Volume 20

CRM
MONOGRAPH
SERIES

Centre de Recherches Mathématiques
Université de Montréal

Riemann Surfaces of Infinite Genus

Joel Feldman
Horst Knörrer
Eugene Trubowitz

The Centre de Recherches Mathématiques (CRM) of the Université de Montréal was created in 1968 to promote research in pure and applied mathematics and related disciplines. Among its activities are special theme years, summer schools, workshops, postdoctoral programs, and publishing. The CRM is supported by the Université de Montréal, the Province of Québec (FCAR), and the Natural Sciences and Engineering Research Council of Canada. It is affiliated with the Institut des Sciences Mathématiques (ISM) of Montréal, whose constituent members are Concordia University, McGill University, the Université de Montréal, the Université du Québec à Montréal, and the Ecole Polytechnique. The CRM may be reached on the Web at www.crm.umontreal.ca.

American Mathematical Society
Providence, Rhode Island USA

Research supported in part by the Natural Sciences and Engineering Research Council of Canada, the Schweizerischer Nationalfonds zur Förderung der wissenschaftlichen Forschung and the Forschungsinstitut für Mathematik, ETH Zürich.

The production of this volume was supported in part by the Fonds pour la Formation de Chercheurs et l'Aide à la Recherche (Fonds FCAR) and the National Sciences and Engineering Research Council of Canada (NSERC).

2000 *Mathematics Subject Classification.* Primary 30F99.

For additional information and updates on this book, visit
www.ams.org/bookpages/crmm-20

Library of Congress Cataloging-in-Publication Data

Feldman, Joel S., 1949–
Riemann surfaces of infinite genus / Joel Feldman, Horst Knörrer, Eugene Trubowitz.
p. cm. — (CRM monograph series, ISSN 1065-8599 ; v. 20)
Includes bibliographical references.
ISBN 0-8218-3357-X (alk. paper)
1. Riemann surfaces. I. Knörrer, Horst. II. Trubowitz, Eugene. III. Title. IV. Series.

QA333.F45 2003
515′.93—dc21 2003045110

∞ The paper used in this book is acid-free and falls within the guidelines
established to ensure permanence and durability.
This volume was submitted to the American Mathematical Society
in camera ready form by the Centre de Recherches Mathématiques.
Visit the AMS home page at `http://www.ams.org/`

10 9 8 7 6 5 4 3 2 1 08 07 06 05 04 03

This work is dedicated to
Henry P. McKean

Contents

Preface vii

Chapter 1. L^2-Cohomology, Exhaustions with Finite Charge and Theta Series 1
Introduction 1
1. Square Integrable One Forms 3
2. Exhaustion Functions with Finite Charge 16
3. Parabolic Surfaces 25
4. Theta Functions 42
A. Summary of Results from Chapter 1 57

Chapter 2. The Torelli Theorem 63
Introduction 63
5. Geometric Hypotheses 66
6. Pointwise Bounds on Differential Forms 79
7. Zeroes of the Theta Function 110
8. Riemann's Vanishing Theorem 124
9. The Geometry of the Theta Divisor 134
10. Hyperelliptic Theta Divisors 152
11. The Torelli Theorem 162

Chapter 3. Examples 169
Introduction 169
12. Hyperelliptic Surfaces 172
13. Heat Curves: Basic Properties 180
14. Heat Curves: Asymptotics 194
15. Heat Curves: Verification of the Geometric Hypotheses 216
16. Fermi Curves: Basic Properties 221
17. Fermi Curves: Asymptotics 223
18. Fermi Curves: Verification of the Geometric Hypotheses 240
B. A Quantitative Morse Lemma 245

Chapter 4. The Kadomcev–Petviashvilli Equation 249
Introduction 249
19. The Formula for the Solution 250
20. Approximations 254
21. Real Periodic Potentials 277

Bibliography 295

Preface

In this book we introduce a class of marked Riemann surfaces $(X; A_1, B_1, \dots)$ of infinite genus that are constructed by pasting plane domains and handles together. Here, $A_1, B_1, \dots$ is a canonical homology basis on X. The asymptotic holomorphic structure is specified by a set of geometric/analytic hypotheses. The analytic hypotheses primarily restrict the distribution and size of the cycles $A_1, A_2, \dots$. The entire classical theory of compact Riemann surfaces up to and including the Torelli Theorem extends to this class. In our generalization, compact surfaces correspond to the special case in which all but finitely many of the A_j, $j \geq 1$, have length zero.

The choice of geometric/analytic hypotheses was guided by two requirements. First, the classical theory of compact Riemann surfaces could be developed in this new context. Secondly, a number of interesting examples satisfy the hypotheses, in particular, the heat curve $\mathcal{H}(q)$ associated to $q \in L^2(\mathbb{R}^2/\Gamma)$. Here, Γ is the lattice

$$\Gamma = (0, 2\pi)\mathbb{Z} \oplus (\omega_1, \omega_2)\mathbb{Z}$$

where $\omega_1 > 0$, $\omega_2 \in \mathbb{R}$ and $\mathcal{H}(q)$ is the set of all points $(\xi_1, \xi_2) \in \mathbb{C}^* \times \mathbb{C}^*$ for which there is a nontrivial distributional solution $\psi(x_1, x_2)$ in $L^\infty_{\mathrm{loc}}(\mathbb{R}^2)$ of the "heat equation"

$$\left(\frac{\partial}{\partial x_1} - \frac{\partial^2}{\partial x_2^2}\right)\psi + q(x_1, x_2)\psi = 0$$

satisfying

$$\psi(x_1 + \omega_1, x_2 + \omega_2) = \xi_1 \psi(x_1, x_2), \quad \psi(x_1, x_2 + 2\pi) = \xi_2 \psi(x_1, x_2).$$

The general theory is used to express the solution of the Kadomcev–Petviashvilli equation with real analytic, periodic initial data q explicitly in terms of the theta function on the infinite-dimensional Jacobian variety corresponding to $\mathcal{H}(q)$. It is evident that the solution is almost periodic in time. In [**Me1, Me2**], this result is improved to finitely differentiable initial data.

This book is divided into four parts. We begin with a discussion, in a very general setting, of L^2-cohomology, exhaustions with finite charge and theta series. In the second part, the geometrical/analytical hypotheses are introduced. Then, an analogue of the classical theory is developed, starting with the construction of a normalized basis of square integrable holomorphic one forms and concluding with the proof of a Torelli theorem. The third part is devoted to a number of examples. Finally, the Kadomcev–Petviashvilli equation is treated in the fourth part.

We speculate that our theory can be extended to surfaces with double points, that a corresponding "Teichmueller theory" can be developed and that there is an infinite-dimensional "Teichmueller space" in which "finite genus" curves are dense.

CHAPTER 1

L^2-Cohomology, Exhaustions with Finite Charge and Theta Series

Introduction

It is an elementary fact of topology that for any oriented, compact surface X of genus g there is a "canonical" basis $A_1, B_1, \ldots, A_g, B_g$ of $H_1(X, \mathbb{Z})$ satisfying

$$A_i \times A_j = 0, \quad A_i \times B_j = \delta_{i,j}, \quad B_i \times B_j = 0$$

for all $1 \leq i, j \leq g$, and that all canonical bases are related by $\mathrm{Sp}(g, \mathbb{Z})$, the group of integral symplectic matrices of order $2g$. For any smooth closed forms ω, η on X the Riemann bilinear relation

$$\int_X \omega \wedge \eta = \sum_{i=1}^{g} \int_{A_i} \omega \int_{B_i} \eta - \int_{A_i} \eta \int_{B_i} \omega$$

is satisfied.

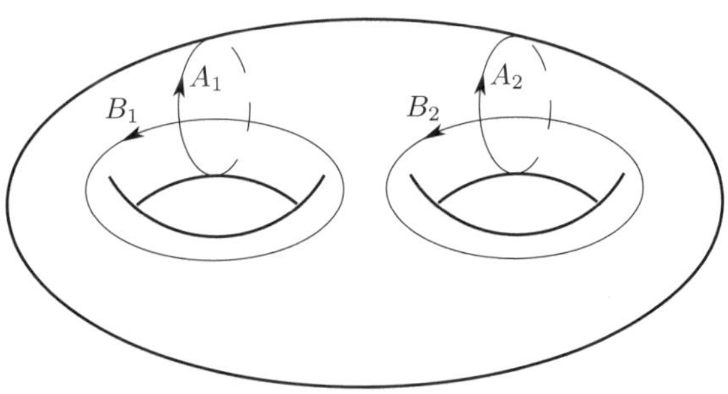

If, in addition, X has a complex structure, in other words, X is a Riemann surface, then it is a basic result that the complex dimension of $\Omega(X)$, the vector space of holomorphic one forms, is g and that for every canonical basis $A_1, B_1, \ldots, A_g, B_g$ of $H_1(X, \mathbb{Z})$, there is a unique "normalized" basis $\omega_1, \ldots, \omega_g$ of $\Omega(X)$ satisfying

$$\int_{A_i} \omega_j = \delta_{i,j}$$

for all $1 \leq i, j \leq g$. It is also easy to show that the "period matrix"

$$\mathcal{R}_X = \left(\int_{B_i} \omega_j \right)$$

is symmetric and $\mathrm{Im}\, \mathcal{R}_X$ is positive definite and that the associated theta series

$$\theta(\mathbf{z}, \mathcal{R}_X) = \sum_{\mathbf{n} \in \mathbb{Z}^g} e^{2\pi i \langle \mathbf{z}, \mathbf{n} \rangle} e^{\pi i \langle \mathbf{n}, R\mathbf{n} \rangle}$$

converges to an entire function of $\mathbf{z} \in \mathbb{C}^g$. The existence of $\omega_1, \dots, \omega_g$ is a simple consequence of elementary Hodge theory. Uniqueness and the properties of the period matrix follow directly from the Riemann bilinear relations.

Suppose X is an open Riemann surface marked with an infinite canonical homology basis $A_1, B_1, A_2, B_2, \dots$, and let $\Omega(X)$ be the Hilbert space of square integrable holomorphic one forms. In §1 we show, again by elementary Hodge theory, that there is a sequence $\omega_j \in \Omega(X)$, $j \geq 1$, with

$$\int_{A_i} \omega_j = \delta_{i,j}$$

for all $i, j \geq 1$. In general, however, one cannot determine whether the sequence $\omega_j, j \geq 1$, is unique or whether its span is $\Omega(X)$. Additional structure is required.

In §2 we introduce (Definition 2.1) the additional structure of an exhaustion function h with finite charge on a marked Riemann surface $(X; A_1, B_1, \dots)$. More precisely, h is a nonnegative, proper, Morse function on X with

$$\int_X |d{*}dh| < \infty$$

that is "compatible" with the marking. A basic result (Theorem 3.8) is that for every marked Riemann surface $(X; A_1, B_1, \dots)$ on which there is an exhaustion function with finite charge one can construct a *unique* sequence $\omega_j \in \Omega(X)$, $j \geq 1$, that spans $\Omega(X)$ and satisfies

$$\int_{A_i} \omega_j = \delta_{i,j}.$$

Furthermore, X is parabolic in the sense of Ahlfors and Nevanlinna. Roughly speaking, an exhaustion function with finite charge allows one to derive enough of the Riemann bilinear relations (Theorem 2.9 and Proposition 3.7) to make the dual sequence ω_j, $j \geq 1$, unique.

In §3, we first review the basic properties of parabolic surfaces. Next, the canonical map from X to $\mathbb{P}(\Omega^*(X))$ is constructed and we show that (Proposition 3.26) it is injective and immersive whenever X has finite ideal boundary and is not hyperelliptic. The Abel–Jacobi map is also discussed (Proposition 3.30).

In §4, we define formal theta series of the argument $\mathbf{z} \in \mathbb{C}^\infty$ for all infinite matrices R with $R^t = R$ and $\operatorname{Im} R > 0$. We prove

THEOREM 4.6. *Suppose that the symmetric matrix R and the sequence $t_j \in (0,1)$, $j \geq 1$, satisfy*

$$\sum_{j \geq 1} t_j^\beta < \infty$$

for some $0 < \beta < \frac{1}{2}$, and

$$\langle \mathbf{n}, \operatorname{Im} R\mathbf{n} \rangle = \sum_{i,j \geq 1} n_i \operatorname{Im} R_{i,j} n_j \geq \frac{1}{2\pi} \sum_{j \geq 1} |\log t_j| n_j^2$$

for all vectors $\mathbf{n} \in \mathbb{Z}^\infty$ with only a finite number of nonzero components. Let B be the Banach space given by

$$B = \left\{ \mathbf{z} = (z_1, z_2, \dots) \in \mathbb{C}^\infty \;\middle|\; \lim_{j \to \infty} \frac{|z_j|}{|\log t_j|} = 0 \right\}$$

with norm

$$\|\mathbf{z}\| = \sup_{j\geq 1} \frac{|z_j|}{|\log t_j|}.$$

Then, for every point $\mathbf{w} \in B$ *the theta series*

$$\theta(\mathbf{z}, R) = \sum_{\substack{\mathbf{n}\in\mathbb{Z}^\infty \\ |\mathbf{n}|<\infty}} e^{2\pi i\langle \mathbf{z},\mathbf{n}\rangle} e^{\pi i\langle \mathbf{n}, R\mathbf{n}\rangle}$$

converges absolutely and uniformly on the ball in B *of radius* $r = 1-2\beta/8\pi$ *centered at* $\mathbf{w}$ *to a holomorphic function.*

Suppose

$$\mathcal{R}_X = \left(\int_{B_i} \omega_j\right)$$

is the period matrix of the marked Riemann surface $(X; A_1, B_1, \dots)$ on which there is an exhaustion function with finite charge. In Proposition 4.4, we put a geometric condition on the cycles $A_1, A_2, \dots,$ that implies the hypothesis of Theorem 4.6 for $\mathcal{R}_X$. We also derive some properties of these theta series that will be used in Chapter 4 to prove that all smooth spatially periodic solutions of the Kadomcev–Petviashvilli equation propagate almost periodically in time.

Chapter 1 is completed by Appendix A, which contains a summary of the results of this chapter.

1. Square Integrable One Forms

In this section we develop L^2-cohomology for an open Riemann surface marked with a canonical homology basis. Our goal (Theorem 1.17) is to derive a criterion for the existence of a unique, normalized basis for the Hilbert space of all square integrable holomorphic one forms. In the next section, the concept of an exhaustion function with finite charge is introduced. It is ultimately shown (Theorem 3.8) that a marked Riemann surface on which there is an exhaustion function with bounded charge satisfies our criterion.

Let λ be a measureable one form on the Riemann surface X. Suppose $z = x^1 + ix^2$ is a coordinate on $U \subset X$ and write

$$\lambda\big|_U = f_1\,dx^1 + f_2\,dx^2.$$

Now set

$$*\lambda\big|_U = -f_2\,dx^1 + f_1\,dx^2.$$

We can use the Cauchy–Riemann equations to derive

PROPOSITION 1.1. *Let* λ *be a measureable one form on the Riemann surface* X. *The local one forms* $*\lambda\big|_U$, (U, z) *a coordinate system on* X, *are consistent and define a global measureable one form* $*\lambda$ *on* X. *Furthermore,*

$$**\lambda = -\lambda$$

and

$$\lambda \wedge \overline{*\lambda}\big|_U = (|f_1|^2 + |f_2|^2)\,dx^1 \wedge dx^2.$$

PROOF. Change coordinates to $w = y^1 + iy^2$ on $V \subset X$. Then,

$$\begin{aligned}\lambda\big|_{U\cap V} &= \left(f_1\frac{\partial x^1}{\partial y^1} + f_2\frac{\partial x^2}{\partial y^1}\right) dy^1 + \left(f_1\frac{\partial x^1}{\partial y^2} + f_2\frac{\partial x^2}{\partial y^2}\right) dy^2 \\ &= g_1\, dy^1 + g_2\, dy^2.\end{aligned}$$

Applying the Cauchy–Riemann equations

$$\begin{aligned}*\lambda\big|_{U\cap V} &= \left(-f_2\frac{\partial x^1}{\partial y^1} + f_1\frac{\partial x^2}{\partial y^1}\right) dy^1 + \left(-f_2\frac{\partial x^1}{\partial y^2} + f_1\frac{\partial x^2}{\partial y^2}\right) dy^2 \\ &= \left(-f_2\frac{\partial x^2}{\partial y^2} - f_1\frac{\partial x^1}{\partial y^2}\right) dy^1 + \left(f_2\frac{\partial x^2}{\partial y^1} + f_1\frac{\partial x^1}{\partial y^1}\right) dy^2 \\ &= -g_2 dy^1 + g_1 dy^2.\end{aligned}$$

□

Let $L^2(X, \mathrm{T}^*X)$ be the Hilbert space of all measureable one forms λ on X satisfying

$$\|\lambda\|^2 = \int_X \lambda \wedge \overline{*\lambda} < \infty$$

with inner product

$$\langle \lambda, \mu\rangle = \int_X \lambda \wedge \overline{*\mu}$$

and let

$$\begin{aligned}\mathcal{C}_X &= \left\{\lambda \in L^2(X, \mathrm{T}^*X) \;\middle|\; \int_X \lambda \wedge d\varphi = 0 \text{ for all } \varphi \in C_0^\infty(X)\right\} \\ &= \text{ the orthogonal complement of } \overline{*dC_0^\infty(X)} \text{ in } L^2(X, \mathrm{T}^*X)\end{aligned}$$

be the closed subspace of all weakly closed forms. Here, $\overline{*dC_0^\infty(X)}$ denotes the closure of $*dC_0^\infty(X)$ in $L^2(X, \mathrm{T}^*X)$.

EXAMPLE 1.2. Recall that

$$d\log r = \frac{dr}{r}, \quad *d\log r = \frac{*dr}{r} = \frac{-x_2\, dx_1 + x_1\, dx_2}{r^2} = d\theta$$

where $r = |z|$ on $\mathbb{C} \setminus \{0\}$. If X is any open subset of $\mathbb{C}$ on which r is bounded away from zero and infinity, then

$$\int_X d\log r \wedge *d\log r = \int_X \frac{dr \wedge d\theta}{r} < \infty.$$

To prepare for the construction of normalized, square integrable holomorphic one forms on noncompact surfaces, we quickly review elementary Hodge theory.

Observe that

$$\overline{dC_0^\infty(X)} = \text{ the closure of } dC_0^\infty(X) \text{ in } L^2(X, \mathrm{T}^*X) \subset \mathcal{C}_X$$

since

$$\int_X d\psi \wedge d\varphi = \int_X d(\psi\, d\varphi) = 0$$

for all $\psi, \varphi \in C_0^\infty(X)$. By definition, the Hodge–Kodaira cohomology space $H^1_{\mathrm{HK}}(X)$ is the orthogonal complement of $\overline{dC_0^\infty(X)}$ in $\mathcal{C}_X$. That is,

$$\mathcal{C}_X = H^1_{\mathrm{HK}}(X) \oplus \overline{dC_0^\infty(X)}$$

or, equivalently,

$$H^1_{\mathrm{HK}}(X) = \mathcal{C}_X / \overline{dC_0^\infty(X)}.$$

THEOREM 1.3 (Hodge Decomposition). *Let X be a Riemann surface. Then*

$$L^2(X, \mathrm{T}^*X) = \overline{dC_0^\infty(X)} \oplus \overline{*dC_0^\infty(X)} \oplus H^1_{\mathrm{HK}}(X)$$

and

$$H^1_{\mathrm{HK}}(X) = \{\lambda \in C^\infty(X, \mathrm{T}^*X) \cap L^2(X, \mathrm{T}^*X) \mid d\lambda = 0 \text{ and } d{*}\lambda = 0\}.$$

PROOF. To verify the first statement, observe that the cohomology space $H^1_{\mathrm{HK}}(X)$ is, by construction, orthogonal to $\overline{dC_0^\infty(X)}$ and $\overline{*dC_0^\infty(X)}$. The latter two subspaces are also orthogonal since

$$\langle d\psi, *d\varphi\rangle = \int_X d\psi \wedge **\overline{d\varphi} = -\int_X d\psi \wedge d\overline{\varphi} = \int_X d(\overline{\varphi}\, d\psi) = 0$$

for all $\psi, \varphi \in C_0^\infty(X)$.

For the second statement, suppose $\lambda \in H^1_{\mathrm{HK}}(X)$. By definition,

$$\langle \lambda, *d\varphi + d\psi\rangle = 0$$

for all $\psi, \varphi \in C_0^\infty(X)$. In particular,

$$\int_U \lambda \wedge (d\varphi - i{*}d\varphi) = 0$$

for all $\overline{\varphi} \in C_0^\infty(U)$ where U is an open subset of X.

In the coordinate system $(U, z = x^1 + ix^2)$ the integrand $\lambda \wedge (d\varphi - i{*}d\varphi)$ becomes,

$$\begin{aligned}
\lambda \wedge (d\varphi - i{*}d\varphi) &= (f_1\, dx^1 + f_2\, dx^2) \wedge \left(\frac{\partial\varphi}{\partial x^1} + i\frac{\partial\varphi}{\partial x^2}\right)(dx^1 - i\, dx^2) \\
&= -i(f_1 - if_2)\left(\frac{\partial\varphi}{\partial x^1} + i\frac{\partial\varphi}{\partial x^2}\right) dx^1 \wedge dx^2 \\
&= -2i(f_1 - if_2)\frac{\partial\varphi}{\partial\bar{z}}\, dx^1 \wedge dx^2
\end{aligned}$$

so that

$$\int_U (f_1 - if_2)\frac{\partial\varphi}{\partial\bar{z}}\, dx^1 \wedge dx^2 = 0$$

for all $\overline{\varphi} \in C_0^\infty(U)$. In other words, $(f_1 - if_2) \in L^2(U)$ is a distributional solution to the elliptic equation

$$\frac{\partial}{\partial\bar{z}} u = 0.$$

It follows that $f_1, f_2 \in C^\infty(U)$. Integrating by parts,

$$\langle d\lambda, \psi\rangle = 0, \quad \langle d{*}\lambda, \varphi\rangle = 0$$

for all $\psi, \varphi \in C_0^\infty(X)$. Consequently, $d\lambda = d{*}\lambda = 0$. □

REMARK 1.4. Suppose $\lambda \in H^1_{\mathrm{HK}}(X)$. By construction, $d\lambda = 0$ and $d{*}\lambda = 0$. Let (U, z) be a coordinate patch on X. Applying Poincaré's lemma there is a $\varphi \in C^\infty(U)$ with $\lambda\big|_U = d\varphi$. We have

$$0 = d{*}\lambda\big|_U = d{*}d\varphi = \left(\left(\frac{\partial}{\partial x^1}\right)^2 \varphi + \left(\frac{\partial}{\partial x^2}\right)^2 \varphi\right) dx^1 \wedge dx^2.$$

That is, φ is a harmonic function on U. Thus, $H^1_{\mathrm{HK}}(X)$ is the Hilbert space of square integrable, harmonic one forms on X.

REMARK 1.5. Let σ be an oriented, simple closed curve on X. Choose a thin, annular region $\mathcal{R} \subset X$ lying just to the left of σ. Precisely, $\mathcal{R}$ is diffeomorphic to

$$\left\{ z \in \mathbb{C} \mid \tfrac{1}{2} < |z| \leq 1 \right\}$$

and, after this diffeomorphism,

$$\sigma(\theta) = e^{i\theta}$$

for all $0 \leq \theta < 2\pi$. Let φ be a smooth, nonnegative function on $\mathcal{R}$ satisfying

$$\varphi(z) = \begin{cases} 1, & \frac{7}{8} < |z| \leq 1 \\ 0, & \frac{1}{2} < |z| < \frac{5}{8} \end{cases}.$$

Observe that,

$$\eta'_\sigma = \begin{cases} d\varphi, & x \in \mathcal{R} \\ 0, & x \in (X \setminus \mathcal{R}) \cup \sigma \end{cases}$$

is a smooth, closed, compactly supported, real valued one form on X. For every smooth, closed one form $\omega \in L^2(X, \mathrm{T}^*X)$ and any oriented, simple closed curve τ that intersects σ transversally, we have

$$\begin{aligned} \langle \omega, *\eta'_\sigma \rangle &= \int_X \eta'_\sigma \wedge \omega = \int_{\mathcal{R}} d\varphi \wedge \omega \\ &= \int_{\mathcal{R}} d(\varphi\omega) - \int_{\mathcal{R}} \varphi\, d\omega \\ &= \int_{\mathcal{R}} d(\varphi\omega) = \int_{\partial\mathcal{R}} \varphi\omega \\ &= \int_\sigma \omega \end{aligned}$$

and

$$\begin{aligned} \langle \eta'_\sigma, *\eta'_\tau \rangle = \int_\tau \eta'_\sigma &= \sum_{\substack{p \in \sigma\cap\tau \\ \tau \text{ crosses } \sigma \text{ from left to right}}} \int_{\substack{\text{a segment of } \tau \\ \text{left of } p}} d\varphi \\ &\quad + \sum_{\substack{p \in \sigma\cap\tau \\ \tau \text{ crosses } \sigma \text{ from right to left}}} \int_{\substack{\text{a segment of } \tau \\ \text{left of } p}} d\varphi \\ &= \sum_{\substack{p \in \sigma\cap\tau \\ \text{left to right}}} 1 + \sum_{\substack{p \in \sigma\cap\tau \\ \text{right to left}}} -1 \\ &= \tau \times \sigma \end{aligned}$$

where $\tau \times \sigma$ is the intersection index of τ and σ.

REMARK 1.6. Let η_σ be the orthogonal projection of η'_σ onto $H^1_{\mathrm{HK}}(X)$. Because $*$ is unitary and leaves $H^1_{\mathrm{HK}}(X)$ invariant, it also leaves the orthogonal complement of $H^1_{\mathrm{HK}}(X)$ invariant. Hence, for all $\omega \in H^1_{\mathrm{HK}}(X)$, $*(\eta_\sigma \eta'_\sigma)$ is perpendicular to ω and

$$\begin{aligned} \langle \omega, *\eta_\sigma \rangle &= \langle \omega, *\eta'_\sigma + *(\eta_\sigma \eta'_\sigma) \rangle \\ &= \langle \omega, *\eta'_\sigma \rangle \\ &= \int_\sigma \omega. \end{aligned}$$

By Schwarz's inequality,

$$\left|\int_\sigma \omega\right| = |\langle \omega, *\eta_\sigma\rangle| \leq \|*\eta_\sigma\|\|\omega\| = \|\eta_\sigma\|\|\omega\|.$$

Thus,

$$\omega \in H^1_{\mathrm{HK}}(X) \to \int_\sigma \omega \in \mathbb{C}$$

defines a bounded linear functional on $H^1_{\mathrm{HK}}(X)$ and $*\eta_\sigma$ is the unique square integrable, harmonic one form satisfying

$$\int_\sigma \omega = \langle \omega, *\eta_\sigma\rangle$$

for all $\omega \in H^1_{\mathrm{HK}}(X)$.

REMARK 1.7. By construction, $\eta_\sigma - \eta'_\sigma$ is a smooth closed form, orthogonal to $H^1_{\mathrm{HK}}(X)$. Thus,

$$\eta_\sigma\eta'_\sigma \in \left(\overline{dC_0^\infty(X)} \oplus \overline{*dC_0^\infty(X)}\right) \cap \mathcal{C}_X = \overline{dC_0^\infty(X)}.$$

It follows that for any oriented, simple closed curves σ and τ,

$$\langle \eta_\sigma, *(\eta_\tau\eta'_\tau)\rangle = -\langle *\eta_\sigma, (\eta_\tau\eta'_\tau)\rangle = 0, \quad \langle(\eta_\sigma\eta'_\sigma), *\eta_\tau\rangle = 0$$

since $*\eta_\sigma, *\eta_\tau \in H^1_{\mathrm{HK}}(X)$ and

$$\langle(\eta_\sigma\eta'_\sigma), *(\eta_\tau\eta'_\tau)\rangle = 0$$

since $\eta_\sigma\eta'_\sigma, \eta_\tau\eta'_\tau \in \overline{dC_0^\infty(X)}$. Expanding,

$$\begin{aligned}\langle \eta'_\sigma, *\eta'_\tau\rangle &= \langle \eta_\sigma - (\eta_\sigma\eta'_\sigma), *\eta_\tau - *(\eta_\tau\eta'_\tau)\rangle \\ &= \langle \eta_\sigma, *\eta_\tau\rangle.\end{aligned}$$

Applying the last statement made in Remark 1.5,

$$\langle \eta_\sigma, *\eta_\tau\rangle = \tau \times \sigma.$$

Our review of elementary Hodge theory is finished.

Let X be a Riemann surface. It may have a boundary ∂X. An element σ of the singular homology space $H_1(X,\mathbb{Z})$ is called a dividing cycle if $\sigma \times \tau = 0$ for every $\tau \in H_1(X,\mathbb{Z})$. The set

$$\{\sigma \in H_1(X,\mathbb{Z}) \mid \sigma \times \tau = 0 \text{ for all } \tau \in H_1(X,\mathbb{Z})\}$$

of all dividing cycles is the kernel of the intersection form on $H_1(X,\mathbb{Z})$. Equivalently, a cycle σ is a dividing cycle if for every compact submanifold Y containing a representative of σ, there is an integral linear combination of cycles in ∂Y that is homologous to σ. Loosely speaking, "σ can be pushed out to the boundary of Y". In particular, each component of the boundary ∂X is a dividing cycle. A dividing cycle composed of simple closed curves in the interior of X divides X into at least two components. Some useful properties of dividing cycles are given in the appendix to this section.

Let

$$H_1^\natural(X,\mathbb{Z}) = H_1(X,\mathbb{Z})/(\text{subgroup of dividing cycles})$$

be the quotient of $H_1(X,\mathbb{Z})$ by the subgroup of all dividingrlo cycles. The intersection form induces a nondegenerate, skew-symmetric bilinear form on $H_1^\natural(X,\mathbb{Z})$.

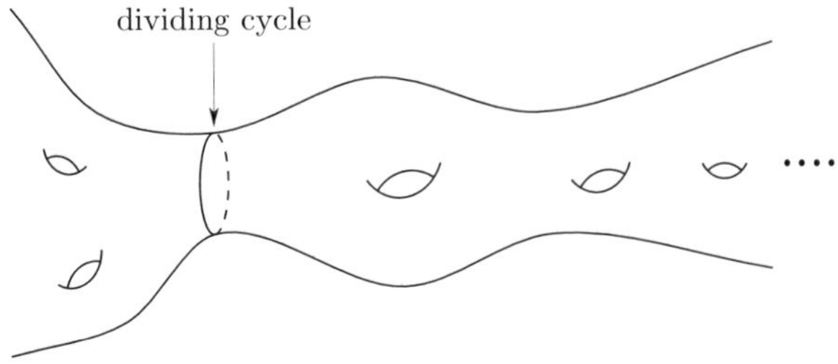

EXAMPLE 1.8. For any $0 < r_1 < r_2 < \infty$, the path

$$\sigma(s) = \frac{r_1 + r_2}{2} e^{is}, \quad 0 \leq 2\pi,$$

represents a dividing cycle on the open annulus $r_1 < |z| < r_2$. Thus,

$$H_1^\natural(r_1 < |z| < r_2, \mathbb{Z}) = 0.$$

Note that,

$$\frac{x\,dy - y\,dx}{x^2 + y^2}$$

is a square integrable, harmonic one form on the annulus with

$$\int_\sigma \frac{x\,dy - y\,dx}{x^2 + y^2} = 2\pi.$$

REMARK 1.9. The period map

$$(\sigma, \omega) \in H_1(X, \mathbb{C}) \times H^1_{\mathrm{HK}}(X) \to \int_\sigma \omega \in \mathbb{C}$$

is a (possibly degenerate) bilinear pairing of the singular homology space $H_1(X, \mathbb{C})$, with complex coefficients, and the cohomology space $H^1_{\mathrm{HK}}(X)$. It does not induce a bilinear pairing of $H_1^\natural(X, \mathbb{C})$ and $H^1_{\mathrm{HK}}(X)$ unless X has the property that

$$\int_\sigma \omega = 0$$

for all dividing cycles $\sigma \in H_1(X, \mathbb{Z})$ and all $\omega \in H^1_{\mathrm{HK}}(X)$.

Let X be a Riemann surface. It may have boundary ∂X.

DEFINITION 1.10. A canonical homology basis for X is a sequence of cycles

$$A_1, B_1, A_2, B_2, \ldots, A_n, B_n, \ldots$$

representing a basis of $H_1^\natural(X, \mathbb{Z})$ that satisfies

(i) For all $i, j \geq 1$,

$$A_i \times B_j = \delta_{i,j}, \quad A_i \times A_j = B_i \times B_j = 0.$$

(ii) For every compact submanifold $Y \subset X$ with boundary there is an $n \geq 1$ such that the range of the canonical map

$$H_1(Y, \mathbb{Z}) \to H_1^\natural(X, \mathbb{Z})$$

is contained in the span of $A_1, B_1, A_2, B_2, \ldots, A_n, B_n$.

It can be shown that there is a canonical homology basis for every Riemann surface. See [**AS**, Chapter I].

A marked Riemann surface $(X; A_1, B_1, \dots)$ is a Riemann surface X with a specified canonical homology basis $A_1, B_1, \dots$.

LEMMA 1.11. *Let A_k, B_k, $k \geq 1$, be a canonical homology basis for X and set*

$$\alpha_k = \eta_{B_k}, \quad \beta_k = -\eta_{A_k}.$$

Then, $\alpha_k, \beta_k \in H^1_{\mathrm{HK}}(X)$, $k \geq 1$, and

$$\int_{A_\ell} \alpha_k = \int_{B_\ell} \beta_k = \delta_{k,\ell}, \quad \int_{B_\ell} \alpha_k = \int_{A_\ell} \beta_k = 0.$$

That is, α_k, β_k, $k \geq 1$ are real, square integrable, harmonic forms dual to the canonical homology basis A_k, B_k, $k \geq 1$.

Here, η_σ is the real, square integrable harmonic form constructed in Remark 1.6 that is characterized by

$$\int_\sigma \omega = \langle \omega, *\eta_\sigma \rangle$$

for all $\omega \in H^1_{\mathrm{HK}}(X)$.

PROOF. By the last statement made in Remark 1.7,

$$\int_{A_\ell} \alpha_k = \langle \alpha_k, *\eta_{A_\ell} \rangle = \langle \eta_{B_k}, *\eta_{A_\ell} \rangle = A_\ell \times B_k = \delta_{k,\ell},$$

$$\int_{B_\ell} \alpha_k = \langle \alpha_k, *\eta_{B_\ell} \rangle = \langle \eta_{B_k}, *\eta_{B_\ell} \rangle = B_\ell \times B_k = 0.$$

Similarly,

$$\int_{B_\ell} \beta_k = \langle \beta_k, *\eta_{B_\ell} \rangle = -\langle \eta_{A_k}, *\eta_{B_\ell} \rangle = \delta_{k,\ell}, \quad \int_{A_\ell} \beta_k = 0. \qquad \square$$

EXAMPLE 1.12. For any $0 < r_1 < r_2 < \infty$, slit the real segment $r_1 < x < r_2$ of the open annulus $r_1 < |z| < r_2$ along infinitely many disjoint, closed subintervals. Reflect the slits onto the segment $-r_2 < x < -r_1$ and glue the lips of corresponding pairs together to construct an infinite genus Riemann surface X. The harmonic function $\log r$ extends from the slit annulus to a nonconstant, harmonic function f on X with

$$\|df\|^2 = \int_X df \wedge \overline{*df} < \infty.$$

Suppose A_k, B_k, $k \geq 1$, is a canonical homology basis for X and α_k, β_k, $k \geq 1$, the dual forms of Lemma 1.11. Then, $\alpha_k + df, \beta_k$, $k \geq 1$, is another set of square integrable, harmonic forms dual to the canonical homology basis.

PROPOSITION 1.13. *Suppose X is a Riemann surface with the properties:*

(i) *For all dividing cycles $\sigma \in H_1(X, \mathbb{Z})$ and all $\omega \in H^1_{\mathrm{HK}}(X)$.*

$$\int_\sigma \omega = 0.$$

(ii) *A harmonic function f on X satisfying*

$$\|df\|^2 = \int_X df \wedge \overline{*df} < \infty$$

is constant.

Let A_k, B_k, $k \geq 1$, be a canonical homology basis for X. Then, the dual one forms α_k, β_k, $k \geq 1$, are unique and span the Hilbert space $H^1_{\mathrm{HK}}(X)$.

PROOF. If $\omega \in H^1_{\mathrm{HK}}(X)$ is orthogonal to α_k, β_k, $k \geq 1$, then,

$$0 = \langle \omega, \alpha_k \rangle = \langle *\omega, *\alpha_k \rangle = \langle *\omega, *\eta_{B_k} \rangle = \int_{B_k} *\omega, \quad 0 = \langle \omega, \beta_k \rangle = -\int_{A_k} *\omega.$$

It now follows from property (i) that

$$\int_\sigma *\omega = 0$$

for all $\sigma \in H_1(X, \mathbb{Z})$ and consequently, that

$$f(x) = \int_{x_0}^{x} *\omega$$

is a harmonic function on X with $\|df\| < \infty$. Our second assumption about X implies that $\omega = 0$ and, as a result, the dual forms α_k, β_k, $k \geq 1$, are a basis for the Hilbert space $H^1_{\mathrm{HK}}(X)$.

That α_k, β_k, $k \geq 1$, are unique is proven by repeating the second half of the above argument. □

Let ω be a measureable one form and suppose $z = x + iy$ is a coordinate on $U \subset X$. We have

$$\begin{aligned} \omega\big|_U &= f\, dx + g\, dy \\ &= \tfrac{1}{2}(f - ig)(dx + i\, dy) + \tfrac{1}{2}(f + ig)(dx\, i\, dy) \\ &= \tfrac{1}{2}(f\, ig)\, dz + \tfrac{1}{2}(f + ig)\, d\bar{z}. \end{aligned}$$

By definition, ω is a holomorphic one form on U, if

$$f + ig = 0$$

and

$$f = \tfrac{1}{2}(f\, ig)$$

is a holomorphic function of z on U. That is,

$$\omega\big|_U = f(z)\, dz$$

and

$$\frac{\partial}{\partial \bar{z}} f(z) = 0.$$

The form ω is holomorphic on X, if it is holomorphic in every coordinate system (U, z).

REMARK 1.14. Suppose ω is a holomorphic one form on X. Then,

$$d\omega = 0, \quad *\omega = -i\omega$$

since

$$d\omega\big|_U = df \wedge dz = 2i \frac{\partial}{\partial \bar{z}} f\, dx \wedge dy = 0, \quad *\omega\big|_U = f * dz = -if\, dz.$$

Conversely, suppose ω is a closed form satisfying $*\omega = -i\omega$. Then,

$$f\, dx + g\, dy = \omega\big|_U = i*\omega\big|_U = -ig\, dx + if\, dy$$

or equivalently,

$$\omega = f\, dz$$

and

$$0 = d\omega\big|_U = df \wedge dz = 2i\frac{\partial}{\partial \bar{z}} f\, dx \wedge dy.$$

Consequently, ω is holomorphic.

We are particularly interested in the space $\Omega(X)$ of all square integrable, holomorphic one forms. By Remark 1.14,

$$\Omega(X) = \{\omega \in H^1_{\mathrm{HK}}(X) \mid *\omega = -i\omega\}.$$

It follows that $\Omega(X)$ is a closed subspace of $H^1_{\mathrm{HK}}(X)$ since $*$ acts as a bounded linear operator on $H^1_{\mathrm{HK}}(X)$. Therefore, $\Omega(X)$ is a Hilbert space with inner product

$$\langle \lambda, \mu \rangle = \int_X \lambda \wedge \overline{*\mu} = i \int_X \lambda \wedge \bar{\mu}.$$

Let A_k, B_k, $k \geq 1$, be a canonical homology basis for X and let $\alpha_k, \beta_k \in H^1_{\mathrm{HK}}(X)$, $k \geq 1$, be the dual forms constructed in Lemma 1.11. Also, let $\mathcal{B}$ be the closed linear span of β_k, $k \geq 1$. Observe that $\mathcal{B}$ and $*\mathcal{B}$ are orthogonal subspaces of $H^1_{\mathrm{HK}}(X)$ since

$$\langle \beta_k, *\beta_\ell \rangle = \langle \eta_{A_k}, *\eta_{A_\ell} \rangle = A_\ell \times A_k = 0.$$

LEMMA 1.15. *For each $k \geq 1$, set*

$$\omega_k = \pi_{*\mathcal{B}}\alpha_k + i{*}\pi_{*\mathcal{B}}\alpha_k$$

where $\pi_{*\mathcal{B}}$ is the orthogonal projection from $H^1_{\mathrm{HK}}(X)$ onto $*\mathcal{B}$. Then, $\omega_k \in \Omega(X)$, $k \geq 1$, and

$$\int_{A_\ell} \omega_k = \delta_{k,\ell}.$$

PROOF. First of all ω_k is a closed form satisfying

$$*\omega_k = *\pi_{*\mathcal{B}}\alpha_k - i\pi_{*\mathcal{B}}\alpha_k = -i\omega_k.$$

Therefore, by Remark 1.14, ω_k, $k \geq 1$, is a holomorphic form in $\Omega(X)$. Dropping the subscript on $\pi_{*\mathcal{B}}$,

$$\begin{aligned}
\int_{A_\ell} \omega_k &= \langle \omega_k, *\eta_{A_\ell} \rangle \\
&= -\langle \pi\alpha_k + i{*}\pi\alpha_k, *\beta_\ell \rangle \\
&= -\langle \pi\alpha_k, *\beta_\ell \rangle \\
&= -\langle \pi\alpha_k + (\mathbb{1}\pi)\alpha_k, *\beta_\ell \rangle \\
&= -\langle \alpha_k, *\beta_\ell \rangle \\
&= -\langle \eta_{B_k}, -*\eta_{A_\ell} \rangle \\
&= \delta_{k,\ell}.
\end{aligned}$$

□

PROPOSITION 1.16. *Suppose X is a Riemann surface with canonical homology basis A_k, B_k, $k \geq 1$. Let X satisfy*

(i) *For all dividing cycles $\sigma \in H_1(X, \mathbb{Z})$ and all $\omega \in H^1_{\mathrm{HK}}(X)$*

$$\int_\sigma \omega = 0.$$

(ii) *A harmonic function f on X satisfying*

$$\|df\|^2 = \int_X df \wedge \overline{*df} < \infty$$

is constant.

(iii) *If $\omega \in \Omega(X)$ satisfies*

$$\int_{A_k} \omega = 0$$

for all $k \geq 1$, then $\omega = 0$.

Then, the normalized holomorphic forms ω_k, $k \geq 1$, are unique and span the Hilbert space $\Omega(X)$.

PROOF. Let $\omega \in \Omega(X)$ be orthogonal to ω_k, $k \geq 1$. Then,

$$\begin{aligned} 2\langle \pi_{*\mathcal{B}}\omega, \alpha_k \rangle &= 2\langle \omega, \pi_{*\mathcal{B}}\alpha_k \rangle \\ &= \langle \omega, \pi_{*\mathcal{B}}\alpha_k \rangle + \langle *\omega, *\pi_{*\mathcal{B}}\alpha_k \rangle \\ &= \langle \omega, \pi_{*\mathcal{B}}\alpha_k \rangle + \langle i\omega, *\pi_{*\mathcal{B}}\alpha_k \rangle \\ &= \langle \omega, \omega_k \rangle \\ &= 0 \end{aligned}$$

for all $k \geq 1$. Recalling that the subspaces $\mathcal{B}$ and $*\mathcal{B}$ are orthogonal,

$$\langle \pi_{*\mathcal{B}}\omega, \beta_k \rangle = 0$$

for all $k \geq 1$. So by Proposition 1.13, $\pi_{*\mathcal{B}}\omega = 0$.

Now,

$$\begin{aligned} \int_{A_k} \omega &= \langle \omega, *\eta_{A_k} \rangle \\ &= -\langle \omega, *\beta_k \rangle \\ &= -\langle \pi_{*\mathcal{B}}\omega + (\mathbb{1}\pi_{*\mathcal{B}})\omega, *\beta_k \rangle \\ &= -\langle (\mathbb{1}\pi_{*\mathcal{B}})\omega, *\beta_k \rangle \\ &= 0. \end{aligned}$$

Therefore, by property (iii), $\omega = 0$ and the forms ω_k, $k \geq 1$ span $\Omega(X)$. □

We now summarize the contents of this section in

THEOREM 1.17. *Let X be a Riemann surface with a canonical homology basis A_k, B_k, $k \geq 1$. Suppose the marked surface $(X; A_1, B_1, \dots)$ has the three properties:*

(i) *For all dividing cycles $\sigma \in H_1(X, \mathbb{Z})$ and all $\omega \in H^1_{\mathrm{HK}}(X)$*

$$\int_\sigma \omega = 0.$$

(ii) *A harmonic function f on X satisfying*

$$\|df\|^2 = \int_X df \wedge \overline{*df} < \infty$$

is constant.

(iii) *If $\omega \in \Omega(X)$ satisfies*

$$\int_{A_k} \omega = 0$$

for all $k \geq 1$, then $\omega = 0$.

Then, $(X; A_1, B_1, \dots)$ *has a unique basis* ω_k, $k \geq 1$, *of square integrable holomorphic one forms dual to* A_k, $k \geq 1$, *and the period map*

$$(\sigma, \omega) \in H_1^\natural(X, \mathbb{C}) \times H^1_{\mathrm{HK}}(X) \to \int_\sigma \omega \in \mathbb{C}$$

is a well-defined, nondegenerate pairing of $H_1^\natural(X, \mathbb{C})$ *and* $H^1_{\mathrm{HK}}(X)$.

In the next section we will introduce certain exhaustion functions on the marked surface $(X; A_1, B_1, \dots)$ that ultimately force properties (i), (ii), (iii) to hold, and therefore the conclusion of Theorem 1.17 as well. This goal is finally realized in Theorem 3.8.

Appendix: Some Properties of Dividing Cycles.

In this appendix we derive some technical properties of dividing cycles that will be used in the next section.

Let Y be a compact, connected submanifold of X with boundary ∂Y. Let B_Y be the set of all components of ∂Y. Each component is given its natural orientation as a submanifold of X. If $\eta, \eta' \in B_Y$, then by definition, $\eta \approx \eta'$ if and only if η and η' both lie in the same component of $\overline{X \setminus Y}$. The relation $\approx$ is an equivalence relation on B_Y. A component $\eta \in B_Y$ that lies in a compact component of $\overline{X \setminus Y}$ is called an accidental component. In the figure below, $\eta_1 \approx \eta_2 \approx \eta_3$, $\eta_4 \approx \eta_5$ and $\eta_6 \approx \eta_7 \approx \eta_8$. The components η_4 and η_5 are accidental.

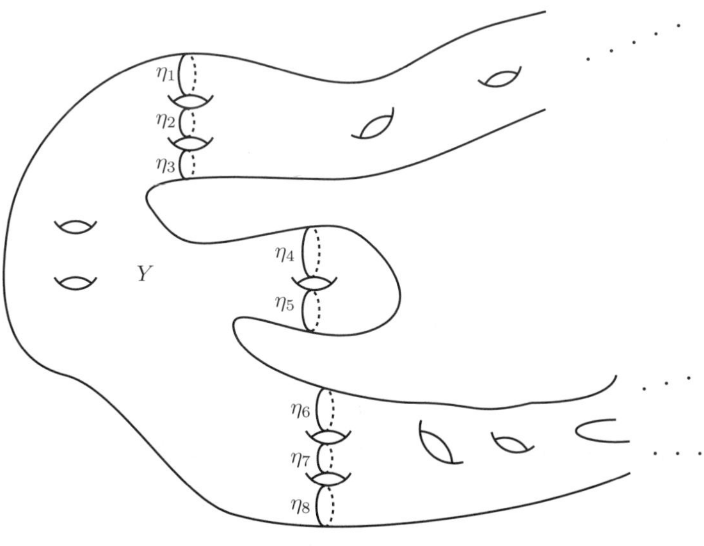

Let $\mathcal{E}(Y)$ be the set of equivalence classes in $(B_Y, \approx)$. Observe that either all elements of a class are accidental components or no element of a class is accidental. Let $\mathcal{E}_a(Y)$ be the set of equivalence classes containing only accidental boundary components. For each $E \in \mathcal{E}(Y)$, set

$$\beta_E = \sum_{\eta \in E} \eta.$$

If $E \in \mathcal{E}_a(Y)$, then β_E is homologous to zero in X.

LEMMA A1.1. *Let σ be a cycle in Y that represents a dividing cycle in $H_1(X, \mathbb{Z})$. Then, σ is homologous in Y to a cycle of the form*

$$\sum_{E \in \mathcal{E}(Y)} n_E \beta_E$$

where $n_E \in \mathbb{Z}$.

PROOF. First of all, σ is homologous in Y to a cycle of the form

$$\sum_{\eta \in B_Y} \nu_\eta \eta.$$

Suppose, $\eta \approx \eta'$. Choose points $p \in \eta$, $p' \in \eta'$ and paths $\alpha_1 \subset Y$, $\alpha_2 \subset \overline{X \setminus Y}$, as in the figure below, connecting p to p'.

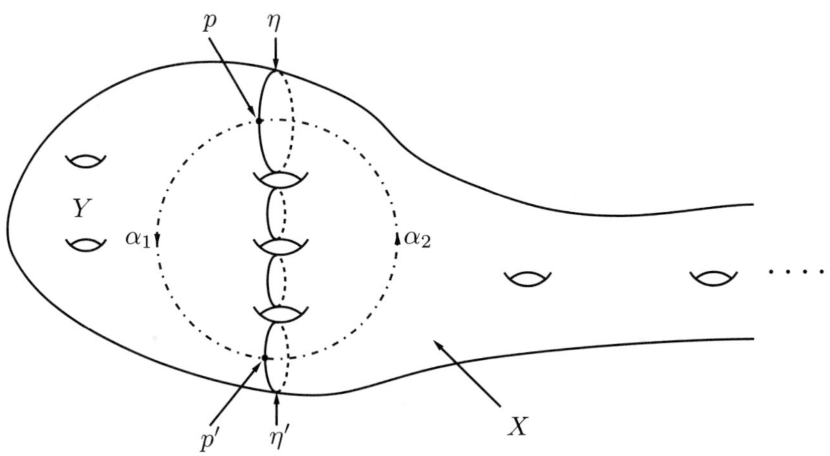

The composition α of α_1 and α_2 is a closed curve on X which, when properly oriented, satisfies

$$\alpha \times \eta = 1, \quad \alpha \times \eta' = -1$$

and

$$\alpha \times \eta'' = 0$$

for all $\eta'' \in B_Y$, $\eta'' \neq \eta, \eta'$. It now follows from the definition of dividing cycle that

$$0 = \sigma \times \alpha = \nu_\eta - \nu_{\eta'}.$$

In other words, $\eta \approx \eta'$ implies $\nu_\eta = \nu_{\eta'}$. □

LEMMA A1.2. *Let Y' be a compact, connected submanifold of X with boundary such that $\overline{Y} \subset \operatorname{int} Y'$. Then, for each $E \in \mathcal{E}(Y) \setminus \mathcal{E}_a(Y)$ there is a subset $S(E)$ of $\mathcal{E}(Y') \setminus \mathcal{E}_a(Y')$ such that β_E is homologous in X to*

$$\sum_{E' \in S(E)} \beta_{E'}.$$

Furthermore, $S(E) \cap S(F) = \varnothing$ for all pairs $E, F \in \mathcal{E}(Y) \setminus \mathcal{E}_a(Y)$ with $E \neq F$.

PROOF. For each $E \in \mathcal{E}(Y) \setminus \mathcal{E}_a(Y)$, let $B_{Y'}(E)$ be the set of all $\eta' \in B_{Y'}$ such that there is a component K of $\overline{Y' \setminus Y}$ with $\eta' \subset \partial K$ and $E \cap \partial K \neq \varnothing$. Then, the cycle $\beta_E = \sum_{\eta \in E} \eta$ is homologous in Y' to $\sum_{\eta' \in B_{Y'}(E)} \eta'$.

Suppose $\eta' \in B_{Y'}(E)$ and $\eta'' \approx \eta'$. We have $\eta'' \in B_{Y'}(F)$ for some $F \in \mathcal{E}(Y)$. Let K and L be components of $\overline{Y' \setminus Y}$ with $\eta' \subset \partial K$, $E \cap \partial K \neq \varnothing$ and $\eta'' \subset \partial L$, $F \cap \partial L \neq \varnothing$. Also, let M be a component of $\overline{X \setminus Y'}$ such that η', $\eta'' \subset \partial M$. Connecting, K and L to M along η' and η'' we see that some element of E and some element of F belong to the boundary of the same connected component of $\overline{X \setminus Y'}$. That is, $E = F$ and $\eta'' \in B_{Y'}(E)$.

It follows from the conclusion of the last paragraph that

$$\sum_{\eta' \in B_{Y'}(E)} \eta' = \sum_{\substack{E' \in \mathcal{E}(Y') \text{ such that there is a component} \\ K \text{ of } \overline{Y' \setminus Y} \text{ with } E' \cap \partial K \neq \varnothing \text{ and } E \cap \partial K \neq \varnothing}} \sum_{\eta' \in E'} \eta'$$
$$= \sum_{\substack{E' \in \mathcal{E}(Y') \text{ such that there is a component} \\ K \text{ of } \overline{Y' \setminus Y} \text{ with } E' \cap \partial K \neq \varnothing \text{ and } E \cap \partial K \neq \varnothing^{c}}} \beta_{E'}.$$

Now, let $S(E)$ be the set of all $E' \in \mathcal{E}(Y') \setminus \mathcal{E}_a(Y')$ such that there is a component K of $\overline{Y' \setminus Y}$ with $E' \cap \partial K \neq \varnothing$ and $E \cap \partial K \neq \varnothing$. The cycle β_E is homologous in X to $\sum_{E' \in S(E)} \beta_{E'}$. □

An exhaustion of X is an increasing sequence

$$X_1 \subset X_2 \subset \cdots$$

of compact submanifolds of X with smooth boundary such that

$$\overline{X_n} \subset \operatorname{int} X_{n+1}$$

for all $n \geq 1$. The following technical lemma will be needed in §2.

LEMMA A1.3. *Let $\sigma \in H_1(X, \mathbb{Z})$ be a dividing cycle on X. Then, there is a positive constant $C(\sigma)$ such that for every exhaustion $X_1 \subset X_2 \subset \cdots$ of X the condition below is satisfied.*

Let $\Gamma_n^1, \ldots, \Gamma_n^{\nu_n}$ be the components of ∂X_n. Then, for all sufficiently large $n \geq 1$ there are integers $s_1, \ldots, s_{\nu_n}$ with

$$|s_i| \leq C(\sigma)$$

such that σ is homologous in X to

$$s_1 \Gamma_n^1 + \cdots + s_{\nu_n} \Gamma_n^{\nu_n}.$$

PROOF. Pick a submanifold Y with smooth boundary that contains a representative of σ. By Lemma A1.1, σ is homologous in X to

$$\sum_{E \in \mathcal{E}(Y) \setminus \mathcal{E}_a(Y)} n_E \beta_E$$

where $n_E \in \mathbb{Z}$. Set

$$C(\sigma) = \max_{E \in \mathcal{E}(Y) \setminus \mathcal{E}_a(Y)} n_E.$$

It follows from Lemma A1.2 that for any n with $\overline{Y} \subset \operatorname{int} X_n$ the condition of Lemma A1.3 is fulfilled. □

2. Exhaustion Functions with Finite Charge

Suppose h is a twice continuously differentiable function on a Riemann surface X. In any coordinate system (U, z),

$$d{*}dh\big|_U = \Delta h \, dx^1 \wedge dx^2$$

where, Δ is the Laplace operator. By analogy with the first of the Maxwell's equations we may regard Δh as the "charge density" corresponding to the "electric field" dh.

Let X be a Riemann surface. If h is any nonnegative function on X and $t \geq 0$, set

$$X_t = h^{-1}([0, t]).$$

DEFINITION 2.1. Let X be a Riemann surface without boundary. An exhaustion function h with finite charge on X is a proper, nonnegative Morse function on X that satisfies:

(i)
$$\int_X |d{*}dh| < \infty.$$

Suppose A_k, B_k, $k \geq 1$, is a canonical homology basis on X. An exhaustion function h with finite charge on the marked Riemann surface $(X; A_1, B_1, \ldots)$ is an exhaustion function with finite charge on X that in addition satisfies:

(ii) For all sufficiently large $t > 0$, there is an $n \geq 1$ such that the cycles

$$A_1, B_1, \ldots, A_n, B_n$$

are homologous in X to cycles of a canonical homology basis for X_t.

(iii) For all sufficiently large $t > 0$, every component of ∂X_t is homologous to a finite linear combination of $A_1, A_2, \ldots$ and dividing cycles.

REMARK 2.2. For all $t > s$, the charge of the electric field dh on the difference $X_t \setminus X_s$ is

$$\int_{X_t \setminus X_s} d{*}dh.$$

For any fixed $s > 0$, the function h exhausts $X \setminus X_s$ by compact subsets $X_t \setminus \operatorname{int} X_s$, $t > s$, with uniformly bounded charge.

EXAMPLE 2.3. Let $p = (y, z)$ be a point on the smooth, algebraic plane curve

$$C = \left\{ (y, z) \in \mathbb{C}^2 \;\middle|\; \sum_{0 \leq i,j \leq n} c_{i,j} y^i z^j = 0 \right\}$$

and let $\pi(p) = z$ be the holomorphic projection from C onto the z-axis in $\mathbb{C}^2$. Then,

$$f(p) = \begin{cases} 0, & |z| \leq 1 \\ \log|z| = \log|\pi(p)|, & |z| > 1 \end{cases}$$

is a proper, nonnegative function on C that is harmonic on the complement

$$C \setminus f^{-1}([0, 1])$$

of the compact subset $f^{-1}([0, 1])$. We have

$$\int_{C \setminus C_1} |d{*}df| = 0$$

since f is harmonic. By Proposition A2.2 there is an exhaustion function h with finite charge on C arbitrarily close to f outside a compact set.

EXAMPLE 2.4. Let $c_i(z)$, $0 \le i \le n$, be entire functions, not all of which are polynomials, and suppose that the transcendental plane curve

$$C = \left\{ (y, z) \in \mathbb{C}^2 \;\middle|\; \sum_{i=0}^{n} c_i(z) y^i = 0 \right\}$$

is smooth. As above,

$$f(p) = \begin{cases} 0, & |z| \le 1 \\ \log |z|, & |z| > 1 \end{cases}$$

is a proper, nonnegative function that is harmonic on the complement of the compact subset $f^{-1}([0,1])$. As above, there is an exhaustion function h with finite charge on C arbitrarily close to f outside a compact set. Using Morse theory one can construct a canonical homology basis $A_1, B_1, \ldots,$ for C such that h becomes an exhaustion function with finite charge on the marked surface $(C; A_1, B_1, \ldots)$.

EXAMPLE 2.5. If φ is a proper, nonconstant holomorphic function on a (necessarily noncompact) Riemann surface X, set

$$K = \{x \in X \mid |\varphi(x)| \le 1\}.$$

Once again,

$$f(x) = \begin{cases} 0, & x \in K \\ \log |\varphi(x)|, & x \notin K \end{cases}$$

is a proper, nonnegative function that is harmonic on the complement of K, and there is an exhaustion function h with finite charge on X that is arbitrarily close to f outside a compact set. We can mark X so that h becomes an exhaustion function with finite charge on the marked surface. In more generality, let $\mathcal{I}$ be a nonempty, finite subset of $\mathbb{P}^1(\mathbb{C})$ and suppose that φ is a proper holomorphic map from X to $\mathbb{P}^1(\mathbb{C}) \setminus \mathcal{I}$. Intuitively, $\mathcal{I}$ is the set of points at "infinity". Choose a rational function ψ on $\mathbb{P}^1(\mathbb{C})$ with poles exactly at the points of $\mathcal{I}$. Then, $\psi \circ \varphi$ is a proper holomorphic function on X.

A linear density $g(x)$ on a Riemann surface X is a piecewise continuous section of the second symmetric tensor product of the real cotangent bundle of X such that $g(x)$ is positive semidefinite for all $x \in X$. The length of curves and the areas of surfaces with respect to the density g is defined just as for a metric. The linear density g_h on X associated to an exhaustion function h is, by definition,

$$g_h(x) = dh \otimes dh + {*dh} \otimes {*dh}.$$

Observe that (x, y) is a multiple of $(-y, x)$ if and only if $x = y = 0$. Consequently $g_h(x)$ is positive definite when $dh(x) \ne 0$.

Let λ be a smooth one form on X. If $dh(x) \ne 0$, write

$$\lambda(x) = \lambda_1 dh(x) + \lambda_2 {*dh(x)}$$

and set

$$|\lambda(x)|^2 = \begin{cases} (|\lambda_1|^2 + |\lambda_2|^2)\big((dh)^2 + (*dh)^2\big), & dh(x) \ne 0 \\ 0, & \text{otherwise} \end{cases}$$

where we suppress the dependence on h. If $z = x_1 + ix_2$ is a coordinate on a open set $U \subset X$ where $dh \neq 0$, then

$$dx_1 = \frac{h_{x_1} dh - h_{x_2} {*dh}}{h_{x_1}^2 + h_{x_2}^2}, \qquad dx_2 = \frac{h_{x_2} dh + h_{x_1} {*dh}}{h_{x_1}^2 + h_{x_2}^2}$$

Substituting,

$$\lambda\big|_U = f_1\, dx_1 + f_2\, dx_2 = \frac{f_1 h_{x_1} + f_2 h_{x_2}}{h_{x_1}^2 + h_{x_2}^2} dh + \frac{-f_1 h_{x_2} + f_2 h_{x_1}}{h_{x_1}^2 + h_{x_2}^2} {*dh}.$$

If $x \in X$ is a Morse critical point of h, there is a coordinate system $(V, w = y_1 + iy_2)$ centered at x with

$$h(y_1, y_2) = h(0) \pm y_1^2 \pm y_2^2$$

for $y_1^2 + y_2^2 < r^2$. We have

$$\lambda\big|_{V \setminus \{x\}} = g_1\, dy_1 + g_2\, dy_2 = \frac{\pm y_1 g_1 \pm y_2 g_2}{2(y_1^2 + y_2^2)} dh + \frac{\mp y_2 g_1 \pm y_1 g_2}{2(y_1^2 + y_2^2)} {*dh}.$$

For any noncritical level set

$$\Gamma_t = \partial X_t$$

of h we have

$$\begin{aligned} \int_{\Gamma_t} |\lambda| &= \int_{\Gamma_t} \sqrt{|\lambda_1|^2 + |\lambda_2|^2} \sqrt{(dh)^2 + ({*dh})^2} \\ &= \int_{\Gamma_t} \sqrt{|\lambda_1|^2 + |\lambda_2|^2} {*dh} \end{aligned}$$

since the pull back of dh to Γ_t vanishes. Here, Γ_t is oriented so that

$${*dh}(\text{an oriented tangent vector to } \Gamma_t) \geq 0.$$

Suppose t is a critical value for h and x is a singular point on Γ_t. Then, with the notation above, set

$$V_\varepsilon = \{w \in V \mid y_1^2 + y_2^2 \geq \varepsilon\}.$$

Now,

$$\begin{aligned} \int_{\Gamma_t \cap V} |\lambda| &= \lim_{\varepsilon \downarrow 0} \int_{\Gamma_t \cap V_\varepsilon} |\lambda| \\ &= \lim_{\varepsilon \downarrow 0} \int_{\Gamma_t \cap V_\varepsilon} \sqrt{|\lambda_1|^2 + |\lambda_2|^2} {*dh} \\ &= \lim_{\varepsilon \downarrow 0} \int_{\Gamma_t \cap V_\varepsilon} \frac{\sqrt{|y_1 g_1 \pm y_2 g_2|^2 + |y_2 g_1 \mp y_1 g_2|^2}}{y_1^2 + y_2^2} (\mp_2\, dy_1 \pm y_1\, dy_2). \end{aligned}$$

If x is a local maximum or minimum, the limit vanishes. If, for example, x is the saddle point

$$h(y_1, y_2) = h(0) + y_1^2 - y_2^2$$

then on Γ_t we have $y_2 = \pm y_1$ and

$$\int_{\Gamma_t \cap V} |\lambda| = \int_{\Gamma_t \cap V} \sqrt{|\lambda_1|^2 + |\lambda_2|^2} {*dh} = 2 \int_{-r/\sqrt{2}}^{r/\sqrt{2}} \sqrt{|g_1 + g_2|^2 + |g_1 - g_2|^2}\, dy_1 < \infty.$$

Also, note that

$$\lambda \wedge \overline{{*\lambda}} = (|\lambda_1|^2 + |\lambda_2|^2)\, dh \wedge {*dh}$$

and

$$\begin{aligned}\left|\int_\gamma \lambda\right| &= \left|\int_\gamma \lambda_1\, dh + \lambda_2 {*dh}\right| \\ &\le \int_a^b \left|\lambda_1(\gamma(s))\, dh(\gamma'(s)) + \lambda_2(\gamma(s)) {*dh}(\gamma'(s))\right| ds \\ &\le \int_\gamma \sqrt{|\lambda_1|^2 + |\lambda_2|^2}\sqrt{(dh)^2 + (*dh)^2} \\ &= \int_\gamma |\lambda|\end{aligned}$$

for any curve γ.

LEMMA 2.6. *Let h be an exhaustion function with finite charge on the Riemann surface X and let*

$$\Gamma_t = \partial X_t$$

where $X_t = h^{-1}([0,t])$, $t \ge 0$. Suppose, λ is a smooth form in $L^2(X, \mathrm{T}^*X)$. Then,

$$\int_0^\infty \left(\int_{\Gamma_s} |\lambda|\right)^2 ds \le \mathrm{const}\|\lambda\|^2 < \infty.$$

PROOF. By Schwarz's inequality,

$$\left(\int_{\Gamma_s} |\lambda|\right)^2 = \left(\int_{\Gamma_s} \sqrt{|\lambda_1|^2 + |\lambda_2|^2 {*dh}}\right)^2 \le \int_{\Gamma_s} {*dh} \int_{\Gamma_s} (|\lambda_1|^2 + |\lambda_2|^2){*dh}$$

However, for all $s' > s$,

$$\int_{\Gamma_{s'}} {*dh} - \int_{\Gamma_s} {*dh} = \int_{X_{s'} \setminus X_s} d{*dh} = O(1).$$

Thus,

$$\int_{\Gamma_s} {*dh}$$

is a bounded function of s. It follows that

$$\left(\int_{\Gamma_s} |\lambda|\right)^2 \le \mathrm{const} \int_{\Gamma_s} (|\lambda_1|^2 + |\lambda_2|^2){*dh}.$$

Integrating,

$$\begin{aligned}\int_0^t \left(\int_{\Gamma_s} |\lambda|\right)^2 ds &\le \mathrm{const} \int_0^t \int_{\Gamma_s} (|\lambda_1|^2 + |\lambda_2|^2){*dh}\, ds \\ &= \mathrm{const} \int_{X_t \setminus X_0} (|\lambda_1|^2 + |\lambda_2|^2)\, dh \wedge {*dh} \\ &\le \mathrm{const} \int_X \lambda \wedge \overline{*\lambda}.\end{aligned}$$

□

We can use Lemma 2.6 to prove

PROPOSITION 2.7. *If there is an exhaustion function h with finite charge on the Riemann surface X, then*

$$\int_\sigma \omega = 0$$

for all dividing cycles $\sigma \in H_1(X, \mathbb{Z})$ and all smooth, closed forms $\omega \in L^2(X, \mathrm{T}^*X)$.

PROOF. Let ω be a smooth closed form in $L^2(X, \mathrm{T}^*X)$. By Lemma 2.6, we can pick a strictly increasing sequence

$$t_1 < t_2 < t_3 \cdots \to \infty$$

of noncritical values for h with

$$\lim_{n\to\infty} \int_{\Gamma_{t_n}} |\omega| = 0.$$

Let $\Gamma_n^1, \ldots, \Gamma_n^{\nu_n}$ be the components of Γ_{t_n}.

Suppose $\sigma \in H_1(X, \mathbb{Z})$ is a dividing cycle. Then, by Lemma A1.3, there is a positive constant $C(\sigma)$ with the property that for all sufficiently large n there are integers $s_1^1, \ldots, s_n^{\nu_n}$ satisfying

$$|s_i^j| \leq C(\sigma)$$

and such that σ is homologous in X to

$$\sigma_n = s_1^1\Gamma_n^1 + \cdots + s_n^{\nu_n}\Gamma_n^{\nu_n}.$$

Now,

$$\left|\int_\sigma \omega\right| = \left|\int_{\sigma_n} \omega\right| \leq \int_{\sigma_n} |\omega| \leq C(\sigma) \int_{\Gamma_{t_n}} |\omega|.$$

The right-hand side of this inequality tends to zero and consequently, the integral of ω over the dividing cycle σ vanishes. □

If there is an exhaustion function h with finite charge on the Riemann surface X, then, by Proposition 2.7, the first hypothesis of Theorem 1.17 is fulfilled. To verify the third hypothesis, we derive a variant of the Riemann bilinear relations. For this purpose we prove

LEMMA 2.8. *Let S be a smooth, compact surface whose boundary ∂S is the union of the closed curves $\gamma_1, \ldots, \gamma_m$. Let $A_1, B_1, \ldots, A_g, B_g$ be a canonical homology basis for S. Suppose that the smooth closed one forms ω and η on S satisfy*

$$\int_{\gamma_i} \omega = \int_{\gamma_i} \eta = 0$$

for $i = 1, \ldots, g$. Then,

$$\int_S \omega \wedge \eta = \sum_{i=1}^{g} \left(\int_{A_i} \omega \int_{B_i} \eta - \int_{B_i} \omega \int_{A_i} \eta \right) + \int_{\partial S} f\eta$$

where f is any smooth function on a neighborhood U of ∂S with

$$df = \omega\big|_U.$$

PROOF. Set

$$\alpha_i' = \eta_{B_i}', \quad \beta_i' = -\eta_{A_i}'$$

for $i = 1, \ldots, g$, where η_σ' is the one form constructed in Remark 1.5. By construction, the supports of α_i' and β_i' do not intersect ∂S. Recall that

$$A_i(\lambda) = \int_{A_i} \lambda, \quad B_i(\lambda) = \int_{B_i} \lambda$$

for all smooth one forms λ on S. It follows from our hypothesis that

$$\omega = \sum_{i=1}^{g} A_i(\omega)\alpha_i' + \sum_{j=1}^{g} B_j(\omega)\beta_j' + d\phi,$$

$$\eta = \sum_{k=1}^{g} A_k(\eta)\alpha_k' + \sum_{\ell=1}^{g} B_\ell(\eta)\beta_\ell' + d\psi.$$

Here, ϕ and ψ are smooth functions on S. We have

$$\begin{aligned}\omega \wedge \eta = \sum_{i,\ell=1}^{g} A_i(\omega)B_\ell(\eta)\alpha_i' \wedge \beta_\ell' + \sum_{i,k=1}^{g} A_i(\omega)A_k(\eta)\alpha_i' \wedge \alpha_k' \\ + \sum_{j,k=1}^{g} B_j(\omega)A_k(\eta)\,\beta_j' \wedge \alpha_k' + \sum_{j,\ell=1}^{g} B_j(\omega)B_\ell(\eta)\beta_j' \wedge \beta_\ell' \\ + \omega \wedge d\psi + d\phi \wedge \eta - d\phi \wedge d\psi.\end{aligned}$$

By Remark 1.5,

$$\int_S \omega \wedge \eta = \sum_{i=1}^{g}\left(\int_{A_i}\omega \int_{B_i}\eta - \int_{B_i}\omega \int_{A_i}\eta\right) + \int_S (\omega \wedge d\psi + d\phi \wedge \eta - d\phi \wedge d\psi)$$

By Stoke's theorem,

$$\begin{aligned}\int_S (\omega \wedge d\psi + d\phi \wedge \eta - d\phi \wedge d\psi) &= \int_S d(-\psi\omega + \phi\eta + \psi d\phi) \\ &= \int_{\partial S} -\psi\omega + \phi\eta + \psi\, d\phi \\ &= \int_{\partial S} \phi\eta\end{aligned}$$

since $d\phi = \omega$ on a neighborhood of ∂S.

If f is any smooth function on a neighborhood U of ∂S with

$$df = \omega\big|_U$$

then there are constants $c_1, \ldots, c_m$ with

$$f\big|_{\gamma_i} = \phi\big|_{\gamma_i} + c_i$$

for $i = 1, \ldots, m$. Therefore,

$$\int_{\partial S} f\eta = \sum_{i=1}^{m}\int_{\gamma_i} f\eta = \sum_{i=1}^{m}\int_{\gamma_i}\phi\eta + \sum_{i=1}^{m} c_i \int_{\gamma_i}\eta = \sum_{i=1}^{m}\int_{\gamma_i}\phi\eta = \int_{\partial S}\phi\eta.$$

Combining this identity with the result of the last paragraph

$$\int_S \omega \wedge \eta = \sum_{i=1}^{g}\left(\int_{A_i}\omega \int_{B_i}\eta - \int_{B_i}\omega \int_{A_i}\eta\right) + \int_{\partial S} f\eta. \qquad \square$$

THEOREM 2.9 (Riemann Bilinear Relations). *Suppose there is an exhaustion function with finite charge on the marked Riemann surface $(X; A_1, B_1, \ldots)$. Let ω and η be smooth, closed, square integrable one forms on X such that*

$$\int_{A_i}\omega = \int_{A_i}\eta = 0$$

for all but finitely many indices i. Then,

$$\int_X \omega \wedge \eta = \sum_{i=1}^{\infty} \left(\int_{A_i} \omega \int_{B_i} \eta - \int_{B_i} \omega \int_{A_i} \eta \right).$$

PROOF. Let h be an exhaustion function with finite charge on $(X; A_1, B_1, \dots)$. By Definition 2.1(iii), for each sufficiently large regular value $t > 0$ of h, the boundary Γ_t of X_t is a finite union of components each of which is homologous to a linear combination of $A_1, A_2, \dots$ and dividing cycles. Observe that for every $j \geq 1$, there is a $t(j)$ such that A_j does not appear in any component of Γ_t when $t > t(j)$, because B_j can be represented by a closed curve lying in the interior of $X_{t(j)}$. It follows from our hypothesis and Proposition 2.7 that the integral of ω and η around every component of Γ_t vanishes when t is large enough.

For each sufficiently large regular value $t > 0$ of h there is, by Definition 2.1(ii), the conclusion of the preceding paragraph and Lemma 2.8, an $n_t \geq 0$ such that

$$\begin{aligned}\int_{X_t} \omega \wedge \eta &= \sum_{i=1}^{n_t} \left(\int_{A_i} \omega \int_{B_i} \eta - \int_{B_i} \omega \int_{A_i} \eta \right) + \int_{\Gamma_t} f_t \eta \\ &= \sum_{i=1}^{\infty} \left(\int_{A_i} \omega \int_{B_i} \eta - \int_{B_i} \omega \int_{A_i} \eta \right) + \int_{\Gamma_t} f_t \eta\end{aligned}$$

where f_t is any smooth function on a neighborhood U of Γ_t with

$$df_t = \omega\big|_U.$$

We choose f_t so that it has a zero on each component of Γ_t. By the fundamental theorem of calculus,

$$\sup_{x \in \Gamma_t} |f_t(x)| \leq \int_{\Gamma_t} |\omega|.$$

Therefore,

$$\begin{aligned}\left| \int_{\Gamma_t} f_t \eta \right| &\leq \int_{\Gamma_t} |f_t \eta| \\ &\leq \left(\sup_{x \in \Gamma_t} |f_t(x)| \right) \int_{\Gamma_t} |\eta| \\ &\leq \int_{\Gamma_t} |\omega| \int_{\Gamma_t} |\eta|.\end{aligned}$$

By Lemma 2.6,

$$\liminf_{t \to \infty} \left| \int_{\Gamma_t} f_t \eta \right| \leq \liminf_{t \to \infty} \int_{\Gamma_t} |\omega| \int_{\Gamma_t} |\eta| = 0.$$

The proof is completed by the observation that

$$\int_X \omega \wedge \eta = \lim_{t \to \infty} \int_{X_t} \omega \wedge \eta.$$

□

PROPOSITION 2.10. *Suppose there is an exhaustion function with finite charge on the marked Riemann surface $(X; A_1, B_1, \dots)$. Let ω be a square integrable holomorphic one form on X such that*

$$\int_{A_k} \omega = 0$$

for all $k \geq 1$. Then, $\omega = 0$.

PROOF. Observe that ω and $\overline{*\omega} = i\overline{\omega}$ are closed. By hypothesis

$$\int_{A_i} \omega = \int_{A_i} \overline{*\omega} = 0$$

for all $i \geq 1$. It follows from Theorem 2.9 that

$$\begin{aligned}\|\omega\|^2 &= \int \omega \wedge \overline{*\omega} \\ &= \sum_{i=1}^{\infty}\left(\int_{A_i} \omega \int_{B_i} \overline{*\omega} - \int_{B_i} \omega \int_{A_i} \overline{*\omega}\right) \\ &= 0.\end{aligned}$$

□

APPENDIX: AN APPROXIMATION LEMMA.

We state and prove a proposition that is useful for the construction of exhaustion functions with finite charge. A technical lemma is required.

LEMMA A2.1. *Let f be a smooth, proper, positive function on the Riemann surface X without boundary such that*

$$\int_X |d{*}df| < \infty.$$

Then, there is an exhaustion function h with finite charge arbitrarily close to f in the Whitney topology on $C^\infty(X)$. If, in addition, V and U are open subsets of X with $\overline{V} \subset U$ and f is Morse on U, then there is an exhaustion function h with finite charge arbitrarily close to f in the Whitney topology on $C^\infty(X)$ such that

$$h\big|_V = f\big|_V.$$

PROOF. Let $Y_n, n \geq 1$, be a sequence of relatively compact open subsets of X satisfying $\overline{Y_n} \subset Y_{n+1}$ for all $n \geq 1$, and $\bigcup_{n\geq 1} Y_n = X$. Fix an open neighborhood $\mathcal{N}$ of f in the Whitney topology (see, for example, [**GG**, II:§3]) on $C^\infty(X)$ that contains only proper, positive functions on X. If we can inductively construct a sequence of functions $f_n \in \mathcal{N}$, $n \geq 0$, such that $f_0 = f$ and for all $n \geq 1$,

(i) $f_n\big|_{Y_n}$ is a Morse function.
(ii) $\int_X |d{*}df_n| \leq \int_X |d{*}df| + \sum_{j=1}^n \frac{1}{2^j}$.
(iii) $f_n\big|_{X\setminus\overline{Y}_{n+1}} = f$.
(iv) $f_n\big|_{Y_{n-2}} = f_{n-1}\big|_{Y_{n-2}}$.
(v) $f_n\big|_V = f\big|_V$.
(vi) $f_n\big|_U$ is a Morse function

then,

$$h = \lim_{n\to\infty} f_n$$

is an exhaustion function with finite charge belonging to $\mathcal{N}$ with $h\big|_V = f\big|_V$.

Suppose the functions $f_0, f_1, \ldots, f_n$, for some $n \geq 0$, have been constructed satisfying conditions (i) through (vi). To construct f_{n+1} choose a nonnegative, smooth function χ on X such that

$$\chi(x) = \begin{cases} 0, & x \in \overline{Y}_{n-1} \cup (X \setminus Y_{n+2}) \cup \overline{V} \\ 1, & x \in \overline{Y}_{n+1} \setminus (Y_n \cup U) \end{cases}.$$

Here, $Y_{-1} = Y_0 = \varnothing$. By construction, χ has compact support.

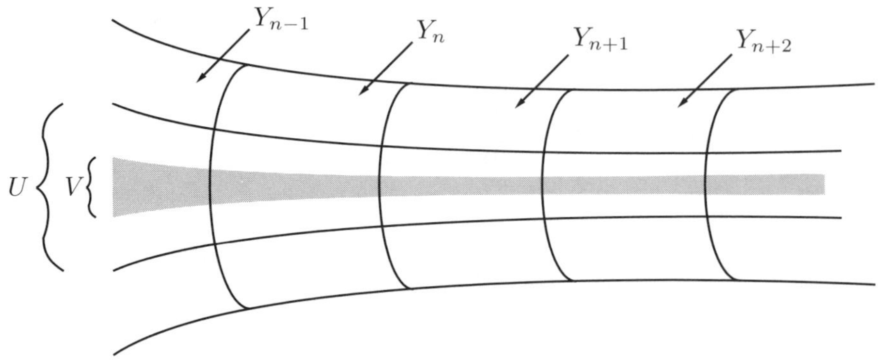

We can pick a Morse function g on $Y_{n+2}\backslash\overline{Y}_{n-1}$ arbitrarily close to $f_n\big|_{Y_{n+2}\backslash\overline{Y}_{n-1}}$, since Morse functions are open and dense in $C^\infty(Y_{n+2}\setminus\overline{Y}_{n-1})$ [**GG**, Chapter II, Theorem 6.2]. Now, set

$$f_{n+1} = (1-\chi)f_n + \chi g.$$

Clearly, f_{n+1} has properties (iii), (iv) and (v). To verify (i), (ii), (vi) and $f_{n+1}\in\mathcal{N}$, observe that

$$f_{n+1} - f_n = \chi(g - f_n).$$

If g is close enough to f_n in $C^\infty(Y_{n+2}\setminus\overline{Y}_{n-1})$, then (ii) holds and f_{n+1} belongs to $\mathcal{N}$, since, by hypothesis, $f_n\in\mathcal{N}$. Property (vi) holds because Morse functions are dense in $C^\infty(U)$ and f_n is a Morse function. For the same reason $f_{n+1}\big|_{Y_n}$ is a Morse function. By construction, $f_{n+1}\big|_{Y_{n+1}\backslash Y_n} = g\big|_{Y_{n+1}\backslash Y_n}$. It follows that $f_{n+1}\big|_{Y_{n+1}}$ is also a Morse function. □

PROPOSITION A2.2. *Let f be a continuous, proper function on X that is smooth outside a compact subset Y. Suppose,*

$$\int_{X\backslash Y} |d{*}df| < \infty.$$

Let V and U be open subsets of X with $\overline{V}\subset U$ such that f is Morse on U. Then, there is a compact subset Z and an exhaustion funtion h with finite charge on X such that $h\big|_{X\backslash Z}$ is arbitrarily close to $|f|\big|_{X\backslash Z}$ in $C^\infty(X\setminus Z)$ and

$$h\big|_{V\backslash Z} = |f|\big|_{V\backslash Z}$$

PROOF. Let $c > \max_{x\in Y}|f(x)|$ and set $Z = f^{-1}([-2c, 2c])$. The set Z is compact since f is proper. Pick any strictly positive smooth function ϕ on an open neighborhood of Z and let χ be a smooth nonnegative function that vanishes on $X\setminus Z$ and is identically one on $f^{-1}([-c,c])$. Observe that the function $\widetilde{f}$ defined by

$$\widetilde{f}(x) = \begin{cases} \phi(x), & f(x)\in[-c,c] \\ \chi(x)\phi(x) + (1-\chi(x))|f(x)|, & |f(x)| > c \end{cases}$$

satisfies the hypotheses of Lemma A2.1. □

3. Parabolic Surfaces

In this section, we review the basic properties of Riemann surfaces that are parabolic in the sense of Ahlfors–Nevanlinna. It is also shown that every Riemann surface with an exhaustion function with finite charge is parabolic.

DEFINITION 3.1. A harmonic exhaustion function h on a Riemann surface X without boundary is a continuous, proper, nonnegative function on X that is harmonic on the complement of a compact subset. A Riemann surface X is parabolic, in the sense of Ahlfors and Nevanlinna [**Ah, Ne**], if there exists a harmonic exhaustion function on X.

REMARK 3.2. A smooth harmonic exhaustion function h with Morse critical points is an exhaustion function with finite charge (see, Definition 2.1 since $d{*}dh = 0$ outside a compact set.

EXAMPLE 3.3. The complex plane is parabolic since

$$h(z) = \begin{cases} 0, & |z| \leq 1 \\ \log|z|, & |z| > 1 \end{cases}$$

is a continuous proper, nonnegative function that is harmonic on the complement of the unit disk. The name "parabolic" derives from the fact that noncompact, simply connected Riemann surfaces satisfying the condition of Definition 3.1 are biholomorphic to $\mathbb{C}$. However, the present use of the term parabolic should not be confused with the use of the same term in the context of universal covering surfaces.

The notion of a parabolic surface is clarified by

THEOREM 3.4. *A Riemann surface X is parabolic if and only if one of the following four equivalent conditions is satisfied*:

(i) *There is no nonconstant, negative, subharmonic function on X.*

(ii) (*The harmonic measure of the ideal boundary vanishes.*) *Suppose,*

$$\{X_n \mid n \geq 0\}$$

is any exhaustion of X by nonempty compact submanifolds $X_n \subset X$ with smooth boundary ∂X_n such that $X_n \subset \operatorname{int} X_{n+1}$, and u_n, $n \geq 1$, is the unique harmonic function on $X_n \setminus \operatorname{int} X_0$ satisfying

$$u_n(x) = \begin{cases} 0, & x \in \partial X_0 \\ 1, & x \in \partial X_n \end{cases}.$$

Then, for each $x \in X_n \setminus X_0$

$$\lim_{n\to\infty} u_n(x) = 0.$$

(iii) (*Nonexistence of Green's functions.*) *Suppose,*

$$\{X_n \mid n \geq 0\}$$

is any exhaustion of X as in (ii). *Let $x_0 \in X_0$ and let (U, z) be a coordinate system centered at x_0. If g_n, $n \geq 0$, is the unique harmonic function on $X_n \setminus \{x_0\}$ vanishing on ∂X_n such that $g_n + \log|z|$ is harmonic on U, then*

$$\lim_{n\to\infty} g_n(x) = \infty$$

for all $x \in X_n \setminus \{x_0\}$.

(iv) (*Maximum principle.*) *Suppose v is a bounded subharmonic function on a region $Y \subset X$ such that for all $x \in \partial Y$,*

$$\limsup_{\substack{y \in Y \\ y \to x}} v(y) \leq M.$$

Then,

$$\sup_{y \in Y} v(y) \leq M.$$

PROOF. See [**AS**, IV.6] and [**Na**]. □

We now review the criterion of extremal distance. It will be used to show that a Riemann surface X with an exhaustion function with finite charge is parabolic.

Recall that a linear density g on a Riemann surface X is a piecewise continuous section of the second symmetric tensor product of the real cotangent bundle of X such that $g(x)$ is positive semidefinite for all $x \in X$. The length of curves and the areas of surfaces with respect to the density g is defined just as for a metric. Suppose, $X_t \subset X$, $t \geq 0$, is a family of compact submanifolds with piecewise smooth boundary such that $X_s \subset X_t$ for all $t > s > 0$. Set

$$L(g : X_s, X_t) = \inf_{\substack{\text{paths } \gamma \text{ connecting} \\ \partial X_s \text{ to } \partial X_t \text{ in } X}} (\text{length of } \gamma \text{ with respect to } g)$$

and

$$A(g; X_t \setminus X_s) = \text{ the area of } X_t \setminus X_s \text{ with respect to } g$$

for all $t > s > 0$.

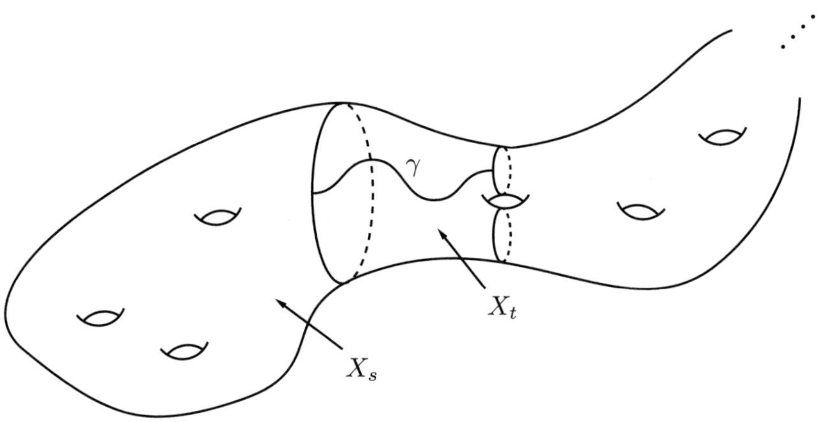

PROPOSITION 3.5 (Criterion of Extremal Distance). *Let X be a Riemann surface. Suppose $X_t \subset X$, $t \geq 0$ is a family of compact submanifolds with piecewise smooth boundary such that $X_s \subset X_t$ for all $t > s > 0$. If there exists a linear density g with*

$$\lim_{t \to \infty} \frac{L(g : X_s, X_t)^2}{A(g; X_t \setminus X_s)} = \infty$$

for some fixed $s > 0$, then X is parabolic.

PROOF. See [**AS**, IV.15B]. □

PROPOSITION 3.6. *There is an exhaustion function with finite charge on the Riemann surface X without boundary if and only if X is parabolic.*

PROOF. Suppose, h is an exhaustion function with finite charge on X. As in Section 2, $X_t = h^{-1}([0,t])$, for all $t \geq 0$, and the linear density g_h on X associated to h is

$$g_h(x) = dh \otimes dh + *dh \otimes *dh.$$

Observe that for any path γ joining Γ_s to Γ_t,

$$(\text{length of } \gamma \text{ with respect to } g_h) = \int_\gamma \sqrt{(dh)^2 + (*dh)^2} \geq \int_\gamma |dh| \geq t - s$$

and consequently,

$$L(g_h; X_s, X_t) \geq t - s.$$

Fix $s > 0$. By Stoke's theorem, for all $t > s$,

$$\begin{aligned} A(g_h; X_t \setminus X_s) &= \int_{X_t \setminus X_s} dh \wedge *dh \\ &= \int_{X_t \setminus X_s} d(h*dh) - hd*dh \\ &= \int_{\Gamma_t} h*dh - \int_{\Gamma_s} h*dh - \int_{X_t \setminus X_s} hd*dh \\ &= t\int_{\Gamma_t} *dh - s\int_{\Gamma_s} *dh - \int_{X_t \setminus X_s} htd*dh \end{aligned}$$

where, as before, $\Gamma_t = \partial X_t$ for all $t \geq 0$. We have

$$\int_{\Gamma_t} *dh - \int_{\Gamma_s} *dh = \int_{X_t \setminus X_s} d*dh = O(1)$$

and

$$\left| \int_{X_t \setminus X_s} hd*dh \right| \leq \int_{X_t \setminus X_s} h|d*dh|X_t \setminus X_s|d*dh| = O(t)$$

since, by hypothesis, h has finite charge. It follows that

$$A(g_h; X_t \setminus X_s) = (t-s)\int_{\Gamma_s} *dh + 0(t) - \int_{X_t \setminus X_s} hd*dh = O(t).$$

We see from the conclusions of the preceding two paragraphs that the criterion

$$\lim_{t\to\infty} \frac{L(g_h; X_s, X_t)^2}{A(g_h; X_t \setminus X_s)} = \infty$$

of extremal distance is satisfied. Therefore, X is parabolic.

Conversely, suppose that X is parabolic. One can use Proposition A2.2 to construct an exhaustion function with finite charge from any harmonic exhaustion function. □

An important fact about parabolic surfaces is

PROPOSITION 3.7. *Let X be a parabolic Riemann surface. A harmonic function f on X satisfying*

$$\|df\|^2 = \int_X df \wedge \overline{*df} < \infty$$

is constant.

PROOF. See [**SN**, §III, 1.3B], or the appendix to this section. □

We now combine Theorem 1.17, Proposition 2.7, Proposition 2.10, Proposition 3.6 and Proposition 3.7 to obtain the basic

THEOREM 3.8. *Let h be an exhaustion function with finite charge on the marked Riemann surface $(X; A_1, B_1, \dots)$. Then, X is parabolic and $(X; A_1, B_1, \dots)$ has a unique, normalized basis ω_k, $k \geq 1$, of square integrable holomorphic one forms dual to A_k, $k \geq 1$. Furthermore, the period map*

$$(\sigma, \omega) \in H_1^\natural(X, \mathbb{C}) \times H^1_{\mathrm{HK}}(X) \to \int_\sigma \omega \in \mathbb{C}$$

is a well-defined, nondegenerate pairing of $H_1^\natural(X, \mathbb{C})$ and $H^1_{\mathrm{HK}}(X)$.

The next topic is the canonical map. Several facts are required. Let f be a meromorphic function on a Riemann surface X. For each $a \in \mathbb{P}^1(\mathbb{C})$, (counting with multiplicity) set

$$n_f(a) = |\{x \in X \mid f(x) = a\}|.$$

LEMMA 3.9. *Let X be a noncompact, parabolic Riemann surface. Let f be a nonconstant meromorphic function on X with $1 \leq m < \infty$ (counted with multiplicity) poles that is bounded on the complement of each neighborhood of its poles. Then,*

$$n_f(a) \leq m$$

for all $a \in \mathbb{C}$, and the set

$$\{a \in \mathbb{C} \mid n_f(a) < m\}$$

has Lebesque measure zero.

PROOF. Let h be a harmonic exhaustion function on X. It can be assumed, by scaling, that the restriction of h to $h^{-1}([1, \infty))$ is harmonic. For all $r > 0$ and $a \in \mathbb{P}^1(\mathbb{C})$, (counting with multiplicity) set

$$n(a, r) = |\{x \in X \mid f(x) = a \text{ and } h(x) \leq r\}|,$$

$$v(r) = \frac{1}{\pi} \int_{\{x \in X \mid h(x) \leq r\}} f^*(\Phi)$$

where, Φ is the volume form of the Fubini–Study metric on $\mathbb{P}^1(\mathbb{C})$. For all $r \geq 1$, set

$$N(a, r) = \int_1^r n(a, t)\, dt, \quad T(r) = \int_1^r v(t)\, dt.$$

By Nevanlinna theory [**W**, p. 47], we have

$$\limsup_{r \to \infty} \frac{N(a, r)}{T(r)} \leq 1$$

for all $a \in \mathbb{P}^1(\mathbb{C})$, and

$$\limsup_{r \to \infty} \frac{N(a, r)}{T(r)} = 1$$

for all a in a subset of full measure of $\mathbb{P}^1(\mathbb{C})$.

Cover the poles of f by the union U of small open disks. By hypothesis,

$$n_f(\infty) = m$$

and

$$\sup_{x \in X \setminus U} |f(x)| \leq \text{const}_U < \infty.$$

It follows from the argument principle that $n_f(a) = m$ for all $a \in \mathbb{P}^1(\mathbb{C})$ sufficiently near ∞, and, as a result, $n(a, r) = m$ for all $a \in \mathbb{P}^1(\mathbb{C})$ sufficiently near ∞ and all sufficiently large r. Integrating,

$$N(a, r) = rm + \text{const}$$

for all $a \in \mathbb{P}^1(\mathbb{C})$ sufficiently near ∞ and all sufficiently large r. Combining this result with the conclusion of the last paragraph, we obtain

$$1 = \limsup_{r \to \infty} \frac{N(a, r)}{T(r)} = \limsup_{r \to \infty} \frac{rm + \text{const}}{T(r)}$$

for all $a \in \mathbb{P}^1(\mathbb{C}) \setminus$(set of Lebesque measure zero) sufficiently near ∞. Consequently,

$$\limsup_{r \to \infty} \frac{r}{T(r)} = \frac{1}{m}.$$

Now, suppose $n_f(a) \geq m + 1$ for some $a \in \mathbb{P}^1(\mathbb{C})$. Then, $n(a, t) \geq m + 1$ for all sufficiently large t and $N(a, r) \geq (m + 1)r + \text{const}$ for all sufficiently large r. These inequalities lead to the contradiction

$$\limsup_{r \to \infty} \frac{N(a, r)}{T(r)} \geq \limsup_{r \to \infty} \frac{(m + 1)r + \text{const}}{T(r)} = \limsup_{r \to \infty} \frac{(m + 1)r}{T(r)} = \frac{m + 1}{m} > 1.$$

Thus, $n_f(a) \leq m$ for all $a \in \mathbb{P}^1(\mathbb{C})$. The same calculation shows that the set of all $a \in \mathbb{P}^1(\mathbb{C})$ with $n_f(a) \leq m - 1$ is contained

$$\left\{ a \in \mathbb{P}^1(\mathbb{C}) \,\middle|\, \limsup_{r \to \infty} \frac{N(a, r)}{T(r)} < 1 \right\}$$

and hence has measure zero. □

THEOREM 3.10 (Royden–Riemann–Roch). *Let X be a parabolic Riemann surface. Let $x_1, \ldots, x_n$, be distinct points on X and $m_1, \ldots, m_n$ nonnegative integers. Then,*

$$\begin{aligned}
&\dim_{\mathbb{C}} \begin{pmatrix} \textit{the vector space of all meromorphic functions on } X \textit{ that} \\ \textit{have a pole at } x_i \textit{ of order at most } m_i, \textit{ for each } i = 1, \ldots, n \\ \textit{and are bounded on the complement of each neighborhood} \\ \textit{of } x_1, \ldots, x_n \end{pmatrix} \\
&\quad = m_1 + \cdots + m_n + 1 \\
&\qquad - \dim_{\mathbb{C}} \begin{pmatrix} \textit{orthogonal complement in } \Omega(X) \textit{ of the closed linear sub-} \\ \textit{space of all } \omega \textit{ that vanish at } x_i \textit{ to order at least } m_i, \textit{ for} \\ \textit{each } i = 1, \ldots, n \end{pmatrix}
\end{aligned}$$

PROOF. See [**R**, Proposition 4 and Theorem 1] □

REMARK 3.11. By Theorem 3.4 (i), a bounded holomorphic function on X is constant. Therefore, a nonconstant meromorphic function belonging to the vector space on the left-hand side of the Royden–Riemann–Roch theorem must have at least one pole.

PROPOSITION 3.12. *Let X be a parabolic Riemann surface that is not biholomorphic to an open subset of $\mathbb{P}^1(\mathbb{C})$. Then, for every $x \in X$ there is a square integrable holomorphic one form ω such that $\omega(x) \neq 0$.*

PROOF. Suppose every $\omega \in \Omega(X)$ vanishes at the point x. It follows from Theorem 3.10 that there is a nonconstant meromorphic function f on X that has a simple pole at x and is bounded on the complement of every neighborhood of x. By Lemma 3.9, $n_f(a) \leq 1$ for all $a \in \mathbb{P}^1(\mathbb{C})$. Therefore, f is a biholomorphic map between X and an open subset of $\mathbb{P}^1(\mathbb{C})$. □

Fix $x \in X$. Let z be a coordinate on a neighborhood U of x. For each $\omega \in \Omega(X)$ set

$$\delta_{x,z}(\omega) = \frac{\omega\big|_U}{dz}(x).$$

LEMMA 3.13. *Let z be a coordinate on an open subset U of X. For all $x \in U$, $\delta_{x,z}$ is a bounded linear functional on $\Omega(X)$. Furthermore, the map*

$$x \in U \to \delta_{x,z} \in \Omega^*(X)$$

is holomorphic. Here, $\Omega^*(X)$ is the dual space of $\Omega(X)$.

PROOF. Write $\omega\big|_U = f\,dz$. We have the representation

$$\delta_{x,z}(\omega) = f\big(z(x)\big).$$

Let $r_0(x) > 0$ be the radius of the largest closed disk centered at $z(x)$ that is contained in the image of U. For all $r \leq r_0(x)$,

$$\delta_{x,z}(\omega) = \frac{1}{2\pi}\int_0^{2\pi} f\big(z(x) + re^{i\theta}\big)\,d\theta = \frac{1}{\pi r_0^2}\int_{|\xi+i\eta-z(x)|\leq r_0(x)} f(\xi+i\eta)\,d\xi\,d\eta.$$

By Schwarz's inequality,

$$|\delta_{x,z}(\omega)| \leq \frac{1}{\sqrt{\pi}r_0}\left(\int_{|\xi+i\eta-z(x)|\leq r_0(x)} |f(\xi+i\eta)|^2\,d\xi\,d\eta\right)^{1/2} \leq \frac{1}{\sqrt{\pi}r_0(x)}\|\omega\|.$$

Thus, $\delta_{x,z} \in \Omega^*(X)$.

It follows from the representation above that $\delta_{x,z}$ is a weakly analytic map from U into $\Omega^*(X)$. By [**RS**, Theorem VI.4], it is holomorphic. □

By Lemma 3.13 and Proposition 3.12, we can make

DEFINITION 3.14. The canonical map κ_X from the parabolic Riemann surface X to the projectivized dual space $\mathbb{P}\big(\Omega^*(X)\big)$ is given by

$$\kappa_X(x) = [\delta_{x,z}]$$

where z is any coordinate on a neighborhood of $x \in X$.

REMARK 3.15. (i) Suppose z and z' are coordinates on a neighbourhood of $x \in X$, then $[\delta_{x,z}] = [\delta_{x,z'}]$ in $\mathbb{P}\big(\Omega^*(X)\big)$. Consequently, $\kappa_X(x)$ is well defined.

(ii) Suppose η_j, $j \geq 1$ is an orthonormal basis for $\Omega(X)$. Then,

$$\delta_{x,z}(\omega) = \sum_{j\geq 1} \delta_{x,z}(\eta_j)\langle\omega,\eta_j\rangle$$

where

$$\big(\delta_{x,z}(\eta_1), \delta_{x,z}(\eta_2), \dots\big) = \left(\frac{\eta_1}{dz}(x), \frac{\eta_2}{dz}(x), \dots\right) \in \ell^2(1,2,\dots)$$

We may write

$$\kappa_X(x) = \left[\frac{\eta_1}{dz}(x); \frac{\eta_2}{dz}(x); \dots\right] \in \mathbb{P}\big(\ell^2(1,2,\dots)\big).$$

Note that for a general independent, spanning sequence η_j, $j \geq 1$ of forms in $\Omega(X)$, say, the forms ω_j, $j \geq 1$ of Theorem 3.8,

$$\left(\frac{\eta_1}{dz}(x), \frac{\eta_2}{dz}(x), \dots\right)$$

may not lie in any recognizable topological vector space. However, see Remark 4.9.

(iii) The image of the canonical map κ_X does not lie in any hyperplane. To see this, let H be a hyperplane in $\mathbb{P}\big(\Omega^*(X)\big)$. Then, there is a nonzero $\eta \in \Omega(X)$, such that

$$H = \big\{[\lambda] \in \mathbb{P}\big(\Omega^*(X)\big) \mid \lambda(\eta) = 0\big\}.$$

If $\kappa_X(x) \in H$ for all $x \in X$, then

$$\delta_{x,z}(\eta) = 0$$

for all $x \in X$ and all coordinates z. It follows that $\eta = 0$.

We now generalize the fact that the canonical map for a compact Riemann surface is an embedding if and only if the surface is not hyperelliptic.

DEFINITION 3.16. A Riemann surface is hyperelliptic if there is a finite subset $\mathcal{I}$ of $\mathbb{P}^1(\mathbb{C})$, a discrete subset S of $\mathbb{P}^1(\mathbb{C}) \setminus \mathcal{I}$ and a proper holomorphic map τ from X to $\mathbb{P}^1(\mathbb{C}) \setminus \mathcal{I}$ of degree two that ramifies precisely over the points of S.

The map τ is called the hyperelliptic projection for X. The holomorphic map $\imath_X$ from X to X characterized by

$$\tau\big(\imath_X(x)\big) = \tau(x),$$

$$\big(x, \imath(x)\big) = \begin{cases} (x, x), & x \in \tau^{-1}(S) \\ \big(x, \imath_X(x) \neq x\big), & x \notin \tau^{-1}(S) \end{cases}$$

is called the hyperelliptic involution. By Example 2.5 one can use τ to construct an exhaustion function with finite charge. Thus, by Proposition 3.6, hyperelliptic surfaces are parabolic.

REMARK 3.17. Suppose $\mathcal{I} \neq \varnothing$. Then, there is a holomorphic function f_S on $\mathbb{P}^1(\mathbb{C}) \setminus \mathcal{I}$ with a simple root at each point of S and no other roots. The hyperelliptic Riemann surface X is biholomorphic to the plane curve

$$\big\{(y, z) \in \mathbb{C} \times \big(\mathbb{P}^1(\mathbb{C}) \setminus \mathcal{I}\big) \mid y^2 = f_S(z)\big\}.$$

In this representation, the hyperelliptic projection and involution are given by $\tau(y, z) = z$ and $\imath_X(y, z) = (-y, z)$.

PROPOSITION 3.18. *Let X be a hyperelliptic Riemann surface. For all $\omega \in \Omega(X)$,*

$$\imath_X^*(\omega) = -\omega.$$

If, in addition, $\dim_{\mathbb{C}} \Omega(X) \geq 2$, *then there is an injective, immersive holomorphic map* φ_X *from* $\mathbb{P}^1(\mathbb{C}) \setminus \mathcal{I}$ *to* $\mathbb{P}\big(\Omega^*(X)\big)$ *such that the diagram*

$$\begin{array}{ccc} X & \xrightarrow{\kappa_X} & \mathbb{P}\big(\Omega^*(X)\big) \\ {\scriptstyle\tau}\big\downarrow & \nearrow{\scriptstyle\varphi_X} & \\ \mathbb{P}^1(\mathbb{C}) \setminus \mathcal{I} & & \end{array}$$

commutes. Here, τ *is the hyperelliptic projection. In other words, the canonical map factors over the hyperelliptic involution.*

PROOF. Let γ and δ be any cycles on X. To compute the intersection index $\delta \times \big(\gamma + \imath_X(\gamma)\big)$, we may assume that δ intersects $\gamma + \imath_X(\gamma)$ transversely and that for all x on γ the cycle δ does not pass through both x and $\imath_X(x)$. One can see that $\tau(\delta) \times \tau(\gamma) = \tau(\delta) \times \tau\big(\imath_X(\gamma)\big) = 0$ which implies that

$$0 = \tau(\delta) \times \tau(\gamma) + \tau(\delta) \times \tau\big(\imath_X(\gamma)\big) = \delta \times \big(\gamma + \imath_X(\gamma)\big).$$

It follows that for all cycles on X, $[\imath_X(\gamma)] = -[\gamma]$ in $H_1^\natural(X, \mathbb{Z})$. That is, $[\imath_X(\gamma)] = -[\gamma]$ in $H_1(X, \mathbb{Z})$ modulo dividing cycles.

Let $\omega \in \Omega(X)$ and let γ be any cycle. Then,

$$-\int_\gamma \omega + \imath_X^*(\omega) = \int_{\imath_X(\gamma)} \omega + \imath_X^*(\omega)$$

since, by Proposition 2.7, the integral of any closed, square integrable form over a dividing cycle vanishes. Since $\imath_X^*$ is an involution on $\Omega(X)$ we have

$$-\int_\gamma \omega + \imath_X^*(\omega) = \int_\gamma \imath_X^*(\omega) + \omega = \int_\gamma \omega + \imath_X^*(\omega)$$

and therefore,

$$\int_\gamma \omega + \imath_X^*(\omega) = 0$$

for all cycles γ. It follows that $\omega + \imath_X^*(\omega) \in \Omega(X)$ is the differential of a holomorphic function on X. This function is constant by Proposition 3.7 so that

$$\imath_X^*(\omega) = -\omega.$$

In other words, $\imath_X^*$ acts on $\Omega(X)$ by multiplication with -1.

The first statement of the proposition, proved immediately above, implies that there is a holomorphic map φ_X from $\mathbb{P}^1(\mathbb{C}) \setminus \mathcal{I}$ to $\mathbb{P}\big(\Omega^*(X)\big)$ such that

$$\kappa_X = \varphi_X \circ \tau.$$

To prove that it is injective we must show that for all pairs x_1 and $x_2 \neq x_1, \imath_X(x_1)$ of points on X there is a $\omega \in \Omega(X)$ with $\omega(x_1) = 0$ and $\omega(x_2) \neq 0$. It is possible to choose a nonzero $\eta \in \Omega(X)$ with $\eta(x_1) = 0$, since $\dim_{\mathbb{C}}\Omega(X) \geq 2$. Let $m \geq 0$ be the order to which η vanishes at x_2. Observe that, by the first part of the proposition, m is even when x_2 is a ramification point of τ_X. Set

$$\omega(x) = \begin{cases} \left(\frac{\tau(x) - \tau(x_1)}{\tau(x) - \tau(x_2)}\right)^m \eta(x), & x_2 \text{ is not a ramification point,} \\ \left(\frac{\tau(x) - \tau(x_1)}{\tau(x) - \tau(x_2)}\right)^{m/2} \eta(x), & x_2 \text{ is a ramification point.} \end{cases}$$

The proof that φ_X is immersive goes in the same way. $\square$

REMARK 3.19. (i) Unravelling the definitions, the injectivity of φ_X is equivalent to the statement that for all $x_1, x_2 \in X$ and coordinates z_1, z_2 in neighbourhoods of x_1, x_2, there exists a constant λ such that $(\omega/dz_1)(x_1) = \lambda(\omega/dz_2)(x_2)$ for all $\omega \in \Omega(X)$ if and only if $\tau(x_1) = \tau(x_2)$. Immersiveness is equivalent to the statement that, for all $x \in X$, there is an $\omega \in \Omega(X)$ with a simple zero at x if and only if $\imath(x) \neq x$. If $\imath(x) = x$ there is an $\omega \in \Omega(X)$ with a zero of order two at x.

(ii) It is a direct consequence of Proposition 3.18 that the hyperelliptic projection is solely determined (up to automorphisms of $\mathbb{P}^1(\mathbb{C})$) by the complex structure on the surface.

DEFINITION 3.20. A pre-end of a Riemann surface is a closed, noncompact, connected submanifold with compact boundary whose complement is connected. An end of a Riemann surface is a pre-end that does not contain two disjoint pre-ends.

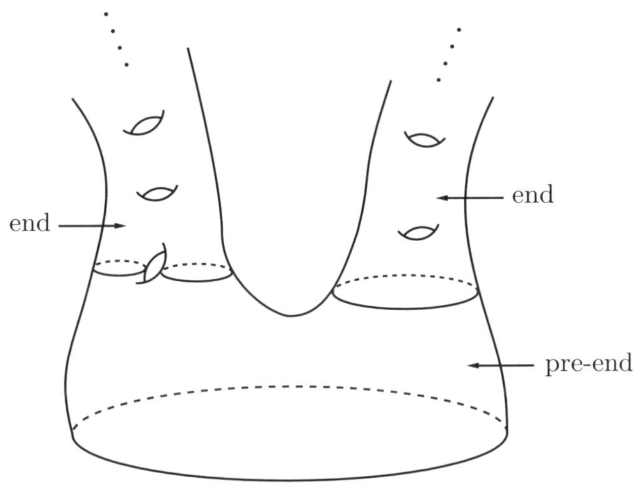

Two ends that contain a common end are called equivalent. An equivalence class of ends is called an ideal boundary point.

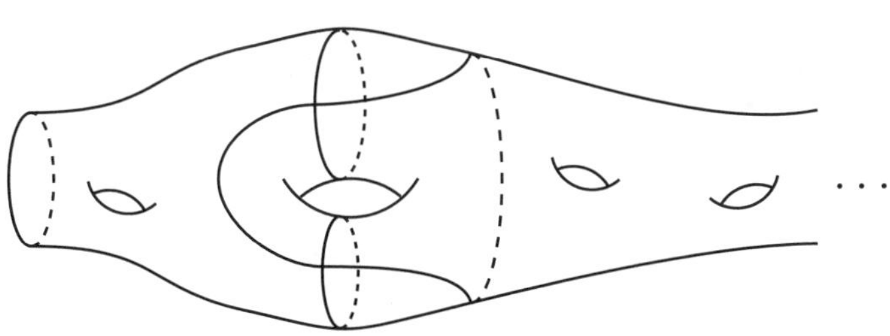

and are boundaries
of two equivalent ends

A Riemann surface has finite ideal boundary if there is a compact subset whose complement is the union of a finite number of ends.

REMARK 3.21. A Riemann surface with finite ideal boundary has a finite number of ideal boundary points. The "infinitely branching" surface below has no ends and hence no ideal boundary points, but still does not have finite ideal boundary.

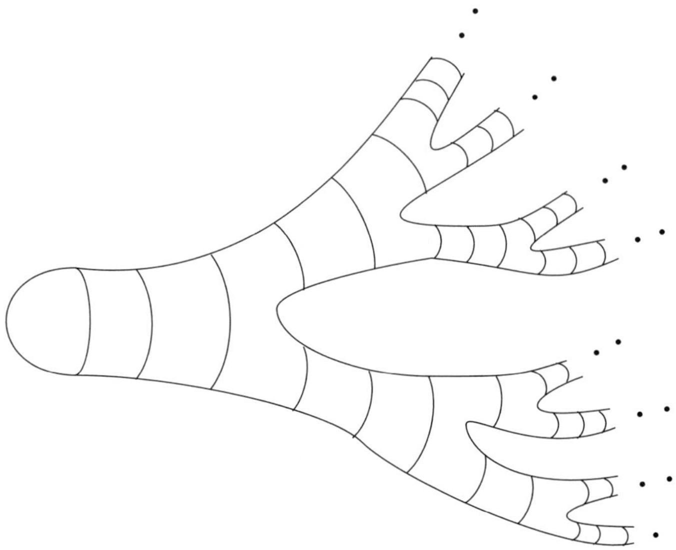

REMARK 3.22. If X is a hyperelliptic surface and τ is the hyperelliptic projection from X to $\mathbb{P}^1(\mathbb{C}) \setminus \mathcal{I}$, then $\mathcal{I}$ is in one to one correspondence with the ideal boundary points of X.

Let E be an end of the Riemann surface X. A function f on E has a limit w at the ideal boundary point represented by E if for every $\varepsilon > 0$ there is an end $E' \subset E$ such that

$$\sup_{x \in E'} |f(x) - w| \leq \varepsilon.$$

If the intersection form on $H_1(E, \mathbb{Z})$ vanishes, the end E is called planar.

THEOREM 3.23 (Heins). *Let p be an ideal boundary point of the Riemann surface X. Supppose f is analytic and bounded on an end E representing p. Then, f has a limit $w_p \in \mathbb{C}$ at p. Furthermore, there is a positive integer d_p and an end $E' \subset E$ such that the restriction of f to E' is a proper map of degree d_p from E' to a punctured neighborhood of w_p in $\mathbb{C}$. If, in addition, p cannot be represented by a planar end, then $d_p \geq 2$.*

PROOF. See, [**H**, pp. 300, 301]. □

COROLLARY 3.24. *Let X be a parabolic Riemann surface with finite ideal boundary. Suppose there is a meromorphic function f on X with $m < \infty$ (counted with multiplicity) poles that is bounded on the complement of any neighborhood of its poles. Then, for each ideal boundary point p of X there is a $w_p \in \mathbb{C}$ and a positive integer d_p such that w_p is the limit of f at p and*

$$n_f(a) + \sum_{\substack{\textit{ideal boundary points } q \\ \textit{such that } w_q = a}} d_q = m$$

for all $a \in \mathbb{P}^1(\mathbb{C})$. Furthermore, if p cannot be represented by a planar end, then $d_p \geq 2$.

PROOF. It is enough to prove that

$$\{a \in \mathbb{P}^1(\mathbb{C}) \mid n_f(a) < m\} \subset \{w_p \in \mathbb{P}^1(\mathbb{C}) \mid p \text{ is an ideal boundary point}\}.$$

Suppose $a \in \mathbb{P}^1(\mathbb{C})$. By Lemma 3.9, there is a sequence a_i, $i \geq 1$, converging to a such that $n_f(a_i) = m$ for all $i \geq 1$. If the preimage

$$f^{-1}(\{a_1, a_2, \dots\})$$

is bounded in X, then every accumulation point is mapped to a and the multiplicities must add up to m, so $n_f(a) = m$. If the preimage is unbounded, there is a subsequence converging to an ideal boundary point p. By Theorem 3.23, $a = w_p$. □

COROLLARY 3.25. *Let X be a parabolic Riemann surface with finite ideal boundary such that no ideal boundary point can be represented by a planar end. Let f be a meromorphic function on X with either two simple poles or one double pole and no other singularities. If f is bounded on the complement of any neighborhood of its poles, then X is hyperelliptic.*

PROOF. Let $\mathcal{I}$ be the set of limits of f at the ideal boundary points of X. By Theorem 3.23 and Corollary 3.24, f is a proper holomorphic map of degree two from X to $\mathbb{P}^1(\mathbb{C}) \setminus \mathcal{I}$. □

PROPOSITION 3.26. *Let X be a parabolic Riemann surface with finite ideal boundary such that no ideal boundary point can be represented by a planar end. Suppose, $\dim_{\mathbb{C}} \Omega(X) \geq 2$. Then, the canonical map κ_X is an injective immersion of X into $\mathbb{P}(\Omega^*(X))$ if and only if X is not hyperelliptic.*

PROOF. Suppose that X is not hyperelliptic. To prove that κ_X is an injective immersion we must show that for every pair of distinct points $x_1, x_2 \in X$ there are $\omega, \eta \in \Omega(X)$ such that $\omega(x_1) = 0$, but $\omega(x_2) \neq 0$ and η vanishes simply at x_1. If every $\omega \in \Omega(X)$ that vanishes at x_1 also vanishes at x_2, then, by Theorem 3.10, there is a meromorphic function with simple poles at x_1 and/or x_2 and no other singularities that is bounded on the complement of every neighborhood of its poles. If there are poles at both x_1 and x_2 then, by Corollary 3.24, X must be hyperelliptic contradicting our hypothesis. If there is a meromorphic function with a simple pole at x_1 or x_2, but not both, that is bounded on the complement of every neighborhood of this pole, then X is biholomorphic to an open subset of $\mathbb{P}^1(\mathbb{C})$ and $\dim_{\mathbb{C}} \Omega(X) = 0$. The case that every $\eta \in \Omega(X)$ that vanishes at x_1 vanishes with multiplicity at least two is treated in the same way.

If X is hyperelliptic, then, by Proposition 3.18, the canonical map is not injective. □

The last topic of this section is the Abel–Jacobi map. Fix a base point x_0 on the Riemann surface X. For any $x \in X$ and any smooth path γ joining x_0 to x, set

$$j(x, \gamma)(\omega) = \int_\gamma \omega$$

for all $\omega \in \Omega(X)$.

LEMMA 3.27. *For each $x \in X$ and each path γ connecting x_0 to x. Then,*

(i) *$j(x, \gamma)$ is a bounded linear functional on $\Omega(X)$.*

(ii) *Let Y be a simply connected open neighborhood of x. For each $y \in Y$, let γ_y be path joining $x_0$to y that is obtained by composing γ with a path inside Y that connects x to y. Then, the map*

$$y \in Y \to j(y, \gamma_y) \in \Omega^*(X)$$

is holomorphic. Its derivative with respect to a local coordinate z centered at x is

$$\frac{d}{dz} j(z, \gamma_z)\big|_{z=0} = \delta_{x,z}$$

where, as before, $\delta_{x,z}(\omega) = \omega\big|_Y / dz(x)$.

PROOF. The proof is similar to that of Lemma 3.13. □

DEFINITION 3.28. Let X be a Riemann surface and $\pi : \widetilde{X} \to X$ its universal cover. Fix a base point $\widetilde{x}_0$ in $\widetilde{X}$. The Abel–Jacobi map j_X from $\widetilde{X}$ to $\Omega^*(X)$ is given by

$$j_X(\widetilde{x}) = j\big(\pi(\widetilde{x}), \pi(\widetilde{\gamma})\big)$$

where $\widetilde{\gamma}$ is any path on $\widetilde{X}$ connecting $\widetilde{x}_0$ to $\widetilde{x}$.

Let $\widetilde{x}_1, \widetilde{x}_2 \in \widetilde{X}$ with $\pi(\widetilde{x}_1) = \pi(\widetilde{x}_2)$. Then, there is a cycle σ on X such that

$$j_X(\widetilde{x}_1)(\omega) - j_X(\widetilde{x}_2)(\omega) = \int_\sigma \omega$$

for all $\omega \in \Omega(X)$. We shall prove the converse for parabolic Riemann surfaces. A prerequisite is

THEOREM 3.29. *Let X be a parabolic Riemann surface. Let $x_1, \dots, x_n$ and $y_1, \dots, y_n$ be points on X with $x_i \neq y_i$ for $i = i, \dots, n$ and let $\gamma_1, \dots, \gamma_n$ be paths such that γ_i connects x_i to y_i for $i = i, \dots, n$. If*

$$\sum_{i=1}^n \int_{\gamma_i} \omega = 0$$

for all $\omega \in \Omega(X)$, then there is a meromorphic function on X with poles at $x_1, \dots, x_n$, roots at $y_1, \dots, y_n$ and no other poles or roots (counting multiplicity) that is bounded on the complement of every neighborhood of $x_1, \dots, x_n$.

PROOF. By Abel's theorem [**AS**, §V, 22.C], there is a meromorphic function f on X with poles at $x_1, \dots, x_n$, roots at $y_1, \dots, y_n$ and no other poles or roots such that $d \log f$ square integrable on the complement of every neighborhood of $x_1, \dots, x_n, y_1, \dots, y_n$. By [**R**, p. 43], *df* is square integrable on the complement of every neighborhood of $x_1, \dots, x_n$. It now follows from [**R**, Proposition 4], that f is bounded on the complement of every neighborhood of $x_1, \dots, x_n$. □

PROPOSITION 3.30. *Let X be a parabolic Riemann surface that is not biholomorphic to an open subset of $\mathbb{P}^1(\mathbb{C})$:*

(i) *The Abel–Jacobi map j_X from $\widetilde{X}$ to $\Omega^*(X)$ is immersive.*
(ii) *If $\widetilde{x}_1, \widetilde{x}_2 \in \widetilde{X}$ and σ is a cycle on X such that*

$$j_X(\widetilde{x}_1)(\omega) - j_X(\widetilde{x}_2)(\omega) = \int_\sigma \omega$$

for all $\omega \in \Omega(X)$, then, $\pi(\widetilde{x}_1) = \pi(\widetilde{x}_2)$.

PROOF. Part (i) is a direct consequence of Lemma 3.27 (ii) and Proposition 3.12.

For part (ii), suppose $\widetilde{x}_1, \widetilde{x}_2 \in \widetilde{X}$ and σ is a cycle on X such that

$$j_X(\widetilde{x}_1)(\omega) - j_X(\widetilde{x}_2)(\omega) = \int_\sigma \omega$$

for all $\omega \in \Omega(X)$. Then, there is a path γ in X joining $\pi(\widetilde{x}_1)$ to $\pi(\widetilde{x}_2)$ such that

$$\int_\gamma \omega = 0$$

for all $\omega \in \Omega(X)$. By Theorem 3.29, there is a meromorphic function with a simple pole at $\pi(\widetilde{x}_1)$ that is bounded on the complement of any neighborhood of $\pi(\widetilde{x}_1)$. The argument is now completed just as in the proof of Proposition 3.12. □

APPENDIX: A PROOF OF PROPOSITION 3.7.

Let h be a harmonic exhaustion function on the parabolic Riemann surface X. Recall that $X_t = h^{-1}([0,t])$ and $\Gamma_t = \partial X_t$, for all $t \geq 0$. By scaling, we may assume that $t = 1$ is a regular value and that h is harmonic on $X \setminus X_1$. Here, as before, Γ_t is oriented so that

$$*dh(\text{an oriented tangent vector to } \Gamma_t) \geq 0.$$

Of course, $dh(\text{a tangent vector to} \Gamma_t) = 0$.

Let f be a real valued, harmonic function on X with $\|df\| < \infty$. By Stoke's theorem,

$$\begin{aligned}\|df\|^2 &= \int_X df \wedge *df = \lim_{t\to\infty} \int_{X_t} df \wedge *df \\ &= \lim_{t\to\infty} \int_{\Gamma_t} f{*df}.\end{aligned}$$

We will show, in the spirit of [**Ah**], that

$$\liminf_{t\to\infty} \int_{\Gamma_t} f{*df} = 0.$$

Then, $\|df\|^2 = 0$ and f is constant on X.

LEMMA A3.1. *For all* $t > 1$,

$$\left|\int_{\Gamma_t} f{*df}\right| \leq 2\int_{\Gamma_t} |df| \int_{\Gamma_t} |{*df}| + \left|\sum_{\substack{\gamma \text{ component} \\ \text{of } \Gamma_t}} \int_\gamma f c_\gamma {*dh}\right|$$

where,

$$c_\gamma = \left(\int_\gamma *dh\right)^{-1}\left(\int_\gamma *dh\right).$$

PROOF. For each component γ of Γ_t,

$$\int_\gamma (*df - c_\gamma {*dh}) = 0.$$

Therefore,

$$\left|\int_\gamma f(*df - c_\gamma *dh)\right| \le \left(\int_\gamma |df|\right)\left(\int_\gamma |*df| + \left(\int_\gamma *dh\right)^{-1}\int_\gamma |*df| \int_\gamma *dh\right)$$
$$\le 2\left(\int_\gamma |df|\right)\left(\int_\gamma |*df|\right).$$

It follows that

$$\left|\int_{\Gamma_t} f*df - \sum_{\substack{\gamma \text{ a component} \\ \text{of } \Gamma_t}} c_\gamma *dh\right| \le \sum_\gamma \left|\int_\gamma f(*df - c_\gamma *dh)\right|$$
$$\le \sum_\gamma 2\left(\int_\gamma |df|\right)\left(\int_\gamma |*df|\right)$$
$$\le 2\left(\int_{\Gamma_t} |df|\right)\left(\int_{\Gamma_t} |*df|\right).$$

□

Suppose λ is a smooth square integrable one form on X. Observe that the exhaustion function with finite charge in the proof of Lemma 2.6 can be replaced by a harmonic exhaustion function. Thus,

$$\int_0^\infty \left(\int_{\Gamma_s} |\lambda|\right)^2 ds \le \text{const}\|\lambda\|^2 < \infty.$$

In particular, the functions $\int_{\Gamma_t} |df|$ and $\int_{\Gamma_t} |*df|$ of t are square integrable on $[0, \infty)$. Therefore, the product

$$\int_{\Gamma_t} |df| \int_{\Gamma_t} |*df|$$

appearing on the right-hand side of Lemma A3.1 is an integrable function of t and as a result, the measure of

$$\left\{t \in [1, \infty) \;\middle|\; 2\int_{\Gamma_t} |df| \int_{\Gamma_t} |*df| > \varepsilon\right\}$$

is finite for all $\varepsilon > 0$

If we can show that

$$F(t) = \frac{1}{\sqrt{t}} \sum_{\substack{\gamma \text{ a component} \\ \text{of } \Gamma_t}} \int_\gamma f c_\gamma *dh$$

is a square integrable function of $t \ge 2$, then the measure of

$$\left\{t \in [2, \infty) \;\middle|\; \left|\sum_{\substack{\gamma \text{ a component} \\ \text{of} \Gamma_t}} \int_\gamma f c_\gamma *dh\right| > \varepsilon\right\}$$

is also finite for all $\varepsilon > 0$. By Lemma A3.1, this will imply that

$$\liminf_{t \to \infty} \int_{\Gamma_t} f*df = 0.$$

We now prepare for the proof that $F(t)$ is a square integrable function of $t \ge 2$.

Let $t > 1$ be a regular value of h. The leaves of the (possibly singular) foliation of $\overline{X_t \setminus X_1}$ given by the harmonic one form $*dh$ coincide with the orbits of the gradient flow of h with respect to any conformal metric. Each leaf, either connects Γ_t to Γ_1, or connects Γ_t to a singular point of h, or connects a singular point of h to Γ_1. Only a finite number of leaves hit a singular point since h is harmonic.

For each component γ of Γ_t, let $L_\gamma \subset X$ be the union of all leaves that hit γ. By the preceding remarks,

(i) the complement of $\bigcup_{\substack{\gamma \text{ a component} \\ \text{of } \Gamma_t}} L_\gamma$ in $\overline{X_t \setminus X_1}$ has measure zero.

(ii) For each γ, the boundary of L_γ consists of γ, $L_\gamma \cap \Gamma_1$, a finite number of leaves of $*dh$ and a finite number of critical points of h.

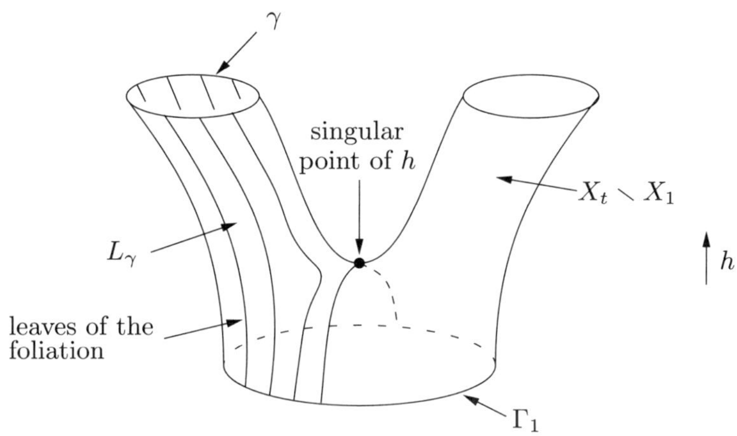

For each $t > 1$, set

$$\omega_t = \sum_{\substack{\gamma \text{a component} \\ \text{of } \Gamma_t}} c_\gamma \mathbb{1}_{L_\gamma} *dh$$

where $\mathbb{1}_{L_\gamma}$ is the indicator function of the set L_γ. Observe that ω_t is a piecewise harmonic one form. By Stoke's theorem, (ii) above, and the fact that $*dh$ vanishes, by definition, on all leaves of the foliation,

$$\begin{aligned} F(t) &= \frac{1}{\sqrt{t}} \sum_{\substack{\gamma \text{ a component} \\ \text{of } \Gamma_t}} \left\{ \int_\gamma f c_\gamma *dh - \int_{L_\gamma \cap \Gamma_1} f c_\gamma *dh \right\} \\ &\qquad + \frac{1}{\sqrt{t}} \sum_{\substack{\gamma \text{ a component} \\ \text{of } \Gamma_t}} \int_{L_\gamma \cap \Gamma_1} f c_\gamma *dh \\ &= \frac{1}{\sqrt{t}} \sum_{\substack{\gamma \text{ a component} \\ \text{of } \Gamma_t}} \int_{L_\gamma} df \wedge c_\gamma *dh + \frac{1}{\sqrt{t}} \sum_{\substack{\gamma \text{ a component} \\ \text{of } \Gamma_t}} \int_{L_\gamma \cap \Gamma_1} f c_\gamma *dh \end{aligned}$$

so that

$$F(t) = \frac{1}{\sqrt{t}} \int_X df \wedge \omega_t + \frac{1}{\sqrt{t}} \int_{\Gamma_1} f \omega_t.$$

LEMMA A3.2. *There is a constant* $\text{const}_{\Gamma_1} > 0$ *such that for all* $t > 2$,

$$\int_{\Gamma_1} |\omega_t| \leq \text{const}_{\Gamma_1} \|\omega_t\|.$$

PROOF. Pick $0 < \delta < 1$ such that h has no critical values in $[1, 1+\delta]$. Let γ be a component of Γ_t. As above, by Stoke's theorem,

$$\int_{L_\gamma \cap \Gamma_1} c_\gamma^2 {*dh} = \int_{L_\gamma \cap \Gamma_s} c_\gamma^2 {*dh}$$

for all $1 \le s \le 1+\delta$. It follows that

$$\begin{aligned}
\delta \int_{L_\gamma \cap \Gamma_1} c_\gamma^2 {*dh} &= \int_1^{1+\delta} ds \int_{L_\gamma \cap \Gamma_s} c_\gamma^2 {*dh} \\
&= \int_{L_\gamma \cap (X_{1+\delta} \setminus X_1)} c_\gamma^2 dh \wedge {*dh} \\
&= \|\omega_t|_{L_\gamma \cap (X_{1+\delta} \setminus X_1)}\|^2.
\end{aligned}$$

By Schwarz's inequality,

$$\begin{aligned}
\left(\int_{\Gamma_1} |\omega_t|\right)^2 &= \left(\int_{\Gamma_1} \left| \sum_{\substack{\gamma \text{ a component} \\ \text{of } \Gamma_t}} c_\gamma \mathbb{1}_{L_\gamma} \right| {*dh}\right)^2 \\
&\le \left(\int_{\Gamma_1} {*dh}\right) \left(\int_{\Gamma_1} \left\{ \sum_{\substack{\gamma \text{ a component} \\ \text{of } \Gamma_t}} c_\gamma \mathbb{1}_{L_\gamma} \right\}^2 {*dh}\right) \\
&= \left(\int_{\Gamma_1} {*dh}\right) \sum_{\substack{\gamma \text{ a component} \\ \text{of } \Gamma_t}} \int_{L_\gamma \cap \Gamma_1} c_\gamma^2 {*dh}.
\end{aligned}$$

Therefore,

$$\begin{aligned}
\left(\int_{\Gamma_1} |\omega_t|\right)^2 &= \frac{1}{\delta} \left(\int_{\Gamma_1} {*dh}\right) \sum_{\substack{\gamma \text{ a component} \\ \text{of } \Gamma_t}} \|\omega_t|_{L_\gamma \cap (X_{1+\delta} \setminus X_1)}\|^2 \\
&= \frac{1}{\delta} \left(\int_{\Gamma_1} {*dh}\right) \|\omega_t|_{(X_{1+\delta} \setminus X_1)}\|^2.
\end{aligned}$$

□

By Schwarz's inequality,

$$\int_2^\infty dt \left| \frac{1}{\sqrt{t}} \int_X df \wedge \omega_t \right|^2 \le \|df\|^2 \int_1^\infty \frac{dt}{t} \|\omega_t\|^2.$$

Also,

$$\begin{aligned}
\int_2^\infty dt \left| \frac{1}{\sqrt{t}} \int_{\Gamma_1} f\omega_t \right|^2 &\le \left(\sup_{x \in \Gamma_1} |f(x)| \right)^2 \int_1^\infty \frac{dt}{t} \left(\int_{\Gamma_1} |\omega_t| \right)^2 \\
&\le (\text{const}_{\Gamma_1} \sup_{x \in \Gamma_1} |f(x)|)^2 \int_1^\infty \frac{dt}{t} \|\omega_t\|^2.
\end{aligned}$$

by Lemma A3.2. Consequently,

$$\int_2^\infty dt |F(t)|^2 \le \text{const} \int_1^\infty \frac{dt}{t} \|\omega_t\|^2.$$

The proof of Proposition 3.7 is now reduced to

LEMMA A3.3.

$$\int_1^\infty \frac{dt}{t}\|\omega_t\|^2 < \infty.$$

PROOF. For each component γ of Γ_t,

$$\begin{aligned}\|c_\gamma{*}dh\big|_{L_\gamma}\|^2 &= c_\gamma^2\int_{L_\gamma} dh\wedge{*}dh\\ &= c_\gamma^2\int_{L_\gamma} d(h{*}dh)\\ &= c_\gamma^2\int_{\partial L_\gamma} h{*}dh\\ &= c_\gamma^2\Big(\int_\gamma h{*}dh - \int_{L_\gamma\cap\Gamma_1} h{*}dh\Big)\end{aligned}$$

since ${*}dh$ vanishes on all the other boundary components. Recall that $h\big|_\gamma = t$ and $h\big|_{L_\gamma\cap\Gamma_1} = 1$. Therefore,

$$\begin{aligned}\|c_\gamma{*}dh\big|_{L_\gamma}\|^2 &= c_\gamma^2\Big(t\int_\gamma {*}dh - \int_{L_\gamma\cap\Gamma_1}{*}dh\Big)\\ &= \Big(\int_\gamma {*}dh\Big)^{-2}\Big(\int_\gamma {*}dh\Big)^2\Big(t\int_\gamma {*}dh - \int_{L_\gamma\cap\Gamma_1}{*}dh\Big)\\ &\le t\Big(\int_\gamma {*}dh\Big)^{-1}\Big(\int_\gamma {*}dh\Big)^2\end{aligned}$$

since

$$\int_{L_\gamma\cap\Gamma_1}{*}dh \ge 0.$$

Summing over components,

$$\|\omega_t\|^2 = \sum_{\substack{\gamma \text{ a component}\\ \text{pof } \Gamma_t}} \|c_\gamma{*}dh\big|_{L_\gamma}\|^2 \le t\sum_{\substack{\gamma \text{ a component}\\ \text{of } \Gamma_t}}\Big(\int_\gamma {*}dh\Big)^{-1}\Big(\int_\gamma {*}dh\Big)^2.$$

The statement of the lemma now follows from Lemma A3.4 below. □

LEMMA A3.4. *Let λ be a smooth one form on X. Then,*

$$\int_1^t ds\sum_{\substack{\gamma \text{ a component}\\ \text{of } \Gamma_s}}\Big(\int_\gamma {*}dh\Big)^{-1}\Big(\int_\gamma |\lambda|\Big)^2 \le \|\lambda\big|_{X_t\setminus X_1}\|^2.$$

In particular,

$$\int_1^\infty ds\sum_{\substack{\gamma \text{ a component}\\ \text{of } \Gamma_s}}\Big(\int_\gamma {*}dh\Big)^{-1}\Big(\int_\gamma {*}dh\Big)^2 < \infty.$$

PROOF. For every component γ of Γ_s,

$$\begin{aligned}\left(\int_\gamma |\lambda|\right)^2 &= \left(\int_\gamma \sqrt{|\lambda_1|^2+|\lambda_2|^2}\sqrt{(dh)^2+(*dh)^2}\right)^2 \\ &= \left(\int_\gamma \sqrt{|\lambda_1|^2+|\lambda_2|^2}*dh\right)^2 \\ &\le \int_\gamma *dh \int_\gamma (|\lambda_1|^2+|\lambda_2|^2)*dh\end{aligned}$$

by Schwarz's inequality. It follows that

$$\begin{aligned}\int_1^t ds \sum_{\substack{\gamma \text{ a component} \\ \text{of } \Gamma_s}} \left(\int_\gamma *dh\right)^{-1}\left(\int_\gamma |\lambda|\right)^2 &\le \int_1^t ds \sum_{\substack{\gamma \text{ a component} \\ \text{of } \Gamma_s}} \int_\gamma (|\lambda_1|^2+|\lambda_2|^2)*dh \\ &= \int_1^t ds \int_{\Gamma_s} (|\lambda_1|^2+|\lambda_2|^2)*dh \\ &= \int_{X_t\setminus X_1} (|\lambda_1|^2+|\lambda_2|^2)dh\wedge *dh. \qquad \square\end{aligned}$$

4. Theta Functions

Suppose there is an exhaustion function with finite charge (see, Definition 2.1) on the marked Riemann surface $(X; A_1, B_1, \dots)$. Then, by Theorem 3.8, the marked surface $(X; A_1, B_1, \dots)$ has a unique basis ω_k, $k \ge 1$, of square integrable holomorphic one forms satisfying

$$\int_{A_i} \omega_j = \delta_{i,j}$$

for all $i, j \ge 1$.

DEFINITION 4.1. Suppose there is an exhaustion function with finite charge on the marked Riemann surface $(X; A_1, B_1, \dots)$. The period matrix $\mathcal{R}_X$ of this marked surface is

$$\mathcal{R}_X = (\mathcal{R}_{i,j}) = \left(\int_{B_i} \omega_j\right).$$

As usual, the Riemann bilinear relations (Theorem 2.9) imply

PROPOSITION 4.2. *Suppose there is an exhaustion function with finite charge on the marked Riemann surface $(X; A_1, B_1, \dots)$. Then, the period matrix $\mathcal{R}_X$ is symmetric and* $\operatorname{Im}\mathcal{R}_X$ *is positive definite. That is,*

$$\langle \mathbf{n}, \operatorname{Im}\mathcal{R}_X\mathbf{n}\rangle = \sum_{i,j\ge 1} n_i \operatorname{Im}\mathcal{R}_{i,j} n_j > 0$$

for all nonzero vectors

$$\mathbf{n} = \begin{pmatrix} n_1 \\ n_2 \\ \vdots \end{pmatrix}$$

in $\mathbb{Z}^\infty$ with only a finite number of nonzero components.

PROOF. For each k and ℓ, the pair of forms ω_k, ω_ℓ satisfies the hypothesis of Theorem 2.9. Consequently,

$$0 = \int_X \omega_k \wedge \omega_\ell = \sum_{i=1}^{\infty} \left(\int_{A_i} \omega_k \int_{B_i} \omega_\ell - \int_{B_i} \omega_k \int_{A_i} \omega_\ell \right) = \mathcal{R}_{k,\ell} - \mathcal{R}_{\ell,k}$$

since $\omega_k \wedge \omega_\ell = 0$. Similarly, for each nonzero $\mathbf{n} \in \mathbb{Z}^\infty$ with only a finite number of nonzero components, the form $\sum_{i\geq 1} n_i\omega_i$ and and its complex conjugate

$$\overline{* \sum_{i\geq 1} n_i\omega_i} = i \sum_{i\geq 1} n_i\bar{\omega}_i$$

satisfy the hypothesis of Theorem 2.9. We have,

$$0 < \left\| \sum_{i\geq 1} n_i\omega_i \right\|^2 = i \int_X \left(\sum_{i\geq 1} n_i\omega_i \right) \wedge \left(\sum_{j\geq 1} n_j\bar{\omega}_j \right) = \sum_{i,j\geq 1} n_i \operatorname{Im} \mathcal{R}_{i,j} n_j. \qquad \square$$

For each $0 < t < 1$, let $H(t) \subset \mathbb{C}^2$ be the model handle given by

$$H(t) = \{(z_1, z_2) \in \mathbb{C}^2 | z_1 z_2 = t, |z_1|, |z_2| \leq 1\}.$$

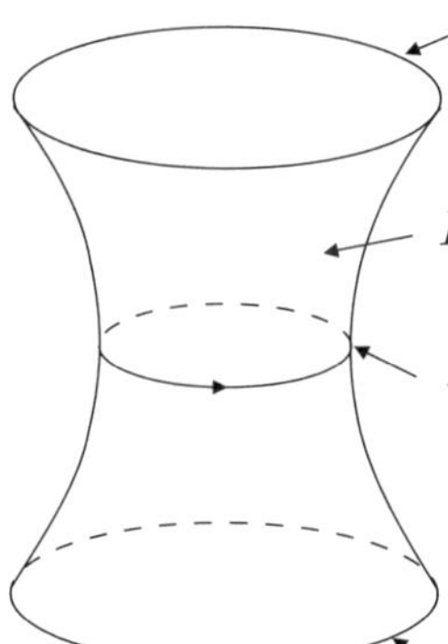

LEMMA 4.3. *Fix* $0 < t < 1$. *Let*

$$A = \{(\sqrt{t}e^{i\theta}, \sqrt{t}e^{-i\theta}) \mid 0 \leq \theta \leq 2\pi\}$$

be the oriented waist on the model handle $H(t)$. *For every holomorphic one form* ω *on* $H(t)$,

$$\|\omega\|_2 \geq \sqrt{\frac{|\log t|}{2\pi}} \left| \int_A \omega \right|.$$

PROOF. The projection

$$(z_1, z_2) \in H(t) \to z_1 \in \{z_1 \mid t \leq |z_1| \leq 1\}$$

from $H(t)$ to the annulus $t \leq |z_1| \leq 1$ is a global coordinate. Write

$$\omega = f(z_1)\, dz_1.$$

For any fixed r

$$\left| \int_A \omega \right|^2 = \left| \int_0^{2\pi} 1 \cdot f(re^{i\theta}) re^{i\theta} d\theta \right|^2 \leq 2\pi \int_0^{2\pi} |rf(re^{i\theta})|^2\, d\theta.$$

Hence

$$\|\omega\|_2^2 = \frac{1}{2}\int_{t\le|z_1|\le 1} |f(z_1)|^2 |\,dz_1\wedge d\bar z_1| = \int_t^1\int_0^{2\pi} |rf(re^{i\theta})|^2 d\theta \frac{dr}{r}$$
$$\ge \frac{1}{2\pi}\left|\int_A \omega\right|^2 \int_t^1 \frac{dr}{r} = \frac{|\log t|}{2\pi}\left|\int_A \omega\right|^2. \qquad \square$$

PROPOSITION 4.4. *Denote by $\mathcal{R}_X$ the period matrix of the marked Riemann surface $(X; A_1, B_1, \dots)$ on which there is an exhaustion function with finite charge. Let $t_j \in (0,1)$, $j \ge 1$. Suppose that for all $j \ge 1$ there exists an injective, holomorphic map*

$$\phi_j : H(t_j) \to X$$

such that

$$\phi_j(\{(\sqrt{t_j}e^{i\theta}, \sqrt{t_j}e^{-i\theta}) | 0 \le \theta \le 2\pi\})$$

is homologous to A_j and

$$\phi_k(H(t_k)) \cap \phi_\ell(H(t_\ell)) = \varnothing$$

for all $k \ne \ell$. Then,

$$\langle \mathbf{n}, \operatorname{Im}\mathcal{R}_X \mathbf{n}\rangle \ge \frac{1}{2\pi}\sum_{j\ge 1} |\log t_j| n_j^2$$

for all vectors $\mathbf{n} \in \mathbb{Z}^\infty$ with only a finite number of nonzero components.

PROOF. Let $\mathbf{n} \in \mathbb{Z}^\infty$ with only a finite number of nonzero components. As in the proof of Proposition 4.2,

$$\langle \mathbf{n}, \operatorname{Im}\mathcal{R}_X \mathbf{n}\rangle = \left\|\sum_{i\ge 1} n_i\omega_i\right\|^2 \ge \sum_{j\ge 1}\left\|\sum_{i\ge 1} n_i\omega_i\Big|_{\phi_j(H(t_j))}\right\|^2.$$

By Lemma 4.3,

$$\langle \mathbf{n}, \operatorname{Im}\mathcal{R}_X \mathbf{n}\rangle \ge \frac{1}{2\pi}\sum_{j\ge 1}|\log t_j| \left|\int_{A_j}\sum_{i\ge 1} n_i\omega_i\right|^2 = \frac{1}{2\pi}\sum_{j\ge 1}|\log t_j| n_j^2. \qquad \square$$

DEFINITION 4.5. Let

$$R = (R_{ij}; i, j \ge 1)$$

be an infinite, symmetric complex matrix such that $\operatorname{Im} R$ is positive definite. The formal theta series $\theta(\mathbf{z}, R), \mathbf{z} \in \mathbb{C}^\infty$, corresponding to R is

$$\theta(\mathbf{z}, R) = \sum_{\mathbf{n}\in\mathbb{Z}^\infty\, |\mathbf{n}|<\infty} e^{2\pi i\langle \mathbf{z}, \mathbf{n}\rangle} e^{\pi i \langle \mathbf{n}, R\mathbf{n}\rangle}$$

where

$$|\mathbf{n}| = |n_1| + |n_2| + \cdots.$$

The formal theta series $\theta(\mathbf{z}, \mathcal{R}_X)$ for a marked Riemann surface $(X; A_1, B_1, \dots)$ on which there is an exhaustion function with finite charge is the formal theta series corresponding to the period matrix $\mathcal{R}_X$.

The generic formal theta series is divergent. The main result of this section is

THEOREM 4.6. *Suppose that the symmetric matrix R and the sequence $t_j \in (0,1)$, $j \geq 1$, satisfy*

$$\sum_{j\geq 1} t_j^{\beta} < \infty \tag{4.1}$$

for some $0 < \beta < \frac{1}{2}$, and

$$\langle \mathbf{n}, \operatorname{Im} R\mathbf{n}\rangle = \sum_{i,j\geq 1} n_i \operatorname{Im} R_{i,j} n_j \geq \frac{1}{2\pi} \sum_{j\geq 1} |\log t_j| n_j^2 \tag{4.2}$$

and all vectors $\mathbf{n} \in \mathbb{Z}^\infty$ with only a finite number of nonzero components. Let B be the Banach space given by

$$B = \left\{ \mathbf{z} = (z_1, z_2, \dots) \in \mathbb{C}^\infty \;\middle|\; \lim_{j\to\infty} \frac{|z_j|}{|\log t_j|} = 0 \right\}$$

with norm

$$\|\mathbf{z}\| = \sup_{j\geq 1} \frac{|z_j|}{|\log t_j|}.$$

Then, for every point $\mathbf{w} \in B$ the theta series

$$\theta(\mathbf{z}, R) = \sum_{\substack{\mathbf{n}\in\mathbb{Z}^\infty \\ |\mathbf{n}|<\infty}} e^{2\pi i \langle \mathbf{z},\mathbf{n}\rangle} e^{\pi i \langle \mathbf{n}, R\mathbf{n}\rangle}$$

converges absolutely and uniformly on the ball in B of radius $r = 1 - 2\beta/8\pi$ centered at $\mathbf{w}$ to a holomorphic function.

PROOF. Fix $\mathbf{w}$ in B. For all $\mathbf{z}$ in the ball of radius r centered at $\mathbf{w}$ and all $\mathbf{n} \in \mathbb{Z}^\infty$ with $|\mathbf{n}| < \infty$,

$$\begin{aligned}
|e^{2\pi i\langle \mathbf{z},\mathbf{n}\rangle} e^{\pi i \langle \mathbf{n}, R\mathbf{n}\rangle}| &\leq e^{2\pi |\langle \operatorname{Im} \mathbf{z}, \mathbf{n}\rangle|} e^{-\pi \langle \mathbf{n}, \operatorname{Im} R\mathbf{n}\rangle} \\
&\leq e^{2\pi |\langle \operatorname{Im} \mathbf{w} + \operatorname{Im} \mathbf{z} - \operatorname{Im} \mathbf{w}, \mathbf{n}\rangle| - 1/2 \sum_{j\geq 1} |\log t_j| n_j^2} \\
&\leq e^{2\pi \sum_{j\geq 1} (|\operatorname{Im} w_j| + |\operatorname{Im} z_j - \operatorname{Im} w_j|)|n_j| - 1/2 \sum_{j\geq 1} |\log t_j| n_j^2} \\
&= e^{1/2 \sum_{j\geq 1} |\log t_j| (4\pi |n_j| (|\operatorname{Im} w_j| + |\operatorname{Im} z_j - \operatorname{Im} w_j| / |\log t_j|) - n_j^2)} \\
&\leq e^{-1/2 \sum_{j\geq 1} |n_j| \, |\log t_j| (|n_j| - 4\pi (|\operatorname{Im} w_j| / |\log t_j| + r))}.
\end{aligned}$$

If the sum of the majorants is finite, then the theta series converges absolutely and uniformly on the ball $\|\mathbf{w} - \mathbf{z}\| \leq r$ to a holomorphic function.

Pick $j_0(r)$ such that

$$\frac{|\operatorname{Im} w_j|}{|\log t_j|} < r$$

for all $j > j_0(r)$. We have

$$\begin{aligned}
\sum_{\substack{\mathbf{n}\in\mathbb{Z}^\infty \\ |\mathbf{n}|<\infty}} & e^{-1/2 \sum_{j\geq 1} |n_j| \, |\log t_j| (|n_j| - 4\pi (|\operatorname{Im} w_j| / |\log t_j| + r))} \\
&\leq \prod_{j\geq 1} \sum_{n\in\mathbb{Z}} e^{-1/2 |n| \, |\log t_j| (|n| - 4\pi (|\operatorname{Im} w_j| / |\log t_j| + r))} \\
&\leq \prod_{1\leq j\leq j_0(r)} \sum_{n\in\mathbb{Z}} e^{-1/2 |n| \, |\log t_j| (|n| - 4\pi (|\operatorname{Im} w_j| / |\log t_j| + r))} \\
&\qquad\qquad \times \prod_{j > j_0(r)} \sum_{n\in\mathbb{Z}} e^{-1/2 |n| |\log t_j| (|n| - 8\pi r)}.
\end{aligned}$$

We have

$$\sum_{n\in\mathbb{Z}} e^{-1/2|\log t_j|(n^2-8\pi|n|r)} \leq \sum_{n\in\mathbb{Z}} e^{-\beta|\log t_j|\,|n|} = 1+2\sum_{n\geq 1} e^{-\beta|\log t_j|n} = 1+2\left(\frac{t_j^\beta}{1-t_j^\beta}\right)$$

since,

$$\tfrac{1}{2}(|n|-8\pi r) = \tfrac{1}{2}\big(|n|-(1-2\beta)\big) = \tfrac{1}{2}(|n|-1)+\beta \geq \beta$$

for $|n| \geq 1$. Consequently, by (4.1),

$$\prod_{j>j_0(r)} \sum_{n\in\mathbb{Z}} e^{-1/2|n|\,|\log t_j|(|n|-8\pi r)} < \infty.$$

The above argument shows that $\theta(x, R)$ is defined and analytic on the subset of the Banach space $\{\mathbf{z} \in \mathbb{C}^\infty \mid \sup_{j\to\infty} |z_j|/|\log t_j| < \infty\}$ consisting of those elements for which

$$\limsup_{j\to\infty} \frac{|\operatorname{Im} z_j|}{|\log t_j|} < \frac{1-2\beta}{4\pi}. \qquad \square$$

REMARK 4.7. Recall that (see, [**RS**, III.2, Example 3]) the dual space B^* of B is given by

$$B^* = \left\{\mathbf{w} = (w_1, w_2, \dots) \in \mathbb{C}^\infty \;\middle|\; \sum_{j\geq 1} |w_j|\,|\log t_j| < \infty\right\}.$$

REMARK 4.8. Supose, f is a map from an open subset U of a complex Banach space E into a complex Banach space F. We recall that f is holomorphic on U, if it is continuously differentiable on U. To be precise, let $L(E, F)$ be the Banach space of bounded linear maps between E and F. Then, f is holomorphic on U, if there is a continuous map

$$x \in U \to d_x f \in L(E, F)$$

such that for each $x \in U$ and every $\varepsilon > 0$ there is a $\delta > 0$ for which

$$\|f(x+h) - f(x) - d_x f(h)\|_F \leq \varepsilon \|h\|_E$$

when $h \in E$ with $\|h\|_E < \delta$. One can show that the limit of a uniformly convergent sequence of holomorphic maps on U is also holomorphic. (See, for example, [**PT**, p. 137].)

Let f be holomorphic on U. One can also show (see, for example, [**PT**, p. 133]), that for each $x \in U$ there are bounded, symmetric, F-valued multilinear forms $f_m(x; h_1, \dots, h_m), m \geq 0$, on E with

$$\|f_m(x; \cdot, \dots, \cdot)\| \leq \operatorname{const}(x)\frac{m!}{\rho^m}$$

for some $\rho > 0$ such that

$$f(x+y) = \sum_{m\geq 0} \frac{1}{m!} f_m(x; y, \dots, y)$$

when $\|y\|_E < \rho$. In our case,

$$\begin{aligned}\theta(\mathbf{z}+\mathbf{h}, R) &= \sum_{\substack{\mathbf{n}\in\mathbb{Z}^\infty \\ |\mathbf{n}|<\infty}} e^{2\pi i\langle \mathbf{z}+\mathbf{h},\mathbf{n}\rangle} e^{\pi i\langle \mathbf{n}, R\mathbf{n}\rangle} \\ &= \sum_{\substack{\mathbf{n}\in\mathbb{Z}^\infty \\ |\mathbf{n}|<\infty}} \sum_{m\geq 0} \frac{1}{m!}(2\pi i\langle \mathbf{h},\mathbf{n}\rangle)^m e^{2\pi i\langle \mathbf{z},\mathbf{n}\rangle} e^{\pi i\langle \mathbf{n}, R\mathbf{n}\rangle} \\ &= \sum_{m\geq 0} \frac{1}{m!}\theta_m(\mathbf{z};\mathbf{h},\ldots,\mathbf{h})\end{aligned}$$

where

$$\theta_m(\mathbf{z};\mathbf{h}_1,\ldots,\mathbf{h}_m) = \sum_{\substack{\mathbf{n}\in\mathbb{Z}^\infty \\ |\mathbf{n}|<\infty}} \left(\prod_{i=1}^m 2\pi i\langle \mathbf{h}_i,\mathbf{n}\rangle\right) e^{2\pi i\langle \mathbf{z},\mathbf{n}\rangle} e^{\pi i\langle \mathbf{n}, R\mathbf{n}\rangle}.$$

We have

$$\begin{aligned}\left|\prod_{i=1}^m 2\pi i\langle \mathbf{h}_i,\mathbf{n}\rangle\right| &\leq \left(2\pi\sum_{j\geq 1}|\log t_j|\,|n_j|\right)^m \prod_{i=1}^m \|\mathbf{h}_i\| \\ &\leq \frac{m!}{\rho^m} e^{2\pi\rho\sum_{j\geq 1}|\log t_j|\,|n_j|} \prod_{i=1}^m \|\mathbf{h}_i\|\end{aligned}$$

and for all $r > \rho > 0$,

$$\begin{aligned}\|\theta_m(\mathbf{z};\cdot,\ldots,\cdot)\| &= \sup_{\substack{\mathbf{h}_i\in B \\ \mathbf{h}_i\neq 0}} \frac{|\theta_m(\mathbf{z};\mathbf{h}_1,\ldots,\mathbf{h}_n)|}{\prod_{i=1}^m \|\mathbf{h}_i\|} \\ &\leq \frac{m!}{\rho^m} \sum_{\substack{\mathbf{n}\in\mathbb{Z}^\infty \\ |\mathbf{n}|<\infty}} e^{2\pi\rho\sum_{j\geq 1}|\log t_j|\,|n_j|} \left|e^{2\pi i\langle \mathbf{z},\mathbf{n}\rangle} e^{\pi i\langle \mathbf{n}, R\mathbf{n}\rangle}\right| \\ &\leq \frac{m!}{\rho^m} \prod_{j\geq 1}\sum_{n\in\mathbb{Z}} e^{-1/2|\log t_j|(n^2-4\pi|n||\operatorname{Im} z_j|/|\log t_j|-4\pi r|n|)} \\ &\leq \operatorname{const}(\mathbf{z},r)\frac{m!}{\rho^m}\end{aligned}$$

since, as in the proof of Theorem 4.6,

$$\prod_{j\geq 1}\sum_{n\in\mathbb{Z}} e^{1/2|\log t_j|(n^2-4\pi|n|\,|\operatorname{Im} z_j|/|\log t_j|-4\pi r|n|)} \leq \operatorname{const}(\mathbf{z},r) < \infty.$$

REMARK 4.9. Let $(X; A_1, B_1, \ldots)$ be a marked Riemann surface satisfying the hypotheses of Proposition 4.4 and . Let $\mathbf{\Omega}$ be the Hilbert space of square holomorphic integrable one forms on X. The linear map

$$\lambda \in \mathbf{\Omega}^* \to \big(\lambda(\omega_j), j\geq 1\big) \in \mathbb{C}^\infty$$

is injective since ω_j, $j \geq 1$, is a basis for $\mathbf{\Omega}$. Here, $\mathbf{\Omega}^*$ is the dual space of bounded linear functionals on $\mathbf{\Omega}$. We have

$$|\lambda(\omega_j)| \leq \|\lambda\|_{\mathbf{\Omega}^*}\|\omega_j\|_{\mathbf{\Omega}}.$$

If $\|\omega_j\|_{\mathbf{\Omega}} = o(|\log t_j|)$, then

$$\lim_{j\to\infty} \frac{|\lambda(\omega_j)|}{|\log t_j|} \le \|\lambda\|_{\mathbf{\Omega}^*} \lim_{j\to\infty} \frac{\|\omega_j\|_\Omega}{|\log t_j|} = 0$$

and

$$\big\|\big(\lambda(\omega_j), j \ge 1\big)\big\|_B = \sup_{j\ge1} \frac{|\lambda(\omega_j)|}{|\log t_j|} \le \left(\sup_{j\ge1} \frac{\|\omega_j\|_\Omega}{|\log t_j|}\right) \|\lambda\|_{\mathbf{\Omega}^*}.$$

That is, $\big(\lambda(\omega_j), j \ge 1\big) \in B$ and

$$\lambda \in \mathbf{\Omega}^* \to i(\lambda) = \big(\lambda(\omega_1), \lambda(\omega_2), \dots\big) \in B$$

is a bounded linear map. In this case, the theta series

$$\theta(\lambda, \mathcal{R}_X) = \sum_{\substack{\mathbf{n}\in\mathbb{Z}^\infty \\ |\mathbf{n}|<\infty}} e^{2\pi i \lambda(\mathbf{n})} e^{\pi i \langle \mathbf{n}, \mathcal{R}_X \mathbf{n}\rangle}$$

with the notation

$$\lambda(\mathbf{n}) = \lambda(n_1\omega_1 + n_2\omega_2 + \cdots) = \big\langle\big(\lambda(\omega_j), j \ge 1\big), \mathbf{n}\big\rangle$$

is holomorphic on $\mathbf{\Omega}^*$.

Proposition 4.10. *Suppose that the symmetric matrix R and the sequence $t_j \in (0,1)$, $j \ge 1$, satisfy* (4.1) *and* (4.2). *For all $\mathbf{n} \in \mathbb{Z}^\infty \cap B$,*

$$\theta(\mathbf{z} + \mathbf{n}, R) = \theta(\mathbf{z}, R).$$

In particular,

$$\theta(\mathbf{z} + \mathbb{1}_j, R) = \theta(\mathbf{z}, R)$$

for all $j \ge 1$, where $\mathbb{1}_j = (\delta_{i,j}, i \ge 1) \in B$ is the jth column of the identity matrix $\mathbb{1}$. Let

$$R_j = \begin{pmatrix} R_{1j} \\ R_{2j} \\ \vdots \end{pmatrix}, \quad j \ge 1,$$

be the columns of R. If R_j belongs to B, then

$$\theta(\mathbf{z} + R_j, R) = e^{-2\pi i (z_j + 1/2 R_{jj})} \theta(\mathbf{z}, R).$$

Proof. The first two statements follow from continuity and the identity

$$e^{2\pi i \langle \mathbf{z} + \mathbb{1}_j, \mathbf{n}\rangle} = e^{2\pi i \langle \mathbf{n}, \mathbf{z}\rangle}$$

for all $\mathbf{n} \in \mathbb{Z}^\infty$ with $|\mathbf{n}| < \infty$. The third is verified by the standard manipulation

$$\begin{aligned}
&\theta(\mathbf{z} + R_j) \\
&\quad= \sum_{\substack{\mathbf{n}\in\mathbb{Z}^\infty \\ |\mathbf{n}|<\infty}} e^{2\pi i \langle \mathbf{z} + R_j, \mathbf{n}\rangle} e^{\pi i \langle \mathbf{n}, R\mathbf{n}\rangle} \\
&\quad= \sum_{\substack{\mathbf{n}\in\mathbb{Z}^\infty \\ |\mathbf{n}|<\infty}} e^{2\pi i \langle \mathbf{z} + R_j, \mathbf{n}\mathbf{1}_j\rangle} e^{\pi i \langle \mathbf{n}\mathbf{1}_j, R\mathbf{n}\mathbf{1}_j\rangle} \\
&\quad= e^{-2\pi i \langle \mathbf{z} + R_j, \mathbf{1}_j\rangle} e^{\pi i \langle \mathbf{1}_j, R\mathbf{1}_j\rangle} \sum_{\substack{\mathbf{n}\in\mathbb{Z}^\infty \\ |\mathbf{n}|<\infty}} e^{2\pi i \langle \mathbf{z}, \mathbf{n}\rangle} e^{2\pi i \langle R_j, \mathbf{n}\rangle - \pi i \langle \mathbf{n}, R\mathbf{1}_j\rangle - \pi i \langle \mathbf{1}_j, R\mathbf{n}\rangle} e^{\pi i \langle \mathbf{n}, R\mathbf{n}\rangle} \\
&\quad= e^{-2\pi i \langle \mathbf{z} + R_j, \mathbf{1}_j\rangle} e^{\pi i \langle \mathbf{1}_j, R\mathbf{1}_j\rangle} \theta(\mathbf{z}).
\end{aligned}$$

The symmetry of R was used to pass from the third to the fourth line. □

REMARK 4.11. Apply Schwarz's inequality for the positive definite bilinear form $\langle \mathbf{x}, \operatorname{Im} R\mathbf{y}\rangle$ to obtain

$$|\operatorname{Im} R_{jk}| = |\langle \mathbb{1}_j, \operatorname{Im} R\mathbb{1}_k\rangle| \leq \langle \mathbb{1}_j, \operatorname{Im} R\mathbb{1}_j\rangle^{1/2}\langle \mathbb{1}_k, \operatorname{Im} R\mathbb{1}_k\rangle^{1/2} = (\operatorname{Im} R_{jj})^{1/2}(\operatorname{Im} R_{kk})^{1/2}.$$

Suppose $\operatorname{Im} R_{jj} = o(|\log t_j|^2)$, then

$$\frac{|\operatorname{Im} R_{jk}|}{|\log t_j|} = \frac{|\operatorname{Im} R_{jk}|}{|\operatorname{Im} R_{jj}|^{1/2}}\frac{|\operatorname{Im} R_{jj}|^{1/2}}{|\log t_j|} < |\operatorname{Im} R_{kk}|^{1/2}\frac{o(|\log t_j|)}{|\log t_j|} \leq |\operatorname{Im} R_{kk}|^{1/2}o(1)$$

and $\operatorname{Im} R_k \in B$ for all $k \geq 1$. If $(X; A_1, B_1, \dots)$ is a marked Riemann surface as in Remark 4.9, then, by Theorem 2.9,

$$\|\omega_j\|^2_{\mathbf{\Omega}} = \frac{i}{2}\int_X \omega_j \wedge \overline{\omega}_j = \operatorname{Im} \mathcal{R}_{j,j}$$

and $\|\omega_j\|_{\mathbf{\Omega}} = o(|\log t_j|)$ implies that the the columns of $\operatorname{Im} \mathcal{R}_X$ belong to B.

Let B_{Re} be the real Banach space given by

$$B_{\mathrm{Re}} = \mathbb{R}^\infty \cap B$$

and let $\mathcal{Z} \subset B_{\mathrm{Re}}$ be the sublattice given by

$$\mathcal{Z} = \mathbb{Z}^\infty \cap B_{\mathrm{Re}}.$$

The quotient

$$\mathcal{T} = B_{\mathrm{Re}}/\mathcal{Z}$$

with metric

$$d(\mathbf{x}, \mathbf{y}) = \inf_{\mathbf{n}\in\mathcal{Z}} \|\mathbf{x} - \mathbf{y} - \mathbf{n}\|$$

is a compact Abelian group. It is isomorphic to the infinite product $\big(\mathbb{R}/\mathbb{Z}\big)^\infty$.

The exponentials $e^{2\pi i\langle \mathbf{n},\mathbf{x}\rangle}$, $\mathbf{n} \in \mathbb{Z}^\infty$, $|\mathbf{n}| < \infty$, are an orthonormal basis for the Hilbert space $L^2(\mathcal{T})$ of all measurable functions $f(\mathbf{x})$ on the infinite-dimensional torus $\mathcal{T}$ with

$$\int_{\mathcal{T}} d\mathbf{x}\, |f(\mathbf{x})|^2 < \infty.$$

Here, $d\mathbf{x}$ is the Haar measure on $\mathcal{T}$. Concretely, for continuous functions

$$\int_{\mathcal{T}} d\mathbf{x}\, f(\mathbf{x}) = \lim_{j\to\infty}\int_{[0,1)^j} dx_1 \cdots dx_j f(x_1, \dots, x_j, 0, \dots).$$

By Proposition 4.10, $\theta(\mathbf{x}, R)$ defines a continuous function on $\mathcal{T}$ and therefore belongs to $L^2(\mathcal{T})$.

PROPOSITION 4.12. *Suppose that the symmetric matrix R and the sequence $t_j \in (0,1)$, $j \geq 1$, satisfy* (4.1) *and* (4.2). *If $\mathbf{n} \in \mathbb{Z}^\infty$ with $|\mathbf{n}| < \infty$, then*

$$\int_{\mathcal{T}} d\mathbf{x}\, \theta(\mathbf{x}, R)e^{-2\pi i\langle \mathbf{n},\mathbf{x}\rangle} = e^{\pi i\langle \mathbf{n}, R\mathbf{n}\rangle}.$$

In other words, the Fourier coefficients of $\theta(\mathbf{x}, R)$ *are* $e^{\pi i\langle \mathbf{n}, R\mathbf{n}\rangle}$. *If, in addition,* R *is pure imaginary, then* $\theta(\mathbf{x}, R)$ *is a real valued function on* $\mathcal{T}$ *with*

$$\inf_{\mathbf{x}\in\mathcal{T}} \theta(\mathbf{x}, R) > 0.$$

PROOF. The domain $[0,1)^\infty$ is a bounded subset of B. In fact, it can be covered by a finite number of balls of radius $r = 1 - 2\beta/8\pi$. By Theorem 4.6 and the bounded convergence theorem

$$\begin{aligned}\int_{\mathcal{T}} d\mathbf{x}\, \theta(\mathbf{x}, R)e^{-2\pi i\langle \mathbf{n},\mathbf{x}\rangle} &= \sum_{\substack{\mathbf{p}\in\mathbb{Z}^\infty\\ |\mathbf{p}|<\infty}} \int_{\mathcal{T}} d\mathbf{x}\, e^{2\pi i\langle \mathbf{z},\mathbf{p}\rangle} e^{\pi i\langle \mathbf{p}, R\mathbf{p}\rangle} e^{-2\pi i\langle \mathbf{n},\mathbf{x}\rangle}\\ &= e^{\pi i\langle \mathbf{n}, R\mathbf{n}\rangle}.\end{aligned}$$

Suppose R is pure imaginary. Observe that

$$\theta(0, tR) = \sum_{\substack{\mathbf{n}\in\mathbb{Z}^\infty\\ |\mathbf{n}|<\infty}} e^{-\pi t\langle \mathbf{n}, \operatorname{Im} R\mathbf{n}\rangle} > 0$$

when R is pure imaginary and $t > 0$. We next show that $\theta(x, R) \geq 0$ for all real vectors $\mathbf{x} \in B$.

For each $m \geq 1$, set

$$\theta_m(\mathbf{x}) = \sum_{\substack{\mathbf{n}\in\mathbb{Z}^\infty\\ n_j=0, j>m}} e^{2\pi i\langle \mathbf{x},\mathbf{n}\rangle} e^{\pi i\langle \mathbf{n}, R\mathbf{n}\rangle} = \sum_{\substack{\mathbf{n}\in\mathbb{Z}^\infty\\ n_j=0, j>m}} e^{2\pi i\langle \mathbf{x},\mathbf{n}\rangle} e^{-\pi\langle \mathbf{n}, \operatorname{Im} R\mathbf{n}\rangle}.$$

By the Poissson summation formula,

$$\theta_m(\mathbf{x}) = \frac{(2\pi)^m}{(\det \operatorname{Im} R_m)^{1/2}} \sum_{\substack{\mathbf{n}\in\mathbb{Z}^\infty\\ n_j=0, j>m}} e^{-\pi\langle(\mathbf{x}+\mathbf{n}), (\operatorname{Im} R)^{-1}(\mathbf{x}+\mathbf{n})\rangle} > 0$$

where R_m is the $m \times m$ principal minor of R. It follows that

$$\theta(\mathbf{x}, R) = \lim_{m\to\infty} \theta_m(\mathbf{x}) \geq 0.$$

The identity,

$$\theta(\mathbf{x}, R) = \int_{\mathcal{T}} d\mathbf{y}\, \theta\big(\mathbf{x} - \mathbf{y}, \tfrac{1}{2}R\big)\theta\big(\mathbf{y}, \tfrac{1}{2}R\big)$$

is easily verified by using the first part of the proposition. Suppose, $\theta(\mathbf{x}, R) = 0$. Then, the identity and the nonnegativity of $\theta\big(\mathbf{x} - \mathbf{y}, \frac{1}{2}R\big)$ and $\theta\big(\mathbf{y}, \frac{1}{2}R\big)$ imply that $\theta\big(\mathbf{x}-\mathbf{y}, \frac{1}{2}R\big)\theta\big(\mathbf{y}, \frac{1}{2}R\big)$ vanishes for all real vectors $\mathbf{y}$. By analyticity, one factor must vanish identically and in particular $\theta\big(0, \frac{1}{2}R\big) = 0$. This contradicts the observation made above. □

Let $U = (U_1, U_2, \dots) \in \mathbb{R}^\infty$. Set

$$\{U\} = (\{U_1\}, \{U_2\}, \dots)$$

where $0 \leq \{U_j\} < 1$ is the fractional part of U_j, $j \geq 1$. Observe that for all $\mathbf{z} \in B$, $U \in \mathbb{R}^\infty$ and $\mathbf{n} \in \mathbb{Z}^\infty$ with $|\mathbf{n}| < \infty$,

$$e^{2\pi i\langle \mathbf{z}+U, \mathbf{n}\rangle} = e^{2\pi i\langle \mathbf{z}+\{U\}, \mathbf{n}\rangle}.$$

It follows that for each $\mathbf{z} \in B$,

$$\theta(\mathbf{z} + U, R) = \theta(\mathbf{z} + \{U\}, R)$$

is well-defined for all $U \in \mathbb{R}^\infty$.

PROPOSITION 4.13. *Suppose that the symmetric matrix R and the sequence $t_j \in (0,1)$, $j \geq 1$, satisfy (4.1) and (4.2). Fix a positive integer k. Let $U_i = (U_{i1}, U_{i2}, \dots) \in \mathbb{R}^\infty$, $i = 0, \dots, p$, satisfy*

$$\sup_{i,j} |U_{i,j}| t_j^{1-2\beta'/2k} < \infty$$

for some $\beta < \beta' < \frac{1}{2}$. Then, for all $\mathbf{z} \in B$,

$$\theta(\mathbf{z} + \xi_0 U_0 + \cdots + \xi_p U_p, R)$$

is a k times continuously differentiable function of $(\xi_0, \dots, \xi_p) \in \mathbb{R}^{p+1}$. If, in addition, R is pure imaginary, then for all real vectors $\mathbf{x} \in B$,

$$\inf_{(\xi_0,\dots,\xi_p)\in\mathbb{R}^p} \theta(\mathbf{x} + \xi_0 U_1 + \cdots + \xi_p U_p, R) > 0.$$

PROOF. We will show that for all $\alpha \in \mathbb{N}^{p+1}$ with $|\alpha| \leq k$, the formal derivatives

$$\begin{aligned} &\frac{\partial^{|\alpha|}}{\partial \xi_0^{\alpha_1} \cdots \partial \xi_p^{\alpha_p}} \theta(\mathbf{z} + \xi_0 U_0 + \cdots + \xi_p U_p) \\ &\qquad = \sum_{\substack{\mathbf{n}\in\mathbb{Z}^\infty \\ |\mathbf{n}|<\infty}} \prod_{\ell=0}^{p} \left(\sum_{j\geq 1} 2\pi i U_{\ell,j} n_j \right)^{\alpha_\ell} e^{2\pi i \langle \mathbf{z}+\xi_0U_0+\cdots+\xi_pU_p, \mathbf{n}\rangle} e^{\pi i \langle \mathbf{n}, R\mathbf{n}\rangle} \end{aligned}$$

converge absolutely and uniformly for $(\xi_0, \dots \xi_p) \in \mathbb{R}^{p+1}$. Fix any $\beta'' \in (\beta, \beta')$. Observe that for all $\mathbf{n} \in \mathbb{Z}^\infty$ with $|\mathbf{n}| < \infty$,

$$\begin{aligned} &\left| \prod_{\ell=0}^{p} \left(\sum_{j\geq 1} 2\pi i U_{\ell,j} n_j \right)^{\alpha_\ell} e^{2\pi i \langle \mathbf{z}+\xi_0U_0+\cdots+\xi_pU_p, \mathbf{n}\rangle} e^{\pi i \langle \mathbf{n}, R\mathbf{n}\rangle} \right| \\ &\quad \leq (2\pi)^{|\alpha|} \prod_{\ell=0}^{p} \left(\sum_{j\geq 1} |U_{\ell,j}|\, |n_j| \right)^{\alpha_\ell} e^{2\pi |\langle \operatorname{Im} \mathbf{z}, \mathbf{n}\rangle|} e^{-1/2 \sum_{j\geq 1} |\log t_j| n_j^2} \\ &\quad \leq (2\pi)^{|\alpha|} \left(\prod_{\ell=0}^{p} \prod_{\substack{j \text{ such that} \\ n_j \neq 0}} |n_j|^{\alpha_\ell} (\max\{|U_{\ell,j}|, 2\})^{\alpha_\ell} \right) \\ &\qquad \times e^{-1/2 \sum_{j\geq 1} |n_j|\, |\log t_j| (|n_j| - 4\pi |\operatorname{Im} z_j| / |\log t_j|)} \\ &\quad = (2\pi)^{|\alpha|} \left(\prod_{\substack{j \text{ such that} \\ n_j \neq 0}} t_j^{1/2(1-2\beta'') n_j^2} |n_j|^{|\alpha|} \prod_{\ell=0}^{p} (\max\{|U_{\ell,j}|, 2\})^{\alpha_\ell} \right) \\ &\qquad \times e^{1/2(1-2\beta'') \sum_{j\geq 1} n_j^2 |\log t_j|} e^{-1/2 \sum_{j\geq 1} |n_j|\, |\log t_j| (|n_j| - 4\pi |\operatorname{Im} z_j| / |\log t_j|)}. \end{aligned}$$

By hypothesis

$$\sup_j t_j^{1/2(1-2\beta')} \prod_{\ell=0}^{p} (\max\{|U_{\ell,j}|, 2\})^{\alpha_\ell} \leq \sup_j \prod_{\ell=0}^{p} (\max\{t_j^{1/2k(1-2\beta')} |U_{\ell,j}|, 2\})^{\alpha_\ell} < \infty.$$

It follows that

$$\begin{aligned} n^{|\alpha|} t_j^{1/2(1-2\beta'')n^2} \prod_{\ell=0}^{p} (\max\{|U_{\ell,j}|, 2\})^{\alpha_\ell} &\leq t_j^{(\beta'-\beta'')n^2} \sup_j \sup_{n>0} n^{|\alpha|} t_j^{1/2(1-2\beta')n^2} \prod_{\ell=0}^{p} (\max\{|U_{\ell,j}|, 2\})^{\alpha_\ell} \\ &\leq \text{const}\, t_j^{(\beta'-\beta'')n^2} \end{aligned}$$

since $t_j^{1/2(1-2\beta')} \geq n^{|\alpha|} t_j^{1/2(1-2\beta')n^2}$ for all sufficiently large j and all $n > 0$, or all j and all sufficiently large $n > 0$. Therefore,

$$\begin{aligned} \sup_{\substack{\mathbf{n}\in\mathbb{Z}^\infty \\ |\mathbf{n}|<\infty}} (2\pi)^{|\alpha|} \left(\prod_{\substack{j \text{ such that} \\ n_j \neq 0}} t_j^{1/2(1-2\beta'')n_j^2} |n_j|^{|\alpha|} \prod_{\ell=0}^{p} (\max\{|U_{\ell,j}|, 2\})^{\alpha_\ell} \right) \\ \leq \sup_{\substack{\mathbf{n}\in\mathbb{Z}^\infty \\ |\mathbf{n}|<\infty}} \prod_{\substack{j \text{ such that} \\ n_j \neq 0}} \text{const}\, t_j^{(\beta'-\beta'')n^2} < \infty. \end{aligned}$$

Combining the last two paragraphs,

$$\begin{aligned} &\left| \prod_{\ell=0}^{p} \left(\sum_{j\geq 1} 2\pi i U_{\ell,j} n_j \right)^{\alpha_\ell} e^{2\pi i \langle \mathbf{z}+\xi_0 U_0+\cdots+\xi_p U_p, \mathbf{n}\rangle} e^{\pi i \langle \mathbf{n}, R\mathbf{n}\rangle} \right| \\ &\quad \leq \text{const}\, e^{1/2(1-2\beta'')\sum_{j\geq 1} n_j^2 |\log t_j|} e^{-1/2\sum_{j\geq 1} |n_j||\log t_j|(|n_j|-4\pi|\operatorname{Im} z_j|/|\log t_j|)} \\ &\quad = \text{const}\, e^{-\beta''\sum_{j\geq 1} |n_j||\log t_j|(|n_j|-2\pi/\beta''|\operatorname{Im} z_j|/|\log t_j|)}. \end{aligned}$$

Pick j_0 such that

$$1 - \frac{2\pi}{\beta''} \frac{|\operatorname{Im} z_j|}{|\log t_j|} \geq \frac{\beta}{\beta''}$$

for all $j \geq j_0$. Summing the majorants,

$$\begin{aligned} \sum_{\substack{\mathbf{n}\in\mathbb{Z}^\infty \\ |\mathbf{n}|<\infty}} & e^{-\beta''\sum_{j\geq 1} |n_j||\log t_j|(|n_j|-2\pi/\beta''|\operatorname{Im} z_j|/|\log t_j|)} \\ &\leq \prod_{j\geq 1} \sum_{n\in\mathbb{Z}} e^{-\beta''|n||\log t_j|(|n|-2\pi/\beta''|\operatorname{Im} z_j|/|\log t_j|)} \\ &\leq \text{const} \prod_{j\geq j_0} \sum_{n\in\mathbb{Z}} e^{-\beta|n||\log t_j|} \\ &\leq \text{const} \prod_{j\geq j_0} \left(1 + 2\frac{t_j^\beta}{1-t_j^\beta}\right) \\ &< \infty \end{aligned}$$

and consequently, $\theta(\mathbf{z} + \xi_0 U_0 + \cdots + \xi_p U_p)$ is k times continuously differentiable.

The second statement follows immediately from Proposition 4.12. □

PROPOSITION 4.14. *Let f be a continuous function on $\mathcal{T}$, $\mathbf{x}$ a point in $\mathcal{T}$ and $U_i = (U_{i1}, U_{i2}, \dots) \in \mathbb{R}^\infty, i = 0, \dots, p$. For all $(\xi_0, \xi_1, \dots, \xi_p) \in \mathbb{R}^{p+1}$, set*

$$f(\mathbf{x} + \xi_0 U_0 + \xi_1 U_1 + \cdots + \xi_p U_p) = f(\{\mathbf{x} + \xi_0 U_0 + \xi_1 U_1 + \cdots + \xi_p U_p\}).$$

Then, for all $\varepsilon > 0$ there is a $\ell(\varepsilon)$ such that in any interval of length $\ell(\varepsilon)$ there is a η for which

$$\sup_{\xi_0} \sup_{(\xi_1,\dots,\xi_p)\in\mathbb{R}^p} |f(\mathbf{x} + (\xi_0+\eta)U_0 + \xi_1 U_1 + \cdots + \xi_p U_p) - f(\mathbf{x} + \xi_0 U_0 + \xi_1 U_1 + \cdots + \xi_p U_p)| \le \varepsilon.$$

In other words, $f(\mathbf{x} + \xi_0 U_0 + \xi_1 U_1 + \cdots + \xi_p U_p)$ is an $L^\infty(\mathbb{R}^p)$ valued almost periodic function of $\xi_0 \in \mathbb{R}$.

PROOF. Fix $\varepsilon > 0$. Recall that f is uniformly continuous on the compact metric space $\mathcal{T}$. Consequently, there is a $\delta > 0$ such that

$$|f(\mathbf{y}) - f(\mathbf{y}')| \le \varepsilon$$

whenever $d(\mathbf{y}, \mathbf{y}') < \delta$.

Pick $j_\delta > 0$ such that

$$d(\mathbf{y}, \mathbf{y}') < \frac{\delta}{2}$$

for all $\mathbf{y}, \mathbf{y}' \in \mathcal{T}$ with $y_j = y'_j$, $1 \le j \le j_\delta$. By elementary arithmetic [**HW**, Theorem 200], there is an $\ell(\varepsilon)$ such that in every interval of length $\ell(\varepsilon)$ there is an η with

$$d\big(0, \eta(U_{0,1}, \dots, U_{0,j_\delta}, 0, \dots)\big) \le \frac{\delta}{2}.$$

Therefore,

$$d(0, \{\eta U_0\}) < \delta.$$

We have

$$\sup_{\xi_0} \sup_{(\xi_1,\dots,\xi_p)\in\mathbb{R}^p} d\big(\mathbf{x} + \{(\xi_0+\eta)U_0 + \xi_1 U_1 + \cdots + \xi_p U_p\}, \mathbf{x} + \{\xi_0 U_0 + \xi_1 U_1 + \cdots + \xi_p U_p\}\big) < \delta. \quad \square$$

REMARK 4.15. We will apply Proposition 4.14 to the function

$$-2\frac{\partial^2}{\partial \xi_1^2} \log \theta(\mathbf{x} + \xi_0 U_0 + \xi_1 U_1 + \xi_2 U_2)$$

to show that all smooth, spatially periodic solutions of the Kadomcev–Petviashvilli equation propagate almost periodically in time.

APPENDIX: A HYPERELLIPTIC SURFACE WITH DIVERGENT THETA SERIES.

In this appendix we explicitly construct a marked hyperelliptic Riemann surface $(X; A_1, B_1, \dots)$ on which there is an exhaustion function with finite charge such that the period matrix $\mathcal{R}_X$ is pure imaginary and

$$\theta(0, \mathcal{R}_X) = \sum_{\substack{\mathbf{n}\in\mathbb{Z}^\infty \\ |\mathbf{n}|<\infty}} e^{-\pi\langle \mathbf{n}, \operatorname{Im} \mathcal{R}_X \mathbf{n}\rangle} = \infty.$$

We will construct this surface using the (convergent!) genus one theta function $\vartheta(z, \tau)$, $\operatorname{Im} \tau > 0$, given by

$$\vartheta(z, \tau) = \sum_{n\in\mathbb{Z}} e^{2\pi i n z} e^{\pi i n^2 \tau}.$$

It is an entire, even function of z that has a simple root at each point of the shifted lattice $1+\tau/2+\mathbb{Z}\oplus\tau\mathbb{Z}$ and no other roots. Moreover,

$$\vartheta(z+1,\tau)=\vartheta(z,\tau)$$
$$\vartheta(z+j\tau)=e^{-\pi i j^2\tau-2\pi i j z}\vartheta(z,\tau)$$

for all $j\in\mathbb{Z}$. It follows that

$$f(\zeta)=\vartheta\left(\frac{1}{2\pi i}\log\zeta+\frac{1+i}{2},i\right)=e^{\pi/4}\sum_{n\in\mathbb{Z}}(-1)^n e^{-\pi(n+1/2)^2}\zeta^n$$

is holomorphic on $\mathbb{C}^*=\mathbb{P}^1\setminus\{0,\infty\}$ and has a simple root at $\zeta=e^{2\pi j}$ for each $j\in\mathbb{Z}$ and no other roots.

Let X be the hyperelliptic Riemann surface given by

$$X=\{(y,\zeta)\in\mathbb{C}\times\mathbb{C}^*\mid y^2=f(\zeta)\}$$

and $\tau_X\colon(y,\zeta)\mapsto\zeta$ the hyperelliptic projection onto $\mathbb{C}^*$.

LEMMA A4.1. *For each $\ell\in\mathbb{Z}$, the inverse image*

$$\tau_X^{-1}(\{e^{(4\ell+1)\pi+2\pi i\alpha}\mid 0\le\alpha<1\})$$

of the oriented circle of radius $e^{(4\ell+1)\pi}$ in $\mathbb{C}^*$ consists of two disjoint oriented cycles.

PROOF. For all $j\in\mathbb{Z}$,

$$\begin{aligned}f(e^{(2j+1)\pi+2\pi i\alpha})&=\vartheta\left(\frac{1}{2\pi i}((2j+1)\pi+2\pi i\alpha)+\frac{1+i}{2},i\right)\\&=\vartheta(\alpha+\tfrac{1}{2}-ij)\\&=e^{\pi j^2+2\pi i j(\alpha+\frac{1}{2})}\vartheta(\alpha+\tfrac{1}{2},i)\end{aligned}$$

Observe that the winding number of $\vartheta(\alpha+\frac{1}{2},it)$ around zero as α runs from zero to one is zero for all $t>0$, since

$$\lim_{t\to\infty}\vartheta(\alpha+\tfrac{1}{2},it)=1$$

and $\vartheta(\alpha+\frac{1}{2},it)$ never vanishes for α,t real. It follows that the winding number of f around zero along the circle of radius $e^{(2j+1)\pi}$ is j, for all $j\in\mathbb{Z}$. Since $j=2\ell$ is even, the inverse image $\tau_X^{-1}(\{e^{(4\ell+1)\pi+2\pi i\alpha}\mid 0\le\alpha<1\})$ consists of two disjoint simple closed curves. □

For each $\ell\in\mathbb{Z}$, let A_ℓ be the component of

$$\tau_X^{-1}(\{e^{(4\ell+1)\pi+2\pi i\alpha}\mid 0\le\alpha<1\})$$

containing the point $(+\sqrt{f(e^{(4\ell+1)\pi})},e^{(4\ell+1)\pi})$ and orient

$$B_\ell=\tau_X^{-1}([e^{4\ell\pi},e^{(4\ell+2)\pi}])$$

such that $A_\ell\times B_\ell=1$.

LEMMA A4.2. (i) *The cycles $A_\ell,B_\ell,\ell\in\mathbb{Z}$ are a canonical homology basis for the hyperelliptic Riemann surface X.*

(ii) *There is a compact subset $Z\subset X$ and an exhaustion function h with finite charge on the marked Riemann surface $(X;A_\ell,B_\ell,\ell\in\mathbb{Z})$ such that*

$$h\big|_{X\setminus Z}=\big|\log|\zeta|\big|$$

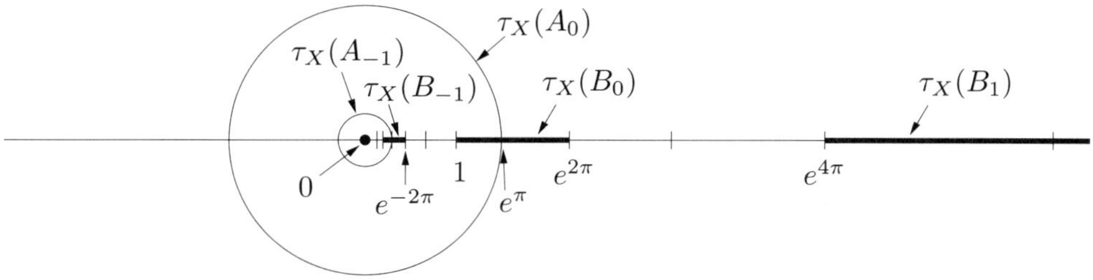

(iii) *The map* $j(y,\zeta) = \overline{(y,\zeta)}$ *is an antiholomorphic involution on* X *with*

$$j(A_\ell) = -A_\ell,$$
$$j(B_\ell) = \quad B_\ell$$

for all $\ell \in \mathbb{Z}$.

(iv) *The map* $s(y,\zeta) = (e^{\pi}\zeta y, e^{4\pi}\zeta)$ *is a holomorphic automorphism of* X *with*

$$s(A_\ell) = A_{\ell+1},$$
$$s(B_\ell) = B_{\ell+1}$$

for all $\ell \in \mathbb{Z}$.

PROOF. It is easy to check that the cycles $A_\ell, B_\ell, \ell \in \mathbb{Z}$, satisfy Definition 1.10 (i) and (ii) and therefore $(X; A_\ell, B_\ell, \ell \in \mathbb{Z})$ is a marked Riemann surface.

To verify (ii), observe that the function $\log|\zeta|$ is harmonic and Morse on X. By Proposition A2.2, there is a compact subset $Z \subset X$ and an exhaustion function h with finite charge on X such that

$$h\big|_{X\setminus Z} = \big|\log|\zeta|\big|.$$

For all $r > 0$, the preimage $\tau_X^{-1}($ circle of radius $r)$ is a dividing cycle that is either connected or is the union of two components one of which is homologous to A_ℓ for some $\ell \in \mathbb{Z}$. By construction, the boundary ∂X_t of $X_t = h^{-1}([0,t])$ is the union

$$\tau_X^{-1}(\text{ circle of radius } e^t) \cup \tau_X^{-1}(\text{ circle of radius} e^{-t})$$

for all sufficiently large $t > 0$. It follows that h satisfies Definition 2.1 (ii) and (iii) and therefore h is an exhaustion function with finite charge on the marked hyperelliptic Riemann surface $(X; A_1, B_1, \dots)$.

Part (iii) is a direct consequence of the reality of f. For part (iv), we use the identity

$$\begin{aligned} f(e^{4\pi}\zeta) &= \vartheta\left(\frac{1}{2\pi i}\log e^{4\pi}\zeta + \frac{1+i}{2}, i\right) \\ &= \vartheta\left(\frac{1}{2\pi i}\log\zeta + \frac{1+i}{2} - 2i, i\right) \\ &= e^{4\pi + 4\pi i(1/2\pi i \log\zeta + 1 + i/2)}\vartheta\left(\frac{1}{2\pi i}\log\zeta + \frac{1+i}{2}, i\right) \\ &= e^{2\pi}\zeta^2 f(\zeta). \quad \square \end{aligned}$$

By Lemma A4.2 (ii) and Theorem 3.8, X is parabolic and $(X; A_\ell, B_\ell, \ell \in \mathbb{Z})$ has a unique, normalized basis $\omega_\ell, \ell \in \mathbb{Z}$, of square integrable holomorphic one forms dual to $A_\ell, \ell \in \mathbb{Z}$.

PROPOSITION A4.3. (i) *The period matrix*

$$\mathcal{R}_X = (\mathcal{R}_{\ell,m}; \ell, m \in \mathbb{Z}) = \left(\int_{B_\ell} \omega_m; \ell, m \in \mathbb{Z} \right)$$

of the marked hyperelliptic Riemann surface $(X; A_\ell, B_\ell, \ell \in \mathbb{Z})$ *is pure imaginary.*

(ii) *For all* $\ell, m \in \mathbb{Z}$,

$$\mathcal{R}_{\ell,m} = \mathcal{R}_{0,m-\ell}$$

(iii) *The evaluation*

$$\theta(0, \mathcal{R}_X) = \sum_{\substack{\mathbf{n} \in \mathbb{Z}^\infty \\ |\mathbf{n}| < \infty}} e^{\pi i \langle \mathbf{n}, \mathcal{R}_X \mathbf{n} \rangle}$$

of the theta series for $(X; A_\ell, B_\ell, \ell \in \mathbb{Z})$ *at* $\mathbf{z} = 0$ *is a divergent sum of positive terms. Here, the sum is over all doubly infinite sequences* $\mathbf{n} = (\ldots, n_{-1}, n_0, n_1, \ldots)$ *of integers with* $|\mathbf{n}| = \cdots + |n_{-1}| + |n_0| + |n_1| + \cdots < \infty$.

PROOF. To verify (i), first observe that $j^*\overline{\omega} \in \Omega(X)$ for all $\omega \in \Omega(X)$. By Lemma A4.2 (iii),

$$\int_{A_\ell} j^*\overline{\omega}_m = \int_{j(A_\ell)} \overline{\omega}_m = -\int_{A_\ell} \overline{\omega}_m = -\delta_{\ell,m}$$

for all $\ell, m \in \mathbb{Z}$. It follows that

$$j^*\overline{\omega}_m = -\omega_m$$

for all $m \in \mathbb{Z}$. Now,

$$\int_{B_\ell} \omega_m = -\int_{B_\ell} j^*\overline{\omega}_m = -\int_{j(B_\ell)} \overline{\omega}_m = -\overline{\int_{B_\ell} \omega_m}.$$

Therefore, the period matrix $\mathcal{R}_X$ of X is pure imaginary.

For (ii), observe that $s^*\omega \in \omega(X)$ for all $\omega \in \Omega(X)$. By Lemma A4.2 (iv),

$$\int_{A_\ell} s^*\omega_m = \int_{s(A_\ell)} \omega_m = \int_{A_{\ell+1}} \omega_m = \delta_{\ell+1,m}$$

for all $\ell, m \in \mathbb{Z}$. It follows that

$$s^*\omega_m = \omega_{m-1}$$

for all $m \in \mathbb{Z}$. Again, by Lemma A4.2 (iv),

$$\int_{B_\ell} \omega_m = \int_{s^\ell(B_0)} \omega_m = \int_{B_0} (s^\ell)^*\omega_m = \int_{B_0} (s^*)^\ell \omega_m = \int_{B_0} \omega_{m-\ell}.$$

Therefore,

$$\mathcal{R}_{\ell,m} = \mathcal{R}_{0,m-\ell}$$

for all $\ell, m \in \mathbb{Z}$.

Finally, by part (i),

$$\theta(0, \mathcal{R}_X) = \sum_{\substack{\mathbf{n} \in \mathbb{Z}^\infty \\ |\mathbf{n}| < \infty}} e^{\pi i \langle \mathbf{n}, \mathcal{R}_X \mathbf{n} \rangle} = \sum_{\substack{\mathbf{n} \in \mathbb{Z}^\infty \\ |\mathbf{n}| < \infty}} e^{-\pi \langle \mathbf{n}, \operatorname{Im} \mathcal{R}_X \mathbf{n} \rangle}$$

is a sum of positive terms. We have

$$\sum_{\substack{\mathbf{n}\in\mathbb{Z}^\infty\\ |\mathbf{n}|<\infty}} e^{-\pi\langle \mathbf{n},\operatorname{Im}\mathcal{R}_X\mathbf{n}\rangle} \geq \sum_{\delta_{\ell,m},\ell\in\mathbb{Z}} e^{-\pi\langle \delta_{\ell,m},\operatorname{Im}\mathcal{R}_X\delta_{\ell,m}\rangle} = \sum_{\ell\in\mathbb{Z}} e^{-\pi\operatorname{Im}\mathcal{R}_{\ell,\ell}}$$

Here, $\delta_{\ell,m}$ is the doubly infinite sequence whose elements are zero for $m \neq \ell$ and one for $m = \ell$. By part (ii),

$$\sum_{\substack{\mathbf{n}\in\mathbb{Z}^\infty\\ |\mathbf{n}|<\infty}} e^{-\pi\langle \mathbf{n},\operatorname{Im}\mathcal{R}_X\mathbf{n}\rangle} \geq e^{-\pi\operatorname{Im}\mathcal{R}_{0,0}} \sum_{\ell\in\mathbb{Z}} 1 = \infty. \quad \square$$

A. Summary of Results from Chapter 1

DEFINITION. Let X be a Riemann surface (possibly with boundary). A system of homology classes $A_1, B_1, A_2, B_2, \ldots$ on X is called a **canonical homology basis** if

(i) $A_i \cdot A_j = B_i \cdot B_j = 0$, $A_i \cdot B_j = \delta_{ij}$.
(ii) If $C \in H_1(X,\mathbb{Z})$ is such that $C \cdot A_i = C \cdot B_i = 0$ for all $i = 1, 2, \ldots$ then $C \cdot D = 0$ for all $D \in H_1(X,\mathbb{Z})$.

DEFINITION. A **marked Riemann surface** is a Riemann surface X, together with a choice of the classes of a canonical homology basis.

A basic concept for our discussion is

DEFINITION. A C^∞ function h on a Riemann surface X that satisfies:

$$\sup_{t>s>0} \left| \int_{X_t\setminus X_s} d{*}dh \right| < \infty$$

is called a **function with bounded charge**. Here, $X_t = h^{-1}\big((-\infty, t]\big)$. A proper Morse function with bounded charge is called an **exhaustion function h with bounded charge**.

An **exhaustion function h with bounded charge on the marked Riemann surface** $(X; A_1, B_1, \ldots)$ is an exhaustion function with bounded charge on X for which there is a bijection $\sigma\colon \mathbb{N} \to \mathbb{N}$ such that for each $t > 0$ there is $n \geq 1$ and $A'_1, B'_1, A'_2, B'_2, \ldots, A'_n, B'_n \in H_1(X_t,\mathbb{Z})$ with the following properties

(i) The homology classes $A'_1, B'_1, A'_2, B'_2, \ldots, A'_n, B'_n$ generate a maximal submodule of $H_1\big(h^{-1}\big((-\infty,t]\big),\mathbb{Z}\big)$ on which the intersection form is nondegenerate.
(ii) If $\iota\colon H_1(X_t,\mathbb{Z}) \to H_1(X,\mathbb{Z})$ denotes the map induced by the inclusion then for $i = 1, \ldots, n$

$$\iota(A'_i) = A_{\sigma(i)},$$
$$\iota(B'_i) = B_{\sigma(i)} \text{ modulo finite linear combinations of the } A_j.$$

LEMMA A.1 (See the Appendix to §2). *Let h be a proper function with bounded charge on X. Then there are exhaustion functions h' with bounded charge arbitrarily close to h in the $C^{(0)}$-topology. If M is a closed subset of X and the restriction of h to a neighborhood of M is already a Morse function, then one can choose h' above such that $h'\big|_M = h\big|_M$.*

THEOREM A.2 (See Proposition 3.6). *A Riemann surface that admits an exhaustion function with bounded charge is parabolic in the sense of Ahlfors Nevanlinna.*

For surveys of properties of parabolic Riemann surfaces see [**Ac1, AS**] [**FK**, §4.3]. We review those properties that are used here. First we discuss the Riemann–Roch theorem and Abel's theorem for finite divisors.

A **divisor** on X is an expression of the form $\sum_{p\in X} m_p p$ with $m_p \in \mathbb{Z}$ for all $p \in X$ and $m_p \neq 0$ for p only in a discrete subset of X. A divisor $\sum m_p p$ is called **finite** if $m_p = 0$ for all but finitely many p. The **degree** of a finite divisor $D = \sum m_p p$ is $\deg D := \sum m_p$. If $D = \sum m_p p$, $D' = \sum m'_p p$ are divisors one says $D \geq D'$ if $m_p \geq m'_p$ for all $p \in X$. A divisor D is called **effective** if $D \geq 0$. If f is a meromorphic function on X then its divisor (f) is

$$(f) = \sum_{p\in X} \operatorname{mult}_p(f)p$$

where $\operatorname{mult}_p(f)$ denotes the order of vanishing of f at p. If f has a pole at p then $\operatorname{mult}_p(f)$ is the negative of the pole order. One defines the divisor of a holomorphic differential form in a similar way.

Denote by Ω the space of square integrable holomorphic differential forms on X. For a divisor D, let

$$\Omega(D) = \{\omega \in \Omega \mid (\omega) \geq -D\}.$$

DEFINITION. A meromorphic function f on X is said to be of class $\mathcal{M}$ if it has only a finite number of poles and is bounded in the complement of each neighbourhood of these poles.

THEOREM A.3 (Riemann–Roch theorem for finite divisors. See [**R**] or Theorem 3.10). *Let D be an effective finite divisor on a parabolic Riemann surface X. Then*

$$\dim\{f \in \mathcal{M} \mid (f) \geq -D\} + \operatorname{codim}_\Omega \Omega(-D) = \deg D + 1.$$

Consequences of the Riemann–Roch theorem include

PROPOSITION A.4 (See Proposition 3.12 and Proposition 3.30). *Let X be a parabolic Riemann surface that is not biholomorphic to an open subset of $\mathbb{P}^1(\mathbb{C})$. Then*

(i) *for every $x \in X$ there is an $\omega \in \Omega$ such that $\omega(x) \neq 0$,*
(ii) *for any two different points $x_1, x_2 \in X$ and any path γ joining x_1 to x_2 there exists $\omega \in \Omega$ such that $\int_\gamma \omega \neq 0$.*

DEFINITION. A Riemann surface X is **hyperelliptic** if there is a finite subset $\mathcal{I}$ of $\mathbb{P}^1(\mathbb{C})$, a discrete subset S of $\mathbb{P}^1(\mathbb{C}) \setminus \mathcal{I}$ and a proper holomorphic map $\tau\colon X \to \mathbb{P}^1 \setminus \mathcal{I}$ of degree 2 that ramifies precisely over the points of S.

DEFINITION. An **end region** of a Riemann surface X is an open connected subset $E \subset X$ that is not relatively compact such that for every compact subset K of X there is a compact subset K' of X such that $K \subset K'$ and $E \cap (X \setminus K')$ is a connected component of $X \setminus K'$. We say that two end regions E_1, E_2 of X are equivalent if there is an end region E_3 contained in $E_1 \cap E_2$. An **end** of X is an equivalence class of end regions. An end is said to be **planar** if it can be represented by an end region E such that the intersection form on $H_1(E, \mathbb{Z})$ is zero. X is said to have only **finitely many ends** if there is a compact subset $K \subset X$ such that $X \setminus K$ is a union of finitely many end regions.

PROPOSITION A.5 (See Proposition 3.26). *Let X be a parabolic Riemann surface that is not biholomorphic to an open subset of $\mathbb{P}^1(\mathbb{C})$. Assume that X has only finitely many ends, none of which is planar. If there is an effective divisor D of degree two on X such that* $\operatorname{codim}_\Omega \Omega(D) \leq 1$ *then X is hyperelliptic.*

To formulate Abel's theorem we use

DEFINITION. A nonzero meromorphic function on a parabolic Riemann surface X is called **quasi-rational** if the set M of zeroes and poles of f is finite and if, for every neighbourhood U of M in X,

$$i \int_{X \setminus U} \frac{df}{f} \wedge \overline{\frac{df}{f}} < \infty.$$

THEOREM A.6 (Abel's theorem for finite divisors. See [**AS**] or Theorem 3.29). *Let $D = p_1 + \cdots + p_n - q_1 - \cdots - q_n$ be a finite divisor of degree zero on a parabolic Riemann surface X. Then D is the divisor of a quasi-rational function on X if and only if there are paths γ_j joining p_j to q_j such that*

$$\sum_{j=1}^{n} \int_{\gamma_j} \omega = 0 \qquad \textit{for all } \omega \in \Omega.$$

To study the analog of the Abel–Jacobi mapping we consider a marked Riemann surface $X; A_1, B_1, A_2, B_2, \ldots$ of infinite genus together with an exhaustion function h of bounded charge.

THEOREM A.7 (Riemann Period Relations. See Theorem 2.9). *Suppose $\alpha, \beta \in \Omega$ are forms with*

$$\int_{A_n} \alpha = \int_{A_n} \beta = 0$$

for all but finitely many $n \geq 1$. Then

$$\int_X \alpha \wedge \overline{\beta} = i \sum_{n \geq 1} \int_{A_n} \alpha \int_{B_n} \bar{\beta} - \int_{A_n} \bar{\beta} \int_{B_n} \alpha.$$

COROLLARY A.8 ([**Ne**]). *If $\omega \in \Omega$ is a form with*

$$\int_{A_n} \omega = 0$$

for $n \geq 1$, then $\omega = 0$.

An important fact for the construction of a theta function is

THEOREM A.9 (See Theorem 3.8). *Let $(X; A_1, B_1, \ldots)$ be a marked Riemann surface that admits an exhaustion function with bounded charge. Then there exists a basis $\omega_1, \omega_2, \ldots$ for the Hilbert space Ω of square integrable holomorphic one forms such that*

$$\int_{A_i} \omega_j = \delta_{i,j}.$$

DEFINITION. The **Riemann period matrix** $R(X)$ is defined by

$$R_{ij} = \int_{B_i} \omega_j.$$

It follows from the period relations that R is symmetric, and its imaginary part is positive definite. Precisely,

$$\sum_{1\le i,j\le n} (\operatorname{Im} R)_{ij} x_i x_j > 0$$

for all $n \ge 1$ and all nonzero real vectors $x = (x_1, x_2, \dots, x_n)$.

The set

$$\Gamma := \left\{ \sum_{j=1}^{\infty} (n_j E_j + m_j R_j) \;\middle|\; n_j, m_j \in \mathbb{Z}, n_j = m_j = 0 \text{ for all but finitely many } j \right\}$$

where R_j denotes the jth column of the period matrix R and E_j is the jth unit vector is called the **period lattice**. It acts by translation on the product $\prod_{j=1}^{\infty} \mathbb{C}$ of infinitely many copies of $\mathbb{C}$. (For now, this product is just considered as a topological vector space, endowed with the product topology). We define the "point set Jacobian" of X as

$$J := \left(\prod_{j=1}^{\infty} \mathbb{C} \right) / \Gamma.$$

For each point $x_0 \in X$ we define the **Abel–Jacobi map** with base point x_0 by

$$X \to J, x \mapsto \int_\gamma \vec{\omega} := \left(\int_\gamma \omega_1, \quad \int_\gamma \omega_2, \dots \right),$$

where γ is a path in X joining x_0 to x. If one replaces γ by another path, $\int_\gamma \vec{\omega}$ only changes by an element of Γ, so the map is well-defined. Part (ii) of the Proposition above shows that the Abel–Jacobi map is injective.

Ω^* can be identified with a subspace of the sequence space $\prod_{j=1}^{\infty} \mathbb{C}$ by the map

$$\Omega^* \hookrightarrow \prod_{j=1}^{\infty} \mathbb{C},$$
$$\varphi \mapsto (\varphi(\omega_1), \varphi(\omega_2), \dots).$$

One can show that $\Gamma \subset \Omega^*$.

If x is a point of X, ξ a local coordinate at x, the componentwise derivative of the Abel–Jacobi map at x with respect to ξ is the vector

$$\kappa_\xi(x) = \left(\left(\frac{\omega_1}{d\xi} \right)(x), \left(\frac{\omega_2}{d\xi} \right)(x), \dots \right) \in \prod_{j=1}^{\infty} \mathbb{C}.$$

Part (i) of the proposition above implies that this vector is never zero. Of course it depends on the choice of the local coordinate ξ, but a change in the local coordinate affects the vector above only by multiplication with a nonzero number. Therefore we let $\mathbb{P}$ be the quotient of $(\prod_{j=1}^{\infty} \mathbb{C}) \setminus \{0\}$ by the action of $\mathbb{C}^*$ given by

$$\lambda \cdot (e_1, e_2, \dots) = (\lambda e_1, \lambda e_2, \dots) \quad (\lambda \in \mathbb{C}^*).$$

By what we said above we get a well-defined map

$$\kappa \colon X \to \mathbb{P}, \quad x \mapsto [\kappa_\xi(x)].$$

This map is called the **canonical map**, we sometimes also write $\kappa(x) = [\vec{\omega}(x)]$.

PROPOSITION A.10 (See Proposition 3.18 and Proposiition 3.26).

(i) *If for every effective divisor D of degree two* $\operatorname{codim}_\Omega \Omega(-D) = 2$ *then* $\kappa\colon X \to \mathbb{P}$ *is injective. Furthermore for each $x \in X$ one has* $\dim \operatorname{span}[\vec{\omega}(x), \dot{\vec{\omega}}(x)] = 1$.

(ii) *If X is hyperelliptic, $\tau\colon X \to \mathbb{P}^1 \setminus M$ the hyperelliptic projection and $M \neq \varnothing$ then for $x_1, x_2 \in X$*

$$\kappa(x_1) = \kappa(x_2) \quad \text{if and only if} \quad \tau(x_1) = \tau(x_2).$$

Furthermore $\dim \operatorname{span}[\vec{\omega}(x), \dot{\vec{\omega}}(x)] = 1$ *if and only if x is not a ramification point of τ.*

Using the Riemann period matrix R one can define the theta function of the marked Riemann surface $(X; A_1, B_1, \dots)$.

THEOREM A.11 (See Theorem 4.6). *Suppose that there is a sequence $t_j \in (0,1)$, $j \geq 1$ such that*

$$\sum_{j \geq 1} t_j^\beta < \infty$$

for some $0 < \beta < \frac{1}{2}$ and such that

$$\langle \mathbf{n}, \operatorname{Im} R\mathbf{n} \rangle = \sum_{i,j \geq 1} n_i \operatorname{Im} R_{i,j} n_j \geq \frac{1}{2\pi} \sum_{j \geq 1} |\log t_j| n_j^2$$

for all vectors

$$\mathbf{n} = \begin{pmatrix} n_1 \\ n_2 \\ \vdots \end{pmatrix}$$

in $\mathbb{Z}^\infty$ with only a finite number of nonzero components. Then the theta series

$$\theta(\mathbf{z}, R) = \sum_{\substack{\mathbf{n} \in \mathbb{Z}^\infty \\ |\mathbf{n}| < \infty}} e^{2\pi i \langle \mathbf{z}, \mathbf{n} \rangle} e^{\pi i \langle \mathbf{n}, R\mathbf{n} \rangle}$$

converges absolutely and uniformly on bounded subsets of the Banach space

$$B = \left\{ z = (z_1, z_2, \dots) \in \mathbb{C}^\infty \;\middle|\; \lim_{j \to \infty} \frac{z_j}{|\log t_j|} = 0 \right\}, \quad \|z\| = \sup_{j \geq 1} \frac{|z_j|}{|\log t_j|}$$

to an entire function that does not vanish identically.

Finally we recall a standard consequence of Stoke's theorem.

PROPOSITION A.12. *Let $\mathcal{X}$ be a compact Riemann surface whose boundary $\partial\mathcal{X}$ has connected components $\partial\mathcal{X}_1, \dots, \partial\mathcal{X}_m$ and with canonical homology basis A_i, B_i, $1 \leq i \leq g$. Let ω, η be closed 1-forms on $\mathcal{X}$ with*

$$\int_{\partial\mathcal{X}_j} \omega = \int_{\partial\mathcal{X}_j} \eta = 0 \quad \text{for} \quad 1 \leq j \leq m.$$

Let f be a single valued Stammfunktion (= primitive) of ω on a neighbourhood of $\bigcup_{j=1}^m \partial\mathcal{X}_j$. Then

$$\int_{\mathcal{X}} \omega \wedge \eta = \sum_{i=1}^{g} \left(\int_{A_i} \omega \int_{B_i} \eta - \int_{B_i} \omega \int_{A_i} \eta \right) + \sum_{j=1}^{m} \int_{\partial\mathcal{X}_j} f\eta.$$

CHAPTER 2

The Torelli Theorem

Proposition II. *Extension is an attribute of God, or God is an extended thing.*
Proof. The proof of this proposition is similar to that of the last.
Benedict De Spinoza, *The Ethics*

Introduction

We introduce a class of marked Riemann surfaces $(X; A_1, B_1, \dots)$ of infinite genus by pasting plane domains and handles together. Here, $A_1, B_1, \dots,$ is a canonical homology basis on X. The asymptotic holomorphic structure is specified by six geometric hypotheses presented in Section 5.

The first important fact, Theorem 5.1, about a surface $(X; A_1, B_1, \dots)$ in this class is the existence of a proper, nonnegative Morse function h on X that satisfies

$$\sup_{t>s>0} \left| \int_{X_t \setminus X_s} d{*}dh \right| < \infty$$

and has the additional property that for each $t > 0$, there is an $n \geq 1$ such that the cycles

$$A_1, B_1, \dots, A_n, B_n$$

generate a maximal submodule of $H_1(X_t, \mathbb{Z})$ on which the intersection form is nondegenerate. Here,

$$X_t = h^{-1}([0,t])$$

We showed in Chapter 1 (see Appendix A, at the end of Chapter 1, for a summary of Chapter 1) that *any* marked Riemann surface supporting a function h with the properties above is "parabolic" in the sense of Nevanlinna–Ahlfors (see, [**AS**]) and has unique holomorphic one forms ω_j, $j \geq 1$, satisfying

$$\|\omega_j\|^2 = \int_X \omega_j \wedge \overline{*\omega_j} < \infty$$

and

$$\int_{A_i} \omega_j = \delta_{ij}.$$

Furthermore, the associated, infinite, Riemann matrix

$$R_X = \left(\int_{B_i} \omega_j \right)$$

is symmetric and $\operatorname{Im} R_X$ is positive definite.

Another important fact (see the proof of Theorem 7.1) about $(X; A_1, B_1, \dots)$ is

$$\langle \mathbf{n}, \operatorname{Im} R_X \mathbf{n} \rangle = \sum_{i,j \geq 1} n_i \operatorname{Im} R_{i,j} n_j \geq \sum_{j \geq 1} |\log t_j| n_j^2$$

for all vectors

$$\mathbf{n} = \begin{pmatrix} n_1 \\ n_2 \\ \vdots \end{pmatrix}$$

in $\mathbb{Z}^\infty$ with only a finite number of nonzero components. Here, t_j, $j \geq g+1$, determines (see (GH2), Section 5) the waists of the handles outside a compact submanifold of X. It follows from our results in Chapter 1 that the theta series

$$\theta(\mathbf{z}, R_X) = \sum_{\substack{\mathbf{n}\in\mathbb{Z}^\infty \text{ with only a finite} \\ \text{number of nonzero components}}} e^{2\pi i\langle \mathbf{z},\mathbf{n}\rangle} e^{\pi i\langle \mathbf{n}, R_X\mathbf{n}\rangle}$$

converges to a nontrivial entire function on the complex Banach space

$$B = \left\{ \mathbf{z} = (z_1, z_2, \dots) \in \mathbb{C}^\infty \;\middle|\; \lim_{j\to\infty} \frac{|z_j|}{|\log t_j|} = 0 \right\}$$

with the norm

$$\|\mathbf{z}\| = \sup_{j\geq 1} \frac{|z_j|}{|\log t_j|}$$

and satisfies the usual tranformation laws.

We see from the preceding discussion that for each surface in our class there is a unique, normalized frame of square integrable holomorphic one forms and an associated theta function that is holomorphic on B. It is now possible to investigate the relationship between the surface $(X; A_1, B_1, \dots)$ and the geometry of its theta function. In this chapter we formulate and prove, among other things, analogues of Riemann's vanishing theorem and Torelli's theorem.

Suppose $(X; A_1, B_1, \dots, A_g, B_g)$ is a marked Riemann surface of genus g and $\theta(\mathbf{z}, R_X)$ its theta function. Fix $e \in \mathbb{C}^g$ and let $\omega_1, \dots, \omega_g$, be the corresponding normalized frame of holomorphic one forms. Then, by a classical theorem of Riemann, the mutilvalued function

$$\theta\left(e + \int_{x_0}^{x} \vec{\omega} \right)$$

of $x \in X$ either vanishes identically or has exactly g roots.

To generalize this fact to our class of infinite genus Riemann surfaces we first show (Proposition 7.3) that for any path joining x_1 to x_2 on X the infinite vector

$$\int_{x_1}^{x_2} \vec{\omega} = \left(\int_{x_1}^{x_2} \omega_1, \int_{x_1}^{x_2} \omega_2, \dots \right)$$

lies in the Banach space B. Next, it is observed that there is one ideal point at infinity for each of the plane domains appearing in the decomposition of X. For the rest of the Introduction it is assumed, for simplicity, that there is just one plane domain. We prove (Proposition 7.6) that there is a path from each point $x \in X$ to ∞ such that

$$\int_{\infty}^{x} \vec{\omega}$$

belongs to B.

Now, let $e \in B$ and suppose $\theta(e) \neq 0$. We show (Theorem 7.11) that there is a compact submanifold Y with boundary such that the multivalued, holomorphic function

$$\theta\Big(e+\int_\infty^x \vec{\omega}, R_X\Big)$$

has exactly genus(Y) roots in Y, exactly one root in each in each handle outside of Y and no other roots. That is, exactly "genus(X)" roots. The solution of the Jacobi inversion problem, formulated in terms of the roots of the theta function, is given in Theorem 7.16.

The proofs of these statements require estimates on the frame $\omega_1, \omega_2, \ldots$. For example, by Theorem 6.4 and formula (6.3), the pull back $w_j(z)\,dz$ of ω_j to any plane domain in the decomposition of X decays quadratically. When there is only one plane domain,

$$\left| w_j(z) - \frac{1}{2\pi i}\left(\frac{1}{z-s_1(j)} - \frac{1}{z-s_2(j)}\right)\right| \le \frac{\text{const}}{|z^2|}$$

where the jth handle is glued into the plane domain close to the points $s_1(j)$, $s_2(j)$. The const is independent of j. We also make a detailed investigation of the pull backs of ω_j to the handles (see, Proposition 6.16).

In Section 8 we introduce the notion of a divisor of degree "genus(X)" on X. This is done by fixing an auxiliary point $\hat{e} \in B$ with $\theta(\hat{e}) \neq 0$ and comparing sequences of points on X to the "genus(X)" many roots $\widehat{x}_1, \widehat{x}_2 \ldots$, of

$$\theta\Big(\hat{e}+\int_\infty^x \vec{\omega}\Big) = 0.$$

Precisely, a sequence $y_j, j \ge 1$, on X represents a divisor of degree "genus(X)" if eventually, y_j lies in the same handle as $\widehat{x}_j$ and the vector

$$\Big(\int_{\widehat{x}_1}^{y_1} \omega_1, \int_{\widehat{x}_2}^{y_2} \omega_2, \ldots\Big)$$

lies in B. The space $W^{(0)}$ of all these sequences is given the structure of a complex Banach manifold modelled on B. The quotient $S^{(0)}$ of $W^{(0)}$ by the group of all finite permutations is the manifold of divisors of degree "genus(X)". The construction is independent of the auxiliary point $\hat{e}$. We similarly construct Banach manifolds $S^{(-n)}$ of divisors of index n, that is, of degree "genus(X) $-\, n$", by deleting the first n components in a sequence $y_1, y_2, \ldots$ belonging to $W^{(0)}$.

Fix $\hat{e}$ as above. The analogue of the Abel–Jacobi map

$$f^{(0)} \colon S^{(0)} \to B$$

is induced by

$$(y_1, y_2, \ldots) \mapsto \hat{e} - \sum_{i\ge 1} \int_{\widehat{x}_i}^{y_i} \vec{\omega}.$$

By Proposition 8.1, $f^{(0)}$ is holomorphic and its derivative is Fredholm of index zero at every point of $S^{(0)}$. Let

$$\Theta = l\{e \in B \mid \theta(e) = 0\}$$

be the theta divisor of X. Then, by the solution of the Jacobi inversion problem, $f^{(0)}$ is a biholomorphism between $f^{(0)-1}(B \setminus \Theta)$ and $B \setminus \Theta$.

Similarly, the map

$$f^{(-1)} \colon S^{(-1)} \to B$$

is induced by

$$(y_2, y_3, \dots) \mapsto \hat{e} - \int_{\widehat{x}_1}^{\infty} \vec{\omega} - \sum_{i \geq 2} \int_{\widehat{x}_i}^{y_i} \vec{\omega}.$$

The analogue of the Riemann vanishing theorem (Theorem 8.4) is that

$$f^{(-1)}(S^{(-1)}) \subset \Theta$$

and

$$\left\{ e \in \Theta \;\middle|\; \theta\left(e - \int_{\infty}^{x} \vec{\omega}\right) \neq 0 \text{ for some } x \text{ in } X \right\} \subset f^{(-1)}(S^{(-1)}).$$

The set

$$\left\{ e \in \Theta \;\middle|\; \theta\left(e - \int_{\infty}^{x} \vec{\omega}\right) = 0 \text{ for all } x \text{ in } X \right\}$$

is stratified and studied in Theorem 9.1.

The Torelli theorem for compact Riemann surfaces states that two Riemann surfaces that have the same period matrices are biholomorphically equivalent. We prove (Theorem 11.1) the same statement for Riemann surfaces in our class. The proof mimics the argument of Andreotti [**An, GH**] for the compact case. We investigate the ramification locus of the Gauss map on the theta divisor. It turns out (Proposition 9.8) that at a generic point e of this ramification locus the kernel of the derivative of the Gauss map is related to the values $\omega_j(x)$, $j = 1, 2, \dots$, of the differentials ω_j at some point $x = x(e)$ of X. If X is not hyperelliptic then almost all points of X occur, while in the hyperelliptic case (Proposition 10.5) only the Weierstrass points occur. Using these observations it is possible to recover the Riemann surface X from Θ, which in turn is completely determined by the period matrix of X.

In Chapter 3 we will show that Fermi curves, spectral curves for the periodic Kadomcev–Petviashvilii equation and spectral curves for periodic ordinary differential operators belong to our class.

5. Geometric Hypotheses

In this section we introduce a class of marked Riemann surfaces $(X; A_1, B_1, \dots)$ that are "asymptotic to" a finite number of complex lines $\mathbb{C}$ joined by infinitely many handles. Here, X is a Riemann surface and $A_1, B_1, \dots$ is a canonical homology basis for X.

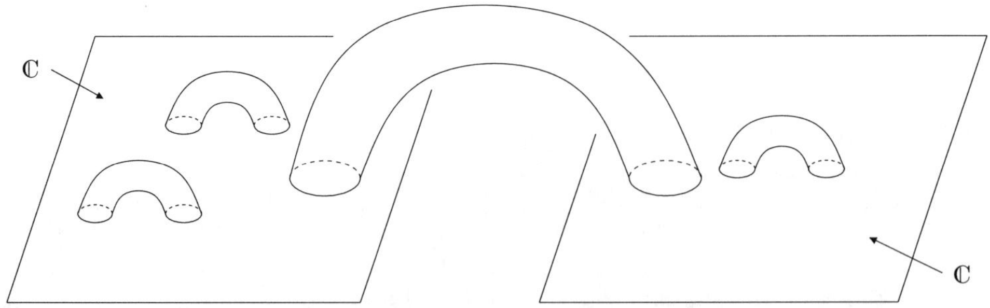

To be precise, we make the

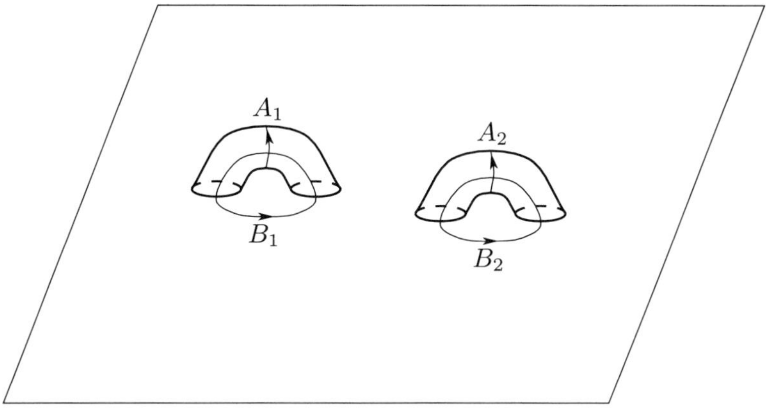

DEFINITION. The notation

$$X = X^{\text{com}} \cup X^{\text{reg}} \cup X^{\text{han}}$$

denotes a marked Riemann surface $(X; A_1, B_1, \dots)$ with a decomposition into a compact, connected submanifold $X^{\text{com}} \subset X$ with smooth boundary and genus $g \geq 0$, a finite number of open "regular pieces" $X^{\text{reg}}_\nu \subset X$, $\nu = 1, \dots, m$,

$$X^{\text{reg}} = \bigcup_{\nu=1}^{m} X^{\text{reg}}_\nu$$

and an infinite number of closed "handles" $Y_j \subset X$, $j \geq g+1$,

$$X^{\text{han}} = \bigcup_{j \geq g+1} Y_j$$

with $X^{\text{com}} \cap (X^{\text{reg}} \cup X^{\text{han}}) \neq \varnothing$, that satisfies the geometric hypotheses (GH1)–(GH6) stated below.

GH1 (Regular Pieces).

(i) For all $1 \leq \mu \neq \nu \leq m$,

$$\overline{X^{\text{reg}}_\mu} \cap \overline{X^{\text{reg}}_\nu} = \varnothing.$$

(ii) For each $1 \leq \nu \leq m$ there is an infinite discrete subset $S_\nu \subset \mathbb{C}$. Furthermore, for each $s \in S_\nu$ there is a compact, simply connected neighborhood $D_\nu(s)$ with smooth boundary $\partial D_\nu(s)$ such that

$$D_\nu(s) \cap D_\nu(s') = \varnothing$$

when $s \neq s'$.

(iii) For each $1 \leq \nu \leq m$, there is a compact simply connected neighborhood $K_\nu \subset \mathbb{C}$ of 0 with smooth boundary and

$$K_\nu \cap D_\nu(s) = \varnothing$$

for all $s \in S_\nu$. Set

$$G_\nu = \mathbb{C} \setminus \left(\text{int } K_\nu \cup \bigcup_{s \in S_\nu} \text{int } D_\nu(s) \right).$$

There is a biholmorphic map Φ_ν,

$$\Phi_\nu \colon G_\nu \to \overline{X_\nu^{\mathrm{reg}}}$$

between G_ν and $\overline{X_\nu^{\mathrm{reg}}}$.

Informally, the closure of the regular piece X_ν^{reg}, $\nu = 1, \ldots, m$, is biholomorphic to a copy of $\mathbb{C}$ minus an open, simply connected neighborhood around each point of S_ν and an additional set K_ν. One end of a closed cylindrical handle will be glued to a closed "annular" region surrounding $D_\nu(s)$ in G_ν. One connected component of $\partial X^{\mathrm{com}}$ will be glued to ∂K_ν.

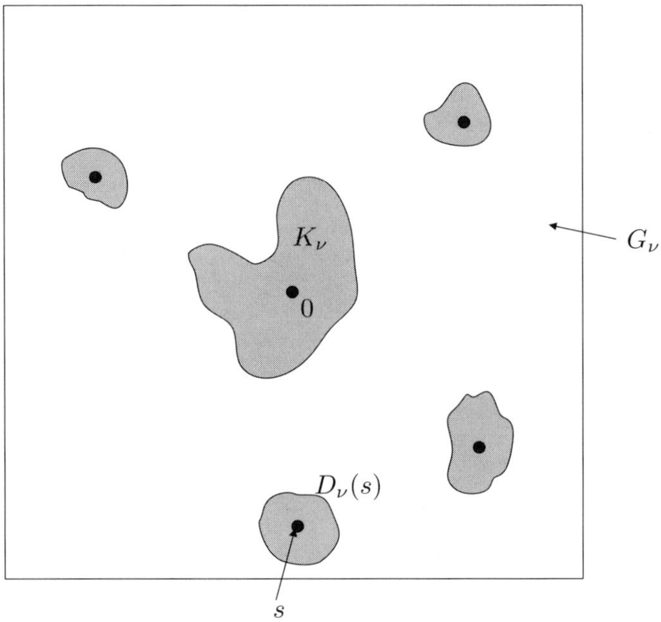

GH2 (Handles).

(i) For all $i \neq j$ with $i, j \geq g+1$,

$$Y_i \cap Y_j = \varnothing.$$

(ii) For each $j \geq g+1$ there is a $0 < t_j < \frac{1}{2}$ and a biholomorphic map ϕ_j

$$\phi_j \colon \mathrm{H}(t_j) \to Y_j$$

between the model handle $\mathrm{H}(t_j)$ and Y_j. Here, the model handle $\mathrm{H}(t)$ is defined by

$$\mathrm{H}(t) = \{(z_1, z_2) \in \mathbb{C}^2 \mid z_1 z_2 = t \text{ and } |z_1|, |z_2| \leq 1\}$$

for all $0 < t < \frac{1}{2}$. It is clearly diffeomorphic to $S^1 \times [0, 1]$. In particular, Y_j, is diffeomorphic to a closed cylinder.

(iii) For all $j \geq g+1$ the cylinder Y_j represents the cycle A_j in homology. Exactly, A_j is the homology class represented by the oriented loop

$$\phi_j\big(\{(\sqrt{t_j}e^{i\theta}, \sqrt{t_j}e^{-i\theta}) \mid 0 \leq \theta \leq 2\pi\}\big).$$

(iv) For every $\beta > 0$

$$\sum_{j\geq g+1} t_j^\beta < \infty$$

In other words, the handle Y_j, $j \geq g+1$, is biholomorphic to the smooth deformation $\mathrm{H}(t_j)$ of the ordinary double point singularity

$$\{(z_1, z_2) \in \mathbb{C}^2 \mid z_1 z_2 = 0 \text{ and } |z_1|, |z_2| \leq 1\}.$$

The size of the handle is determined by t_j.

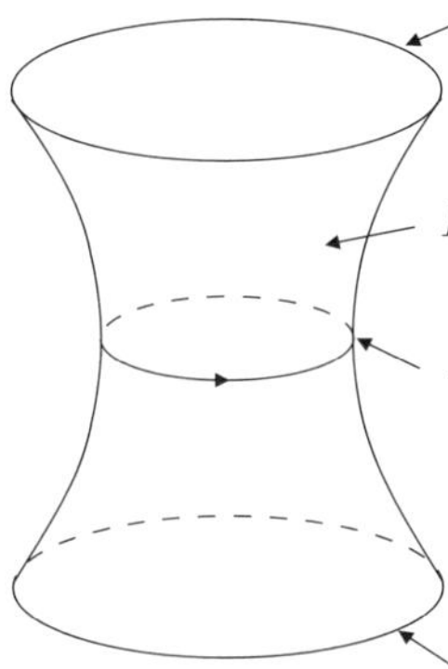

Intuitively, condition (iv) is fulfilled when the waists of Y_j, $j \geq g+1$ are rapidly decreasing.

GH3 (Glueing handles and regular pieces together).

(i) For each $j \geq g+1$ the intersection $Y_j \cap X^{\mathrm{reg}}$ consists of two components Y_{j1}, Y_{j2}:

$$Y_j \cap X^{\mathrm{reg}} = Y_{j1} \cup Y_{j2}.$$

For each pair (j, μ) with $j \geq g+1$ and $\mu = 1, 2$ there is a radius

$$\tau_\mu(j) \in (\sqrt{t_j}, 1)$$

and a sheet number

$$\nu_\mu(j) \in \{1, \ldots, m\}$$

such that

$$Y_{j\mu} = \phi_j\big(\{(z_1, z_2) \in H(t_j) \mid \tau_\mu(j) \leq |z_\mu| \leq 1\}\big)$$

and

$$Y_{j\mu} \subset X^{\mathrm{reg}}_{\nu_\mu(j)}.$$

There is a bijective map

$$(j, \mu) \in \{j \in \mathbb{Z} \mid j \geq g+1\} \times \{1, 2\} \mapsto s_\mu(j) \in \bigcup_{\nu=1}^{m} S_\nu \quad \text{(disjoint union)}$$

such that

$$\phi_j\big(\{(z_1, z_2) \in H(t_j) \mid |z_\mu| = \tau_\mu(j)\}\big) = \Phi_{\nu_\mu(j)}\big(\partial D_{\nu_\mu(j)}(s_\mu(j))\big).$$

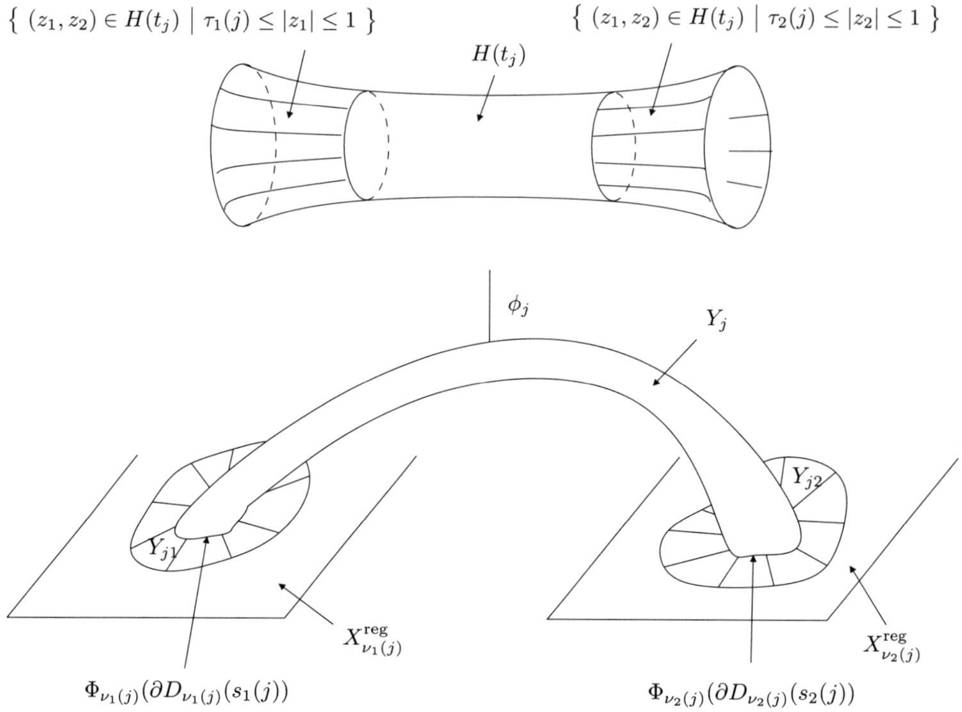

(ii) For each $j \ge g+1$ and $\mu = 1,2$ there are

$$R_\mu(j) > 4r_\mu(j) > 0$$

such that the biholomorphic map

$$g_{j\mu}\colon \mathcal{A}_{j\mu} = \{z \in \mathbb{C} \mid \tau_\mu(j) \le |z| \le 1\} \to \mathbb{C}$$

defined by

$$g_{j\mu}(z) = \begin{cases} \Phi^{-1}_{\nu_1(j)} \circ \phi_j\big(z, \frac{t_j}{z}\big), & \mu = 1 \\ \Phi^{-1}_{\nu_2(j)} \circ \phi_j\big(\frac{t_j}{z}, z\big), & \mu = 2 \end{cases}$$

satisfies

$$\big|g_{j\mu}\big(4\tau_\mu(j)e^{i\theta}\big) - s_\mu(j)\big| < r_\mu(j)$$

and

$$|g_{j\mu}(e^{i\theta}) - s_\mu(j)| > R_\mu(j) > |g_{j\mu}(e^{i\theta}/2) - s_\mu(j)|, \quad |g_{j\mu}(e^{i\theta}/4) - s_\mu(j)| > R_\mu(j)/4$$

for all $0 \le \theta \le 2\pi$.

REMARK. Informally, the map $s_\mu(j)$ enumerates the points of $\bigcup_{\nu=1}^m S_\nu$ and specifies that the end of the handle $\mathrm{H}(t_j)$ containing

$$\{(z_1,z_2) \in \mathrm{H}(t_j) \mid |z_\mu| = 1\}$$

is glued to the annular region $\Phi^{-1}_{\nu_\mu(j)}(Y_{j\mu})$ in $G_{\nu_\mu(j)}$.

REMARK. The map $g_{j,\mu}$ describes how $Y_{j\mu}$ is glued to $G_{\nu_\mu(j)}$.

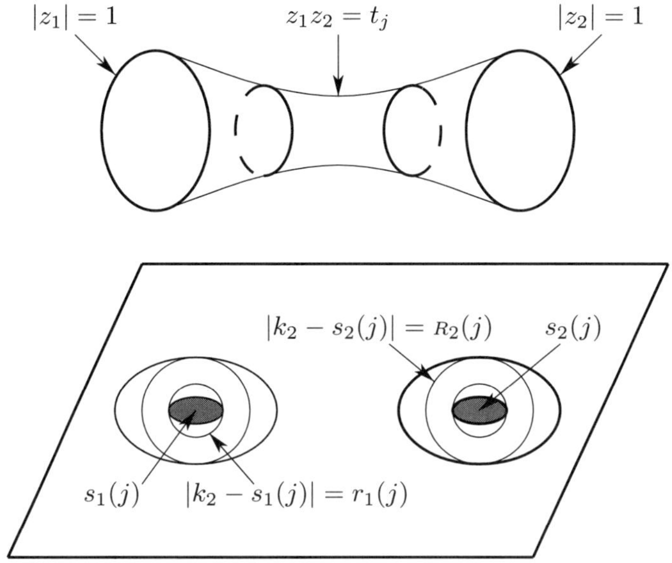

The parameters that control the overlap $Y_{j1} \cup Y_{j2}$ between X^{reg} and the handle Y_j are introduced in hypothesis (GH3). The component $Y_{j\mu}$ of the overlap is specified in two charts; in the plane region G_ν, and on the model handle $H(t_j)$. The preimage $\phi_j^{-1}(Y_{j\mu})$ in $H(t_j)$ of the overlap $Y_{j\mu}$ is $\{(z_1, z_2) \in H(t_j) \mid \tau_\mu(j) \leq |z_\mu| \leq 1\}$.

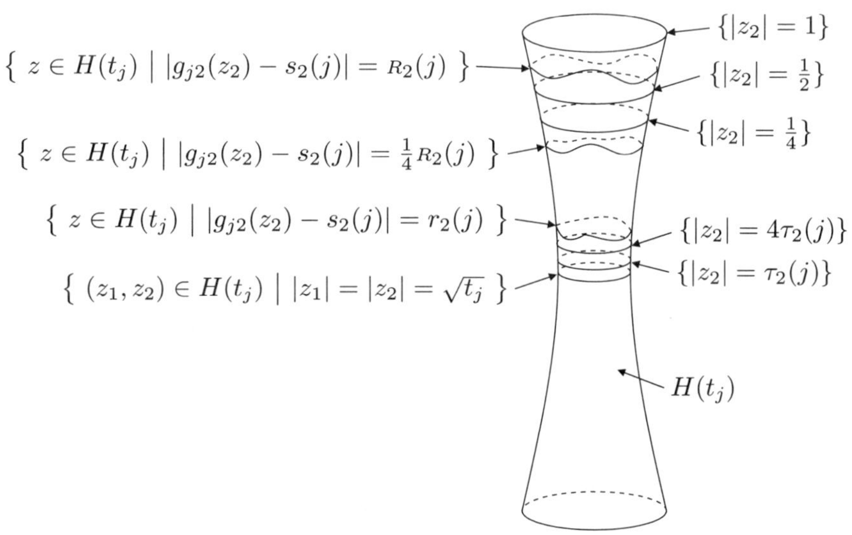

We imagine, as in the figure above, that $\tau_\mu(j)$ is relatively close to the radius of the waist $\sqrt{t_j}$ so that the overlap on $H(t_j)$ is large.

The preimage $\Phi^{-1}_{\nu_\mu(j)}(Y_{j\mu})$ in the other chart $G_{\nu_\mu(j)}$ is the annular plane region surrounding $s_\mu(j)$ whose inner boundary is $\partial D_{\nu_\mu(j)}\big(s_\mu(j)\big)$ and whose outer boundary is $\Phi^{-1}_{\nu_\mu(j)} \circ \phi_j\big(\{(z_1, z_2) \in H(t_j) \mid |z_\mu| = 1\}\big)$. It contains $\{z \in \mathbb{C} \mid r_\mu(j) \leq |z - s_\mu(j)| \leq R_\mu(j)\}$. We will assume, (GH5)(ii), that $r_\mu(j)$ and $r_\mu(j/R_\mu(j)$

are both asymptotically small. Consequently, the "holes" $D_\nu(s)$, $s \in S_\nu$, in G_ν are also asymptotically small and the overlap is asymptotically big.

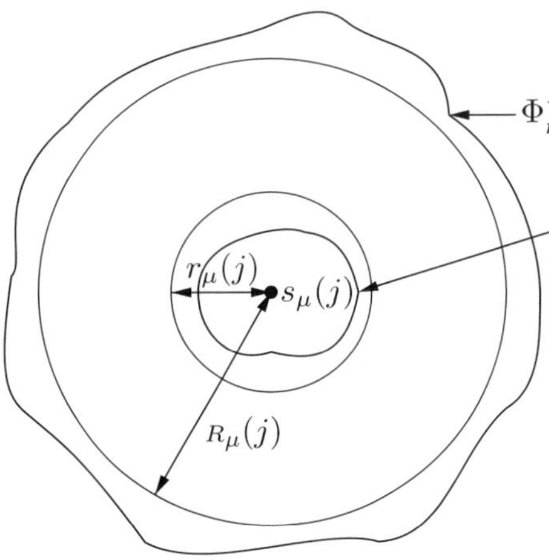

Passing from the chart $G_{\nu_\mu(j)}$ to $H(t_j)$, the image of the circle $|z - s_\mu(j)| = R_\mu(j)$ lies near

$$\{(z_1, z_2) \in H(t_j) \mid |z_\mu| = \tfrac{1}{2}\}$$

and the image of the circle $|z - s_\mu(j)| = r_\mu(j)$ lies outside

$$\{(z_1, z_2) \in H(t_j) \mid |z_\mu| = 4\tau_\mu(j)\}$$

REMARK. If $\partial D_{\nu_\mu(j)}$ is counterclockwise oriented, then, by construction, $\Phi(\partial D_{\nu_\mu(j)})$ is homologous to $(-1)^{\mu+1} A_j$.

GH4 (Glueing in the Compact Piece).

$$\partial X^{\mathrm{com}} = \Phi_1(\partial K_1) \cup \cdots \cup \Phi_m(\partial K_m).$$

Furthermore $A_1, B_1, \ldots, A_g, B_g$ is the image of a canonical homology basis of X^{com} under the map $H_1(X^{\mathrm{com}}, \mathbb{Z}) \to H_1(X, \mathbb{Z})$ induced by inclusion.

GH5 (Estimates on the Glueing Maps).

(i) For each $j \geq g+1$ and $\mu = 1, 2$

$$R_\mu(j) < \tfrac{1}{4} \min_{\substack{s \in S_{\nu_\mu(j)} \\ s \neq s_\mu(j)}} |s - s_\mu(j)|,$$

$$R_\mu(j) < \tfrac{1}{4} \operatorname{dist}\left(s_\mu(j), K_{\nu_\mu(j)}\right).$$

(ii) There are $0 < \delta < d$ such that

$$\sum_{j,\mu} \frac{1}{|s_\mu(j)|^{d-4\delta-2}} < \infty$$

and such that, for all $j \geq g+1$ and $\mu = 1, 2$

$$r_\mu(j) < \frac{1}{|s_\mu(j)|^d}, \quad R_\mu(j) > \frac{1}{|s_\mu(j)|^\delta}, \quad |s_1(j) - s_2(j)| > \frac{1}{|s_\mu(j)|^\delta}.$$

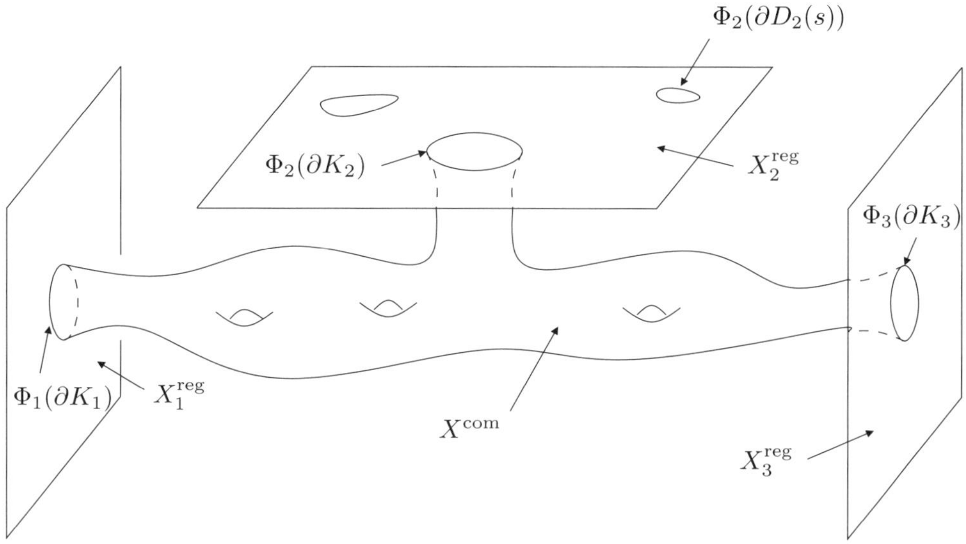

(iii) For all $j \geq g+1$

$$\big||s_1(j)| - |s_2(j)|\big| \leq \tfrac{1}{4} \min_{\mu=1,2} \min_{\substack{s \in S_{\nu_\mu(j)} \\ s \neq s_\mu(j)}} |s - s_\mu(j)|.$$

For $\mu = 1, 2$

$$\sum_j \frac{||s_1(j)| - |s_2(j)||}{|s_\mu(j)|} < \infty.$$

(iv) For $\mu = 1, 2$

$$\lim_{j \to \infty} \frac{\log |s_\mu(j)|}{|\log t_j|} = 0.$$

(v) For $\mu = 1, 2$

$$\lim_{j \to \infty} \frac{R_\mu(j)}{\min_{\substack{s \in S_{\nu_\mu(j)} \\ s \neq s_\mu(j)}} |s - s_\mu(j)|} \log |s_\mu(j)| = 0.$$

(vi) For each $j \geq g+1$ and $\mu = 1, 2$ we define $\alpha_{j,\mu}(z)$ by

$$\alpha_{j,\mu}(z)\, dz = (g_{j,\mu})_* \left(\frac{1}{2\pi i} \frac{dz_1}{z_1} \right) - \frac{(-1)^{\mu+1}}{2\pi i} \frac{1}{z - s_\mu(j)}\, dz.$$

We assume

$$\sup_{j,\mu} \|\alpha_{j,\mu}(z)\, dz\big|_{\{z \in \mathbb{C} | r_\mu(j) < |z - s_\mu(j)| < R_\mu(j)\}}\|_2 < \infty$$

and, for $\mu = 1, 2$

$$\lim_{j \to \infty} R_\mu(j) \sup_{|z - s_\mu(j)| = R_\mu(j)} |\alpha_{j,\mu}(z)| = 0.$$

Morally, (GH5)(i) says that the holes $D_\nu(s)$ are separated. The first and last parts of (GH5)(ii) and (GH5)(v) bound their density at infinity. The second part of (GH5)(ii) implies that $r_\mu(j)$ and $r_\mu(j)/R_\mu(j)$ are both asymptotically small. The condition (GH5)(iii) forces the two ends of a handle to be attached at approximately the same distance from the origin on the regular pieces. (GH5)(iv) relates the size of the waist of the handle Y_j to the distance from the origin at which it is attached to the regular piece. Finally, (GH5)(vi) measures the derivative of the glueing map $g_{j,\mu}$ by pulling back the holomorphic form dz_1/z_1 on $H(t_j)$ and comparing it to the meromorphic form $1/z - s_\mu(j)\,dz$. The intuition is that both of these forms have the same A_j period and should be the leading part of ω_j.

GH6 (Distribution of s_ν). For all $\nu = 1, \dots, m$ such that

$$\#\{(j, \mu) \mid \nu_\mu(j) = \nu, \nu_1(j) \neq \nu_2(j)\} < \infty$$

that is, such that the sheet X_ν^{reg} is joined to other sheets by only finitely many handles, one has

$$\limsup_{\substack{j\to\infty \\ \nu_1(j)=\nu_2(j)=\nu}} |s_1(j) - s_2(j)| = \infty.$$

This ends the statment of the hypotheses that specify the class of Riemann surfaces that we consider. The first four hypotheses are essentially topological in nature. The estimates in (GH5) control the analytic structure of X. Hypothesis (GH6) is used only in the proof of the Torelli theorem in §11.

We should point out that the results of this paper apply more generally than under hypotheses (GH1)–(GH6). It suffices to assume (GH1)–(GH4), (GH5)(i), (GH5)(iv) and the conclusions of Lemmas 5.2, 6.1, 6.2, 6.3, 6.19, 7.5 and 7.7 and Corollary 6.17 below.

The first important consequence of the hypotheses is

THEOREM 5.1. *A marked Riemann surface $X = X^{\mathrm{com}} \cup X^{\mathrm{reg}} \cup X^{\mathrm{han}}$ as above admits an exhaustion function with bounded charge.*

By (A.2) the theorem implies that these surfaces are parabolic in the sense of Ahlfors–Nevanlinna. By construction they have only finitely many ends.

To prepare for the proof of Theorem 5.1 we first note

LEMMA 5.2.

(a) *For all $j \geq g+1$ and $\mu = 1, 2$*

$$\frac{r_\mu(j)}{|s_\mu(j)|} < \frac{1}{2}$$

(b)
$$\sum_{\substack{j\geq g+1\\ \mu=1,2}} \left(\frac{1 + |\log \tau_\mu(j)|}{|\log t_j|} \left| \log \frac{|s_1(j)|}{|s_2(j)|} \right| + \frac{r_\mu(j)}{|s_\mu(j)|} \right) < \infty.$$

PROOF. (a) Since $0 \in K_\nu$ for $\nu = 1, \dots, m$ we have by (GH5)(i)

$$r_\mu(j) < \tfrac{1}{4} R_\mu(j) < \tfrac{1}{16} |s_\mu(j)|.$$

(b) This part follows from

$$\frac{1 + |\log \tau_\mu(j)|}{|\log t_j|} = O(1),$$

$$\sum_j \left| \log \frac{|s_1(j)|}{|s_2(j)|} \right| \leq \sum_{j,\mu} \log\left(1 + \frac{||s_1(j)| - |s_2(j)||}{|s_\mu(j)|} \right) < \infty$$

which is a consequence of (GH5)(iii) and

$$\sum_j \frac{r_\mu(j)}{|s_\mu(j)|} \le \sum_j \frac{1}{|s_\mu(j)|^{d+1}} < \infty$$

which is a consequence of (GH5)(ii). □

The estimates of Lemma 5.2 are motivated by the following simple Ansatz for an exhaustion function with bounded charge. Fix a smooth monotone function $\chi(t)$ satisfying

$$\chi(t) = \begin{cases} 0, & t \le 2 \\ 1, & t \ge 3 \end{cases}$$

and set

$$c_\mu(j) = \frac{\log |s_\mu(j)|}{\log t_j}$$

for $\mu = 1, 2$. Notice that $\log |\Phi_\nu^{-1}(x)|$ is a proper, harmonic function on X_ν^{reg} and that

$$c_2(j) \log |z_1| + c_1(j) \log |z_2|$$

is a proper, harmonic function of $x = \phi_j(z_1, z_2)$ on Y_j. We now introduce our Ansatz

$$h_X(x) = \begin{cases} \log |\Phi_\nu^{-1}(x)|, & x \in X_\nu^{\mathrm{reg}} \setminus \bigcup_{j \ge g+1} Y_{j1} \cup Y_{j2} \\ c_2(j) \log |z_1| + c_1(j) \log |z_2|, & x = \phi_j(z_1, z_2) \in Y_j \setminus (Y_{j1} \cup Y_{j2}) \\ \left(1\chi\left(\frac{|z_\mu|}{\tau_\mu(j)}\right)\right) \left(c_2(j) \log |z_1| + c_1(j) \log |z_2|\right) & \\ \qquad + \chi\left(\frac{|z_\mu|}{\tau_\mu(j)}\right) \log |\Phi_{\nu_\mu(j)}^{-1}(x)|, & x = \phi_j(z_1, z_2) \in Y_{j\mu} \end{cases}$$

for an exhaustion function with bounded charge on X. Suppose, for example, that $x = \phi_j(z_1, z_2) \in Y_{j1}$. Then

$$h_X(x) = h_X(z_1, z_2) = h_X\left(z_1, \frac{t_j}{z_1}\right)$$

on $\mathcal{A}_{j1}$. By construction, h_X is smooth and proper on $X \setminus X^{\mathrm{com}}$ and harmonic on

$$X \setminus \left(X^{\mathrm{com}} \cup \bigcup_{j \ge g+1} Y_{j1} \cup Y_{j2}\right).$$

Let $C = \min_{1 \le \nu \le m} \inf\{\log |z| \mid z \in G_\nu\}$. Let $\widetilde{X}^{\mathrm{com}} \subset X^{\mathrm{com}}$ be a submanifold of X^{com} which is the complement of a small neighbourhood of the boundary $\partial X^{\mathrm{com}}$. For $x \in X^{\mathrm{com}}$ define $h_X(x) = C - 1$ if $x \in \widetilde{X}^{\mathrm{com}}$ and interpolate smoothly in the neighbourhood of $\partial X^{\mathrm{com}}$.

LEMMA 5.3.

$$\sum_{\substack{j \ge g+1 \\ \mu = 1,2}} \int_{\mathcal{A}_{j\mu}} |\Delta h_X(\Phi_{\nu_\mu(j)}(z))| |dz \wedge d\bar{z}| < \infty.$$

PROOF. Again, suppose that $x = \phi_j(z_1, z_2) \in Y_{j1}$. Collecting terms,

$$\begin{aligned} h_X(x) &= \left(1\chi\left(\frac{|z_1|}{\tau_1(j)}\right)\right)\left(c_2(j)\log|z_1| + c_1(j)\log|z_2|\right) + \chi\left(\frac{|z_1|}{\tau_1(j)}\right)\log|\Phi^{-1}_{\nu_1(j)}(x)| \\ &= c_2(j)\log|z_1| + c_1(j)\log\left|\frac{t_j}{z_1}\right| + \chi\left(\frac{|z_1|}{\tau_1(j)}\right)h_{j1}(x) \end{aligned}$$

with

$$\begin{aligned} h_{j1}(x) &= \log|g_{j1}(z_1)| - \left(c_2(j)\log|z_1| + \log|s_1(j)| - c_1(j)\log|z_1|\right) \\ &= \log\left|\frac{g_{j1}(z_1)}{s_1(j)}\right| - \frac{1}{\log t_j}\log\left|\frac{s_2(j)}{s_1(j)}\right|\log|z_1|. \end{aligned}$$

For all $x \in Y_{j1}$, we have

$$\Delta h_X(x) = h_{j1}(z_1)\Delta\chi\left(\frac{|z_1|}{\tau_1(j)}\right) + 2\langle\nabla\chi, \nabla h_{j1}\rangle(z_1)$$

since h_{j1} is harmonic. Integrating the absolute value,

$$\begin{aligned} &\frac{1}{2}\int_{\mathcal{A}_{j1}} |\Delta h_X||dz\wedge d\bar{z}| \\ &\qquad \le \frac{1}{2}\int_{\mathcal{A}_{j1}} \left|h_{j1}(z)\Delta\chi\left(\frac{|z|}{\tau_1(j)}\right)\right||dz\wedge d\bar{z}| + \int_{\mathcal{A}_{j1}} |\langle\nabla\chi, \nabla h_{j1}\rangle||dz\wedge d\bar{z}| \end{aligned}$$

To estimate the first term, observe that in polar coordinates,

$$\begin{aligned} &\frac{1}{2}\int_{A_{j1}} \left|h_{j1}(z)\Delta\chi\left(\frac{|z|}{\tau_1(j)}\right)\right||dz\wedge d\bar{z}| \\ &\quad = \int_{\tau_1(j)}^{1}\int_0^{2\pi} \left|h_{j1}(re^{i\theta})r\Delta\chi\left(\frac{r}{\tau_1}\right)\right| dr\,d\theta \\ &\quad \le \int_{2\tau_1}^{3\tau_1} \left|\frac{r}{\tau_1^2}\chi''\left(\frac{r}{\tau_1}\right) + \frac{1}{\tau_1}\chi'\left(\frac{r}{\tau_1}\right)\right| \int_0^{2\pi} \left|\log\left|\frac{g_{j1}(re^{i\theta})}{s_1(j)}\right|\right| d\theta\,dr \\ &\qquad + \int_{2\tau_1}^{3\tau_1} \left|\frac{r}{\tau_1^2}\chi''\left(\frac{r}{\tau_1}\right) + \frac{1}{\tau_1}\chi'\left(\frac{r}{\tau_1}\right)\right| \int_0^{2\pi} \frac{|\log r|}{|\log t_j|}\left|\log\left|\frac{s_1(j)}{s_2(j)}\right|\right| d\theta\,dr \end{aligned}$$

since, $\chi(r/\tau_1)$ is supported between $2\tau_1$ and $3\tau_1$. It follows that,

$$\begin{aligned} &\frac{1}{2}\int_{A_{j1}} \left|h_{j1}(z)\Delta\chi\left(\frac{|z|}{\tau_1(j)}\right)\right||dz\wedge d\bar{z}| \\ &\qquad \le \text{const} \sup_{2\tau_1\le r\le 3\tau_1} \int_0^{2\pi} \left|\log\left|\frac{g_{j1}(re^{i\theta})}{s_1(j)}\right|\right| d\theta + \text{const}\frac{|\log\tau_j|}{|\log t_j|}\left|\log\frac{|s_1(j)|}{|s_2(j)|}\right|. \end{aligned}$$

By hypothesis (GH3)(iii) on the glueing map g_{j1},

$$\begin{aligned} \sup_{2\tau_1\le r\le 3\tau_1} \int_0^{2\pi} \left|\log\left|\frac{g_{j1}(re^{i\theta})}{s_1(j)}\right|\right| d\theta &\le \text{const} \sup_{\substack{2\tau_1\le r\le 3\tau_1 \\ 0\le\theta\le 2\pi}} \left|\frac{g_{j1}(re^{i\theta}) - s_1}{s_1}\right| \\ &\le \text{const}\,\frac{\rho_1^-(j)}{|s_1(j)|}. \end{aligned}$$

Therefore,

$$\frac{1}{2}\int_{A_{j1}} \left|h_{j1}(z)\Delta\chi\left(\frac{|z|}{\tau_1(j)}\right)\right||dz\wedge d\bar{z}| \le \text{const}\left(\frac{|\log\tau_j|}{|\log t_j|}\left|\log\frac{|s_1(j)|}{|s_2(j)|}\right| + \frac{\rho_1^-(j)}{|s_1(j)|}\right).$$

For the second term, we evaluate the inner product to obtain

$$2\langle \nabla\chi, \nabla h_{j1}\rangle = \frac{\chi'(|z|/\tau_1)}{\tau_1|z||g_{j1}(z)|^2}\big(\operatorname{Re} z \operatorname{Re}(g_{j1z}\overline{g}_{j1}) - \operatorname{Im} z \operatorname{Im}(g_{j1z}\overline{g}_{j1})\big) - \frac{1}{\log t_j}\log\left|\frac{s_2(j)}{s_1(j)}\right|\frac{\chi'(|z|/\tau_1)}{\tau_1|z|}.$$

By Schwarz's inequality,

$$2|\langle \nabla\chi, \nabla h_{j1}\rangle| \leq \frac{|\chi'(|z|/\tau_1)|}{\tau_1|z||g_{j1}(z)|^2}|z||g_{j1z}\overline{g}_{j1}| + \frac{1}{|\log t_j|}\left|\log\left|\frac{s_2(j)}{s_1(j)}\right|\right|\frac{|\chi'(|z|/\tau_1)|}{\tau_1|z|}.$$

Integrating,

$$\begin{aligned}\int_{\mathcal{A}_{j1}} &|\langle \nabla\chi, \nabla h_{j1}\rangle|\,|dz\wedge d\bar{z}| \\ &\leq \frac{1}{|\log t_j|}\left|\log\frac{|s_2(j)|}{|s_1(j)|}\right|\int_{2\tau_1}^{3\tau_1}\frac{|\chi'(r/\tau_1)|}{\tau_1}\,dr\,d\theta + \int_{2\tau_1}^{3\tau_1}\frac{|\chi'(|z|/\tau_1)|}{\tau_1|g_{j1}|}|g_{j1z}|r\,dr\,d\theta \\ &< \text{const}\frac{1}{|\log t_j|}\left|\log\frac{|s_2(j)|}{|s_1(j)|}\right| + \text{const}\,\tau_1 \sup_{2\tau_1\leq|z|\leq 3\tau_2}\frac{|g_{j1z}|}{|g_{j1}|}.\end{aligned}$$

Now observe that for all $2\tau_1 \leq |z| \leq 3\tau_1$,

$$\tfrac{1}{2}|s_1(j)| \leq |s_1(j)| - \rho_1^-(j) \leq \big||s_1(j)| - |g_{j1}(z) - s_1(j)|\big| \leq |g_{j1}(z)|$$

and by Cauchy's formula,

$$\sup_{2\tau_1\leq|z|\leq 3\tau_2}|g_{j1z}| = \sup_{2\tau_1\leq|z|\leq 3\tau_2}\big|\big(g_{j1} - s_1(j)\big)_z\big| \leq \frac{1}{\tau_1}\sup_{|z|=\tau_1, 4\tau_2}|g_{j1} - s_1(j)| < \frac{\rho_1^-(j)}{\tau_1}.$$

Combining this remark with the result of the last paragraph we obtain

$$\begin{aligned}\int_{\mathcal{A}_{j1}} &|\langle \nabla\chi, \nabla h_{j1}\rangle||dz\wedge d\bar{z}| \\ &< \text{const}\frac{1}{|\log t_j|}\left|\log\frac{|s_2(j)|}{|s_1(j)|}\right| + \text{const}\frac{\tau_1}{|s_1(j)|}\sup_{2\tau_1\leq|z|\leq 3\tau_2}|g_{j1z}| \\ &\leq \text{const}\left(\frac{1}{|\log t_j|}\left|\log\frac{|s_1(j)|}{|s_2(j)|}\right| + \frac{\rho_1^-(j)}{|s_1(j)|}\right).\end{aligned}$$

The proof is completed by applying the estimate of Lemma 5.2(b). □

PROOF OF THEOREM 5.1. Lemma 5.3 implies that h_X is a proper function with bounded charge. By (A.1) there exists an exhaustion function h' with bounded charge arbitrary close to h. We may assume that h' coincides with h on $X^{\text{reg}} \setminus \bigcup_{j=1}^m Y_j$ and that for each j the restriction $h'\big|_{\partial Y_j}$ is a Morse function. We put $X_t = h^{(-1)}\big((-\infty, t]\big)$. It remains to show that h' is an exhaustion function of bounded charge on the marked Riemann surface $(X; A_1, B_1, A_2, B_2, \dots)$, i.e., that the exhaustion function is compatible with the chosen homology basis.

For each $j \geq g+1$ the handle Y_j is homeomorphic to a closed cylinder. Therefore its homology $H_1(Y_j, \mathbb{Z})$ is generated by the class of

$$\tilde{A}_j = \phi_j\big(\{(\sqrt{t_j}e^{i\theta}, \sqrt{t_j}e^{-i\theta}) \mid 0 \leq \theta \leq 2\pi\big)$$

and its relative homology $H_1(Y_j, \partial Y_j)$ is generated by the class of

$$\widetilde{B}_j = \phi_j\left(\left\{\frac{1}{r}\sqrt{t_j}r\sqrt{t_j}\;\middle|\;(\sqrt{t_j} \leq r \leq \frac{1}{\sqrt{t_j}}\right\}\right).$$

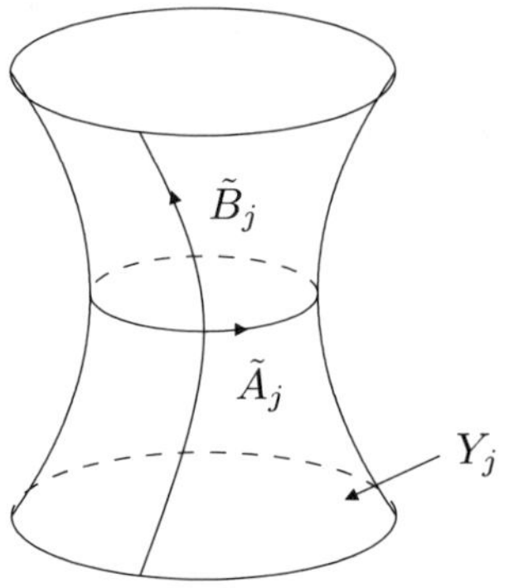

If v is a regular value of h and of $h'|_{\partial Y_j}$ then $Y_j \cap X_v$ is a submanifold with boundary and corners inside Y_j.

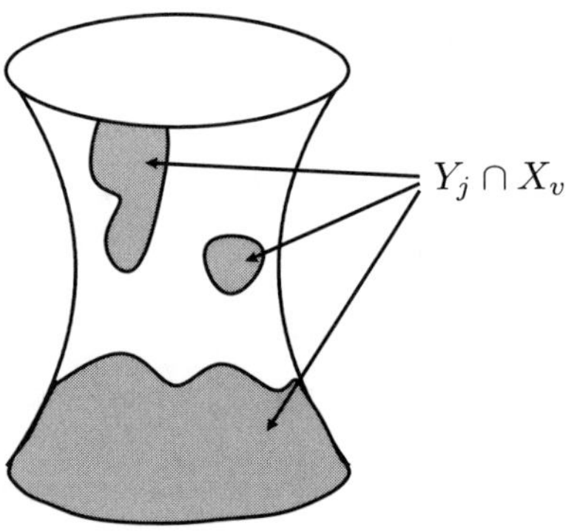

The image of the natural map

$$H_1(Y_j \cap X_v, (\partial Y_j) \cap X_v) \to H_1(Y_j, \partial Y_j)$$

contains the class $[\widetilde{B}_j]$ of $\widetilde{B}_j$ if and only if there are points p_1, p_2 on the two components of ∂Y_j that can be connected by a path that lies completely inside $Y_j \cap X_v$. Similarly $[\tilde{A}_j]$ lies in the image of

$$H_1(Y_j \cap X_v, \mathbb{Z}) \to H_1(Y_j, \mathbb{Z})$$

if and only if it is possible to go around the cylinder on a path that lies completely inside $Y_j \cap X_v$.

Now let $C < v_1' < v_2' < \cdots$ be the crititical values of h' above C. Choose $v_0, v_1, v_2, \ldots$ such that

$$C < v_0 < v_1' < v_1 < v_2' < v_2 < \cdots$$

and such that each v_j is a regular value of $h'|_{\cup_j \partial Y_j}$. We put

$$N_i = \{1, \ldots, g\} \cup \left\{ j \geq g+1 \;\middle|\; \begin{array}{l} \text{the images of the natural maps } H_1(Y_j \cap X_{v_i}, \mathbb{Z}) \to \\ H_1(Y_j, \mathbb{Z}) \text{ resp. } H_1(Y_j \cap X_{v_i}, (\partial Y_j) \cap X_{v_i}) \to \\ H_1(Y_j, \partial Y_j) \text{ contain } [\tilde{A}_j] \text{ resp. } [\widetilde{B}_j] \end{array} \right\}$$

Obviously $N_0 \subset N_1 \subset \cdots$ and $\#|N_{i+1} \setminus N_i| \leq 1$, since h is a Morse function. Now we recursively construct cycles

$$A_j', \quad B_j' j \in N_i$$

such that for each $i \geq g+1$

$(5.1)_i$ A'_j, B'_j, $j \in N_i$ represent a canonical homology basis for X_{v_i} and A'_j lies inside Y_j and is homologous to $\tilde{A}_j$ whenever $j \geq g+1$, $j \in N_i$.

For $i = 0$ the compact piece X^{com} is a deformation retract of X_{v_0}, so $N_0 = \{1, \dots, g\}$. We choose $A'_1, B'_1, \dots, A'_g, B'_g$ as cycles that represent the canonical homology basis for X^{com} of (GH4). Now suppose that $A'_j, B'_j, j \in N_i$ have been constructed such that $(5.1)_i$ holds. If $N_{i+1} = N_i$ then nothing has to be done. Otherwise, by Morse theory, $\dim H_1(X_{v_{i+1}}, X_{v_i}) = 1$ and $N_{i+1} \setminus N_i$ consists of one element, say k. Choose a cycle A'_k in $Y_k \cap X_{v_{i+1}}$ whose image under the the natural map $H_1(Y_k \cap X_{v_{i+1}}, \mathbb{Z}) \to H_1(Y_k, \mathbb{Z})$ is $[\tilde{A}_k]$. Then

$$A'_j \cdot A'_k = 0, \quad B'_j \cdot A'_k = 0 \quad \text{for } j \in N_i.$$

With this choice the second condition of $(5.1)_{i+1}$ is fulfilled. Next choose a curve $\widetilde{B}'_k$ connecting two points p_1, p_2 of different components of ∂Y_k completely inside $Y_k \cap X_{v_{i+1}}$. Connect p_1 and p_2 by paths $B_k^{(1)}$ resp. $B_k^{(2)}$ inside $X^{\text{reg}} \cap X_{v_{i+1}}$ to points q_1 resp. q_2 of ∂X^{com}, and join q_1 and q_2 by a path $B_k^{(3)}$ inside X^{com} which has zero intersection number with $A'_1, \dots, A'_g$. The union of $\widetilde{B}'_k$, $B_k^{(1)}$, $B_k^{(2)}$, $B_k^{(3)}$ is a closed curve B''_k such that with suitably chosen orientation

$$A'_j \cdot B''_k = 0 \quad \text{for } j \neq k,\ j \in N - i,$$
$$A'_k \cdot B''_k = 1.$$

By adding suitable linear combinations of the A'_j, $j \neq k$ to B''_k we get a cycle B'_k such that

$$A'_j \cdot B'_k = \delta_{jk}, \quad B'_j \cdot B'_k = 0 \text{ for } j \in N_i.$$

To prove that A'_j, B'_j, $j \in N_{i+1}$ indeed represent a homology basis for $X_{v_{i+1}}$ we use the exact sequence

$$0 \to H_1(X_{v_i}, \mathbb{Z}) \to H_1(X_{v_{i+1}}, \mathbb{Z}) \to H_1(X_{v_{i+1}}, X_{v_i}) \to 0.$$

By the induction hypothesis the rank of the intersection form on $H_1(X_{v_i}, \mathbb{Z})$ is $2|N_i|$. As observed before, $\dim H_1(X_{v_{i+1}}, X_{v_i}) = 1$. So the rank of the intersection form on $H_1(X_{v_{i+1}}, \mathbb{Z})$ is $2|N_{i+1}| = 2|N_i| + 2$. This completes the construction of the cycles A'_j, B'_j, $j \in \mathbb{N}$.

The cycles A'_j, B'_j, $j \in \mathbb{N}$ represent a canonical homology basis of X for which h' is an exhaustion function of bounded charge. As

$$A_j = [A'_j] \text{ and } B_j = [B'_j] \text{ modulo finite linear combinations of the } A_i$$

h' is also an exhaustion function with bounded charge for the original marked Riemann surface $(X; A_1, B_1, \dots)$. □

6. Pointwise Bounds on Differential Forms

From now on we consider a Riemann surface $X = X^{\text{com}} \cup X^{\text{reg}} \cup X^{\text{han}}$ that fulfills the hypotheses (GH1)–(GH6) of §5. By Theorem 5.1 and A.9 there exist square integrable holomorphic differential forms ω_j with

$$\int_{A_i} \omega_j = \delta_{i,j}.$$

The main purpose of this section is to give pointwise bounds on the forms ω_j.

Before doing that we derive a geometric consequence of the hypotheses which allows us to construct a convenient class of compact submanifolds of X. To simplify the notation, set, for each $s \in S_\nu$,

$$r(s) = r_\mu(j),\ R(s) = R_\mu(j) \quad \text{when } s = s_\mu(j),$$
$$\mathcal{A}(s) = \{z \in \mathbb{C} \mid r(s) < |z - s| < R(s)\}.$$

Define the curves

$$a(s) := \Phi_\nu\big(\{z \in \mathbb{C} \mid |z - s| = R(s)\}\big), \quad s \in S_\nu.$$

Let Y_j' be the cylinder in Y_j bounded by $a(s_1(j))$ and $a(s_2(j))$.

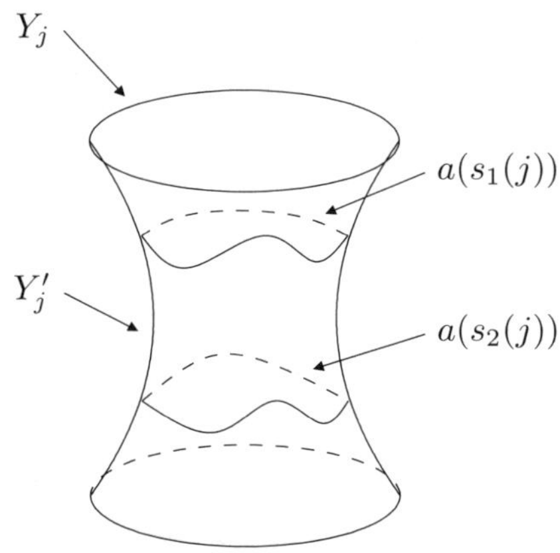

LEMMA 6.1. *For each ε, $N > 0$ there exists a system $\Gamma = (\Gamma_1, \ldots, \Gamma_m)$ of simple closed curves $\Gamma_\nu \subset G_\nu$ such that*

$$\sum_{\nu=1}^{m} \sup_{z \in \Gamma_\nu} \frac{\operatorname{length} \Gamma_\nu}{|z|^2} < \varepsilon, \tag{6.1a}$$

$$\sum_{s \in S_\nu} \big(r(s) \operatorname{length} \Gamma_\nu\big)^2 \sup_{z \in \Gamma_\nu} \frac{1}{|z - s|^4} < \varepsilon^2 \quad \text{for } 1 \le \nu \le m \tag{6.1b}$$

and for all $j \ge g + 1$ and $\mu = 1, 2$

$$\int_{\Gamma_\nu} \frac{|s_1(j) - s_2(j)|}{|z - s_1(j)||z - s_2(j)|} |dz| \le \varepsilon |\log t_j| \quad \text{if } \nu = \nu_1(j) = \nu_2(j), \tag{6.1c}$$

$$\int_{\Gamma_\nu} \frac{|s_\mu(j)|}{|z(z - s_\mu(j))|} |dz| \le \varepsilon |\log t_j| \quad \text{if } \nu = \nu_\mu(j),\ \nu_1(j) \ne \nu_2(j). \tag{6.1d}$$

Furthermore $\bigcup_{\nu=1}^{m} \Phi_\nu(\Gamma_\nu)$ decomposes X into a compact connected component $X(\Gamma)$ containing X^{com} and a noncompact component such that for all $j \ge g + 1$

$$Y_j' \subset X(\Gamma) \text{ or } Y_j' \cap X(\Gamma) = \varnothing,$$
$$Y_i \subset X(\Gamma) \quad \text{for } i \le N.$$

PROOF. Define, for each $j \ge g + 1$ and $\mu = 1, 2$

$$\rho_\mu(j) = \tfrac{1}{4} \min_{\substack{s \in S_{\nu_\mu(j)} \cup \{0\} \\ s \ne s_\mu(j)}} |s - s_\mu(j)|,$$
$$C_\mu(j) = \{z \in \mathbb{C} \mid |z - s_\mu(j)| = \rho_\mu(j)\}.$$

Observe that, by hypothesis (GH5)(iii)

$$\big||s_1(j)| - |s_2(j)|\big| \le \tfrac{1}{4}\min\{\rho_1(j), \rho_2(j)\}. \tag{6.2}$$

Let $R \gg 1$. Consider the circle $\Gamma'(R)$ of radius R around the origin in $\mathbb{C}$. If neither of the circles $C_\mu(j)$, $\mu = 1, 2$ intersects $\Gamma'(R)$ then, by (6.2), $s_1(j)$ and $s_2(j)$ are either both outside or both inside $\Gamma'(R)$. Now construct $\Gamma_\nu(R)$ by modifying $\Gamma'(R)$ as follows. Suppose that $\Gamma'(R)$ meets at least one of the circles $C_1(j)$, $C_2(j)$. If $|s_1(j)|, |s_2(j)| \le R$ replace the arc of $\Gamma'(R)$ joining the two points of $\Gamma'(R) \cap C_\mu(j)$ by the outer part of $C_\mu(j)$ joining the same two points. In the other case use the inner parts of $C_1(j)$, $C_2(j)$.

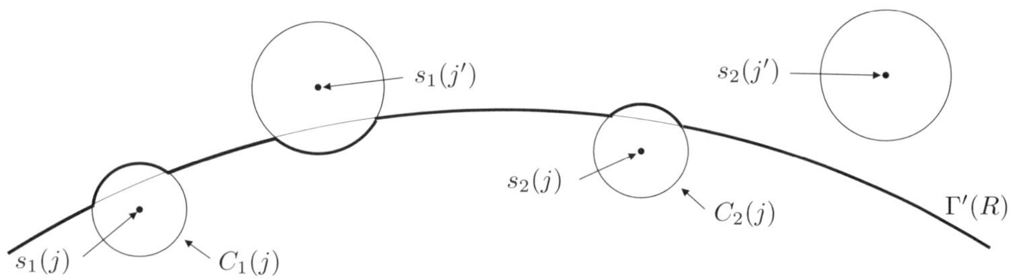

By construction, $\cup_{\nu=1}^m \Phi_\nu(\Gamma_\nu)$ decomposes X into a compact connected component $X(\Gamma)$ containing X^{com} and a noncompact component such that for all $j \ge g+1$

$$Y_j' \subset X(\Gamma) \quad \text{or} \quad Y_j' \cap X(\Gamma) = \varnothing,$$
$$Y_i \subset X(\Gamma) \quad \text{for } i \le N.$$

Also by construction

$$\operatorname{length}\Gamma_\nu(R) \le \operatorname{const} R, \qquad \inf_{z \in \Gamma_\nu(R)} |z| \ge \operatorname{const} R$$

which implies (6.1a).

To prove (6.1b) observe that for $z \in G_\nu$ with $|z - s| \ge \frac{1}{4}R(s)$ for all $s \in S_\nu$

$$\begin{aligned}
\sum_{s \in S_\nu} \frac{r(s)}{|z-s|^2} &= \sum_{\substack{s \in S_\nu \\ |s| \le |z|/2}} \frac{r(s)}{|z-s|^2} + \sum_{\substack{s \in S_\nu \\ |s| > |z|/2}} \frac{r(s)}{|z-s|^2} \\
&\le \sum_{s \in S_\nu} \frac{4r(s)}{|z|^2} + \frac{4}{|z|^2} \sum_{\substack{s \in S_\nu \\ |s| > |z|/2}} \frac{16 r(s)}{R(s)^2}|s|^2 \\
&\le \frac{4}{|z|^2} \sum_{s \in S_\nu} \frac{1}{|s|^d} + \frac{64}{|z|^2} \sum_{s \in S_\nu} \frac{1}{|s|^{d-2\delta-2}}.
\end{aligned}$$

Hence by (GH5)(ii), if $z \in G_\nu$ with $|z - s| \ge \frac{1}{4}R(s)$ for all $s \in S_\nu$ then

$$\sum_{s \in S_\nu} \frac{r(s)}{|z-s|^2} \le \frac{\operatorname{const}}{|z|^2} \tag{6.3}$$

So (6.1b) follows from (6.1a).

We now prove (6.1c) and (6.1d). First observe that by

$$\begin{aligned}\left|\frac{s_1 - s_2}{(z-s_1)(z-s_2)}\right| &= \left|\frac{1}{z-s_1} - \frac{1}{z} + \frac{1}{z} - \frac{1}{z-s_2}\right| \\ &\le \left|\frac{s_1}{(z-s_1)z}\right| + \left|\frac{s_2}{z(z-s_2)}\right|\end{aligned}$$

both (6.1c) and (6.1d) follow from

$$\lim_{R\to\infty} \sup_{j,\mu} \frac{1}{\log t_j} \int_{\Gamma_{\nu_\mu(j)}(R)} \frac{|s_\mu(j)|}{|z(z - s_\mu(j))|} |dz| = 0. \tag{6.4}$$

We now prove (6.4). If $|s| = |s_\mu(j)| > 2R$

$$\int_{\Gamma_\nu(R)} \frac{|s|}{|z(z-s)|} |dz| \le \text{const} \frac{|s|}{R|s|} R = \text{const}$$

Since $\lim_{j\to\infty} t_j = 0$ this implies

$$\lim_{R\to\infty} \sup_{\substack{j,\mu \\ |s_\mu(j)|>2R}} \frac{1}{\log t_j} \int_{\Gamma_{\nu_\mu(j)}(R)} \frac{|s_\mu(j)|}{|z(z - s_\mu(j))|} |dz| = 0.$$

If $2R \ge |s| = |s_\mu(j)| > \frac{1}{2}$

$$\int_{\Gamma_\nu(R)} \frac{|s|}{|z(z-s)|} |dz| \le \text{const} \int_{\Gamma_\nu(R)} \frac{|dz|}{|z-s|} = \text{const} \int_{\Gamma_\nu(R)/s} \frac{|d\zeta|}{|\zeta - 1|}.$$

To bound the last integral observe that

$$\int_{|\zeta-1|=\rho_\mu(j)/|s|} \frac{|d\zeta|}{|\zeta-1|} = 2\pi$$

and

$$\int_{\substack{|\zeta|=R/|s| \\ |\zeta-1|\ge\rho_\mu(j)/|s|}} \frac{|d\zeta|}{|\zeta-1|} \le \text{const} \log \frac{|s|}{\rho_\mu(j)}.$$

Therefore

$$\int_{\Gamma_\nu(R)} \frac{|s|}{|z(z-s)|} |dz| \le \text{const} \log \frac{|s|}{\rho_\mu(j)} \le \text{const} \log \frac{|s|}{R(s)} \le \text{const}(1+\delta) \log |s|$$

by (GH5)(i)–(GH5)(ii). So, by (GH5)(iv)

$$\lim_{R\to\infty} \sup_{\substack{j,\mu \\ 2R\ge|s_\mu(j)|>\frac{1}{2}R}} \frac{1}{\log t_j} \int_{\Gamma_{\nu_\mu(j)}(R)} \frac{|s_\mu(j)|}{|z(z - s_\mu(j))|} |dz| = 0.$$

Finally, if $|s| = |s_\mu(j)| \le \frac{1}{2}R$

$$\int_{\Gamma_\nu(R)} \frac{|s|}{|z(z-s)|} |dz| \le \text{const} \frac{|s|}{R^2} R = \text{const} \frac{|s|}{R}.$$

Let $\varepsilon > 0$. Then

$$\frac{1}{|\log t_j|} \frac{|s_\mu(j)}{R} \le \begin{cases} \varepsilon \log 2 & \text{if } |s_\mu(j)| \le \varepsilon R \\ \frac{1}{2\log t_j} & \text{if } |s_\mu(j)| > \varepsilon R \end{cases}.$$

So

$$\lim_{R\to\infty} \sup_{\substack{j,\mu \\ |s_\mu(j)|\le \frac{1}{2}R}} \frac{1}{\log t_j} \int_{\Gamma_{\nu_\mu(j)}(R)} \frac{|s_\mu(j)|}{|z(z-s_\mu(j))|} |dz| = 0.$$

This finishes the proof of (6.4) and hence of (6.1c), (6.1d). □

We also need the following fact which one deduces immediately from (GH5)(ii).

LEMMA 6.2.

$$\sum_j \left(\frac{\max\{r_1(j), r_2(j)\}}{\min\{R_1(j), R_2(j)\}} \right)^2 < \infty.$$

The strategy for obtaining pointwise bounds on the forms ω_j is to compare $\Phi_\nu^*\omega_j$ with the "Rosenlicht differentials"

$$\left(\frac{\delta_{\nu,\nu_1(j)}}{2\pi i} \frac{1}{z-s_1(j)} - \frac{\delta_{\nu,\nu_2(j)}}{2\pi i} \frac{1}{z-s_2(j)} \right) dz$$

which would play the role of ω_j if all the t_j's were zero and X^{com} were empty. On G_ν the difference between $\Phi_\nu^*\omega_j$ and the Rosenlicht differential can be written as a sum of terms $w_{j,s}^\nu(z)\,dz$, $s \in S_\nu$ and $w_{j,\mathrm{com}}^\nu(z)\,dz$ which are holomorphic and decay quadratically away from the points s and 0 respectively. The coefficient of $1/|z-s|^2$ in $w_{j,s}^\nu(z)$ is related to the norm $\|\omega_j|_{Y_i}\|_2$ if $s \in \{s_1(i), s_2(i)\}$. On the other hand, we use (A.12) to get bounds on $\|\omega_j|_{Y_i}\|_2$ in terms of the expansions $\sum_{s\in S_\nu} w_{j,s}^\nu(z)\,dz + w_{j,\mathrm{com}}^\nu(z)\,dz$. In this way we get a system of inequalities for the norms $\|\omega_j|_{Y_i}\|_2$, which allow us give bounds on $\|\omega_j|_{Y_i}\|_2$ and then pointwise estimates on ω_j in X^{reg} and in the handles Y_i.

In the course of the argument, the following quantities will play an important role.

(a) The numbers

$$\mathfrak{A}_{i,k} = \begin{cases} 6\pi \max_{\mu=1,2} \left[\frac{r_\mu(i)}{R_\mu(i)} + \delta_{\nu_1(i),\nu_2(i)} \frac{2r_{3-\mu}(i)R_\mu(i)}{|s_1(i)-s_2(i)|^2} \right] & \text{if } i = k \\ 24\pi \max_{\mu,\tau=1,2} \delta_{\nu_\mu(i),\nu_\tau(k)} \frac{r_\tau(k)R_\mu(i)}{|s_\mu(i)-s_\tau(k)|^2} & \text{if } i \neq k \end{cases}$$

estimate the influence of $\|\omega_j|_{Y_k}\|_2$ on $\|\omega_j|_{Y_i}\|_2$.

(b) The numbers

$$\aleph_j = \max_{\mu=1,2} \left\| \alpha_{j,\mu}(z)\,dz \big|_{\mathcal{A}(s_\mu(j))} \right\|_2$$

are used to control the difference of ω_j from the Rosenlicht differential in a neighbourhood of Y_j.

(c) When $\nu_1(j) = \nu_2(j)$, that is, when both ends of the handle Y_j belong to the same sheet, define

$$\mathcal{O}^j = 4 \max_{\mu=1,2} \frac{R_\mu(j)^2}{|s_1(j)-s_2(j)|^2} + 4 \sum_{\substack{s\in S_\nu(j) \\ s\neq s_1(j), s_2(j)}} \frac{R(s)^2 |s_1(j)-s_2(j)|^2}{|s-s_1(j)|^2 |s-s_2(j)|^2}$$
$$+ 8\pi^2 \max_{\mu=1,2} R_\mu(j)^2 \sup_{|z-s_\mu(j)|=R_\mu(j)} |\alpha_{j,\mu}(z)|^2.$$

When the jth handle joins distinct sheets define, for $N > 0$,

$$\mathcal{O}^j(N) = 4 \max_{\mu=1,2} \frac{R_\mu(j)^2}{|s_\mu(j)|^2} + 4 \sum_{\substack{i \geq N+1 \\ i \neq j}} \max_{\mu,\tau=1,2} \delta_{\nu_\mu(i),\nu_\tau(j)} \frac{R_\mu(i)^2 |s_\tau(j)|^2}{|s_\mu(i)|^2 |s_\mu(i) - s_\tau(j)|^2}$$
$$+ 8\pi^2 \max_{\mu=1,2} R_\mu(j)^2 \sup_{|z - s_\mu(j)| = R_\mu(j)} |\alpha_{j,\mu}(z)|^2.$$

LEMMA 6.3.

(a)
$$\sum_{\substack{i,k \geq g+1 \\ i \neq k}} \mathfrak{A}_{i,k}^2 < \infty.$$

(b) *For each* j, N,

$$\mathcal{O}^j, \quad \mathcal{O}^j(N) < \infty.$$

(c)
$$\lim_{\substack{j \to \infty \\ \nu_1(j) = \nu_2(j)}} \mathcal{O}^j = 0 \quad \lim_{N \to \infty} \limsup_{\substack{j \to \infty \\ \nu_1(j) \neq \nu_2(j)}} \mathcal{O}^j(N) = 0.$$

(d)
$$\sup_j \aleph_j < \infty.$$

PROOF. (a) For each fixed k, τ

$$\pi \sum_{i \neq k} \sum_{\mu=1,2} \delta_{\nu_\mu(i),\nu_\tau(k)} \frac{R_\mu(i)^2}{|s_\mu(i) - s_\tau(k)|^4} \leq \sum_{\substack{s \in S_{\nu_\tau(k)} \\ s \neq \sigma_\tau(k)}} \frac{R_\mu(s)}{|s - s_\tau(k)|^4}$$
$$\leq 16 \sum_{\substack{s \in S_{\nu_\tau(k)} \\ s \neq \sigma_\tau(k)}} \int_{|y-s| \leq R(s)} \frac{dy}{|y - s_\tau(k)|^4}$$
$$\leq 16 \int_{|y - s_\tau(k)| \geq 1/2 |s_\tau(k)|^\delta} \frac{dy}{|y - s_\tau(k)|^4}$$

since for each $s \in S_{\nu_\tau(k)}$ with $s \neq s_\tau(k)$ and each $y \in \mathbb{R}^2$ with $|y - s| R(s)$

$$|y - s_\tau(k)| \geq \tfrac{1}{2} |s - s_\tau(k)| \geq \tfrac{1}{2|s_\tau(k)|^\delta}$$

by (GH5)(ii) and since the discs around the points s with radii $R(s)$ do not overlap. Therefore

$$\sum_{i \neq k} \mathfrak{A}_{i,k}^2 \leq 24^2 \cdot 16\pi \sum_{k,\tau} r_\tau(k)^2 \int_{|y - s_\tau(k)| \geq 1/2 |s_\tau(k)|^\delta} \frac{dy}{|y - s_\tau(k)|^4}$$
$$\leq \text{const} \sum_{k,\tau} r_\tau(k)^2 |s_\tau(k)|^{2\delta}$$
$$\leq \text{const} \sum_{k,\tau} \frac{1}{|s_\tau(k)|^{2d - 2\delta}} < \infty$$

by (GH5)(ii).

(b) For $s \in S_\nu$ put

$$\rho(s) = \frac{R(s)\log|s|}{\min_{\substack{s'\in S_\nu\\ s'\neq s}} |s'-s|}.$$

By (GH5)(v)

$$\lim_{|s|\to\infty} \rho(s) = 0. \tag{6.5}$$

As above, if $\nu_1(j) = \nu_2(j) = \nu$

$$\begin{aligned}
(6.6)\quad \pi &\sum_{\substack{s\in S_\nu\\ s\neq s_1(j),s_2(j)}} \frac{R(s)^2|s_1(j)-s_2(j)|^2}{|s-s_1(j)|^2|s-s_2(j)|^2} \\
&= \pi \sum_{\substack{s\in S_\nu\\ s\neq s_1(j),s_2(j)}} \rho(s)^2 \frac{(\min_{s'\neq s}|s'-s|)^2|s_1(j)-s_2(j)|^2}{|s-s_1(j)|^2|s-s_2(j)|^2\log^2|s|} \\
&\leq 128 \sum_{\substack{s\in S_\nu(j)\\ s\neq s_1(j),s_2(j)}} \rho(s)^2 \int_{|y-s|\leq 1/2\min_{s'\neq s}|s'-s|} \frac{|s_1(j)-s_2(j)|^2}{|y-s_1(j)|^2|y-s_2(j)|^2\log^2|y|}\, dy.
\end{aligned}$$

By (6.5) this is bounded by

$$\text{const} \int_{|y-s_\mu(j)|\geq 1/2|s_\mu(j)|^\delta} \frac{|s_1(j)-s_2(j)|^2}{|y-s_1(j)|^2|y-s_2(j)|^2\log^2|y|}\, dy$$

which converges. This proves part (b) when $\nu_1(j) = \nu_2(j)$. The case $\nu_1(j) \neq \nu_2(j)$ is treated similarly.

(c) Again we give the proof only for the case $\nu_1(j) = \nu_2(j) = \nu$. The first term in $\mathcal{O}^j$ goes to zero with j since by (GH5)(v)

$$\lim_{j\to\infty} \frac{R_\mu(j)}{|s_1(j)-s_2(j)|} \leq \lim_{j\to\infty} \frac{R_\mu(j)}{\min_{s\neq s_\mu(j)}|s-s_\mu(j)|} = 0.$$

The third term in $\mathcal{O}^j$ goes to zero by (GH5)(vi). We now concentrate on the second term.

Fix $\varepsilon > 0$. By (6.5) there is a finite subset $\widetilde{S} \subset \bigcup_{\nu=1}^m S_\nu$ such that $\rho(s) < \varepsilon$ for all $s \notin \widetilde{S}$. Clearly

$$\lim_{j\to\infty} \sum_{s\in\widetilde{S}} \rho(s)^2 \int_{|y-s|\leq 1/2\min_{s'\neq s}|s'-s|} \frac{|s_1(j)-s_2(j)|^2}{|y-s_1(j)|^2|y-s_2(j)|^2\log^2|y|}\, dy = 0.$$

Hence, by (6.6), it suffices to prove

$$\sup_j \int_{|y-s_\mu(j)|\geq 1/2|s_\mu(j)|^\delta} \frac{|s_1(j)-s_2(j)|^2}{|y-s_1(j)|^2|y-s_2(j)|^2\log^2|y|}\, dy < \infty. \tag{6.7}$$

Clearly the integral in (6.7) over the region $\{y\in\mathbb{C} \mid |y|\leq e\}$ is bounded in j. The integral in (6.7) over the region $\{y\in\mathbb{C} \mid |y|\geq e, |y-s_\mu(j)| \geq \frac{1}{2}|s_1(j)-s_2(j)| \text{ for } \mu=1,2\}$ is bounded by

$$\int_{|y-s_\mu(j)|\geq 1/2|s_1(j)-s_2(j)|} \frac{|s_1(j)-s_2(j)|^2}{|y-s_1(j)|^2|y-s_2(j)|^2}\, dy.$$

By translating and scaling, we see that this is independent of j.

Now, we consider the integral in (6.7) over the region

$$M_j = \left\{ y \in \mathbb{C} \,\middle|\, |y| \geq e, \frac{1}{2|s_1(j)|^\delta} \leq |y - s_1(j)| \leq \frac{1}{2}|s_1(j) - s_2(j)| \right\}.$$

There

$$\frac{|s_1(j) - s_2(j)|}{|y - s_2(j)|} \leq 2$$

so the integral is bounded by

$$4 \int_{M_j} \frac{dy}{|y - s_1(j)|^2 \log^2 |y|}.$$

Clearly

$$\int_{\{y \in M_j \,|\, |y| \leq 1/2|s_1(j)|\}} \frac{dy}{|y - s_1(j)|^2 \log^2 |y|} \leq \frac{4}{|s_1(j)|^2} \int_{\{y \in M_j \,|\, |y| \leq 1/2|s_1(j)|\}} dy \leq \pi.$$

Finally,

$$\begin{aligned}
&\int_{\{y \in M_j \,|\, |y| \geq 1/2|s_1(j)|\}} \frac{dy}{|y - s_1(j)|^2 \log^2 |y|} \\
&\qquad \leq \frac{1}{\log^2 \, (1/2|s_1(j)|^2)} \int_{1/2|s_1(j)|^\delta \leq |y| \leq 1/2|s_1(j) - s_2(j)|} \frac{dy}{|y - s_1(j)|^2} \\
&\qquad = \frac{2\pi(\log 1/2|s_1(j) - s_2(j)| + \log 2|s_1(j)|^\delta)}{\log^2(1/2|s_1(j)|^2)}
\end{aligned}$$

By (GH5)(iii)

$$|s_1(j) - s_2(j)| \leq \text{const}\, |s_1(j)|.$$

This show that the integral over M_j is bounded. The integral in (6.7) over the analogous region centred at $s_2(j)$ is treated in the same way.

(d) is a reformulation of the first part of (GH5)(vi). □

We now come to the statement of the main theorem of this section. By (GH5)(i), there exists a collar T_ν of K_ν such that $|z - s| > \sqrt{2}/\sqrt{2} - 1R(s)$ for all $z \in T_\nu$ and $s \in S_\nu$.

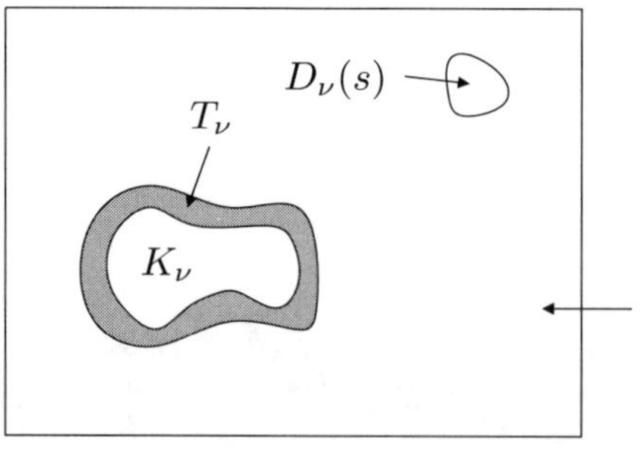

We define functions $w_j^\nu(z)$ on G_ν by

$$w_j^\nu(z)\, dz = \Phi_\nu^* \omega_j.$$

THEOREM 6.4. *There exists a constant C such that, for $1 \le \nu \le m$ and $z \in G_\nu \setminus T_\nu$ with $|z-s| \ge 3r(s)$ for all $s \in S_\nu$*

$$|w_j^\nu(z)| \le C\left(\sum_{s\in S_\nu} \frac{r(s)}{|z-s|^2} + \frac{1}{\operatorname{dist}(z,T_\nu)^2}\right) \qquad \text{for } j \le g,$$

$$\left|w_j^\nu(z) - \frac{\delta_{\nu,\nu_1(j)}}{2\pi i}\left(\frac{1}{z-s_1(j)} - \frac{1}{z}\right) + \frac{\delta_{\nu,\nu_2(j)}}{2\pi i}\left(\frac{1}{z-s_2(j)} - \frac{1}{z}\right)\right|$$
$$\le C\left(\sum_{s\in S_\nu} \frac{r(s)}{|z-s|^2} + \frac{1}{\operatorname{dist}(z,T_\nu)^2}\right) \qquad \text{for } j \ge g+1.$$

Furthermore, for every $\rho > 0$ there is a constant $g_0(\rho)$ such that

$$\limsup_{j\to\infty} \sum_{\substack{i\ge g_0+1\\ i\ne j}} \left\|\omega_j\big|_{Y_i'}\right\|_2^2 + \left\|\left(\omega_j - (\phi_j)_*\left(\frac{1}{2\pi i}\frac{dz_1}{z_1}\right)\right)\Big|_{Y_j'}\right\|_2^2 < \rho.$$

In the single sheet case, $m = 1$, it is possible to choose $g_0(\rho) = g$. More detailed bounds are given in Remark 6.13.

We first outline the proof of Theorem 6.4 in the simple setting of a single sheet, that is $m = 1$, without a compact region, that is $X^{\text{com}} = \emptyset$. We also make the simplifying assumption $\|\mathfrak{A}\| < \frac{1}{2}$. For simplicity of notation we delete the sub and superscripts ν. In this setting the estimate of Theorem 6.4 on $w_j^\nu(z)$ becomes

$$\left|w_j(z) - \frac{1}{2\pi i}\left(\frac{1}{z-s_1(j)} - \frac{1}{z-s_2(j)}\right)\right| \le C\sum_{s\in S}\frac{r(s)}{|z-s|^2} \qquad \text{for } j \ge 1.$$

For each form ω_j and each $s \in S$ we consider the principal part

$$w_{j,s}(z) = -\frac{1}{2\pi i}\int_{|\zeta-s|=r(s)} \frac{w_j(\zeta)}{\zeta - z}\,d\zeta, \qquad |z-s| > r(s)$$

of the Laurent expansion of w_j in the annulus $\mathcal{A}(s)$.

PROPOSITION 6.5. *For $z \in G$ with $|z-s| > r(s)$ for all $s \in S$*

$$w_j(z) = \sum_{s\in S} w_{j,s}(z). \tag{6.8}$$

The series on the right-hand side converges absolutely and uniformly on compact subsets of $\{z \in G \mid |z-s| > r(s)\ \forall s \in S\}$. Furthermore, for $s \in S$ and $|z-s| > 3r(s)$

$$|w_{j,s}(z)| \le \frac{3r(s)}{|z-s|^2}\left\|w_j\,dz\big|_{\mathcal{A}(s)}\right\|_2 \qquad \text{for } s \ne s_1(j), s_2(j),$$

$$\left|w_{j,s}(z) - \frac{(-1)^{\mu+1}}{2\pi i}\frac{1}{z-s}\right| \le \frac{3r(s)}{|z-s|^2}\left\|\left(w_j - \frac{(-1)^{\mu+1}}{2\pi i}\frac{1}{z-s}\right)dz\Big|_{\mathcal{A}(s)}\right\|_2 \qquad \text{for } s = s_\mu(j). \tag{6.9}$$

PROOF. Formula (6.9) is a standard estimate for Laurent series relating pointwise bounds to L^2 bounds on annular regions. It will be stated and proven in a more general setting in Lemma 6.9 below. The absolute and uniform convergence on compacts of the right-hand side of (6.8) follows from the convergence of

$\sum_{s\in S} r(s)/|s|^2$, which is a consequence of Lemma 5.2(b), and from the fact that $\|w_j\,dz|_{\mathcal{A}(s)}\|_2 \le \|\omega_j\|_2$.

We now prove equation (6.8). For $R \gg 0$ we define curves $L(R)$ around 0 similar to the curves $\Gamma(R)$ constructed in the proof of Lemma 6.3. Again we start with the circle $\Gamma'(R)$ of radius R around 0. For each $s \in S$ such that $D(s) \cap \Gamma'(R) \neq \varnothing$ let $p_1(s)$ and $p_2(s)$ be the two extremal points of $D(s) \cap \Gamma'(R)$. Replace the segment of $\Gamma'(R)$ between $p_1(s)$ and $p_2(s)$ by the piece of $\partial D(s)$ that joins $p_1(s)$ and $p_2(s)$ on the outside resp. inside of s if $|s| \le R$ resp. $|s| > R$. Call the resulting curve $L(R)$, and put

$$\tilde{L}(R) = L(R) \setminus \cup_{s\in S} L(R) \cap \partial D(s).$$

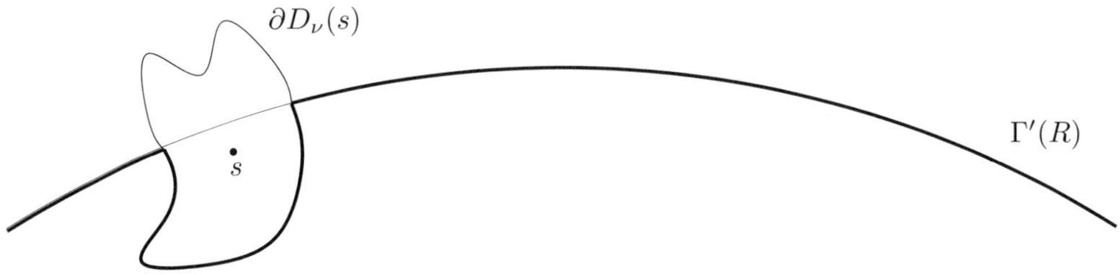

Now let $z \in G_\nu$ with $|z - s| > r(s)$ for all $s \in S_\nu$. Let $R > 2|z|$. By Cauchy's formula

$$w_j(z) = \frac{1}{2\pi i}\int_{L(R)} \frac{w_j(\zeta)}{\zeta - z}\,d\zeta + \sum_{s \text{ inside } L(R)} w_{j,s}(z).$$

Then for $R' > R$

$$w_j(z) = \frac{1}{2\pi i}\frac{1}{R' - R}\int_R^{R'} \left(\int_{L(t)} \frac{w_j(\zeta)}{\zeta - z}\,d\zeta\right) dt \\ + \sum_{R<|s|\le R'} \frac{R' - |s|}{R' - R} w_{j,s}(z) + \sum_{|s|\le R} w_{j,s}(z).$$

Therefore it suffices to show that

$$\lim_{R\to\infty} \inf_{R'>R} \frac{1}{R' - R}\left|\int_R^{R'} \left(\int_{L(t)} \frac{w_j(\zeta)}{\zeta - z}\,d\zeta\right) dt\right| = 0.$$

Now

$$\begin{aligned}
&\frac{1}{R' - R}\left|\int_R^{R'} \left(\int_{\tilde{L}(t)} \frac{w_j(\zeta)}{\zeta - z}\,d\zeta\right) dt\right| \\
&\le \frac{2}{R(R' - R)}\int_R^{R'} \int_{\tilde{L}(t)} |w_j(\zeta)|d|\zeta|\,dt \\
&\le \frac{2}{R(R' - R)}\int_{\cup_{R\le t\le R'}\tilde{L}(t)} |w_j(z)|\,|dz \wedge d\bar{z}| \\
&\le \frac{2}{R(R' - R)}\left(\int_{\cup_{R\le t\le R'}\tilde{L}(t)} |dz \wedge d\bar{z}|\right)^{1/2} \cdot \left(\int_{\cup_{R\le t\le R'}\tilde{L}(t)} |w_j(z)|^2 |dz \wedge d\bar{z}|\right)^{1/2}
\end{aligned}$$

$$\le \frac{\text{const}}{R}\sqrt{\frac{R'+R}{R'-R}}\big\|w_j(z)\,dz\big|_{\cup_{R\le t\le R'}\tilde L(t)}\big\|_2.$$

By construction

$$L(t)\setminus \tilde L(t)\subset \cup_{\{s\in S} t-r(s)\le |s|\le t+r(s)\} L(t)\cap\partial D(s).$$

If $s=s_\mu(i)$ is an element of S such that $L(t)\cap\partial D(s)\ne\varnothing$ then $L(t)\cap\partial D(s)$ is a part of

$$\phi_i\big(\{(z_1,z_2)\in H(t_i)\mid |z_\mu|=\tau_i\}\big)$$

so that

$$\left|\int_{L(t)\cup\partial D(s)}\frac{w_j(\zeta)}{\zeta-z}\,d\zeta\right|\le\left|\int_{|z_\mu|=\tau_i}\frac{\phi_i^*(\omega_j)}{g_{i,\mu}(z_\mu)-z}\right|.$$

In Lemma 6.8 below we derive a general pointwise bound on differentials in handles in terms of the L^2 norm of the differentials in the handle. In the case at hand it implies

$$\left|\frac{\phi_i^*(\omega_j)}{dz_\mu}\right|\le\frac{2}{\sqrt{\pi}}\left(1+\left|\frac{z_{3-\mu}}{z_\mu}\right|\right)\big\|\omega_j\big|_{Y_i}\big\|_2.$$

As $|g_{i,\mu}(z_\mu)-z|\ge \text{const}\,|s|$ and $z_\mu z_{3-\mu}=t_i$

$$\left|\int_{|z_\mu|=\tau_i}\frac{\phi_i^*(\omega_j)}{g_{i,\mu}(z_\mu)-z}\right|\le\text{const}\frac{\tau_i}{|s|}\left(\frac{t_i}{\tau_i^2}+1\right)\big\|\omega_j\big|_{Y_i}\big\|_2.$$

Using this estimate and the fact that $L(t)\cap\partial D(s)=\varnothing$ unless $|s|-r(s)\le t\le |s|+r(s)$ we have

$$\begin{aligned}\frac{1}{R'-R}&\left|\int_R^{R'}\left(\int_{L(t)\setminus\tilde L(t)}\frac{w_j(\zeta)}{\zeta-z}\,d\zeta\right)dt\right|\\&\le\frac{\text{const}}{R'-R}\sum_{i,\mu}\frac{\tau_i r_\mu(i)}{|s_\mu(i)|}\big\|\omega_j\big|_{Y_i}\big\|_2\\&\le\frac{\text{const}}{R'-R}\left(\sum_i\frac{r_\mu(i)^2\tau_i^2}{|s_\mu(i)|^2}\right)^{1/2}\left(\sum_i\big\|\omega_j\big|_{Y_i}\big\|_2^2\right)^{1/2}.\end{aligned}$$

Since $\tau_i\le 1$ and $\sum_{s\in S} r(s)^2/|s|^2\le 4\sum_{s\in S} r(s)^2/R(s)^2<\infty$ by Lemma 6.2 this goes to zero as $R'\to\infty$. □

We want to use (6.8) to derive bounds on the L^2 norms appearing in (6.9). Clearly,

$$\big\|w_j\,dz\big|_{\mathcal{A}(s)}\big\|_2\le\big\|\omega_j\big|_{Y_i'}\big\|_2\quad\text{if } s=s_1(i)\text{ or }s_2(i).\tag{6.10}$$

To estimate $\big\|\omega_j\big|_{Y_i'}\big\|_2$ we use the following special case of (A.12).

LEMMA 6.6. *Let α be a holomorphic differential form on Y_i' with $\int_{A_i}\alpha=0$ and let f_μ be functions in a neighborhood of $\{z\in\mathbb{C}\mid |z-s_\mu(i)|=R_\mu(i)\}$ with $df_\mu=\Phi^*(\alpha)$. Then*

$$\big\|\alpha\big|_{Y_i'}\big\|_2^2=\frac{i}{2}\left(\int_{|z-s_1(i)|=R_1(i)}f_1(z)\overline{\Phi^*(\alpha)}+\int_{|z-s_2(i)|=R_2(i)}f_2(z)\overline{\Phi^*(\alpha)}\right).$$

Observe that, since $\int_{A_i}\alpha=0$, we may choose any f_μ obeying $df_\mu=\Phi^*(\alpha)$. By choosing f_μ to have a zero on $|z-s_\mu(i)|=R_\mu(i)$ we get

COROLLARY 6.7. *Let α be a holomorphic differential form on Y_i' with $\int_{A_i} \alpha = 0$. Write the form $\Phi^*(\alpha) = f(z)\,dz$. Then*

$$\|\alpha|_{Y_i'}\|_2^2 \leq 2\pi^2\left(R_1(i)^2\Big[\sup_{|z-s_1(i)|=R_1(i)} |f(z)|\Big]^2 + R_2(i)^2\Big[\sup_{|z-s_2(i)|=R_2(i)} |f(z)|\Big]^2\right).$$

PROOF OF THEOREM 6.4 (Simple single sheet case). We apply Corollary 6.7 to $\alpha = \omega_j$ with $j \neq i$ and substitute (6.8)

$$\|\omega_j|_{Y_i'}\|_2 \leq 2\pi \sup_{\mu=1,2} R_\mu(i)[\sup_{|z-s_\mu(i)|=R_\mu(i)} |w_j(z)|] \tag{6.11}$$

$$\leq 2\pi \sup_{\mu=1,2} R_\mu(i) \sup_{|z-s_\mu(i)|=R_\mu(i)} \Big|\sum_{s\in S} w_{j,s}(z)\Big|.$$

By (6.9) and hypothesis (GH5)(i)

$$\sup_{|z-s_\mu(i)|=R_\mu(i)} |w_{j,s}(z)| \leq \begin{cases} \frac{6r(s)}{|s_\mu(i)-s|^2}\|w_j\,dz|_{\mathcal{A}(s)}\|_2 & \text{if } s \neq s_1(j),\ s_2(j),\ s_\mu(i) \\ \frac{3r(s)}{R(s)^2}\|w_j\,dz|_{\mathcal{A}(s)}\|_2 & \text{if } s = s_\mu(i) \end{cases}.$$

Since

$$w_{j,s_1(j)}(z) + w_{j,s_2(j)}(z) \tag{6.12}$$
$$= \frac{1}{2\pi i}\frac{s_1(j)-s_2(j)}{(z-s_1(j))(z-\sigma_2(j))} + \sum_{\tau=\pm1}\left(w_{j,s_\tau(j)}(z) - \frac{\tau}{2\pi i}\frac{1}{z-s_\tau(j)}\right)$$

we have

$$\sup_{|z-s_\mu(i)|=R_\mu(i)} |w_{j,s_1(j)}(z) + w_{j,s_2(j)}(z)| \leq \frac{1}{\pi}\frac{|s_1(j)-s_2(j)|}{|s_\mu(i)-s_1(j)||s_\mu(i)-s_2(j)|}$$
$$+ \sum_{\tau=\pm1}\frac{6r_\tau(j)}{|s_\mu(i)-s_\tau(j)|^2}\left\|\left(w_j - \frac{\tau}{2\pi i}\frac{1}{z-s_\tau(j)}\right)dz\Big|_{\mathcal{A}(s_\tau(j))}\right\|_2.$$

Substituting these bounds into (6.11) and using (6.10)

$$\|\omega_j|_{Y_i'}\|_2 \leq 6\pi \sup_{\mu=1,2}\left[\frac{r_\mu(i)}{R_\mu(i)} + \frac{2r_{3-\mu}(i)R_\mu(i)}{|s_1(i)-s_2(i)|^2}\right]\|\omega_j|_{Y_i'}\|_2$$
$$+ 24\pi\sum_{k\neq i,j}\sup_{\mu,\tau=\pm1}\frac{r_\tau(k)R_\mu(i)}{|s_\mu(i)-s_\tau(k)|^2}\|\omega_j|_{Y_k'}\|_2$$
$$+ 2\sup_{\mu=1,2}\frac{R_\mu(i)|s_1(j)-s_2(j)|}{|s_\mu(i)-s_1(j)||s_\mu(i)-s_2(j)|}$$
$$+ 12\pi\sup_{\mu=1,2}\sum_{\tau=\pm1}\frac{r_\tau(j)R_\mu(i)}{|s_\mu(i)-s_\tau(j)|^2}\left\|\left(w_j - \frac{\tau}{2\pi i}\frac{1}{z-s_\tau(j)}\right)dz\Big|_{\mathcal{A}(s_\tau(j))}\right\|_2.$$

To clarify the structure of this system of inequalities, define, for each j and $i \neq j$

$$\Omega_i^j = \|\omega_j|_{Y_i'}\|_2, \quad \widetilde{\Omega}_i^j = 2\sup_{\mu=1,2}\frac{R_\mu(i)|s_1(j)-s_2(j)|}{|s_\mu(i)-s_1(j)||s_\mu(i)-s_2(j)|}. \tag{6.13}$$

Then the system of inequalities reads, for $i \neq j$

$$\text{(6.14)}\quad \Omega_i^j \leq \widetilde{\Omega}_i^j + \sum_{k\neq j} \mathfrak{A}_{i,k}\Omega_k^j + \frac{1}{2}\mathfrak{A}_{i,j}\sum_{\mu=1}^{2}\left\| \left(w_j - \frac{(-1)^{\mu+1}}{2\pi i}\frac{1}{z-s_\mu(j)}\right) dz\Big|_{\mathcal{A}(s_\mu(j))}\right\|_2 .$$

By Lemma 6.3(b) and 6.3(c), the vector $\widetilde{\Omega}^j = (\widetilde{\Omega}_i^j)$ is in ℓ^2, with a norm bounded uniformly in j. By Lemma 6.10(a) below, $\mathfrak{A} = (\mathfrak{A}_{i,k})$ is a bounded operator on ℓ^2. By our additional assumption, its norm is smaller than 1. So, if it weren't for the term involving $\|(w_j - (-1)^{\mu+1}/2\pi i 1/z - s_\mu(j))\, dz|_{\mathcal{A}(s_\mu(j))}\|_2$ we would get a bound on the ℓ^2 norm of Ω^j in terms of the ℓ^2 norm of $\widetilde{\Omega}^j$ and the operator norm of $\mathfrak{A}$.

We now wish to incorporate a term involving $\omega_j|_{Y_j'}$ in the system of inequalities (6.14). The form $(w_j - (-1)^{\mu+1}/2\pi i 1/z - s_\mu(j))\, dz|_{\mathcal{A}(s_\mu(j))}$ is not, in general, the pull-back under Φ of a holomorphic differential form on Y_j. Therefore we cannot apply the analysis based on Lemma 6.6 directly. However, $\omega_j - (\phi_j)_*(1/2\pi i dz_1/z_1)$ is a holomorphic form on Y_j whose pull-backs $\Phi^*\big(\omega_j - (\phi_j)_*(1/2\pi i dz_1/z_1)\big) = w_j(z)\, dz - (g_{j,\mu})_*(1/2\pi i dz_1/z_1)$ can be used in place of $(w_j - (-1)^{\mu+1}/2\pi i 1/z - s_\mu(j))\, dz$.

Put

$$\Omega_j^j = \left\| \left(\omega_j - (\phi_j)_*\left(\frac{1}{2\pi i}\frac{dz_1}{z_1}\right)\right)\Big|_{Y_j'}\right\|_2 .$$

Since

$$\text{(6.15a)}\quad \left(w_j - \frac{(-1)^{\mu+1}}{2\pi i}\frac{1}{z-s_\mu(j)}\right) dz\Big|_{\mathcal{A}(s_\mu(j))} = \left(w_j (g_{j,\mu})_*\left(\frac{1}{2\pi i}\frac{dz_1}{z_1}\right) + \alpha_{j,\mu}\, dz\right)\Big|_{\mathcal{A}(s_\mu(j))}$$

we get

$$\text{(6.15b)}\quad \left\| \left(w_j - \frac{(-1)^{\mu+1}}{2\pi i}\frac{1}{z-s_\mu(j)}\right) dz\Big|_{\mathcal{A}(s_\mu(j))}\right\|_2 \leq \Omega_j^j + \aleph_j$$

and, by substituting into (6.14),

$$\text{(6.16a)}\quad \Omega_i^j \leq (\widetilde{\Omega}_i^j + \mathfrak{A}_{i,j}\aleph_j) + \sum_k \mathfrak{A}_{i,k}\Omega_k^j \quad \text{for } i \neq j.$$

We now derive a similar formula for $i = j$. We apply Corollary 6.7 with $i = j$ and $\alpha = \omega_j - (\phi_j)_*(1/2\pi i dz_1/z_1)$. Observe that

$$\left(w_j (g_{j,\mu})_*\left(\frac{1}{2\pi i}\frac{dz_1}{z_1}\right)\right)\Big|_{\mathcal{A}(s_\mu(j))} = \sum_{s\neq s_\mu(j)} w_{j,s}\, dz + \left(w_{j,s_\mu(j)} - \frac{(-1)^{\mu+1}}{2\pi i}\frac{1}{z-s_\mu(j)}\right) dz - \alpha_{j,\mu}\, dz.$$

Hence

$$\Omega_j^j \le 2\pi \sup_{\mu=1,2} R_\mu(j) \sup_{|z-s_\mu(j)|=R_\mu(j)} \left| \sum_{s\neq s_\mu(j)} w_{j,s}(z) \right|$$
$$+ 2\pi \sup_{\mu=1,2} R_\mu(j) \sup_{|z-s_\mu(j)|=R_\mu(j)} \left| w_{j,s_\mu(j)} - \frac{(-1)^{\mu+1}}{2\pi i}\frac{1}{z-s_\mu(j)} \right|$$
$$+ 2\pi \sup_{\mu=1,2} R_\mu(j) \sup_{|z-s_\mu(j)|=R_\mu(j)} |\alpha_{j,\mu}|$$

By the estimates of Proposition 6.5 and (6.10)

$$\Omega_j^j \le (\widetilde{\Omega}_j^j + \mathfrak{A}_{j,j}\aleph_j) + \sum_k \mathfrak{A}_{j,k}\Omega_k^j. \tag{6.16b}$$

with

$$\widetilde{\Omega}_j^j = 2\pi \sup_{\mu=1,2} R_\mu(j) \left[\frac{\sqrt{2}}{2\pi|s_1(j)-s_2(j)|} + \sup_{|z-s_\mu(j)|=R_\mu(j)} |\alpha_{j,\mu}(z)| \right].$$

Now view $\Omega^j = (\Omega_i^j)_{i\ge 1}$ as a vector in ℓ^2. Observe that

$$\sum_i |\widetilde{\Omega}_i^j|^2 \le \mathcal{O}^j. \tag{6.17}$$

We shall show below, in Lemma 6.11, that $(\mathfrak{A}_{i,j}\aleph_j)_{i\ge1}$ is a vector in ℓ^2 whose norm C^j goes to zero with j. As $\mathfrak{A} = (\mathfrak{A}_{i,k})_{i,k\ge1}$ is an operator on ℓ^2 of norm smaller than one half (6.16) implies

$$\|\Omega^j\| \le \frac{\sqrt{\mathcal{O}^j} + C^j}{1 - \|\mathfrak{A}\|} \le 2(\sqrt{\mathcal{O}^j} + C^j).$$

As $\mathcal{O}^j$ goes to zero with j by Lemma 6.3(c), this proves the last inequality of the Theorem.

Put

$$C = 3\sup_j(\sqrt{\mathcal{O}^j} + C^j + \aleph_j).$$

By Lemma 6.3(c) and 6.3(d), C is finite. Now, by (6.10), (6.15b) and Proposition 6.5

$$\left| w_j(z) - \frac{1}{2\pi i}\frac{1}{z-s_1(j)} + \frac{1}{2\pi i}\frac{1}{z-s_2(j)} \right| \le C \sum_{s\in S} \frac{r(s)}{|z-s|^2}$$

when $|z-s| > 3r(s)$ for all $s \in S$. This concludes the proof of Theorem 6.4, under the special assumptions made above. □

Before proving the technical Lemma 6.8–6.11 used above, we describe how to modify the strategy in the general case. First, one can not expect the operator norm of $\mathfrak{A}$ to be smaller than a half. On the other hand, the projected operator $(\mathfrak{A}_{i,k})_{i,k>N}$ will have small operator norm if N is large enough. This and the problems arising from the existence of X^{com} is taken care of by enclosing X^{com} and the handles Y_i, $i \le N$ in an enlarged compact piece $X(\Gamma)$ that is treated as a unit.

The other main difference is that the cancellation (6.12) is only possible when both ends of Y_j hit the same component of X^{reg}. If only one end of Y_j hits X_ν^{reg} then $\Phi_\nu^*\omega_j$ will have a $\pm 1/z$ decay away from the compact piece and $\mp 1/(z-s)$ decay away from the annulus $\mathcal{A}(s)$ corresponding to the intersection of Y_j' and X_ν^{reg}.

LEMMA 6.8. *Let ω be a holomorphic differential form on $H(t) = \{(z_1, z_2) \in \mathbb{C} \mid z_1 z_2 = t, |z_1|, |z_2| \leq 1\}$ with $t < 1/4$, such that $\int_{|z_1|=\sqrt{t}} \omega = 0$. Then*

$$\left| \frac{\omega}{dz_1/z_1} \right| \leq \frac{2}{\sqrt{\pi}} (|z_1| + |z_2|) \left\| \omega|_{H(t)} \right\|_2$$

for $(z_1, z_2) \in H(t)$, $|z_1|, |z_2| \leq \frac{1}{2}$.

PROOF. We use $z = z_1$ as a coordinate on $H(t)$. Write

$$\omega = f(z) \frac{dz}{z} \quad \text{with} \quad f(z) = \sum_{n=-\infty}^{\infty} a_n z^n, \quad a_0 = 0.$$

Then

$$\begin{aligned}
\left\| \omega|_{H(t)} \right\|_2^2 &= \frac{1}{2i} \int_{H(t)} |f(z)|^2 \frac{dz \wedge d\bar{z}}{|z|^2} \\
&= \sum_{n \neq 0} \int_t^1 \int_0^{2\pi} |a_n|^2 r^{2n-2} r \, dr \, d\phi \\
&= 2\pi \sum_{n \neq 0} \frac{|a_n|^2}{2n} (1 - t^{2n}) \\
&\geq 2\pi \sum_{n \geq 1} \frac{|a_n|^2}{2n} + 2\pi \sum_{n \geq 1} \frac{|a_{-n}|^2}{2n} \frac{1}{t^{2n}}.
\end{aligned}$$

On the other hand, for $|z| = \frac{1}{2}$

$$\begin{aligned}
\left| \frac{f(z)}{z} \right|^2 &\leq 4 \left(\sum_{n \geq 1} |a_n| \frac{1}{2^n} + \sum_{n \geq 1} |a_{-n}| 2^n \right)^2 \\
&\leq 8 \left[\left(\sum_{n \geq 1} \frac{|a_n|}{\sqrt{2^n}} \frac{1}{\sqrt{2^n}} \right)^2 + \left(\sum_{n \geq 1} \sqrt{8^n} |a_{-n}| \frac{1}{\sqrt{2^n}} \right)^2 \right] \\
&\leq 8 \left(\sum_{n \geq 1} \frac{1}{2^n} \right) \left[\sum_{n \geq 1} \frac{|a_n|^2}{2^n} + \sum_{n \geq 1} 8^n |a_{-n}|^2 \right] \\
&\leq 8 \left[\sum_{n \geq 1} \frac{|a_n|^2}{2n} + \sum_{n \geq 1} \frac{|a_{-n}|^2}{2n} \frac{1}{t^{2n}} \right] \leq \frac{4}{\pi} \|\omega\|_2^2.
\end{aligned}$$

Similarly, for $|z| = \sqrt{t}$

$$\begin{aligned}
\left| \frac{f(z)}{z} \right|^2 &\leq \frac{2}{t} \left[\left(\sum_{n \geq 1} \frac{|a_n|}{\sqrt{2^n}} (\sqrt{2t})^n \right)^2 + \left(\sum_{n \geq 1} \frac{|a_{-n}|}{(\sqrt{2}t)^n} (\sqrt{2t})^n \right)^2 \right] \\
&\leq \frac{2}{t} \left(\sum_{n \geq 1} (2t)^n \right) \left[\sum_{n \geq 1} \frac{|a_n|^2}{2^n} + \sum_{n \geq 1} \frac{|a_{-n}|^2}{2^n} \frac{1}{t^{2n}} \right] \\
&\leq \frac{4}{1 - 2t} \frac{1}{2\pi} \|\omega\|_2^2 \leq \frac{4}{\pi} \|\omega\|_2^2.
\end{aligned}$$

So, by the maximum principle,

$$|f(z)| \leq \frac{2}{\sqrt{\pi}} \|\omega\|_2 |z| \quad \text{for } \sqrt{t} \leq |z| \leq \frac{1}{2}$$

and hence

$$\left| \frac{\omega}{dz_1/z_1} \right| \leq \frac{2}{\sqrt{\pi}} |z_1| \|\omega\|_2 \leq \frac{2}{\sqrt{\pi}} (|z_1| + |z_2|) \|\omega\|_2$$

on $\{(z_1, z_2) \in H(t) \mid \sqrt{t} \leq |z_1| \leq \frac{1}{2}\}$. The same estimate on

$$\{(z_1, z_2) \in H(t) \mid t/2 \leq |z_1| \leq \sqrt{t}\} = \{(z_1, z_2) \in H(t) \mid \sqrt{t} \leq |z_2| \leq \tfrac{1}{2}\}$$

is proven in the same way. □

LEMMA 6.9. *Let $J\colon S^1 \times [0,1] \to \mathbb{C}$ be a smooth function with $J(S^1 \times \{t\})$ winding once in the positive direction around $s \in \mathbb{C}$ for each $t \in [0,1]$. Let w be a holomorphic function on $J(S^1 \times [0,1])$. Then*

(a) $$\left| \frac{1}{2\pi i} \int_{J(S^1 \times \{0\})} \frac{w(\zeta)}{\zeta - z} \, d\zeta \right| \leq C_J \left\| w\big|_{J(S^1 \times [0,1])} \right\|_2 \sup_{\zeta \in J(S^1 \times [0,1])} \left| \frac{1}{\zeta - z} \right|$$

where

$$C_J = \frac{1}{2\pi} \sqrt{\int_{S^1 \times [0,1]} \left| \frac{\partial J}{\partial \phi} \right|^2 \frac{d\phi dt}{DJ(\phi, t)}}$$

and DJ is the Jacobian of $J(\phi, t)$.

(b) *If, in addition, $\int_{J(S^1 \times \{0\})} w(\zeta) \, d\zeta = 0$ then*

$$\left| \frac{1}{2\pi i} \int_{J(S^1 \times \{0\})} \frac{w(\zeta)}{\zeta - z} \, d\zeta \right| \leq C_J \left\| w\big|_{J(S^1 \times [0,1])} \right\|_2 \sup_{\zeta \in J(S^1 \times [0,1])} \frac{|s - \zeta|}{|\zeta - z||s - z|}.$$

(c) *If w is a holomorphic function on the annulus $\{\zeta \mid r \leq |\zeta - s| \leq 2r\}$ and $\int_{|\zeta - s| = r} w(\zeta) \, d\zeta = 0$ then, for $|z - s| \geq 3r$,*

$$\left| \frac{1}{2\pi i} \int_{|\zeta - s| = r} \frac{w(\zeta)}{\zeta - z} \, d\zeta \right| \leq \frac{3r}{|z - s|^2} \left\| w\big|_{r \leq |\zeta - s| \leq 2r} \right\|_2.$$

PROOF. (a) By Cauchy's theorem, for $z \notin J(S^1 \times [0,1])$,

$$\begin{aligned}
\int_{J(S^1 \times \{0\})} \frac{w(\zeta)}{\zeta - z} \, d\zeta &= \int_0^1 \int_{J(S^1 \times \{t\})} \frac{w(\zeta)}{\zeta - z} \, d\zeta \, dt \\
&= \int_0^1 \int_{S^1} \frac{w(J(\phi, t))}{J(\phi, t) - z} \frac{\partial J}{\partial \phi}(\phi, t) \, d\phi \, dt \\
&= \int_{S^1 \times [0,1]} w\big(J(\phi, t)\big) \sqrt{DJ(\phi, t)} \left[\frac{1}{(J(\phi, t) - z)\sqrt{DJ(\phi, t)}} \frac{\partial J}{\partial \phi} \right] d\phi dt.
\end{aligned}$$

By the Cauchy–Schwarz inequality

$$\begin{aligned}\left|\int_{J(S^1\times\{0\})}\frac{w(\zeta)}{\zeta-z}\,d\zeta\right|^2 &\le \left[\int_{S^1\times[0,1]}\left|w_j(J(\phi,t))\right|^2 DJ(\phi,t)\,d\phi\,dt\right]\\ &\quad\times\left[\int_{S^1\times[0,1]}\frac{1}{|J(\phi,t)-z|^2 DJ(\phi,t)}\left|\frac{\partial J}{\partial\phi}\right|^2 d\phi\,dt\right]\\ &\le \left\|w|_{J(S^1\times[0,1])}\right\|_2^2\left[\int_{S^1\times[0,1]}\left|\frac{\partial J}{\partial\phi}\right|^2\frac{d\phi dt}{DJ(\phi,t)}\right]\sup_{\zeta\in J(S^1\times[0,1])}\frac{1}{|\zeta-z|^2}.\end{aligned}$$

(b) Since $\int_{J(S^1\times\{0\})} w(\zeta)\,d\zeta = 0$

$$\begin{aligned}\int_{J(S^1\times\{0\})}\frac{w(\zeta)}{\zeta-z}\,d\zeta &= \int\frac{w(\zeta)}{\zeta-z}\,d\zeta-\int\frac{w(\zeta)}{s-z}\,d\zeta\\ &= \int_{J(S^1\times\{0\})} w(\zeta)\frac{s-\zeta}{(\zeta-z)(s-z)}\,d\zeta\end{aligned}$$

Now, continue as in part (a).

(c) Apply (b) with

$$J(\phi,t) = s+(t+1)re^{i\phi}.$$

Since

$$\frac{\partial J}{\partial\phi} = i(t+1)re^{i\phi}$$

and

$$DJ = \begin{vmatrix} r\cos\phi & r\sin\phi\\ -(t+1)r\sin\phi & (t+1)r\cos\phi\end{vmatrix} = (t+1)r^2$$

we have

$$C_J \le \frac{1}{2\pi}\sqrt{\int_0^{2\pi}d\phi\int_0^1 dt(t+1)} \le \frac{\sqrt{3\pi}}{2\pi}.$$

Thus

$$\begin{aligned}\left|\frac{1}{2\pi i}\int_{|\zeta-s|=r}\frac{w(\zeta)}{\zeta-z}\,d\zeta\right| &\le \frac{\sqrt{3\pi}}{2\pi}\sup_{r\le|\zeta-s|\le 2r}\frac{|s-\zeta|}{|\zeta-s||s-z|}\left\|w|_{r\le|\zeta-s|\le 2r}\right\|_2\\ &\le \frac{3r}{|z-s|^2}\left\|w|_{r\le|\zeta-s|\le 2r}\right\|_2.\end{aligned}$$

□

LEMMA 6.10.

(a) $(\mathfrak{A}_{i,k})_{i,k\ge g+1}$ *is a bounded operator on* ℓ^2.

(b) *There is an* N *such that the norm of the operator* $(\mathfrak{A}_{i,k})_{i\ge N+1,k\ge g+1}$ *is bounded by* $\frac{1}{5}$.

(c) *The vector* $\left(\max_{\mu=1,2} R_\mu(i)/\mathrm{dist}(s_\mu(i),T)^2\right)_{i\ge g+1}$ *is in* ℓ^2.

PROOF. (a)–(b) The norm of the operator is bounded by

$$\sqrt{\sum_{\substack{i\ge N+1\\ k\ge g+1\\ i\ne k}}\mathfrak{A}_{i,k}^2} + \sup_{j\ge N+1}\mathfrak{A}_{j,j}.$$

The first term is finite for all $N \geq g+1$ and goes to zero as $N \to \infty$ by Lemma 6.3(a). Since

$$\mathfrak{A}_{j,j} \leq 6\pi \frac{\max\{r_1(j), r_2(j)\}}{\min\{R_1(j), R_1(j)\}} \left(1 + \sup_{\mu=\pm 1} \frac{R_\mu(j)^2}{|s_1(j) - s_2(j)|^2}\right)$$

(GH5)(i) and Lemma 6.3(b), 6.3(c) imply that

$$\lim_{j\to\infty} \mathfrak{A}_{j,j} = 0.$$

(c) The convergence of $\sum_s R(s)^2/|s|^4$ is equivalent to the convergence of the first term of $\mathcal{O}^j$ for any fixed j. □

LEMMA 6.11. *For each $j \geq g+1$, the vector $l(\mathfrak{A}_{i,j}\aleph_j)_{i\geq g+1}$ is in ℓ^2 and*

$$\lim_{j\to\infty} \|(\mathfrak{A}_{i,j}\aleph_j)_{i\geq g+1}\|_2 = 0.$$

PROOF. We bound

$$\|(\mathfrak{A}_{i,j}\aleph_j)_{i\geq g+1}\|_2 \leq \|(\mathfrak{A}_{i,j})_{i\geq g+1}\|_2 \sup_j \aleph_j.$$

Lemma 6.3(d) states that $\aleph_j$ is bounded uniformly in j. As in Lemma 6.10, $(\mathfrak{A}_{i,j})_{i\geq g+1}$ is in ℓ^2 for all j and its norm converges to zero as j tends to infinity. □

PROOF OF THEOREM 6.4 (General single sheet case). We now prove Theorem 6.4, still in the single sheet case $m = 1$, but allowing X^{com} to be nonempty and deleting the simplifying assumption $\|\mathfrak{A}\| < \frac{1}{2}$. Again, we supress the subscript ν.

We define

$$w_{j,\text{com}}(z) = -\frac{1}{2\pi i}\int_{\partial K} \frac{w_j(\zeta)}{\zeta - z}\, d\zeta, \quad z \in \mathbb{C}\setminus K.$$

By Lemma 6.9(b), 6.9(c) we have the following analog of Proposition 6.5.

PROPOSITION 6.12. *For $z \in G$ with $|z - s|$ for all $s \in S$*

$$w_j(z) = w_{j,\text{com}}(z) + \sum_{s\in S} w_{j,s}(z).$$

The series on the right-hand side converges absolutely and uniformly on compact subsets of $\{z \in G \mid |z-s| > r(s)\ \forall s \in S\}$. Furthermore, for $s \in S$ and $|z-s| > 3r(s)$

$$|w_{j,s}(z)| \leq \frac{3r(s)}{|z-s|^2}\left\|w_j\, dz\big|_{\mathcal{A}(s)}\right\|_2 \quad \text{for } s \neq s_1(j),\ s_2(j),$$

$$\left|w_{j,s}(z) - \frac{(-1)^{\mu+1}}{2\pi i}\frac{1}{z-s}\right| \leq \frac{3r(s)}{|z-s|^2}\left\|\left(w_j - \frac{(-1)^{\mu+1}}{2\pi i}\frac{1}{z-s}\right)dz\Big|_{\mathcal{A}(s)}\right\|_2$$
$$\text{for } s = s_\mu(j).$$

Finally there is a constant γ, independent of j, such that, for all $z \in \mathbb{C}\setminus K \cup T$,

$$|w_{j,\text{com}}(z)| \leq \frac{\gamma}{\text{dist}(z,T)^2}\left\|w_j\, dz\big|_T\right\|_2$$

where $\text{dist}(z,T)$ *is the distance from z to T.*

PROOF. Proposition 6.12 is proven just like Proposition 6.5. The last estimate follows immediately from Lemma 6.9(b). □

Observe that Proposition 6.12 implies a bound on $w_j(z)$ as claimed in Theorem 6.4, but with a possibly j dependent constant

$$C_j^2 = \gamma^2 \big\| w_j\, dz\big|_T \big\|_2^2 + 9 \sum_{\substack{i \geq g+1 \\ i \neq j}} \big\| \omega_j \big|_{Y_i'} \big\|_2^2 + 9 \sum_{\mu=1,2} \left\| \left(w_j - \frac{(-1)^{\mu+1}}{2\pi i} \frac{1}{z - s_\mu(j)} \right) dz \right|_{\mathcal{A}(s_\mu(j))} \Bigg\|_2^2 .$$

We wish to improve this to a j independent bound. Define

$$\Omega_{\rm com}^j = \big\| w_j\, dz\big|_T \big\|_2$$

and as above, for $i \geq g+1$

$$\Omega_i^j = \begin{cases} \big\| \omega_j \big|_{Y_i'} \big\|_2 & \text{if } i \neq j \\ \Big\| \Big(\omega_j - (\phi_j)_* \Big(\frac{1}{2\pi i} \frac{dz_1}{z_1} \Big) \Big) \Big|_{Y_j'} \Big\|_2 & \text{if } i = j \end{cases} .$$

In the same way as (6.16) one shows

$$\Omega_i^j \leq (\widetilde{\Omega}_i^j + \mathfrak{A}_{i,j} \aleph_j) + \sum_{k \geq g+1} \mathfrak{A}_{i,k} \Omega_k^j + \mathfrak{A}_{i,\rm com} \Omega_{\rm com}^j \quad \text{for } i \geq g+1 \tag{6.18}$$

where $\widetilde{\Omega}_i^j$ is as in (6.13), $\mathfrak{A}_{i,k}$ is defined as above and

$$\mathfrak{A}_{i,\rm com} = 4\pi \sup_{\mu=1,2} \frac{\gamma R_\mu(i)}{\operatorname{dist}(s_\mu(i), T)^2} .$$

By Lemma 6.10(b), 6.10(c), there is an N such that the operator $\mathfrak{A}_{i,k}$ with i running over $N+1, N+2, \dots$ and k running over $\text{com}, g+1, g+2, \dots$ has norm at most $\frac{1}{4}$. Fix any such N and also fix any $b < \frac{1}{2}$. As before, we define Ω^j to be the vector $(\Omega_{\rm com}^j, \Omega_{g+1}^j, \Omega_{g+2}^j, \dots)$ in ℓ^2. Denote by $\underline{\Omega}^j$ the components $(\Omega_{\rm com}^j, \Omega_{g+1}^j, \Omega_{g+2}^j, \dots, \Omega_N^j)$ and by $\overline{\Omega}^j$ the components $(\Omega_{N+1}^j, \dots)$. Define

$$a_i = \begin{cases} 0 & i = \text{com or } i \leq N \\ \Big(\mathfrak{A}_{i,\rm com}^2 + \sum_{k=g+1}^N \mathfrak{A}_{i,k}^2 \Big)^{1/2} & i \geq N+1 \end{cases} .$$

Then (6.18) implies that, for $i \geq N+1$,

$$\Omega_i^j \leq (\widetilde{\Omega}_i^j + \mathfrak{A}_{i,j} \aleph_j) + \sum_{k \geq N+1} \mathfrak{A}_{i,k} \Omega_k^j + a_i \| \underline{\Omega}^j \| \tag{6.19a}$$

and, in particular,

$$\| \overline{\Omega}^j \| \leq \| \widetilde{\Omega}^j + \mathfrak{A} \aleph_j \| + \tfrac{1}{4} (\| \overline{\Omega}^j \| + \| \underline{\Omega}^j \|). \tag{6.19b}$$

Let Γ be a curve as in Lemma 6.1 with $\varepsilon = b / \sqrt{\gamma^2 + 1}$ such that $X(\Gamma)$ contains all handles Y_i with $i \leq N$. Observe that for any j

$$(\Omega_{\rm com}^j)^2 + \sum_{i \leq N} (\Omega_i^j)^2 \leq \big\| \omega_j \big|_{X(\Gamma)} \big\|_2^2 . \tag{6.20}$$

Consider any j such that $s_1(j)$, $s_2(j)$ are outside of Γ. Since $\int_\Gamma w_j(z)\, dz = 0$ and $\int_{A_i} \omega_j = 0$ for all i such that $Y_i \subset X(\Gamma)$, (A.12) implies that

$$\big\| \omega_j \big|_{X(\Gamma)} \big\|_2^2 = \frac{i}{2} \int_\Gamma f(z) \overline{w_j(z)\, dz}$$

where $f(z)$ is a Stammfunktion of w_j on Γ. As in the proof of Corollary 6.7 we may choose a Stammfunktion that is zero at some point of Γ and conclude that

$$\big\| \omega_j \big|_{X(\Gamma)} \big\|_2 \le (\text{length } \Gamma) \sup_{z\in\Gamma} |w_j(z)|. \tag{6.21}$$

By (6.10), (6.15b) and Proposition 6.12, for $z \in \Gamma$

$$\begin{aligned} |w_j(z)| &\le \Big| w_{j,\text{com}}(z) + \sum_{s\in S} w_{j,s}(z) \Big| \\ &\le \frac{\gamma}{\text{dist}(z,T)^2} \Omega^j_{\text{com}} + \sum_{i\ge g+1} \sum_{\mu=1,2} \frac{3r_\mu(i)}{|z-s_\mu(i)|^2} \Omega^j_i + \sum_{\mu=1,2} \frac{3r_\mu(j)}{|z-s_\mu(j)|^2} \aleph_j \\ &\qquad + \frac{1}{2\pi} \frac{1}{|z-s_1(j)|} + \frac{1}{2\pi} \frac{1}{|z-\sigma_2(j)|}. \end{aligned}$$

By (6.20), (6.21) and Cauchy–Schwarz

$$\begin{aligned} \|\underline{\Omega}^j\| \le \big\| \omega_j \big|_{X(\Gamma)} \big\|_2 &\le \widetilde{\Omega}^j_\Gamma + b\|\Omega^j\| \\ &\le \widetilde{\Omega}^j_\Gamma + b\|\overline{\Omega}^j\| + b\|\underline{\Omega}^j\| \end{aligned} \tag{6.22}$$

where

$$\begin{aligned} \widetilde{\Omega}^j_\Gamma &= (\text{length } \Gamma) \sup_{z\in\Gamma} \left(\sum_{\mu=1,2} \frac{3r_\mu(j)}{|z-s_\mu(j)|^2} \aleph_j + \frac{1}{2\pi} \frac{1}{|z-s_1(j)|} + \frac{1}{2\pi} \frac{1}{|z-\sigma_2(j)|} \right) \\ &\le \text{const}(\Gamma) \frac{1}{\min_{\mu=1,2} |s_\mu(j)|} \end{aligned}$$

since by Lemma 6.1 and the choice of Γ

$$(\text{length } \Gamma) \sup_{z\in\Gamma} \sqrt{\frac{\gamma^2}{\text{dist}(z,T)^4} + \sum_s \frac{18 r(s)^2}{|z-s|^4}} \le b.$$

By Lemma 6.3(d), $\aleph_j$ is uniformly bounded. Since the series $\sum_s r(s)/|s|^2$ converges and $\lim_{j\to\infty} 1/s_\mu(j) = 0$ we have

$$\lim_{j\to\infty} \widetilde{\Omega}^j_\Gamma = 0. \tag{6.23}$$

Since $b < \frac{1}{2}$, (6.22) implies

$$\|\underline{\Omega}^j\| \le 2\widetilde{\Omega}^j_\Gamma + 2b\|\overline{\Omega}^j\|.$$

By (6.19b)

$$\|\overline{\Omega}^j\| \le \tfrac{1}{2}\widetilde{\Omega}^j_\Gamma + \|\widetilde{\Omega}^j + \mathfrak{A}\aleph_j\| + \tfrac{1}{4}\|\overline{\Omega}^j\| + \tfrac{b}{2}\|\overline{\Omega}^j\|$$

which implies that

$$\|\overline{\Omega}^j\| \le \widetilde{\Omega}^j_\Gamma + 2\|\widetilde{\Omega}^j + \mathfrak{A}\aleph_j\|, \quad \|\underline{\Omega}^j\| \le 3\widetilde{\Omega}^j_\Gamma + 4b\|\widetilde{\Omega}^j + \mathfrak{A}\aleph_j\|. \tag{6.24}$$

By (6.23), (6.17), Lemma 6.3(c) and Lemma 6.11, $\|\Omega^j\|$ goes to zero with j. By (6.22) and (6.23), $\big\| \omega_j \big|_{X(\Gamma)} \big\|_2$ also goes to zero with j. The Theorem follows as above. □

REMARK 6.13. We have actually proven more detailed bounds than claimed in the theorem. Pick any N sufficient for Lemma 6.10(b) and any $b > 0$. Define, for $k \geq N+1$

$$a_k = \left(\mathfrak{A}^2_{k,\text{com}} + \sum_{n=g+1}^{N} \mathfrak{A}^2_{k,n} \right)^{1/2}$$

Then there exists a constant $V = V(N, b)$ such that the following bounds hold.

Define,

$$\widehat{\Omega}^j_i \begin{cases} \frac{V}{\min_{\mu=1,2} s_\mu(j)} + b & \text{if } i \leq N \text{ or } i = \text{com} \\ \sum_{k \geq N+1} (\mathbb{1} - \mathfrak{A}^r)^{-1}_{i,k} \left(\widetilde{\Omega}^j_k + \mathfrak{A}_{k,j} \aleph_j + b a_k + \frac{V}{\min_{\mu=1,2} s_\mu(j)} a_k \right) & \text{if } i \geq N+1 \end{cases}$$

where the superscript r denotes the resriction to $i, k \geq N+1$. The vectors $\widehat{\Omega}^j$ are uniformly bounded in ℓ^2. The expansion of $(\mathbb{1} - \mathfrak{A}^r)^{-1}$ converges, since $\|\mathfrak{A}^r\| \leq 1/4$. Then, for all $z \in G_\nu$,

$$\begin{aligned} &\left| w^\nu_j(z) - \frac{\delta_{\nu,\nu_1(j)}}{2\pi i} \left(\frac{1}{z - s_1(j)} - \frac{1}{z} \right) + \frac{\delta_{\nu,\nu_2(j)}}{2\pi i} \left(\frac{1}{z - s_2(j)} - \frac{1}{z} \right) \right| \\ &\qquad \leq \sum_{s \in S_\nu} \frac{3r(s) \widehat{\Omega}^j_{i(s)}}{|z - s|^2} + \sum_{\mu=1,2} \delta_{\nu,\nu_\mu(j)} \frac{3r_\mu(j) \aleph_j}{|z - s_\mu(j)|^2} + \frac{\gamma \widehat{\Omega}^j_{\text{com}}}{\text{dist}(z, T_\nu)^2} \end{aligned}$$

where we put $\nu_1(j) = \nu_2(j) = 0$ for $j \leq g$. Define

$$\Omega^j_i = \begin{cases} \left\| \omega_j \big|_{Y'_i} \right\|_2 & \text{if } i \neq j \\ \left\| \left(\omega_j - (\phi_j)_* \left(\frac{1}{2\pi i} \frac{dz_1}{z_1} \right) \right) \Big|_{Y'_j} \right\|_2 & \text{if } i = j \end{cases}$$

$$\Omega^j_{\text{com}} = \| w_j \, dz \big|_T \|_2.$$

Then, for all $i \geq g+1$ and $i = \text{com}$,

$$\Omega^j_i \leq \widehat{\Omega}^j_i.$$

Choosing Γ sufficiently large we have that, for any compact subset $\mathcal{K}$ of X,

$$\lim_{j \to \infty} \|\omega_j \big|_{\mathcal{K}}\|_2 = 0.$$

PROOF OF THEOREM 6.4 (General, multiple sheet case). Recall

$$\mathfrak{A}_{i,k} = \begin{cases} 6\pi \max_{\mu=1,2} \left[\frac{r_\mu(i)}{R_\mu(i)} + \delta_{\nu_1(i),\nu_2(i)} \frac{2 r_{3-\mu}(i) R_\mu(i)}{|s_1(i) - s_2(i)|^2} \right] & \text{if } i = k \\ 24\pi \max_{\mu,\tau=1,2} \delta_{\nu_\mu(i),\nu_\tau(k)} \frac{r_\tau(k) R_\mu(i)}{|s_\mu(i) - s_\tau(k)|^2} & \text{if } i \neq k \end{cases}$$

$$\mathfrak{A}_{i,\text{com}} = 4\pi \sup_{\mu=1,2} \frac{\gamma R_\mu(i)}{\text{dist}(s_\mu(i), T_{\nu_\mu(i)})^2},$$

$$\aleph_j = \max_{\mu=1,2} \left\| \alpha_{j,\mu}(z) \, dz \big|_{A(s_\mu(j))} \right\|_2$$

and define

$$\widetilde{\Omega}_i^j = \begin{cases} 2\max_{\mu=1,2} \delta_{\nu_\mu(i),\nu_1(j)} \frac{R_\mu(i)|s_1(j)-s_2(j)|}{|s_\mu(i)-s_1(j)||s_\mu(i)-s_2(j)|} \\ \qquad\qquad\qquad\qquad\qquad \text{if } i \neq j, \nu_1(j) = \nu_2(j) \\ 2\max_{\mu,\tau=1,2} \delta_{\nu_\mu(i),\nu_\tau(j)} \frac{R_\mu(i)|s_\tau(j)|}{|s_\mu(i)||s_\mu(i)-s_\tau(j)|} \\ \qquad\qquad\qquad\qquad\qquad \text{if } i \neq j,\ \nu_1(j) \neq \nu_2(j) \\ \max_{\mu=1,2}\left[\frac{\sqrt{2}R_\mu(j)}{|s_1(j)-s_2(j)|} + \sup_{|z-s_\mu(j)|=R_\mu(j)} 2\pi R_\mu(j)|\alpha_{j,\mu}(z)|\right] \\ \qquad\qquad\qquad\qquad\qquad \text{if } i = j,\ \nu_1(j) = \nu_2(j) \\ \max_{\mu=1,2}\left[\frac{\sqrt{2}R_\mu(j)}{|s_\mu(j)|} + \sup_{|z-s_\mu(j)|=R_\mu(j)} 2\pi R_\mu(j)|\alpha_{j,\mu}(z)|\right] \\ \qquad\qquad\qquad\qquad\qquad \text{if } i = j,\ \nu_1(j) \neq \nu_2(j) \end{cases}.$$

Observe that for all $N > 0$

$$\text{(6.25)} \qquad \sum_{i\geq N+1} (\widetilde{\Omega}_i^j)^2 \leq \begin{cases} \mathcal{O}^j(N) & \text{if } \nu_1(j) \neq \nu_2(j) \\ \mathcal{O}^j & \text{if } \nu_1(j) = \nu_2(j) \end{cases}.$$

Recall that in Theorem 6.4 we wish to get bounds dependent on a small number ρ. Select first, any positive $b < \min(\frac{1}{16}\rho, \frac{1}{2})$; second, any N sufficient for Lemma 6.10(b) and such that

$$\text{(6.26)} \qquad \limsup_{\substack{j\to\infty \\ \nu_1(j)\neq\nu_2(j)}} \mathcal{O}^j(N) < \frac{1}{16}\rho.$$

This is possible by Lemma 6.3(c). Third, choose a system Γ of curves as in Lemma 6.1 with the selected N and $\varepsilon = b/(1+\gamma)$. Here γ is defined as in Proposition 6.12. It is determined purely by the collars T_ν. Finally select, for each pair (ν_1, ν_2) of sheets with $\nu_1 \neq \nu_2$, a handle $I(\nu_1, \nu_2)$ with $\nu_1\big(I(\nu_1,\nu_2)\big) = \nu_1$, $\nu_2\big(I(\nu_1,\nu_2)\big) = \nu_2$ and $Y_{I(\nu_1,\nu_2)} \cap X(\Gamma) = \varnothing$. Furthermore, choose these handles far enough out that for all (ν_1, ν_2), $I = I(\nu_1, \nu_2)$ obeys

$$\text{(6.27a)} \qquad \sum_\nu (\operatorname{length}\Gamma_\nu) \times \sup_{z\in\Gamma_\nu}\left(\sum_{\mu=1,2} \delta_{\nu_\mu(I),\nu} \frac{3r_\mu(I)}{|z-s_\mu(I)|^2}\aleph_I + \sum_\mu \delta_{\nu_\mu(I),\nu}\frac{1}{2\pi}\frac{1}{|z-s_\mu(I)|}\right) < b,$$

$$\text{(6.27b)} \qquad \|(\mathfrak{A}_{i,I}\aleph_I)_{i\geq g+1}\|_2 < b,$$

$$\text{(6.27c)} \qquad \mathcal{O}^I(N) < \tfrac{1}{16}\rho.$$

Requirement (6.27a) is possible by the uniform boundedness of $\aleph_j$ (Lemma 6.3(d)), the convergence of $\sum_s r(s)/|s|^2$ and by $\lim_{j\to\infty} 1/s_\mu(j) = 0$. Requirement (6.27b) is possible by Lemma 6.11. Requirement (6.27c) is possible by (6.26).

We will develop bounds on

$$\Omega_{\text{com}}^j = \left[\sum_{\nu=1}^m \|(w_j^\nu - w_{I(j)}^\nu)\, dz\big|_{T_\nu}\|_2^2\right]^{1/2},$$

$$\Omega_i^j = \left\|\left(\omega_j - \omega_{I(j)} - (\delta_{i,j} - \delta_{i,I(j)})(\phi_i)_*\left(\frac{1}{2\pi i}\frac{dz_1}{z_1}\right)\right)\Big|_{Y_i'}\right\|_2$$

where, by convention $I(j) = I\big(\nu_1(j), \nu_2(j)\big)$ and $\omega_{I(\nu,\nu)} = \delta_{i,I(\nu,\nu)} = 0$ for all $1 \le \nu \le m$.

As before we define, for each form ω_j, each $1 \le \nu \le m$, and each $s \in S_\nu$

$$w^\nu_{j,s}(z) = -\frac{1}{2\pi i}\int_{|z-s|=r(s)} \frac{w^\nu_j(\zeta)}{\zeta - z}\,d\zeta, \quad |z-s| > r(s),\ z \in G_\nu$$

and

$$w^\nu_{j,\mathrm{com}}(z) = -\frac{1}{2\pi i}\int_{\partial K_\nu} \frac{w^\nu_j(\zeta)}{\zeta - z}\,d\zeta, \quad z \in G_\nu.$$

The analog of Propositions 6.5 and 6.12 is

PROPOSITION 6.14. *For $z \in G_\nu$ with $|z-s| > r(s)$ for all $s \in S_\nu$*

$$w^\nu_j(z) = w^\nu_{j,\mathrm{com}}(z) + \sum_{s\in S_\nu} w^\nu_{j,s}(z).$$

The series on the right-hand side converges absolutely and uniformly on compact subsets of $\{z \in G_\nu \mid |z-s| > r(s)\ \forall s \in S_\nu\}$. Furthermore, for $s \in S_\nu$ and $|z-s| > 3r(s)$

$$|f(z)_s| \le \frac{3r(s)}{|z-s|^2}\|f(z)\,dz\big|_{\mathcal{A}(s)}\|_2$$

for both

$$f(z) = w^\nu_j - \sum_{\mu=1,2} \delta_{s,s_\mu(j)} \frac{(-1)^{\mu+1}}{2\pi i}\frac{1}{z-s},$$

$$f(z) = w^\nu_j - w^\nu_{I(j)} - \sum_{\mu=1,2} \big(\delta_{s,s_\mu(j)} - \delta_{s,s_\mu(I(j))}\big)\frac{(-1)^{\mu+1}}{2\pi i}\frac{1}{z-s}.$$

Finally there is a constant γ, independent of j, $I(j)$, ν, such that, for all $z \in \mathbb{C} \setminus K_\nu \cup T_\nu$,

$$\left|w^\nu_{j,\mathrm{com}}(z) + \frac{\delta_{\nu,\nu_1(j)}}{2\pi i z} - \frac{\delta_{\nu,\nu_2(j)}}{2\pi i z}\right| \le \frac{\gamma}{\mathrm{dist}(z,T_\nu)^2}\left\|\left(w^\nu_j(z) + \frac{\delta_{\nu,\nu_1(j)}}{2\pi i z} - \frac{\delta_{\nu,\nu_2(j)}}{2\pi i z}\right) dz\big|_{T_\nu}\right\|_2,$$

$$\left|w^\nu_{j,\mathrm{com}}(z) - w^\nu_{I(j),\mathrm{com}}(z)\right| \le \frac{\gamma}{\mathrm{dist}(z,T_\nu)^2}\left\|\left(w^\nu_j(z) - w^\nu_{I(j)}(z)\right) dz\big|_{T_\nu}\right\|_2$$

where $\mathrm{dist}(z, T_\nu)$ *is the distance from z to T_ν.*

PROOF. Proposition 6.14 is proven just like Proposition 6.12. □

Again, Proposition 6.14 implies a bound on $w^\nu_j(z)$ as claimed in Theorem 6.4, but with a possibly j dependent constant C_j.

By Corollary 6.7, for all $i \geq g+1$ and $j \neq I(j)$,

$$\begin{aligned}
\Omega_i^j &= \left\| \left(\omega_j - \omega_{I(j)} - (\delta_{i,j} - \delta_{i,I(j)})(\phi_i)_* \left(\frac{1}{2\pi i} \frac{dz_1}{z_1} \right) \right) \Big|_{Y_i'} \right\|_2 \\
&\leq 2\pi \sup_{\mu=1,2} R_\mu(i) \\
&\quad \times \sup_{|z-s_\mu(i)|=R_\mu(i)} \left| w_j^{\nu_\mu(i)}(z) - w_{I(j)}^{\nu_\mu(i)}(z) - (\delta_{i,j} - \delta_{i,I(j)})(g_{i,\mu})_* \left(\frac{1}{2\pi i} \frac{dz_1}{z_1} \right) \frac{1}{dz} \right| \\
&= 2\pi \sup_{\mu=1,2} R_\mu(i) \\
&\quad \times \sup_{|z-s_\mu(i)|=R_\mu(i)} \left| w_j^{\nu_\mu(i)} - w_{I(j)}^{\nu_\mu(i)} - (\delta_{i,j} - \delta_{i,I(j)}) \left(\frac{(-1)^{\mu+1}}{2\pi i} \frac{1}{z - s_\mu(i)} + \alpha_{i,\mu}(z) \right) \right| \\
&= 2\pi \sup_{\mu=1,2} R_\mu(i) \\
&\quad \times \sup_{|z-s_\mu(i)|=R_\mu(i)} \Bigg| w_{j,\mathrm{com}}^{\nu_\mu(i)}(z) - w_{I(j),\mathrm{com}}^{\nu_\mu(i)}(z) + \sum_{s \in S_{\nu_\mu(i)}} (w_{j,s}^{\nu_\mu(i)} - w_{I(j),s}^{\nu_\mu(i)})(z) \\
&\qquad\qquad - (\delta_{i,j} - \delta_{i,I(j)}) \left(\frac{(-1)^{\mu+1}}{2\pi i} \frac{1}{z - s_\mu(i)} + \alpha_{i,\mu}(z) \right) \Bigg|.
\end{aligned}$$

Using Lemma 6.9(c), the terms with $s \neq s_1(j), s_2(j), s_1\big(I(j)\big), s_2\big(I(j)\big)$ are bounded by

$$\sup_{|z-s_\mu(i)|=R_\mu(i)} |(w_{j,s}^{\nu_\mu(i)} - w_{I(j),s}^{\nu_\mu(i)})(z)| \leq \begin{cases} \frac{6r(s)}{|s_\mu(i)-s|^2} \Omega_k^j & \text{if } s \neq s_\mu(i) \\ \frac{3r(s)}{R_\mu(i)^2} \Omega_i^j & \text{if } s = s_\mu(i) \end{cases}$$

with k chosen so that $s \in \{s_1(k), s_2(k)\}$ and, using Lemma 6.9(b), the "com" terms are bounded by

$$\sup_{|z-s_\mu(i)|=R_\mu(i)} |(w_{j,\mathrm{com}}^{\nu_\mu(i)} - w_{I(j),\mathrm{com}}^{\nu_\mu(i)})(z)| \leq \frac{2\gamma}{\mathrm{dist}(s_\mu(i), T_{\nu_\mu(i)})^2} \Omega_{\mathrm{com}}^j.$$

If $\nu_1(j) \neq \nu_2(j)$ then, since $\nu_\tau(j) = \nu_\tau\big(I(j)\big)$, the remaining terms are bounded using

$$\begin{aligned}
&\Bigg| \sum_{\tau=1,2} \delta_{\nu_\mu(i),\nu_\tau(j)} (w_{j,s_\tau(j)}^{\nu_\mu(i)} - w_{I(j),s_\tau(j)}^{\nu_\mu(i)} + w_{j,s_\tau(I(j))}^{\nu_\mu(i)} - w_{I(j),s_\tau(I(j))}^{\nu_\mu(i)})(z) \\
&\qquad\qquad - (\delta_{i,j} - \delta_{i,I(j)}) \left(\frac{(-1)^{\mu+1}}{2\pi i} \frac{1}{z - s_\mu(i)} + \alpha_{i,\mu}(z) \right) \Bigg| \\
&\leq \Bigg| \frac{\delta_{\nu_\mu(i),\nu_1(j)}}{2\pi i} \left(\frac{1}{z - s_1(j)} - \frac{1}{z - s_1(I(j))} \right) \\
&\qquad - \frac{\delta_{\nu_\mu(i),\nu_2(j)}}{2\pi i} \left(\frac{1}{z - s_2(j)} - \frac{1}{z - s_2(I(j))} \right) - \frac{(-1)^{\mu+1}}{2\pi i} \frac{\delta_{i,j} - \delta_{i,I(j)}}{z - s_\mu(i)} \Bigg| \\
&\quad + \sum_{\tau=1,2} \delta_{\nu_\mu(i),\nu_\tau(j)} \left| w_{j,s_\tau(j)}^{\nu_\mu(i)}(z) - w_{I(j),s_\tau(j)}^{\nu_\mu(i)}(z) - \frac{1}{2\pi i} \frac{(-1)^{\tau+1}}{z - s_\tau(j)} \right| \\
&\quad + \sum_{\tau=1,2} \delta_{\nu_\mu(i),\nu_\tau(j)} \left| w_{j,s_\tau(I(j))}^{\nu_\mu(i)}(z) - w_{I(j),s_\tau(I(j))}^{\nu_\mu(i)}(z) + \frac{(-1)^{\tau+1}}{2\pi i} \frac{1}{z - s_\tau(I(j))} \right| \\
&\quad + (\delta_{i,j} + \delta_{i,I(j)}) |\alpha_{i,\mu}(z)|
\end{aligned}$$

$$\leq \frac{1}{2\pi}\left|\delta_{\nu_\mu(i),\nu_1(j)}\frac{s_1(j)-s_1(I(j))}{(z-s_1(j))(z-s_1(I(j)))} - \delta_{\nu_\mu(i),\nu_2(j)}\frac{s_2(j)-s_2(I(j))}{(z-s_2(j))(z-s_2(I(j)))} - \frac{(-1)^{\mu+1}}{2\pi i}\frac{\delta_{i,j}-\delta_{i,I(j)}}{z-s_\mu(i)}\right|$$
$$+ \sum_{\tau=1,2}\delta_{\nu_\mu(i),\nu_\tau(j)}(\Omega_j^j+\aleph_j)\begin{cases}\frac{6r_\tau(j)}{|s_\mu(i)-s_\tau(j)|^2} & \text{if } (i,\mu)\neq(j,\tau)\\ \frac{3r_\tau(j)}{R_\mu(i)^2} & \text{if } (i,\mu)=(j,\tau)\end{cases}$$
$$+ \sum_{\tau=1,2}\delta_{\nu_\mu(i),\nu_\tau(j)}(\Omega_{I(j)}^j+\aleph_{I(j)})\begin{cases}\frac{6r_\tau(I(j))}{|s_\mu(i)-s_\tau(I(j))|^2} & \text{if } (i,\mu)\neq\big(I(j),\tau\big)\\ \frac{3r_\tau(I(j))}{R_\mu(i)^2} & \text{if } (i,\mu)=\big(I(j),\tau\big)\end{cases}$$
$$+ (\delta_{i,j}+\delta_{i,I(j)})|\alpha_{i,\mu}(z)|$$

on $|z-s_\mu(i)|=R_\mu(i)$. If $\nu_1(j)=\nu_2(j)$ the remaining terms are bounded by

$$\left|\sum_{\tau=1,2}\delta_{\nu_\mu(i),\nu_\tau(j)}w_{j,s_\tau(j)}^{\nu_\mu(i)}(z) - \delta_{i,j}\left(\frac{(-1)^{\mu+1}}{2\pi i}\frac{1}{z-s_\mu(i)}+\alpha_{i,\mu}(z)\right)\right|$$
$$\leq \left|\frac{\delta_{\nu_\mu(i),\nu_1(j)}}{2\pi i}\left(\frac{1}{z-s_1(j)}-\frac{1}{z-s_2(j)}\right) - \frac{(-1)^{\mu+1}}{2\pi i}\frac{\delta_{i,j}}{z-s_\mu(i)}\right|$$
$$+ \sum_{\tau=1,2}\delta_{\nu_\mu(i),\nu_\tau(j)}\left|w_{j,s_\tau(j)}^{\nu_\mu(i)}(z) - \frac{1}{2\pi i}\frac{(-1)^{\tau+1}}{z-s_\tau(j)}\right| + \delta_{i,j}|\alpha_{i,\mu}(z)|$$
$$\leq \frac{1}{2\pi}\left|\delta_{\nu_\mu(i),\nu_1(j)}\frac{s_1(j)-s_2(j)}{(z-s_1(j))(z-s_2(j))} - \frac{(-1)^{\mu+1}}{2\pi i}\frac{\delta_{i,j}}{z-s_\mu(i)}\right|$$
$$+ \sum_{\tau=1,2}\delta_{\nu_\mu(i),\nu_\tau(j)}(\Omega_j^j+\aleph_j)\begin{cases}\frac{6r_\tau(j)}{|s_\mu(i)-s_\tau(j)|^2} & \text{if } (i,\mu)\neq(j,\tau)\\ \frac{3r_\tau(j)}{R_\mu(i)^2} & \text{if } (i,\mu)=(j,\tau)\end{cases}$$
$$+ \delta_{i,j}|\alpha_{i,\mu}(z)|,$$

on $|z-s_\mu(i)|=R_\mu(i)$.

So, for $i\geq g+1$ and $j\neq I(j)$,

$$\Omega_i^j \leq (\widetilde{\Omega}_i^j+\widetilde{\Omega}_i^{I(j)}+\mathfrak{A}_{i,j}\aleph_j+\mathfrak{A}_{i,I(j)}\aleph_{I(j)}) + \sum_{k\geq g+1}\mathfrak{A}_{i,k}\Omega_k^j + \mathfrak{A}_{i,\mathrm{com}}\Omega_{\mathrm{com}}^j. \tag{6.28}$$

This system of equations is supplemented by the following bound on the first N handles. Consider any j such that $s_1(j)$, $s_2(j)$ are outside of Γ.

$$(\Omega_{\mathrm{com}}^j)^2 + \sum_{i\leq N}(\Omega_i^j)^2 \leq \big\|(\omega_j-\omega_I)\big|_{X(\Gamma)}\big\|_2^2.$$

Clearly,

$$\int_{A_i}(\omega_j-\omega_I)=0$$

for all i such that $Y_i'\subset X(\Gamma)$. We claim that

$$\int_{\Gamma_\nu}(w_j^\nu-w_I^\nu)(z)\,dz=0 \tag{6.29}$$

for all ν. To prove this, let $\varepsilon'>0$ and $\Gamma'=(\Gamma_1',\dots,\Gamma_m')$ a system of curves as in Lemma 6.1 with this ε' such that

$$X(\Gamma)\subset \operatorname{int}X(\Gamma') \quad\text{and}\quad Y_j\subset X(\Gamma'), Y_I\subset X(\Gamma').$$

Then $\Gamma_\nu - \Gamma'_\nu$ is homologous to $\sum_{i\in P(\Gamma_\nu)} A_i - \sum_{i\in N(\Gamma_\nu)} A_i$ where

$$P(\Gamma_\nu) = \{i \mid Y'_i \cap X(\Gamma) = \varnothing, Y'_i \subset X(\Gamma'), \nu_1(i) \neq \nu,\ \nu_2(i) = \nu\},$$
$$N(\Gamma_\nu) = \{i \mid Y'_i \cap X(\Gamma) = \varnothing, Y'_i \subset X(\Gamma'), \nu_1(i) = \nu,\ \nu_2(i) \neq \nu\}.$$

Therefore

$$\int_{\Gamma_\nu} (w_j^\nu - w_I^\nu)(z)\,dz = \int_{\Gamma'_\nu} (w_j^\nu - w_I^\nu)(z)\,dz.$$

By Proposition 6.14

$$\begin{aligned}\left|\int_{\Gamma'_\nu} (w_j^\nu - w_I^\nu)(z)\,dz\right| &\le \int_{\Gamma'_\nu} \left(\sum_{\substack{i,\mu \\ \nu_\mu(i)=\nu}} 3r_\mu(i) \frac{\|\omega_j|_{Y_i}\|_2 + \|\omega_I|_{Y_i}\|_2}{|z - s_\mu(i)|^2} + \frac{\text{const}}{|z^2|} \right) |dz| \\ &\le \text{const}\big(1 + \|\omega_j\|_2 + \|\omega_I\|_2\big) \int_{\Gamma'_\nu} \sqrt{\frac{1}{|z|^4} + \sum_{s\in S_\nu} \frac{r(s)^2}{|z-s|^4}}\,|dz| \\ &\le \text{const}\,\varepsilon'.\end{aligned}$$

This shows (6.29).

Now (A.12) implies that

$$\|(\omega_j - \omega_I)\big|_{X(\Gamma)}\|_2^2 = \sum_{\nu=1}^{m} \frac{i}{2} \int_{\Gamma_\nu} f_\nu(z) \overline{(w_j^\nu - w_I^\nu)(z)\,dz}$$

where, for each $1 \le \nu \le m$, $f_\nu(z)$ is a Stammfunktion of $w_j^\nu - w_I^\nu$ on Γ_ν. As in the proof of Corollary 6.7, we may choose f_ν to be zero at some point of Γ_ν and conclude that

$$\begin{aligned}\|(\omega_j - \omega_I)\big|_{X(\Gamma)}\|_2 &\le \sqrt{\sum_\nu (\text{length}\,\Gamma_\nu)^2 \sup_{z\in\Gamma_\nu} |w_j^\nu(z) - w_I^\nu(z)|^2} \\ &\le \sum_\nu (\text{length }\Gamma_\nu) \sup_{z\in\Gamma_\nu} |w_j^\nu(z) - w_I^\nu(z)|.\end{aligned}$$

By Proposition 6.14, for $z \in \Gamma_\nu$

$$\begin{aligned}|w_j^\nu(z) - w_I^\nu(z)| &\le \left| w_{j,\text{com}}^\nu(z) - w_{I,\text{com}}^\nu(z) + \sum_{s\in S_\nu} \big(w_{j,s}^\nu(z) - w_{I,s}^\nu(z)\big) \right| \\ &\le \frac{\gamma}{\text{dist}(z, T_\nu)^2} \Omega_{\text{com}}^j + \sum_{i\ge g+1} \sum_{\mu=1,2} \delta_{\nu_\mu(i),\nu} \frac{3r_\mu(i)}{|z - s_\mu(i)|^2} \Omega_i^j \\ &\quad + \sum_{\mu=1,2} \delta_{\nu_\mu(j),\nu} \frac{3r_\mu(j)}{|z - s_\mu(j)|^2} \aleph_j \\ &\quad + \sum_{\mu=1,2} \delta_{\nu_\mu(I),\nu} \frac{3r_\mu(I)}{|z - s_\mu(I)|^2} \aleph_I + \sum_\mu \delta_{\nu_\mu(j),\nu} \frac{1}{2\pi} \frac{1}{|z - s_\mu(j)|} \\ &\quad + \sum_\mu \delta_{\nu_\mu(I),\nu} \frac{1}{2\pi} \frac{1}{|z - s_\mu(I)|}.\end{aligned}$$

As before denote by $\underline{\Omega}^j$ the components $(\Omega^j_{\text{com}}, \Omega^j_{g+1}, \Omega^j_{g+2}, \dots, \Omega^j_N)$ and by $\overline{\Omega}^j$ the components $(\Omega^j_{N+1}, \dots)$. By Cauchy–Schwarz

$$\begin{aligned} \|\underline{\Omega}^j\| &\le \|(\omega_j - \omega_I)\big|_{X(\Gamma)}\|_2 \le \widetilde{\Omega}^j_\Gamma + b\|\Omega^j\| \\ &\le \widetilde{\Omega}^j_\Gamma + b\|\overline{\Omega}^j\| + b\|\underline{\Omega}^j\| \end{aligned} \tag{6.30}$$

where

$$\begin{aligned} \widetilde{\Omega}^j_\Gamma &= \sum_\nu (\operatorname{length}\Gamma_\nu) \\ &\quad\times \sup_{z\in\Gamma_\nu}\Bigg(\sum_{\mu=1,2}\delta_{\nu_\mu(j),\nu}\frac{3r_\mu(j)}{|z-s_\mu(j)|^2}\aleph_j + \sum_{\mu=1,2}\delta_{\nu_\mu(I),\nu}\frac{3r_\mu(I)}{|z-s_\mu(I)|^2}\aleph_I \\ &\qquad\qquad + \sum_\mu \delta_{\nu_\mu(j),\nu}\frac{1}{2\pi}\frac{1}{|z-s_\mu(j)|} + \sum_\mu \delta_{\nu_\mu(I),\nu}\frac{1}{2\pi}\frac{1}{|z-s_\mu(I)|}\Bigg) \\ &\le \operatorname{const}(\Gamma)\frac{1}{\min_{\mu=1,2}|s_\mu(j)|} \end{aligned}$$

since by Lemma 6.1 and the choice of Γ

$$\sum_\nu (\operatorname{length}\Gamma_\nu)\sup_{z\in\Gamma_\nu}\sqrt{\frac{\gamma^2}{\operatorname{dist}(z,T_\nu)^4} + \sum_{s\in S_\nu}\frac{18r(s)^2}{|z-s|^4}} \le b.$$

By Lemma 6.3(d), $\aleph_j$ is uniformly bounded. Since the series $\sum_s r(s)/|s|^2$ converges and $\lim_{j\to\infty} 1/s_\mu(j) = 0$ we have, by (6.27a)

$$\limsup_{j\to\infty}\widetilde{\Omega}^j_\Gamma < b. \tag{6.31}$$

Since $b < \frac{1}{2}$, (6.30) implies

$$\|\underline{\Omega}^j\| \le 2\widetilde{\Omega}^j_\Gamma + 2b\|\overline{\Omega}^j\|.$$

By (6.28) and Lemma 6.10(b)

$$\|\overline{\Omega}^j\| \le \frac{1}{2}\widetilde{\Omega}^j_\Gamma + \|(\widetilde{\Omega}^j_i + \widetilde{\Omega}^{I(j)}_i + \mathfrak{A}_{i,j}\aleph h_j + \mathfrak{A}_{i,I(j)}\aleph_{I(j)})_{i\ge N+1}\| + \frac{1}{4}\|\overline{\Omega}^j\| + \frac{b}{2}\|\underline{\Omega}^j\|$$

which implies that

$$\begin{aligned} \|\overline{\Omega}^j\| &\le \widetilde{\Omega}^j_\Gamma + 2\|(\widetilde{\Omega}^j_i + \widetilde{\Omega}^{I(j)}_i + \mathfrak{A}_{i,j}\aleph p h_j + \mathfrak{A}_{i,I(j)}\aleph_{I(j)})_{i\ge N+1}\|, \\ \|\underline{\Omega}^j\| &\le 3\widetilde{\Omega}^j_\Gamma + 4b\|(\widetilde{\Omega}^j_i + \widetilde{\Omega}^{I(j)}_i + \mathfrak{A}_{i,j}\aleph_j + \mathfrak{A}_{i,I(j)}\aleph_{I(j)})_{i\ge N+1}\| \end{aligned} \tag{6.32}$$

By (6.25), Lemma 6.3(c), (6.26) and (6.27)(c), $\limsup_{j\to\infty}\|(\widetilde{\Omega}^j_i + \widetilde{\Omega}^{I(j)}_i)_{i\ge N+1}\| < 2/16\rho$. By Lemma 6.11 and (6.27b), $\limsup_{j\to\infty}\|(\mathfrak{A}_{i,j}\aleph_j + \mathfrak{A}_{i,I(j)}\aleph_{I(j)})_{i\ge N+1}\| < \frac{1}{16}\rho$. Using this, (6.31) and (6.32) we conclude that

$$\limsup_{j\to\infty}\|\Omega^j\| < \rho.$$

Recall that the components of Ω^j are given by

$$\Omega^j_{\text{com}} = \left[\sum_{\nu=1}^{m} \|(w_j^\nu - w_{I(j)}^\nu)\, dz\big|_{T_\nu}\|_2^2\right]^{1/2}$$

$$\Omega^j_i = \left\|\left(\omega_j - \omega_{I(j)} - (\delta_{i,j} - \delta_{i,I(j)})(\phi_i)_*\left(\frac{1}{2\pi i}\frac{dz_1}{z_1}\right)\right)\big|_{Y_i'}\right\|_2, \qquad i \geq g+1.$$

Since

$$\sum_{\nu=1}^{m}\left\|\left(w_{I(j)}^\nu + \frac{\delta_{\nu,\nu_1(j)}}{2\pi i z} - \frac{\delta_{\nu,\nu_2(j)}}{2\pi i z}\right) dz\Big|_{T_\nu}\right\|_2^2$$
$$+ \sum_{i\geq g+1}\left\|\left(\omega_{I(j)} - \delta_{i,I(j)}(\phi_i)_*\left(\frac{1}{2\ pii}\frac{dz_1}{z_1}\right)\right)\Big|_{Y_i'}\right\|_2^2 < \infty$$

there is a g_0 such that

$$\limsup_{j\to\infty} \sum_{i\geq g_0+1}\left\|\left(\omega_j - \delta_{i,j}(\phi_i)_*\left(\frac{1}{2\pi i}\frac{dz_1}{z_1}\right)\right)\Big|_{Y_i'}\right\|_2^2 < \rho$$

and, since $\aleph_j$ is uniformly bounded,

$$\sup_{\nu,j}\left[\left\|\left(w_j^\nu + \frac{\delta_{\nu,\nu_1(j)}}{2\pi i z} - \frac{\delta_{\nu,\nu_2(j)}}{2\pi i z}\right) dz\Big|_{T_\nu}\right\|_2^2\right.$$
$$\left.+ \sum_{s\in S_\nu}\left\|\left(w_j^\nu - \sum_{\mu=1,2}\delta_{s,s_\mu(j)}\frac{(-1)^{\mu+1}}{2\pi i}\frac{1}{z-s}\right)\Big|_{\mathcal{A}(s)}\right\|_2^2\right] < \infty.$$

The Theorem follows by Proposition 6.14. □

REMARK 6.15. Choosing Γ sufficiently large we have by (6.30) and (6.31) that, for any compact subset $\mathcal{K}$ of X,

$$\sup_j \|\omega_j\big|_{\mathcal{K}}\|_2 < \infty.$$

Using Lemma 6.9(a) one sees that for each $x_0 \in X$ and each $i_0 \in \mathbb{N}$ with $\omega_{i_0}(x_0) \neq 0$ there is a neighbourhood U of x_0 in X such that

$$\sup_{x\in U, i\in\mathbb{N}}\left|\frac{\omega_i(x)}{\omega_{i_0}(x)}\right| < \infty.$$

We now state three corollaries of the preceding discussion. Applying Lemma 6.8, we have

PROPOSITION 6.16. *Let j be sufficiently large. Then for $z = (z_1, z_2) \in H(t_j)$ with $|z_1|, |z_2| \leq \frac{1}{4}$*

$$\left|\frac{\phi_j^*\omega_j(z)}{dz_1/z_1} - \frac{1}{2\pi i}\right| \leq \frac{2}{5\pi}(|z_1| + |z_2|) \leq \frac{1}{4\pi}$$

and for $i \neq j$,

$$\left|\frac{\phi_j^*\omega_i(z)}{\phi_j^*\omega_j(z)}\right| \leq 16\sqrt{\pi}(|z_1| + |z_2|)\|\omega_i\big|_{Y_j'}\|_2.$$

PROOF. The map

$$H(4t_j) \to Y_j, \quad (\zeta_1, \zeta_2) \mapsto \phi_j\left(\tfrac{1}{2}\zeta_1, \tfrac{1}{2}\zeta_2\right)$$

parametrizes the part of Y_j that lies betwen the curves $\phi_j(|z_1| = \frac{1}{2})$ and $\phi_j(|z_2| = \frac{1}{2})$. By hypothesis (GH3) this part of Y_j is contained in Y_j'. So we can apply Lemma 6.8 using the bounds of Theorem 6.4. $\square$

COROLLARY 6.17.

$$0 < \liminf_{j\to\infty} \frac{\|\omega_j\|_2^2}{|\log t_j|} < \limsup_{j\to\infty} \frac{\|\omega_j\|_2^2}{|\log t_j|} < \infty.$$

Furthermore

$$\lim_{j\to\infty} \frac{\|\omega_j|_{X\setminus Y_j'}\|_2^2}{|\log t_j|} = 0.$$

PROOF. By Theorem 6.4 for all sufficiently big j

$$\frac{1}{2}\sqrt{|\log t_j|} \leq \left\|(\phi_j)_*\left(\frac{1}{2\pi i}\frac{dz_1}{z_1}\right)\Big|_{Y_j'}\right\|_2 - \frac{1}{10\sqrt{\pi}}$$

$$\leq \|\omega_j|_{Y_j'}\|_2 \leq \left\|(\phi_j)_*\left(\frac{1}{2\pi i}\frac{dz_1}{z_1}\right)\Big|_{Y_j'}\right\|_2 + \frac{1}{10\sqrt{\pi}} \leq 2\sqrt{|\log t_j|}.$$

We show that this is the dominant contribution to $\|\omega_j\|_2$. By Theorem 6.4

$$\|\omega_j|_{\cup_{i\neq j} Y_i'}\|_2$$

is bounded uniformly in j. Now fix a compact subset $\mathcal{K}$ of X such that $X^{\mathrm{com}} \cup_{\nu=1}^m T_\nu \subset \mathcal{K}$. By Remark 6.13

$$\lim_{j\to\infty} \|\omega_j|_{\mathcal{K}}\|_2 = 0.$$

Furthermore by Proposition 6.14 for $\nu = 1, \dots, m$

$$\|\omega_j|_{X_\nu^{\mathrm{reg}}\setminus\mathcal{K}\cup_j Y_j'}\|_2$$

$$\leq \sum_{\mu=1,2} \delta_{\nu,\nu_\mu(j)} \left\| \frac{1}{2\pi i}\left(\frac{1}{z - s_\mu(j)} - \frac{1}{z}\right) dz\Big|_{\{z\in\mathbb{C}||z-s_\mu(j)|\geq R_\mu(j), |z|\geq \mathrm{const}\}}\right\|_2$$

$$+ \left\|\left(w_{j,\mathrm{com}}^\nu(z) + \sum_{s\in S_\nu} w_{j,s}^\nu\right) dz\Big|_{\{z\in G_\nu||z-s|\geq R(s) \text{ for all } s\in S_\nu\}}\right\|_2.$$

The first term is bounded by

$$\mathrm{const}\sqrt{\log R_1(j) + \log R_2(j) + \log|s_1(j)| + \log|s_2(j)|}.$$

So it is irrelevant by (GH5)(i), (GH5)(iv). By Proposition 6.14 the second term is bounded by

$$\left\|\left(\sum_{\substack{i,\mu\\ \nu_\mu(i)=\nu}} \frac{r_\mu(i)\Omega_i^j}{|z - s_\mu(i)|^2} + \frac{1}{|z|^2}\Omega_{\mathrm{com}}^j\right) dz\Big|_{\{z\in G_\nu||z-s|\geq R(s) \text{ for all } s\in S_\nu\}}\right\|_2.$$

Using Cauchy–Schwarz and the fact that $(\Omega_{\mathrm{com}}^j)^2 + \sum_i(\Omega_i^j)^2 \leq C$ we get a bound

$$\mathrm{const}\left(1 + \sum_{s\in S_\nu}\left\|\frac{r(s)}{|z-s|^2} d|z|\Big|_{|z-s|\geq R(s)}\right\|_2^2\right)^{1/2} \leq \mathrm{const}\left(1 + \sum_{s\in S_\nu}\frac{r(s)^2}{R(s)^2}\right)^{1/2}.$$

This last expression is bounded by Lemma 6.2. □

To conclude we state one more property of the forms ω_j. It will be used only in the proof of Torelli's theorem in Section 11.

For $j \geq g+1$ and $\nu = 1, \ldots, m$ define

$$\sigma_\nu(j) := \begin{cases} 0 & \text{if } \nu_1(j) \neq \nu, \nu_2(j) \neq \nu \\ (-1)^{\mu+1} s_\mu(j) & \text{if } \nu_\mu(j) = \nu, \nu_1(j) \neq \nu_2(j) \\ s_1(j) - s_2(j) & \text{if } \nu_1(j) = \nu_2(j) = \nu \end{cases}.$$

PROPOSITION 6.18. *Let $(x_n)_{n\in\mathbb{N}}$ be a sequence of points in X. Suppose that there is $i_0 \geq g+1$ such that $\omega_{i_0}(x_n) \neq 0$ for all $n \in \mathbb{N}$ and*

$$\sup_{i,n\in\mathbb{N}} \left| \frac{\omega_i(x_n)}{\omega_{i_0}(x_n)} \right| < \infty. \tag{6.33}$$

Assume furthermore that for each $i \geq g+1$ the limit $\lim_{n\to\infty} \omega_i(x_n)/\omega_{i_0}(x_n)$ exists, and that there is no $\nu \in \{1, \ldots, m\}$ such that

$$\lim_{n\to\infty} \frac{\omega_i(x_n)}{\omega_{i_0}(x_n)} = \frac{\sigma_\nu(i)}{\sigma_\nu(i_0)} \quad \textit{for all } i \geq g+1. \tag{6.34}$$

Then the sequence $(x_n)_{n\in\mathbb{N}}$ has an accumulation point.

To prepare for the proof we state

LEMMA 6.19. *For every $\nu = 1, \ldots, m$ and every $\widetilde{C} > 0$ the set*

$$\Bigg\{ z \in G_\nu \,\Bigg|\, \widetilde{C} \sum_{s\in S_\nu \nu} \frac{r(s)}{|z-s|^2} \geq \sup_{\substack{j\geq g+1 \\ \nu_1(j)=\nu_2(j)=\nu}} \left| \frac{1}{z-s_1(j)} - \frac{1}{z-s_2(j)} \right| + \sup_{\substack{j\geq g+1, \mu=1,2 \\ \nu_\mu(j)=\nu \\ \nu_1(j)\neq\nu_2(j)}} \left| \frac{1}{z-s_\mu(j)} - \frac{1}{z} \right| \text{ and } |z-s| \geq \tfrac{1}{4}R(s) \text{ for all } s \in S_\nu \Bigg\}$$

is compact in $\mathbb{C}$.

PROOF. For $z \in G_\nu$ with $|z-s| \geq \frac{1}{4}R(s)$ for all $s \in S_\nu$

$$\sum_{s\in S_\nu} \frac{r(s)}{|z-s|^2} \leq \frac{\text{const}}{|z|^2}$$

by (6.3). The Lemma follows from (GH6) and the facts that

$$\lim_{z\to\infty} |z|^2 \left| \frac{1}{z-s_1(j)} - \frac{1}{z-s_2(j)} \right| = |s_1(j) - s_2(j)| \quad \text{if } \nu_1(j) = \nu_2(j) = \nu,$$

$$\lim_{z\to\infty} |z|^2 \left| \frac{1}{z-s_\mu(j)} - \frac{1}{z} \right| = |s_\mu(j)| \quad \text{if } \nu_\mu(j) = \nu \text{ and } \nu_1(j) \neq \nu_2(j).$$

Observe that, if X_ν^{reg} is joined to other sheets by infinitely many handles then

$$\lim_{\substack{j,\mu \\ \nu_\mu(j)=\nu \nu_1(j)\neq\nu_2(j)}} |s_\mu(j)| = \infty$$

since S_ν is discrete. □

PROOF OF PROPOSITION 6.18. Suppose the sequence has no accumulation point. Now, Proposition 6.16 and (6.33) imply that the sequence meets only finitely many handles $\phi_j(\{(z_1, z_2) \in H(t_j) \mid |z_1|, |z_2| \leq \frac{1}{4}\})$. By going to a subsequence we may therefore assume that there is $\nu \in \{1, \ldots, \mathbf{m}\}$ and a sequence of points z_n with $|z_n - s| \geq \frac{1}{4}R(s)$ for all $s \in S_\nu$ such that $x_n = \Phi_\nu(z_n)$. Write

$$\Phi_\nu^*(\omega_j) = w_j(z)\, dz.$$

Clearly

$$\lim_{n\to\infty} z_n = \infty.$$

Put

$$\rho_j(z) = \frac{\delta_{\nu,\nu_1(j)}}{2\pi i}\left(\frac{1}{z - s_1(j)} - \frac{1}{z}\right) - \frac{\delta_{\nu,\nu_2(j)}}{2\pi i}\left(\frac{1}{z - s_2(j)} - \frac{1}{z}\right),$$

$$R(z) = \sum_{s\in S_\nu} \frac{r(s)}{|z - s|^2}.$$

By Theorem 6.4 there is $C > 0$ such that for $j \geq g + 1$

$$|w_j(z) - \rho_j(z)| \leq C|R(z)|.$$

Now fix any $\widetilde{C} > 0$. By (GH6) we may assume that none of the points z_n lies in

$$\{z \in \mathbb{C} \mid \widetilde{C}|R(z)| \geq \sup_{j\geq g+1} |\rho_j(z)|\}.$$

Therefore for each $n \in \mathbb{N}$ there is $j(n) \geq g + 1$ such that

$$|\rho_{j(n)}(z_n)| \geq \widetilde{C}|R(z_n)|$$

By assumption (6.33), there is a (possibly small) constant $K > 0$ such that for all $n \in \mathbb{N}$

$$\left|\frac{w_{j(n)}(z_n)}{w_{i_0}(z_n)}\right| \leq \frac{1}{K}$$

so that

$$|w_{i_0}(z_n)| \geq K\big(|\rho_{j(n)}(z_n)| - C|R(z_n)|\big) \geq K(\widetilde{C} - C)|R(z_n)|.$$

Since $|\rho_{i_0}(z_n)| \geq |w_{i_0}(z_n)| - C|R(z_n)|$ we have

$$|R(z_n)| \leq \frac{1}{K\widetilde{C} - (K+1)C}|\rho_{i_0}(z_n)| \quad \text{for all } n \in \mathbb{N}. \tag{6.35}$$

Hence for each j there is $n_0(\widetilde{C}, j)$ such that for $n \geq n_0(\widetilde{C}, j)$

$$\begin{aligned}
\left|\frac{w_j(z_n)}{w_{i_0}(z_n)} - \frac{\rho_j(z_n)}{\rho_{i_0}(z_n)}\right| &= \left|\frac{w_j(z_n)\rho_{i_0}(z_n) - \rho_j(z_n)w_{i_0}(z_n)}{w_{i_0}(z_n)\rho_{i_0}(z_n)}\right| \\
&\leq \frac{C|R(z_n)|\big|\,|\rho_{i_0}(z_n)| + |\rho_j(z_n)|\,\big|}{(|\rho_{i_0}(z_n)| - C|R(z_n)|)|\rho_{i_0}(z_n)|} \\
&\leq \frac{C}{K\widetilde{C} - (K+1)C}\,\frac{|\rho_{i_0}(z_n)|\,\big|\,|\rho_{i_0}(z_n)| + |\rho_j(z_n)|\,\big|}{(1 - C/K\widetilde{C} - (K+1)C)|\rho_{i_0}(z_n)|^2} \\
&= \frac{C}{K\widetilde{C} - (K+2)C}\,\big|\,|\rho_{i_0}(z_n)| + |\rho_j(z_n)|\,\big|\,\big|\rho_{i_0}(z_n)\big|
\end{aligned}$$

Here we used (6.35) twice. The inequality above holds for every $\widetilde{C}$. As

$$\rho_j(z) = \frac{1}{2\pi i}\frac{\sigma_\nu(j)}{z^2} + O\left(\frac{1}{|z|^3}\right) \text{ as } |z| \to \infty$$

the inequality above implies that

$$\lim_{n\to\infty}\frac{w_j(z_n)}{w_{i_0}(z_n)} = \lim_{n\to\infty}\frac{\rho_j(z_n)}{\rho_{i_0}(z_n)} = \frac{\sigma_\nu(j)}{\sigma_\nu(i_0)}$$

□

7. Zeroes of the Theta Function

We continue considering a Riemann surface $X = X^{\rm com}\cup X^{\rm reg}\cup X^{\rm han}$ that fulfills the hypotheses (GH1)–(GH6) of §5. Its period matrix

$$\mathcal{R} = (\mathcal{R}_{i,j}) = \int_{B_i}\omega_j$$

is symmetric and has positive definite imaginary part. Select any numbers $t_1,\ldots,$ $t_g \in (0,\frac{1}{2})$.

THEOREM 7.1. *The theta series*

$$\theta(\mathbf{z}) = \theta(\mathbf{z},\mathcal{R}) = \sum_{\substack{\mathbf{n}\in\mathbb{Z}^\infty\\ |\mathbf{n}|<\infty}} e^{2\pi i\langle\mathbf{z},\mathbf{n}\rangle}e^{\pi i\langle\mathbf{n},\mathcal{R}\mathbf{n}\rangle}$$

associated with the Riemann period matrix $\mathcal{R}$, *converges absolutely and uniformly on bounded subsets of the Banach space*

$$B = \left\{\mathbf{z} = (z_1,z_2,\ldots)\in\mathbb{C}^\infty\;\middle|\;\lim_{j\to\infty}\frac{z_j}{|\log t_j|} = 0\right\} \text{ with norm } \|\mathbf{z}\| = \sup_{j\geq 1}\frac{|z_j|}{|\log t_j|}$$

to an entire function that does not vanish identically.

This theorem is proven using (S.11). By way of preparation we show

LEMMA 7.2. *Let* ω *be a holomorphic* 1*-form on* $H(t) = \{(z_1,z_2)\in\mathbb{C}^2 \mid z_1z_2 = t,$ $|z_1|,|z_2|\leq 1\}$. *Then*

$$\|\omega\|_2 \geq \sqrt{\frac{|\log t|}{2\pi}}\left|\int_A\omega\right|$$

where $A = \{(z_1,z_2)\in H(t) \mid |z_1| = |z_2| = \sqrt{t}\}$.

PROOF. We use z_1 as coordinate on $H(t)$. Write

$$\omega = f(z_1)\,dz_1.$$

For any fixed r

$$\left|\int_A\omega\right|^2 = \left|\int_0^{2\pi}1\cdot f(re^{i\phi})re^{i\phi}\,d\phi\right|^2 \leq 2\pi\int_0^{2\pi}|rf(re^{i\phi})|^2\,d\phi.$$

Hence

$$\begin{aligned}\|\omega\|_2^2 = \frac{1}{2}\int_{t\leq|z_1|\leq 1}|f(z_1)|^2|\,dz_1\wedge d\bar{z}_1| &= \int_t^1\int_0^{2\pi}|rf(re^{i\phi})|^2\,d\phi\frac{dr}{r}\\ &\geq \frac{1}{2\pi}\left|\int_A\omega\right|^2\int_t^1\frac{dr}{r} = \frac{|\log t|}{2\pi}\left|\int_A\omega\right|^2.\end{aligned}$$

□

PROOF OF THEOREM 7.1. We show that the hypotheses of (S.11) are fulfilled. The condition that $\sum t_j^\beta < \infty$ for some $\beta < \frac{1}{2}$ is an immediate consequence of (GH2)(iv).

Choose nonintersecting representatives of the cycles $A_1, \ldots, A_g$ and let $U_1, \ldots, U_g$ be small tubular neghbourhoods of these representatives which are biholomorphic to some $H(t)$'s. By Lemma 7.2, if ω is any square integrable form on X, then

$$\|\omega\|_2^2 \geq \sum_{j=1}^{g} \|\omega\big|_{U_j}\|_2^2 + \sum_{j\geq g+1} \|\omega\big|_{Y_j}\|_2^2 \geq \frac{1}{2\pi}\left(\sum_{j=1}^{\infty} |\log t_j| \left|\int_{A_j} \omega\right|^2\right). \tag{7.1}$$

Now let

$$\mathbf{n} = \begin{pmatrix} n_1 \\ n_2 \\ \vdots \end{pmatrix}$$

be a vector in $\mathbb{Z}^\infty$ with only finitely many nonzero components. By the Riemann period relation (S.7) and formula (7.1)

$$\langle \mathbf{n}, \operatorname{Im} R\mathbf{n}\rangle = \left\|\sum_j n_j\omega_j\right\|_2^2 \geq \frac{1}{2\pi}\sum_{j\geq 1} |\log t_j| n_j^2. \qquad \square$$

In this section we consider the analog of the Abel–Jacobi map from X to its Jacobian and the zero set of the Theta function of the image of X. More precisely, let $\pi\colon \widetilde{X} \to X$ be the universal covering of X. The analog of the Abel–Jacobi map is the map from $\widetilde{X}$ to B described in part (ii) of

PROPOSITION 7.3.

(i) *Let $x_0 \in X$, ξ a local coordinate on a neighbourhood U of x_0 and write $\omega_j\big|_U = f_j(\xi)\,d\xi$. Then the map*
$$\xi \mapsto (f_1(\xi), f_2(\xi), \ldots)$$
is an analytic map from U to B.

(ii) *Let $\widetilde{x}_0 \in \widetilde{X}$. Then the map*
$$\widetilde{X} \to B, \quad \widetilde{x} \mapsto \left(\int_{\widetilde{x}_0}^{\widetilde{x}} \pi^*\omega_1, \int_{\widetilde{x}_0}^{\widetilde{x}} \pi^*\omega_2, \ldots\right)$$
is analytic.

PROOF. Let $x_0 \in X$. If U is chosen small enough then it follows from Remark 6.15 and Lemma 6.9(a) that

$$\sup_j \sup_{\xi\in U} |f_j(\xi)| < \infty.$$

Then the claims follow from the Weierstrass convergence theorem. $\square$

For any path γ in X we put

$$\int_\gamma \vec{\omega} = \left(\int_\gamma \omega_1, \int_\gamma \omega_2, \ldots\right).$$

By Proposition 7.3, $\int_\gamma \vec{\omega}$ lies in B for all compact γ and depends analytically on the end points of γ.

PROPOSITION 7.4. *There is a constant* const *such that for all systems of closed curves* $\Gamma = (\Gamma_1, \dots, \Gamma_m)$ *fulfilling the conclusions of Lemma* 6.1

$$\left\| \int_{\Gamma_\nu} |\vec{\omega}| \right\| \leq \text{const}\, \varepsilon$$

where $|\vec{\omega}| = (|\omega_1|, |\omega_2|, \dots)$.

PROOF. By Theorem 6.4 and Proposition 6.14 there exists a constant C and, for each j, constants Ω_s^j, $s \in \cup S_\nu$ (disjoint union) and Ω_{com}^j such that

$$\left| w_j^\nu(z) \frac{\delta_{\nu,\nu_1(j)}}{2\pi i} \left(\frac{1}{z - s_1(j)} \frac{1}{z} \right) \frac{\delta_{\nu,\nu_2(j)}}{2\pi i} \left(\frac{1}{z - s_2(j)} \frac{1}{z} \right) \right| \leq \frac{\Omega_{\text{com}}^j}{\text{dist}(z, T_\nu)^2} + \sum_{s \in S_\nu} \frac{\Omega_s^j r(s)}{|z - s|^2}$$

and

$$(\Omega_{\text{com}}^j)^2 + \sum_s (\Omega_s^j)^2 \leq C^2.$$

For all j,

$$\begin{aligned} \int_{\Gamma_\nu} \left| \frac{\Omega_{\text{com}}^j}{\text{dist}(z, T_\nu)^2} + \sum_{s \in S_\nu} \frac{\Omega_s^j r(s)}{|z - s|^2} \right| |dz| &\leq C \int_{\Gamma_\nu} \sqrt{\frac{1}{\text{dist}(z, T_\nu)^4} + \sum_{s \in S_\nu} \frac{r(s)^2}{|z - s|^4}}\, |dz| \\ &\leq \sqrt{2} C \varepsilon \end{aligned}$$

by (6.1b).

If $\nu_1(j) = \nu_2(j) = \nu$ then

$$\int_{\Gamma_\nu} \left| \frac{1}{z - s_1(j)} - \frac{1}{z - s_2(j)} \right| |dz| = \int_{\Gamma_\nu} \frac{|s_1(j) - s_2(j)|}{|z - s_1(j)||z - s_2(j)|} |dz| \leq \varepsilon |\log t_j|$$

by (6.1c). If $\nu = \nu_1(j) \neq \nu_2(j)$ then, by (6.1d)

$$\int_{\Gamma_\nu} \left| \frac{1}{z - s_1(j)} - \frac{1}{z} \right| |dz| = \int_{\Gamma_\nu} \frac{|s_1(j)|}{|z||z - s_1(j)|} |dz| \leq \varepsilon |\log t_j|.$$

This implies the Proposition. □

We choose standard paths to infinity in

LEMMA 7.5. *There is a point* $x^{(0)} \in X^{\text{com}}$ *and there are non self-intersecting paths* $P_\nu \colon [0, \infty) \to X$, $\nu = 1, \dots, m$ *such that* $P_\nu(0) = x^{(0)}$, P_ν *and* P_μ *intersect only at* $x^{(0)}$, $P_\nu(t)$ *lies in* $X_\nu^{\text{reg}} \setminus h_X^{-1}([0, t])$ *for all sufficiently big* t, $P_\nu^{-1}(X^{\text{reg}})$ *is connected and*

(7.2a) $$\lim_{t \to \infty} \sup_{\substack{j \\ \nu_1(j) = \nu_2(j) = \nu}} \frac{1}{|\log t_j|} \left| \log \frac{\Phi_\nu^{-1} P_\nu(t) - s_1(j)}{\Phi_\nu^{-1} P_\nu(t) - s_2(j)} \right| = 0,$$

(7.2b) $$\lim_{t \to \infty} \sup_{\substack{\nu_1(j) \neq \nu_2(j) \\ \nu_\mu(j) = \nu}} \frac{1}{|\log t_j|} \left| \log \left(1 - \frac{s_\mu(j)}{\Phi_\nu^{-1} P_\nu(t)} \right) \right| = 0,$$

(7.2c) $$\int_{\Phi^{-1} P_\nu([t_0, \infty))} \sqrt{\frac{1}{|z|^4} + \sum_{s \in S_\nu} \frac{r(s)^2}{|z - s|^4}}\, |dz| < \infty$$

for some t_0.

PROOF. Choose $\zeta_\nu \in G_\nu$ sufficiently large and obeying $|\zeta_\nu - s| \geq \frac{1}{4}R(s)$ for all $s \in S_\nu$. Join ζ_ν to ∞ by a straight line that goes radially outward. Replace each line segment that intersects one of the discs

$$\{|z-s| < \tfrac{1}{4}R(s)\}, \quad s \in S_\nu$$

by the shorter arc of the circle $\{|z-s| = \frac{1}{4}R(s)\}$. Call the resulting path $\widetilde{P}_\nu$.

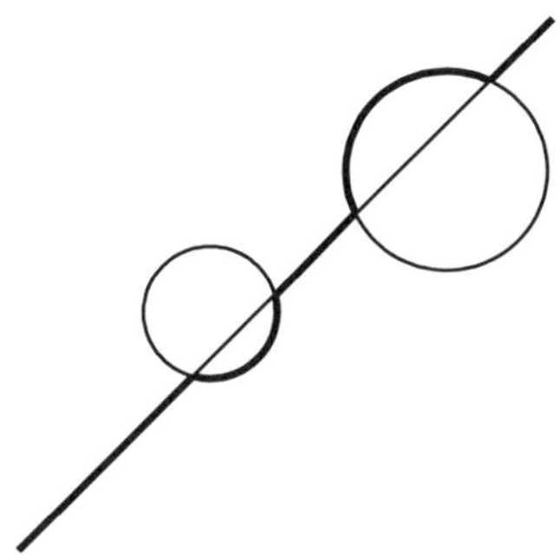

Choose any $x^{(0)} \in X^{\mathrm{com}}$. Connect $x^{(0)}$ to $\Phi_\nu(\zeta_\nu)$, $1 \leq \nu \leq m$ by paths that are nonself-intersecting and that intersect each other only at $x^{(0)}$. Call P_ν the path obtained by composing the path connecting $x^{(0)}$ to $\Phi_\nu(\zeta_\nu)$ with $\Phi_\nu(\widetilde{P}_\nu)$.

We now prove (7.2a). Pick any $z \in \widetilde{P}_\nu$ and let j obey $\nu_1(j) = \nu_2(j) = \nu$. We may assume without loss of generality that $|z - s_1(j)| \geq |z - s_2(j)|$. Using (GH5)(iii), we have

$$\begin{aligned}\frac{1}{|\log t_j|}\left|\log\frac{z-s_1(j)}{z-s_2(j)}\right| &\leq \frac{1}{|\log t_j|}\log\left(1+\left|\frac{s_1(j)-s_2(j)}{z-s_2(j)}\right|\right)\\ &\leq \frac{1}{|\log t_j|}\log\left(1+\frac{\mathrm{const}|s_2(j)|}{R_2(j)}\right) \leq \frac{\mathrm{const}(1+\delta)}{|\log t_j|}\log|s_2(j)|.\end{aligned}$$

This is independent of z and goes to zero as j goes to infinity, by (GH5)(iv). Furthermore, for each fixed j

$$\lim_{\substack{|z|\to\infty \\ z\in\widetilde{P}_\nu}} \log\left|\frac{z-s_1(j)}{z-s_2(j)}\right| = 0.$$

This proves (7.2a). The proof of (7.2b) is similar.

To prove (7.2c), notice that by (6.3)

$$\int_{\widetilde{P}_\nu}\sqrt{\frac{1}{|z|^4}+\sum_{s\in S_\nu}\frac{r(s)^2}{|z-s|^4}}\,|dz| \leq \mathrm{const}\int_{\widetilde{P}_\nu}\frac{1}{|z|^2}|dz| < \infty. \qquad \square$$

PROPOSITION 7.6. *For* $\nu = 1, \ldots, m$

$$\hat{e}_\nu := \lim_{t\to\infty}\int_{P_\nu([0,t])}\vec{\omega}$$

exists in B.

PROOF. We must show that

$$\lim_{t\to\infty}\sup_{t'>t}\left\|\int_{P_\nu([t,t'])}\vec{\omega}\right\| = 0.$$

As in the proof of Proposition 7.4

$$\left| w_j^\nu(z) \frac{\delta_{\nu,\nu_1(j)}}{2\pi i}\left(\frac{1}{z-s_1(j)}\frac{1}{z}\right)\frac{\delta_{\nu,\nu_2(j)}}{2\pi i}\left(\frac{1}{z-s_2(j)}\frac{1}{z}\right)\right| \leq \frac{\Omega^j_{\mathrm{com}}}{\mathrm{dist}(z,T_\nu)^2} + \sum_{s\in S_\nu}\frac{\Omega^j_s r(s)}{|z-s|^2}$$

with

$$(\Omega^j_{\mathrm{com}})^2 + \sum_s (\Omega^j_s)^2 \leq C^2.$$

The integral

$$\int_{\Phi_\nu^{-1}P_\nu([t,t'])}\left|\frac{\Omega^j_{\mathrm{com}}}{\mathrm{dist}(z,T_\nu)^2} + \sum_{s\in S_\nu}\frac{\Omega^j_s r(s)}{|z-s|^2}\right||dz|$$
$$\leq \int_{\Phi_\nu^{-1}P_\nu([t,t'])}\sqrt{\frac{1}{\mathrm{dist}(z,T_\nu)^4} + \sum_{s\in S_\nu}\frac{r(s)^2}{|z-s|^4}}\,|dz|$$

converges to zero as $t\to\infty$ by (7.2c).

If $\nu_1(j)=\nu_2(j)$ then

$$\int_{\Phi_\nu^{-1}P_\nu([t,t'])}\left(\frac{1}{z-s_1(j)}-\frac{1}{z-s_2(j)}\right)dz = \log\frac{z-s_1(j)}{z-s_2(j)}\Bigg|_{\Phi_\nu^{-1}P_\nu(t)}^{\Phi_\nu^{-1}P_\nu(t')}$$

so that, by (7.2a),

$$\lim_{t\to\infty}\sup_{t'>t}\sup_j\frac{1}{|\log t_j|}\left|\int_{\Phi_\nu^{-1}P_\nu([t,t'])}\left(\frac{1}{z-s_1(j)}-\frac{1}{z-s_2(j)}\right)dz\right| = 0.$$

Also by (7.2b), if $\nu_1(j)\neq\nu_2(j)$ and $\nu=\nu_1(j)$ then

$$\lim_{t\to\infty}\sup_{t'>t}\sup_j\frac{1}{|\log t_j|}\left|\int_{\Phi_\nu^{-1}P_\nu([t,t'])}\left(\frac{1}{z-s_1(j)}-\frac{1}{z}\right)dz\right| = 0. \qquad \square$$

LEMMA 7.7.

$$\lim_{\substack{j\to\infty\\ \nu_1(j)=\nu_2(j)}}\frac{1}{|\log t_j|}\log\left(\frac{|s_1(j)-s_2(j)|}{\min(R_1(j),R_2(j))}\right) = 0,$$

and

$$\lim_{\substack{j\to\infty\\ \nu_1(j)\neq\nu_2(j)}}\frac{1}{|\log t_j|}\left(\log\frac{|s_1(j)|}{R_1(j)} + \log\frac{|s_2(j)|}{R_2(j)}\right) = 0.$$

PROOF. We consider the case $\nu_1(j)=\nu_2(j)$. The other case is similar. By (GH5)(ii)

$$\min\big(|s_1(j)|^{-1-\delta},|s_2(j)|^{-1-\delta}\big) \leq \frac{|s_1(j)-s_2(j)|}{\min(R_1(j),R_2(j))} \leq 2\max\big(|s_1(j)|^{1+\delta},|s_2(j)|^{1+\delta}\big)$$

So the claim follows from (GH5)(iv). $\square$

PROPOSITION 7.8. *For every $\varepsilon>0$ and $T>0$ there is a $t>T$ such that for every point x in $X_\nu^{\mathrm{reg}}\cap h_X^{-1}\big([t,\infty)\big)$ with $|\Phi_\nu^{-1}(x)-s|\geq\frac14 R(s)$ for all $s\in S_\nu$ there exists a path γ in $X_\nu^{\mathrm{reg}}\cap h_X^{-1}\big([T,\infty)\big)$ with the following properties:*

(i) *γ joins x to a point of $P_\nu\big([0,\infty)\big)$.*

(ii) $\sup_j |1/|\log t_j|\int_\gamma \omega_j| < \varepsilon$.

PROOF. Let $\zeta \in G_\nu$ be sufficiently large and obey $|\zeta - s| \geq \frac{1}{4}R(s)$ for all $s \in S_\nu$. Join ζ to ∞ by a straight line that goes radially outward. Replace each line segment that intersects one of the discs

$$\{|z-s| < \tfrac{1}{4}R(s)\}, \quad s \in S_\nu$$

by the shorter arc of the circle $\{|z-s| = \frac{1}{4}R(s)\}$. Call the resulting path $\widetilde{\gamma}$.

Using the notation as in the proofs of Propositions 7.4, 7.6 we have for each $\rho > 0$

$$\int_{\widetilde{\gamma}_\zeta} \sum_{\substack{s \in S_\nu \\ |s|>\rho}} \frac{\Omega_s^j r(s)}{|z-s|^2} |dz| \leq \sum_{\substack{s \in S_\nu \\ |s|>\rho}} \Omega_s^j \frac{r(s)}{R(s)} \leq C \sqrt{\sum_{\substack{s \in S_\nu \\ |s|>\rho}} \frac{r(s)^2}{R(s)^2}}. \tag{7.3}$$

To see the the last inequality observe that

$$\int_{\widetilde{P}_\nu} \frac{1}{|z-s|^2} |dz| < \frac{\text{const}}{R(s)} \qquad \forall s \in S_\nu.$$

When $\widetilde{P}_\nu$ is a straight line this is obvious from

$$\int_{-\infty}^{\infty} \frac{dx}{x^2+\sigma^2} = \frac{\pi}{\sigma} \leq \frac{4\pi}{R(s)} \qquad \forall \sigma \geq \tfrac{1}{4}R(s).$$

The process of substituting circular arcs for line segments just changes the constant in the bound.

If ρ is sufficiently large, then the right-hand side of (7.3) is smaller than ε by Lemma 6.2. Furthermore

$$\int_{\widetilde{\gamma}_\zeta} \sum_{\substack{s \in S_\nu \\ |s|\leq\rho}} \frac{\Omega_s^j r(s)}{|z-s|^2} |dz| \leq C \int_{\widetilde{\gamma}_\zeta} \sum_{\substack{s \in S_\nu \\ |s|\leq\rho}} \frac{r(s)}{|z-s|^2} |dz|$$

goes to zero as $|\zeta| \to \infty$ uniformly in j.

Next, if $\nu_1(j) = \nu_2(j) = \nu$ then for any $\tau > 0$ and $j \geq g+1$

$$\begin{aligned} &\left| \frac{1}{\log t_j} \int_{\widetilde{\gamma}_\zeta([0,\tau])} \left(\frac{1}{z - s_1(j)} - \frac{1}{z - s_2(j)} \right) dz \right| \\ &\qquad \leq \frac{1}{|\log t_j|} \left| \log \frac{\zeta - s_1(j)}{\zeta - s_2(j)} - \log \frac{\widetilde{\gamma}_\zeta(\tau) - s_1(j)}{\widetilde{\gamma}_\zeta(\tau) - s_2(j)} \right| \\ &\qquad \leq \frac{1}{|\log t_j|} \left[2 \log \left(1 + \frac{4|s_1(j) - s_2(j)|}{\min(R_1(j), R_2(j))} \right) + 4\pi \right]. \end{aligned}$$

This goes to zero with j by Lemma 7.7. Also, for any fixed j with $\nu_1(j) = \nu_2(j)$

$$\lim_{|\zeta|\to\infty} \sup_{\tau>0} \left| \frac{1}{\log t_j} \int_{\widetilde{\gamma}_\zeta([0,\tau])} \left(\frac{1}{z - s_1(j)} - \frac{1}{z - s_2(j)} \right) dz \right| = 0.$$

Finally, if $\nu_1(j) = \nu$ with $\nu_1(j) \neq \nu_2(j)$ then for any $\tau > 0$

$$\left| \frac{1}{\log t_j} \int_{\widetilde{\gamma}_\zeta([0,\tau])} \left(\frac{1}{z - s_1(j)} - \frac{1}{z} \right) dz \right| \leq \frac{1}{|\log t_j|} \left[2 \log \left(1 + \frac{|4 s_1(j)|}{R_1(j)} \right) + 4\pi \right]$$

goes to zero with j by Lemma 7.7 and, for any fixed j as above

$$\lim_{|\zeta|\to\infty} \sup_{\tau>0} \left| \frac{1}{\log t_j} \int_{\widetilde{\gamma}_\zeta([0,\tau])} \left(\frac{1}{z - s_1(j)} - \frac{1}{z} \right) dz \right| = 0.$$

By the estimates above

$$\lim_{|\zeta|\to\infty} \sup_j \sup_{\tau>0} \left| \frac{1}{\log t_j} \int_{\widetilde{\gamma}_\zeta([0,\tau])} w_j^{(\nu)}(z)\, dz \right| = 0. \tag{7.4}$$

Given ε and T as in the proposition choose $t > T$ such that the expression of (7.5) is smaller than ε for all ζ with $h_X(\Phi_\nu(\zeta)) \geq t$ and $|\zeta - s| \geq \frac{1}{4}R(s)$ for all $s \in S_\nu$. Consider x such that $\zeta = \Phi_\nu^{-1}(x)$ has these properties. By Proposition 7.4 there exists $\Gamma = (\Gamma_1, \ldots, \Gamma_m)$ such that $x \in X(\Gamma)$, $h_X^{-1}([0,T]) \subset X(\Gamma)$ and

$$\left\| \int_\Gamma |\vec{\omega}| \right\| \leq \varepsilon.$$

Let $\Phi_\nu(\widetilde{\gamma}_\zeta(\tau))$ be an intersection point of $\Phi_\nu \widetilde{\gamma}_\zeta([0,\infty))$ with Γ_ν. Define γ as the composition of $\Phi(\widetilde{\gamma}_\zeta([0,\tau]))$ and a piece $\widehat{\Gamma}$ of Γ_ν that joins $\Phi_\nu(\widetilde{\gamma}_\zeta(\tau))$ to a point of $P_\nu([0,\infty))$. Since

$$\frac{1}{|\log t_j|} \left| \int_{\widehat{\Gamma}} \omega_j \right| \leq \left\| \int_{\widehat{\Gamma}} |\vec{\omega}| \right\|$$

we have

$$\sup_j \frac{1}{|\log t_j|} \left| \int_{\widehat{\Gamma}} \omega_j \right| \leq 2\varepsilon.$$

□

If x_1, x_2 are points of the same regular piece X_ν^{reg} and γ, γ' are paths in X_ν^{reg} joining x_1 to x_2 then the homology classes of γ and γ' differ by a linear combination of A-cycles. So, by the periodicity of the theta function, for any $e \in B$

$$\theta\Big(e + \int_\gamma \vec{\omega}\Big) = \theta\Big(e + \int_{\gamma'} \vec{\omega}\Big).$$

In this situation we define

$$\theta\Big(e + \int_{x_1}^{x_2} \vec{\omega}\Big) = \theta\Big(e + \int_\gamma \vec{\omega}\Big).$$

If x is a point of X_ν^{reg} and $e \in B$ we define $\theta(e + \int_{\infty_\nu}^x \vec{\omega})$ as follows: Choose an auxiliary point $x' = P_\nu(t')$ in X_ν^{reg} on the path $P_\nu([0,\infty))$ and put

$$\theta\Big(e + \int_{\infty_\mu}^x \vec{\omega}\Big) := \theta\Big(e - \hat{e}_\mu + \int_{w_\nu([0,t'])} \vec{\omega} + \int_{x'}^x \vec{\omega}\Big).$$

If $x'' = P_\nu(t'')$ is another choice of the auxiliary point then by construction the path $P_\nu([t', t'']) \subset X_\nu^{\text{reg}}$. As above one sees that the definition is independent of the choice of auxiliary point. Put

$$\theta\Big(e + \int_\infty^x \vec{\omega}\Big) := \theta\Big(e + \int_{\infty_1}^x \vec{\omega}\Big).$$

LEMMA 7.9. *For each $e \in B$ and $\nu = 1, \ldots, m$*

$$\lim_{|\Phi_\nu^{-1}(x)|\to\infty} \theta\Big(e + \int_\infty^x \vec{\omega}\Big) = \theta\Big(e + \hat{e}_\nu - \hat{e}_1\Big).$$

The limit is over all x in $\{x \in X_\nu^{\text{reg}} \mid |\Phi_\nu^{-1}(x) - s| \geq \frac{1}{4}R(s)\ \forall s \in S_\nu\}$

PROOF. Let $\varepsilon > 0$. By Propositions 7.8 and 7.6 respectively there exists, for each point of $\{x \in X_\nu^{\mathrm{reg}} \mid |\Phi_\nu^{-1}(x) - s| \geq \frac{1}{4}R(s)\ \forall s \in S_\nu\}$ with $\Phi_\nu^{-1}(x)$ sufficiently large, a path γ_x in X_ν^{reg} joining x to a point $P_\nu\big(t(x)\big)$ such that

$$\sup_j \frac{1}{|\log t_j|}\left|\int_{\gamma_x} \omega_j\right| < \varepsilon \quad \text{and} \quad \left\|\int_{P_\nu([t(x),\infty))} \vec{\omega}\right\| < \varepsilon.$$

We have

$$\theta\left(e + \int_\infty^x \vec{\omega}\right) = \theta\left(e + \hat{e}_\nu - \hat{e}_1 - \int_{P_\nu([t(x),\infty))} \vec{\omega} - \int_{\gamma_x} \vec{\omega}\right).$$

The claim now follows from the continuity of the theta function. □

We use the previous Lemma to discuss the integral of $d\log\theta(e + \int_\infty^x \vec{\omega})$ around the curves

$$a(s) = \Phi_\nu\big(\{z \in \mathbb{C} \mid |z - s| = R(s)\}\big) \quad s \in S_\nu$$

$a(s)$ is a special representative of $\pm A_j$, $s = s_1(j)$ or $s = s_2(j)$.

COROLLARY 7.10. *Let $e \in B$ and $\nu \in \{1, \ldots, m\}$ such that*

$$\theta(e - \hat{e}_1 + \hat{e}_\nu) \neq 0.$$

(i) *For all sufficiently large $s \in S_\nu$*

$$\int_{a(s)} d\log\theta\left(e + \int_\infty^x \vec{\omega}\right) = 0.$$

(ii) *For all curves $\Gamma = (\Gamma_1, \ldots, \Gamma_m)$ as in Lemma 6.1 with ε sufficiently small*

$$\int_{\Gamma_\nu} d\log\theta\left(e + \int_\infty^x \vec{\omega}\right) = 0.$$

PROOF. Whenever s is sufficiently large then, by the previous lemma

$$\left|\theta\left(e + \int_\infty^x \vec{\omega}\right) - \theta(e - \hat{e}_1 + \hat{e}_\nu)\right| < \frac{1}{2}|\theta(e - \hat{e}_1 + \hat{e}_\nu)|$$

for all $x \in a(s)$ so that there is a single valued branch of $\log\theta(e + \int_\infty^x \vec{\omega})$ defined on $a(s)$. The same argument applies to Γ_ν. □

We are now going to discuss zeroes of the theta function on X. Observe that, by the transformation properties of the theta function, for any two points $x_1, x_2 \in X$ and any two paths γ, γ' joining x_1 to x_2,

$$\theta\left(e + \int_\gamma \vec{\omega}\right) = 0 \iff \theta\left(e + \int_{\gamma'} \vec{\omega}\right) = 0.$$

In this case we say that $\theta(e + \int_{x_1}^{x_2} \vec{\omega}) = 0$. Otherwise we say $\theta(e + \int_{x_1}^{x_2} \vec{\omega}) \neq 0$. Observe that, in general, the actual value depends on the homology class of the joining path. Similarly we say that $\theta(e + \int_{\infty_\mu}^x \vec{\omega}) = 0$ if $\theta(e - \hat{e}_\mu + \int_{x(0)}^x \vec{\omega}) = 0$.

THEOREM 7.11. *Let $e \in B$ such that $\theta(e - \hat{e}_1 + \hat{e}_\nu) \neq 0$ for $\nu = 1, \ldots, m$. Then there is a compact subset K of X such that*

(i) *$\theta(e + \int_\infty^x \vec{\omega})$ has exactly* $\operatorname{genus}\big(X(\Gamma)\big)$ *zeroes in $X(\Gamma)$ if Γ is as in Lemma 6.1 with $K \subset X(\Gamma)$.*
(ii) *$\theta(e + \int_\infty^x \vec{\omega})$ has no zeroes in $\bigcup_\nu \{x \in X_\nu^{\mathrm{reg}} \setminus K \mid |\Phi_\nu^{-1}(x) - s| \geq \frac{1}{4}R(s)\ \forall s \in S_\nu\}$.*

(iii) *For each j such that the handle Y_j is contained in $X \setminus K$ there is exactly one zero x_j of $\theta(e + \int_\infty^x \vec{\omega})$ in Y_j. There is $\eta > 0$ such that for all such j*

$$x_j \in Y_j(\eta) := \phi_j\{(z_1, z_2) \in H(t_j) \mid |z_\mu| < |t_j|^\eta \text{ for } \mu = 1, 2\}.$$

PROOF. Part (ii) of the theorem is an immediate consequence of Lemma 7.9. We now discuss points of $Y_j' - Y_j(\eta)$ for a priori arbitrary η. Let $y = \phi(\xi_1, \xi_2)$ be such a point with, say, $|\xi_1| \geq \exp(-\eta|\log t_j|)$. Let σ be the image of the line segment

$$\{(z_1, z_2) \in H(t_j) \mid \arg z_1 = \arg \xi_1,\ |\xi_1| \leq |z_1| \leq 1\}$$

under ϕ_j and let $x := \phi_j(\xi_1/|\xi_1|, |\xi_1|\xi_2)$.

Let j be sufficiently large. Then, by Proposition 6.16

$$\sup_i \frac{1}{\log t_i} \int_\sigma |\omega_i| \leq \text{const}\, \eta$$

and, as x lies in one of the sets X_ν^{reg}, there is, by Propositions 7.8 and 7.6, a path γ_x joining x to a point $P_\nu\big(t(x)\big)$ such that

$$\sup_i \frac{1}{\log t_i} \int_{\gamma_x} |\omega_i| + \left\| \int_{P_\nu([t(x),\infty))} \vec{\omega} \right\| < \eta.$$

If η was chosen such that for all e' with $\sup_j |e_j'|/|\log t_j| \leq (\text{const} + 2)\eta$ one has $\theta(e - \hat{e}_1 + \hat{e}_\nu + e') \neq 0$ then

$$\theta\left(e + \int_\infty^y \vec{\omega}\right) = \theta\left(e - \hat{e}_1 + \hat{e}_\nu - \left(\int_{P_\nu([t(x),\infty))} \vec{\omega} + \int_{\gamma_x} \vec{\omega} + \int_\sigma \vec{\omega}\right)\right) \neq 0.$$

To count the number of zeroes of $\theta(e + \int_\infty^x \vec{\omega})$ we use

LEMMA 7.12. *Let $e \in B$ be such that $\theta(e - \hat{e}_1 + \hat{e}_\nu) \neq 0$ for $\nu = 1, \ldots, m$. Let X' be a compact submanifold with boundary in X containing X^{com} such that*

(i) *$X' \cap X_\nu^{\text{reg}}$ is connected for $\nu = 1, \ldots, m$.*
(ii) *$\partial X' \subset X_\nu^{\text{reg}}$.*
(iii) *$\int_{C'} d\log\theta(e + \int_\infty^x \vec{\omega}) = 0$ for each component C' of $\partial X'$.*
(iv) *Then $\theta(e + \int_\infty^x \vec{\omega})$ has precisely* genus(X') *zeroes in X'.*

PROOF. Let $C_1, \ldots, C_n$ be the components of $\partial X'$. There is $i_1 < i_2 < \cdots < i_p$ such that $A_{i_1}, \ldots, A_{i_p}, B_{i_1}, \ldots, B_{i_p}, C_1, \ldots, C_n$ generate the homology of X'. This basis can be represented by closed curves $a_{i_1}, \ldots, a_{i_p}, b_{i_1}, \ldots, b_{i_p}, C_1 \ldots, C_n$ in X' obeying $a_{i_k} \cap a_{i_l} = b_{i_k} \cap b_{i_l} = a_{i_k} \cap C_j = b_{i_k} \cap C_j = \varnothing$ for all k, l, j. Furthermore $a_{i_k} \cap b_{i_l} = \varnothing$ if $k \neq l$ and $a_{i_k} \cap b_{i_k}$ consists of exactly one point. Cutting up X' along this system of paths we get a Riemann surface $\widetilde{X}'$ whose homology is generated by $A_{i_j} B_{i_j} A_{i_j}^{-1} B_{i_j}^{-1}$, $1 \leq j \leq p$ and C_j, $1 \leq j \leq n$.

Each C_j is a closed curve in X_ν^{reg} and hence an integer linear combination of A cycles. Thus $\int_{C_j} \vec{\omega}$ is an integer linear combination of $\mathbb{1}_k$'s. Clearly $\int_{A_{i_j} B_{i_j} A_{i_j}^{-1} B_{i_j}^{-1}} \vec{\omega} = 0$. By the periodicity of the theta function, $\theta(e + \int_\infty^x \vec{\omega})$ is a single valued holomorphic function on $\widetilde{X}'$. Here, for $x \in X'$, the vector $\int_\infty^x \vec{\omega}$ is defined as $-\hat{e}_1 + \int_\gamma \vec{\omega}$, where γ is a path in $\widetilde{X}'$ joining $x^{(0)}$ to x. We now verify that, this definition coincides with the previous definition whenever $x \in X^{\text{reg}}$.

Let $P_\nu^{\text{com}} = P_\nu\big([0, \infty)\big) \cap X^{\text{com}}$. We can choose the representative curves a_{i_k}, b_{i_k} so that they do not intersect any P_ν^{com}. Also we can view $\widetilde{X}' \cap X_\nu^{\text{reg}}$ as follows.

Take $X' \cap X_\nu^{\rm reg}$, which by hypothesis is connected, and delete from it the curves b_{i_k} with $i_k \geq g+1$. Whenever $\nu_1(i_k) = \nu_2(i_k) = \nu$ this involves deleting a line segment joining the "holes" around $s_1(i_k)$ and $s_2(i_k)$. When $\nu = \nu_1(i_k) \neq \nu_2(i_k)$ this involves deleting a line segment joining the "hole" around $s_1(i_k)$ to $X^{\rm com}$. Consequently, $X' \cap X_\nu^{\rm reg} \setminus \{b_{i_k}\}$ is connected and we can choose, for any $x \in X' \cap X_\nu^{\rm reg}$ the path γ to consist of $P_\nu^{\rm com}$ composed with some path in $X' \cap X_\nu^{\rm reg} \setminus \{b_{i_k}\}$. This path was also allowed under the previous definition.

By Stokes' theorem the number of zeroes of $\theta(e + \int_\infty^x \vec\omega)$ in X' is

$$\frac{1}{2\pi i} \int_{\partial \tilde X'} d \log \theta\left(e + \int_\infty^x \vec\omega \right).$$

By the assumption of the temma, $\int_{C_j} d\log\theta\left(e + \int_\infty^x \vec\omega\right) = 0$ so we are left with

$$\begin{aligned}\frac{1}{2\pi i}\sum_{k=1}^{p}\Bigg\{ \int_{a_{i_k}} d\log\theta\left(e+\int_\infty^x \vec\omega\right) + \int_{b_{i_k}} d\log\theta\left(e+\int_\infty^x\vec\omega\right) \\ - \int_{a_{i_k}} d\log\theta\left(e+\int_\infty^x \vec\omega + R_{i_k}\right) - \int_{b_{i_k}} d\log\theta\left(e+\int_\infty^x\vec\omega + \mathbb{1}_{i_k}\right)\Bigg\} \\ = \frac{1}{2\pi i}\sum_{k=1}^{p}\int_{a_{i_k}} d\left[2\pi i\left(e_{i_k} + \int_\infty^x \omega_{i_k} + \frac{1}{2}R_{i_k i_k}\right)\right] = p. \quad \square\end{aligned}$$

To continue the proof of Theorem 7.11 observe that by Corollary 7.10 the surface $X(\Gamma)$ fulfils the hypothesis of Lemma 7.12 if Γ is sufficiently far out. This gives part (i) of the theorem. To prove the remaining part of (iii) recall that Y_j' is the cylinder in Y_j bounded by $a\big(s_1(j)\big)$ and $a\big(s_2(j)\big)$. Let Γ be far enough out that $Y_j \subset X(\Gamma)$, and put $X' := X(\Gamma) - Y_j'$. By Corollary 7.10 the hypotheses of Lemma 7.12 are fulfilled if j is big enough. Since $\operatorname{genus}(X') = \operatorname{genus}\big(X(\Gamma)\big) - 1$ there must be a zero of $\theta\left(e + \int_\infty^x \vec\omega\right)$ inside Y_j'. $\square$

REMARK 7.13. The proof shows that there exist $E > 0$, $\eta > 0$ and $j_0 > 0$ such that for all $j \geq j_0$ and all $\|e'\| < E$, there is exactly one zero x_j' of $\theta(e + e' + \int_\infty^x \vec\omega)$ in $Y_j(\eta)$. This is the unique zero of $\theta(e + e' + \int_\infty^x \vec\omega)$ in Y_j.

To compare the zero sets of $\theta(e + \int_\infty^x \vec\omega)$ and $\theta(e' + \int_\infty^x \vec\omega)$ for different e, e' we will use

LEMMA 7.14. *Let $x_{g+1}, x_{g+2}, \dots$ and $x'_{g+1}, x'_{g+2}, \dots$ be sequences of points in X such that $x_j, x_j' \in Y_j$ for all j and let γ_j be paths in Y_j connecting x_j to x_j'. Assume that there is an $\eta > 0$ such that $x_j, x_j' \in Y_j(\eta)$ for all sufficiently big j. Assume further that there is an increasing sequence of finite subsets J_n of $\{g+1, g+2, \dots\}$ such that $\cup_{n\geq 0} J_n = \{g+1, g+2, \dots\}$ and*

$$\lim_{n\to\infty} \sum_{j\in J_n} \int_{\gamma_j} \vec\omega$$

exists in B. Then

$$\left(0, \dots, 0, \int_{\gamma_{g+1}} \omega_{g+1}, \int_{\gamma_{g+2}} \omega_{g+2}, \dots\right) \in B.$$

PROOF. We may assume that $x_j, x_j' \in Y_j(\eta)$ and $\gamma_j \subset Y_j(\eta)$ for all j. By Proposition 6.16

$$\left|\int_{\gamma_j} \omega_i\right| \le 48\sqrt{\pi}\|\omega_i|_{Y_j'}\|_2 t_j^\eta \left|\int_{\gamma_j} \omega_j\right|$$

for all $i \neq j$. Therefore, for each finite subset J of $\{g+1, g+2, \dots\}$

$$\begin{aligned}
&\left\|\sum_{j\in J}\int_{\gamma_j} (\vec{\omega} - \mathbb{1}_j\omega_j)\right\| \\
&\qquad \le 48\sqrt{\pi}\sup_i\left(\frac{1}{|\log t_i|}\sum_{j\in J}\|\omega_i|_{Y_j'}\|_2 t_j^\eta\left|\int_{\gamma_j}\omega_j\right|\right) \\
&\qquad \le 48\sqrt{\pi}\sup_i\left(\frac{1}{|\log t_i|}\sum_{j\in J}\|\omega_i|_{Y_j'}\|_2 t_j^\eta|\log t_j|\right)\sup_{j\in J}\frac{1}{|\log t_j|}\left|\int_{\gamma_j}\omega_j\right| \\
&\qquad \le 48\sqrt{\pi}\sup_i\left(\sum_{j\in J}\|\omega_i|_{Y_j'}\|_2^2\right)^{1/2}\left(\sum_{j\in J} t_j^{2\eta}|\log t_j|^2\right)^{1/2}\left\|\sum_{j\in J}\int_{\gamma_j}\mathbb{1}_j\omega_j\right\|.
\end{aligned}$$

Here $\mathbb{1}_j$ denotes the vector $(0,\dots,0,1,0,\dots)$ of B. By Theorem 6.4 the first factor is bounded by a constant independent of J. By (GH2)(iv) there is for each $\varepsilon > 0$ an $N \in \mathbb{N}$ such that

$$\sum_{j\in J} t_j^{2\eta}|\log t_j|^2 < \varepsilon^2$$

whenever $J \subset \{N, N+1, \dots\}$. So we see that for every $\varepsilon > 0$ there is an $N \in \mathbb{N}$ such that for all finite subsets J of $\{g, g+1, \dots\}$

$$\left\|\sum_{\substack{j\in J\\ j\ge N}}\int_{\gamma_j}(\vec{\omega} - \mathbb{1}_j\omega_j)\right\| \le \varepsilon\left\|\sum_{\substack{j\in J\\ j\ge N}}\int_{\gamma_j}\mathbb{1}_j\omega_j\right\|. \tag{7.5}$$

Observe that we have not used the convergence of $\lim_{n\to\infty}\sum_{j\in J_n}\int_{\gamma_j}\vec{\omega}$ yet. We do so now. From (7.5) it follows that there is an N such that for all $n \ge N$ and for all $n' \ge n$

$$\begin{aligned}
\left\|\sum_{j\in J_{n'}\setminus J_n}\mathbb{1}_j\int_{\gamma_j}\omega_j\right\| &\le \left\|\sum_{j\in J_{n'}\setminus J_n}\int_{\gamma_j}\vec{\omega}\right\| + \left\|\sum_{j\in J_{n'}\setminus J_n}\int_{\gamma_j}(\vec{\omega} - \mathbb{1}_j\omega_j)\right\| \\
&\le \left\|\sum_{j\in J_{n'}\setminus J_n}\int_{\gamma_j}\vec{\omega}\right\| + \frac{1}{2}\left\|\sum_{j\in J_{n'}\setminus J_n}\mathbb{1}_j\int_{\gamma_j}\omega_j\right\|.
\end{aligned}$$

Therefore $\left(\sum_{j\in J_n}\mathbb{1}_j\int_{\gamma_j}\omega_j\right)_n$ is a Cauchy sequence in B and

$$\sum_{j=g+1}^{\infty}\mathbb{1}_j\int_{\gamma_j}\omega_j = \left(0,\dots,0,\int_{\gamma_{g+1}}\omega_{g+1},\int_{\gamma_{g+2}}\omega_{g+2},\dots\right)$$

lies in B. □

The proof of the Lemma above also yields the following result, which will be useful later.

LEMMA 7.15. *Let $x_{g+1}, x_{g+2}, \dots$ and $x'_{g+1}, x'_{g+2}, \dots$ be sequences of points in X such that $x_j, x'_j \in Y_j(\eta)$ for all j and let γ_j be paths in $Y_j(\eta)$ connecting x_j to x'_j. Assume further*

$$\left(0, \dots, 0, \int_{\gamma_{g+1}} \omega_{g+1}, \int_{\gamma_{g+2}} \omega_{g+2}, \dots\right) \in B.$$

Then

(i) $\sum_{j=g+1}^{\infty} \int_{\gamma_j} \vec{\omega}$ *converges in B.*

(ii) *For every $\varepsilon > 0$ there is an N, depending only on ε, η, not on the sequences x, x', such that for all $n \geq N$*

$$\left\| \sum_{j=n}^{\infty} \int_{\gamma_j} (\vec{\omega} - \mathbb{1}_j \omega_j) \right\| \leq \varepsilon \left\| \sum_{j=n}^{\infty} \int_{\gamma_j} \mathbb{1}_j \omega_j \right\|.$$

PROOF. The bound (7.5) is also true under the assumptions of this lemma. It implies that

$$\sum_{j=g+1}^{\infty} \int_{\gamma_j} (\vec{\omega} - \mathbb{1}_j \omega_j)$$

converges in B. Therefore

$$\sum_{j=g+1}^{\infty} \int_{\gamma_j} \vec{\omega} = \sum_{j=g+1}^{\infty} \mathbb{1}_j \int_{\gamma_j} \omega_j + \sum_{j=g+1}^{\infty} \int_{\gamma_j} (\vec{\omega} - \mathbb{1}_j \omega_j)$$

also converges. This proves part (i). Part (ii) is again a direct consequence of (7.5). □

THEOREM 7.16. *Let $e, e' \in B$ be such that*

$$\theta(e - \hat{e}_1 + \hat{e}_\nu) \neq 0, \quad \theta(e' - \hat{e}_1 + \hat{e}_\nu) \neq 0 \quad \textit{for } \nu = 1, \dots, m.$$

Let $x_1, x_2, \dots,$ resp. $x'_1, x'_2, \dots,$ be the zeroes of $\theta(e + \int_\infty^x \vec{\omega})$ resp., $\theta(e' + \int_\infty^x \vec{\omega})$ such that $x_j, x'_j \in Y_j$ for all sufficiently big j. Then there are paths γ_j joining x_j to x'_j such that $\gamma_j \subset Y'_j$ for all sufficiently large j

$$\left(\int_{\gamma_1} \omega_1, \int_{\gamma_2} \omega_2, \dots\right) \in B$$

and

$$e - e' = \sum_{j \geq 1} \int_{\gamma_j} \vec{\omega}.$$

To prepare for the proof, we construct a special exhaustion of X by compact submanifolds $X^{(n)}$. Namely, let $\Gamma^{(n)}$, n sufficiently big, be systems of curves as in Lemma 6.1 with $N = n$, $\varepsilon = 1/n$ such that, with $X^{(n)} = X(\Gamma^{(n)})$,

$$X^{(n)} \subset \operatorname{int} X^{(n+1)} \tag{7.6}$$

PROOF OF THEOREM 7.16. Denote $J^{(n)} = \{j \mid Y'_j \subset X^{(n)}\}$ and $C_\nu^{(n)} = P_\nu([0, \infty)) \cap X^{(n)}$. Cut $X^{(n)}$ open as follows and call the resulting surface $\widetilde{X}^{(n)}$. First cut along the $C_\nu^{(n)}$'s.

The result is a manifold with boundary $\sum_{\nu=1}^{m} C_\nu^{(n)} \partial X_\nu^{(n)} (C_\nu^{(n)})^{-1}$ and handles Y_j, $j \in J^{(n)}$. Choose curves a_j, b_j, $j \in J^{(n)}$ representing the A and B cycles in X'

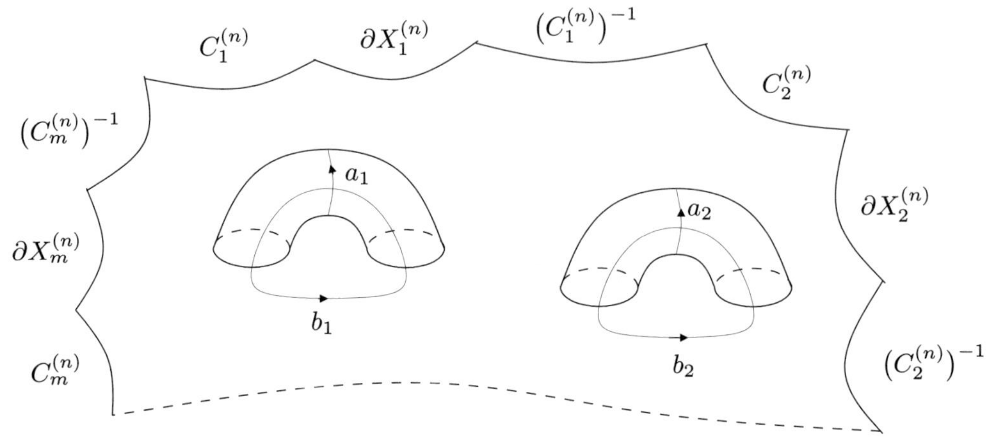

that obey $a_j \cap a_{j'} = b_j \cap b_{j'} = a_j \cap C_\nu^{(n)} = b_j \cap C_\nu^{(n)} = a_j \cap \partial X_\nu^{(n)} = b_j \cap \partial X_\nu^{(n)} = \varnothing$ for all j, j', ν. Furthermore $a_j \cap b_{j'} = \varnothing$ if $j \neq j'$ and $a_j \cap b_j$ consists of exactly one point. Cut along the a_j's and b_j's.

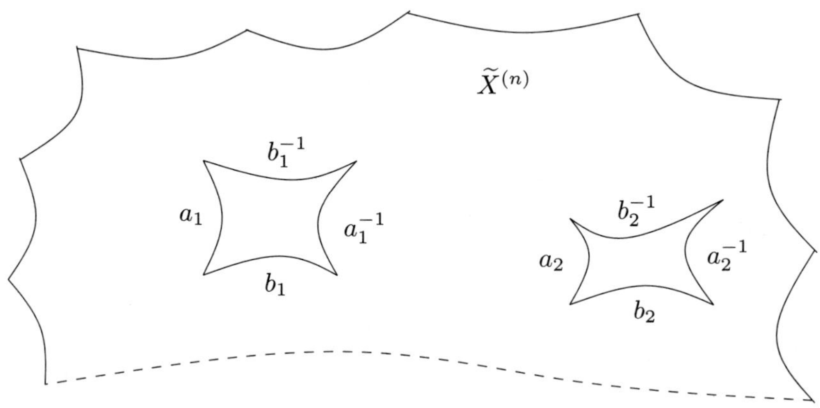

Then $\widetilde{X}^{(n)}$ has outer boundary given by the single curve $\sum_{\nu=1}^m C_\nu^{(n)} \partial X_\nu^{(n)} (C_\nu^{(n)})^{-1}$ and $|J^{(n)}|$ holes each bounded by a $A_j B_j A_j^{-1} B_j^{-1}$, $j \in J^{(n)}$.

We define the holomorphic function $u\colon \widetilde{X}^{(n)} \to B$ by

$$u(x) = \int_{x^{(0)}}^{x} \vec{\omega} - \hat{e}_1.$$

Since

$$\int_{A_j} \vec{\omega} + \int_{B_j} \vec{\omega} + \int_{B_j^{-1}} \vec{\omega} + \int_{A_j^{-1}} \vec{\omega} = 0,$$

$$\sum_{\nu=1}^{m} \int_{C_\nu^{(n)} \partial X_\nu^{(n)} (C_\nu^{(n)})^{-1}} \vec{\omega} = \int_{\partial X^{(n)}} \vec{\omega} = 0$$

the function u is single-valued on $\widetilde{X}^{(n)}$. So, by the residue theorem

$$\sum_{j \in J^{(n)}} u(x_j) = \frac{1}{\pi i} \int_{\partial \widetilde{X}^{(n)}} u(x) d \log \theta\big(e + u(x)\big).$$

A standard calculation like in the case of compact Riemann surfaces (see, e.g., [**Mu1**, p. 149]) shows that the contribution to the right-hand side from the holes $A_j B_j A_j^{-1} B_j^{-1}$ is

$$\sum_{j\in J^{(n)}}\left\{\int_{a_j}(u(x)+R_j)\omega_j-\frac{1}{2\pi i}R_j\int_{a_j}d\log\theta\big(e+u(x)\big)+\frac{1}{2\pi i}\mathbb{1}_j\int_{b_j}d\log\theta\big(e+u(x)\big)\right\}.$$

The contribution from the outer boundary is

$$\frac{1}{2\pi i}\int_{\partial X^{(n)}}u(x)d\log\theta\big(e+u(x)\big)-\sum_{\nu=1}^{m}1\frac{2\pi i}{\int}_{\partial X_\nu^{(n)}}\vec\omega\int_{C_\nu^{(n)}}d\log\theta\big(e+u(x)\big).$$

Substracting the corresponding result for $\sum_{j\in J^{(n)}}u(x'_j)$ we get

$$\begin{aligned}(7.7)\qquad &\sum_{j\in J^{(n)}}\big(u(x_j)-u(x'_j)\big)\\ &=\frac{1}{2\pi i}\sum_{j\in J^{(n)}}\left\{-R_j\int_{a_j}\big[d\log\theta\big(e+u(x)\big)-d\log\theta\big(e'+u(x)\big)\big]\right.\\ &\qquad\qquad\left.+\mathbb{1}_j\int_{b_j}\big[d\log\theta\big(e+u(x)\big)-d\log\theta\big(e'+u(x)\big)\big]\right\}\\ &\quad+\frac{1}{2\pi i}V_n+\frac{1}{2\pi i}W_n\end{aligned}$$

where

$$\begin{aligned}V_n&=\int_{\partial X^{(n)}}u(x)\left(d\log\theta\left(e+\int_\infty^x\vec\omega\right)-d\log\theta\left(e'+\int_\infty^x\vec\omega\right)\right)\\ W_n&=\sum_{\nu=1}^{m}\frac{1}{2\pi i}\int_{\partial X_\nu^{(n)}}\vec\omega\int_{C_\nu^{(n)}}\big(d\log\theta\big(e'+u(x)\big)-d\log\theta\big(e+u(x)\big)\big)\end{aligned}$$

We now show that both V_n and W_n converge to zero as $n\to\infty$. By Corollary 7.10(ii) $\log\theta(e+\int_\infty^x\vec\omega)$ and $\log\theta(e'+\int_\infty^x\vec\omega)$ are single valued functions on $\partial X^{(n)}$ whenever n is big enough. In this case we can apply partial integration and get

$$V_n=-\int_{\partial X^{(n)}}\left[\log\theta\left(e+\int_\infty^x\vec\omega\right)-\log\theta\left(e'+\int_\infty^x\vec\omega\right)\right]\vec\omega$$

yielding the bound

$$\|V_n\|\le\sup_{x\in\partial X^{(n)}}\left|\log\theta\left(e+\int_\infty^x\vec\omega\right)-\log\theta\left(e'+\int_\infty^x\vec\omega\right)\right|\left\|\int_{\partial X^{(n)}}|\vec\omega|\right\|.$$

By Lemma 7.9, the supremum remains bounded as $n\to\infty$. By Proposition 7.4, the integral converges to zero. Observe that $\partial X_\nu^{(n)}$ is a linear combination, with coefficients ±1 or 0, of A_j's with $j\notin J^{(n)}$. Thus $|\int_{\partial X_\nu^{(n)}}\omega_i|\le1$ and is zero if $i\in J^{(n)}$ so that

$$\lim_{n\to\infty}\left\|\int_{\partial X_\nu^{(n)}}\vec\omega\right\|=0.$$

By Lemma 7.9, the integral $\int_{C_\nu^{(n)}}\big(d\log\theta\big(e'+u(x)\big)-d\log\theta\big(e+u(x)\big)\big)$ is uniformly bounded and we conclude that

$$\lim_{n\to\infty}V_n=\lim_{n\to\infty}W_n=0\quad\text{in } B.$$

To continue the discussion of (7.7) put

$$m_j := \frac{1}{2\pi i} \int_{a_j} \big(d \log \theta\big(e + u(x)\big) - d \log \theta\big(e' + u(x)\big)\big).$$

Clearly $m_j \in \mathbb{Z}$. Since a_j is represented by one $a(s)$ for $j \geq g+1$ we have by Corollary 7.10(i)

$$m_j = 0 \quad \text{for all but finitely many } j.$$

Next

$$\begin{aligned} \frac{1}{2\pi i} \int_{b_j} &\big[d \log \theta\big(e + u(x)\big) - d \log \theta\big(e' + u(x)\big)\big] \\ &= \frac{1}{2\pi i}\big[\log \theta\big(e + u(x^{(j)}) + R_j\big) - \log \theta\big(e + u(x^{(j)})\big) \\ &\qquad - \log \theta\big(e' + u(x^{(j)}) + R_j\big) - \log \theta\big(e' + u(x^{(j)})\big)\big] + n_j \end{aligned}$$

with an integer n_j depending on the choices of the branches of the logarithms. By the "periodicity" rules for the theta function it is equal to

$$\frac{1}{2\pi i}\big[-\pi i\big(2u_j(x^{(j)}) + 2e_j + R_{jj}\big) + \pi i\big(2u_j(x^{(j)}) + 2e'_j + R_{jj}\big)\big] + n_j = e'_j - e_j + n_j.$$

Putting everything together we get

$$\sum_{j \in J^{(n)}} \big(u(x_j) - u(x'_j)\big) = \sum_{j \in J^{(n)}} \big(\mathbb{1}_j(e'_j - e_j) - m_j R_j + n_j \mathbb{1}_j\big) + \frac{1}{2\pi i} V_n + \frac{1}{2\pi i} W_n.$$

As $\widetilde{X}^{(n)}$ can be considered as a subset of $\widetilde{X}^{(n+1)}$ the integers m_j, n_j do not depend on n. Now choose paths γ_j from x_j to x'_j such that

$$\int_{\gamma_j} \vec{\omega} = u(x'_j) - u(x_j) - m_j R_j + n_j \mathbb{1}_j.$$

By Theorem 7.11, there is an $\eta > 0$ such that x_j and x'_j are in $Y_j(\eta)$ for all but finitely many j. As $m_j = 0$ for all but finitely many j, we can have $\gamma_j \subset Y_j(\eta)$ for all but finitely many j. Then

$$\sum_{j \in J^{(n)}} \int_{\gamma_j} \vec{\omega} = \sum_{j \in J^{(n)}} \mathbb{1}_j(e_j - e'_j) - \frac{1}{2\pi i} V_n - \frac{1}{2\pi i} W_n.$$

As $\lim_{n\to\infty} V_n = \lim_{n\to\infty} W_n = 0$ and $\lim_{n\to\infty} \sum_{j \in J^{(n)}} \mathbb{1}_j(e_j - e'_j) = e - e'$ the theorem now follows from Lemma 7.14. □

8. Riemann's Vanishing Theorem

We fix a point $\hat{e} \in B$ such that

$$\theta(\hat{e} - \hat{e}_1 + \hat{e}_\nu) \neq 0 \text{ for } \nu = 1, \dots, m.$$

Let $\widehat{x}_1, \widehat{x}_2, \dots$ be the zeroes of $\theta(\hat{e} + \int_\infty^x \vec{\omega})$ such that $\widehat{x}_j \in Y_j$ for all sufficiently large j. We wish to mimic the classical construction of the genus(X)-fold symmetric product of X using the points $\widehat{x}_j$ as reference points.

Denote by $\pi\colon \widetilde{X} \to X$ the universal cover of X and choose $\widetilde{x}_j \in \pi^{-1}(\widehat{x}_j)$. Put

$$Y''_j = \phi_j\big(\{(z_1, z_2) \in H(t_j) \mid |z_1|, |z_2| \leq \tfrac{1}{4}\}\big) \subset Y_j.$$

Furthermore let $\widetilde{Y}_j$ be a component of $\pi^{-1}(Y_j'')$ such that $\widetilde{x}_j \in \widetilde{Y}_j$, whenever $\widehat{x}_j \in Y_j''$. Let $\widetilde{Y}_j(\eta)$ be the preimage of $Y_j(\eta)$ in $\widetilde{Y}_j$. By abuse of notation we retain the symbol ω_j for the differential form $\pi^*(\omega_j)$ on $\widetilde{X}$. Put

$$W^{(-n)} := \Big\{ (y_{n+1}, y_{n+2}, \dots) \mid y_j \in \widetilde{X}, y_j \in \widetilde{Y}_j \text{ for all sufficiently big } j,\\ \Big(0,\dots,0, \int_{\widetilde{x}_{n+1}}^{y_{n+1}} \omega_{n+1}, \int_{\widetilde{x}_{n+2}}^{y_{n+2}} \omega_{n+2}, \dots \Big) \in B \Big\}.$$

The group $\mathfrak{S}$ of permutations of $\{r \in \mathbb{Z} \mid r \geq n+1\}$, that leave all but finitely many numbers fixed, acts naturally on $W^{(-n)}$. Put

$$S^{(-n)} := W^{(-n)}/\mathfrak{S}.$$

We call elements of $S^{(-n)}$ divisors of index n.

We give $W^{(-n)}$ the structure of a B–Banach manifold as follows. For $N \geq n+1$ put

$$W_N^{(-n)} := \{(y_j) \in W^{(-n)} \mid y_j \in \widetilde{Y}_j \text{ for } j \geq N\}.$$

The $W_N^{(-n)}$ form an exhaustion of $W^{(-n)}$ by open sets. The map

$$\begin{aligned} W_N^{(-n)} &\to \widetilde{X}^{N-n-1} \times \{e \in B \mid e_1 = \dots = e_{N-1} = 0\} \\ y &\mapsto \Big((y_{n+1}, \dots, y_{N-1}), \Big(0,0,\dots,0, \int_{\widetilde{x}_N}^{y_N} \omega_N, \int_{\widetilde{x}_{N+1}}^{y_{N+1}} \omega_{N+1}, \dots \Big)\Big) \end{aligned} \tag{8.1}$$

is injective, since, as we now show, the map

$$\widetilde{Y}_j \to \mathbb{C}, \quad y \mapsto \int_{\widetilde{x}_j}^{y} \omega_j$$

is biholomorphic to its image. First observe that the derivative of this map at y is $\omega_j(y)$, which is nonzero by Proposition 6.16. Furthermore

$$\begin{aligned} \pi_j \colon \{\xi \in \mathbb{C} \tfrac{1}{2} \log t_j + \log 2 \leq \mathrm{Re}\xi \leq \tfrac{1}{2} |\log t_j| - \log 2\} &\to Y_j'', \\ \xi &\mapsto \phi_j(\sqrt{t_j} e^{\xi}, \sqrt{t_j} e^{-\xi}) \end{aligned}$$

is the universal cover of Y_j''. Again, by Proposition 6.16, in $\widetilde{Y}_j$

$$\pi_j^*(\omega_j) = \frac{1}{2\pi i}(1+g)\, d\xi \text{ with } |g(\xi)| \leq \tfrac{1}{2}.$$

So for any two points $y, y' \in \widetilde{Y}_j$

$$\frac{1}{4\pi}|y-y'| \leq \Big| \int_{\widetilde{x}_j}^{y} \omega_j - \int_{\widetilde{x}_j}^{y'} \omega_j \Big| \leq \frac{3}{4\pi}|y-y'|. \tag{8.2}$$

Second, we verify that the image of the map (8.1) is open in the product of $\widetilde{X}^{N-n-1}$ and the appropriate subspace of B. Theorem 7.11(iii) implies that there exists an $\eta > 0$ such that for all sufficiently big j

$$|\,\mathrm{Re}\,\widetilde{x}_j| < (\tfrac{1}{2} - \eta)|\log t_j|.$$

Now let e be a point in the image of (8.1), and y its preimage. Since

$$\lim_{j\to\infty} \frac{1}{|\log t_j|} \Big| \int_{\widetilde{x}_j}^{y_j} \omega_j \Big| = 0$$

it follows from (8.2) that there is $\eta' > 0$ such that

$$|\operatorname{Re} y_j| < (\tfrac{1}{2} - \eta')|\log t_j|$$

for all j big enough. Using (8.2) again one sees that the image of (8.1) contains a neighbourhood of e.

So the maps (8.1) give coordinates on $W^{(-n)}$. The quotient $S^{(-n)} = W^{(-n)}/\mathfrak{S}$ inherits the structure of an B–Banach manifold as follows. Let $y = (y_{n+1}, \dots) \in W^{(-n)}$. Then there is $N \geq n+1$ such that for $j \geq N$ one has $y_i \in Y_j$ if and only if $i = j$. By (8.1) a neighbourhood of y in $W^{(-n)}$ is isomorphic to a neighbourhood U of its image in $\widetilde{X}^{N-n-1} \times B^{(-N+1)}$, where

$$B^{(-k)} := \{e \in B \mid e_j = 0 \text{ for } j \leq k\}.$$

The symmetric group $\mathfrak{S}_{N-n-1}$ acts by permutation on $\widetilde{X}^{N-n-1}$, and the quotient $\widetilde{X}^{N-n-1}/\mathfrak{S}_{N-n-1}$ is again a manifold (see [**GH**, p. 236]). Let U' be the image of U under the natural map

$$\widetilde{X}^{N-n-1} \times B^{(-N+1)} \to \widetilde{X}^{N-n-1}/\mathfrak{S}_{N-n-1} \times B^{(-N+1)}.$$

Then there is a homeomorphism from U' to a neighbourhood of the image of y in $S^{(-n)}$ such that the following diagram commutes

$$\begin{array}{ccc} \widetilde{X}^{N-n-1} \times B^{(-N+1)} \supset U & \longrightarrow & W^{(-n)} \\ \downarrow & & \downarrow \\ \widetilde{X}^{N-n-1}/\mathfrak{S}_{N-n-1} \times B^{(-N+1)} \supset U' & \longrightarrow & S^{(-n)}. \end{array}$$

Here the top row is the inverse map to (8.1). The maps constructed this way are coordinates on $S^{(-n)}$.

PROPOSITION 8.1.

(a) *The map*

$$\mu^{(-n)} \colon W^{(-n)} \to B, \quad y \mapsto \sum_{j=n+1}^{\infty} \int_{\widetilde{x}_j}^{y_j} \vec{\omega}$$

is holomorphic for each $n \geq g+1$.

(b) *Denote by P_n the projection*

$$\big(P_n(e)\big)_j = \begin{cases} e_j & \text{for } j \geq n+1 \\ 0 & \text{for } j \leq n \end{cases}$$

from B to $B^{(-n)}$. Then, for each η such that $\widetilde{x}_j \in \widetilde{Y}_j(\eta)$ for j big enough there is an $n_0 > 0$ such that for all $n > n_0$ the map

$$P_n \circ \mu^{(-n)} \colon \{y \in W^{(-n)} \mid y_j \in \widetilde{Y}_j(\eta) \text{ for } j \geq n+1\} \to B^{(-n)}$$

is injective.

(c) *At every point of $W^{(-n)}$ the derivative of $\mu^{(-n)}$ is Fredholm of index n. That is $d\mu^{(-n)}(g)$ has kernel of finite dimension and range of finite codimension with*

$$\operatorname{codim}\big(\operatorname{range} d\mu^{(-n)}(y)\big) - \dim\big(\ker d\mu^{(-n)}(y)\big) = n.$$

(d) *Let $\eta > 0$. If n is big enough, the derivative of $P_n \circ \mu^{(-n)}$ is boundedly invertible at every point of $\{y \in W^{(-n)} \mid y_j \in \widetilde{Y}_j(\eta) \text{ for } j \geq n+1\}$.*

COROLLARY 8.2. *$\mu^{(-n)}$ induces a holomorphic map $j^{(-n)}: S^{(-n)} \to B$ whose derivative is Fredholm of index n.*

PROOF OF PROPOSITION 8.1(a)–(b). (a) Fix any η such that $\widetilde{x}_j \in \widetilde{Y}_j(2\eta)$ for all j sufficiently large. For each $\widehat{y} \in W^{(-n)}$ there is an $N > 0$ and a neighbourhood U of $\widehat{y}$ such that $y_j \in \widetilde{Y}_j(\eta)$ for all $y \in U$, $j \geq N+1$. On U write

$$\mu^{(-n)}(y) = f(y) + \sum_{j=N+1}^{\infty} \mathbb{1}_j \int_{\widetilde{x}_j}^{y_j} \omega_j + g(y)$$

with

$$f(y) = \sum_{j=n+1}^{N} \int_{\widetilde{x}_j}^{y_j} \vec{\omega}, \quad g(y) = \sum_{j=N+1} \int_{\widetilde{x}_j}^{y_j} (\vec{\omega} - \mathbb{1}_j \omega_j).$$

By Proposition 7.3, f is a holomorphic map into B which is a function of $y_{n+1}, \dots,$ y_N only. By Lemma 7.15, for every $\varepsilon > 0$ there is an n_ε such that if $m > n > n_\varepsilon$

$$\left\| \sum_{j=n}^{m} \int_{\widetilde{x}_j}^{y_j} (\vec{\omega} - \mathbb{1}_j \omega_j) \right\| \leq \varepsilon \left\| \sum_{j=n}^{m} \int_{\widetilde{x}_j}^{y_j} \mathbb{1}_j \omega_j \right\|.$$

Thus g converges uniformly on bounded subsets of $W^{(-n)}$ and is holomorphic by the Weierstrass convergence theorem. In particular, if N was chosen big enough

$$\begin{aligned} \|g(y)\| &\leq \frac{1}{2} \left\| \sum_{j=N+1}^{\infty} \mathbb{1}_j \int_{\widetilde{x}_j}^{y_j} \omega_j \right\| \quad \forall y \in U, \\ \|g(y) - g(y')\| &\leq \frac{1}{2} \left\| \sum_{j=N+1}^{\infty} \mathbb{1}_j \int_{\widetilde{y}'_j}^{y_j} \omega_j \right\| \quad \forall y, y' \in U. \end{aligned} \tag{8.3}$$

As $y \to \sum_{j=N+1}^{\infty} \mathbb{1}_j \int_{\widetilde{x}_j}^{y_j} \omega_j$ gives (some of the) coordinates on U, $\mu^{(-n)}(y)$ is holomorphic on U.

(b) If n is big enough we can choose $N = n$. Then

$$P_n \circ \mu^{(-n)}(y) = \sum_{j=n+1}^{\infty} \mathbb{1}_j \int_{\widetilde{x}_j}^{y_j} \omega_j + P_n \circ g(y)$$

and the statement follows from (8.3). □

For the proofs of Proposition 8.1(c)–(d) we use the coordinates on $\widetilde{Y}_j$ defined just before (8.2). In these coordinates $\omega_i = w_i(y_j)\, dy_j$ with $|w_j(y_j) - 1| \leq \frac{1}{2}$. Pick any point $(y_{n+1}, y_{n+2}, \dots) \in W^{(-n)}$. Pick any coordinate patch in a neighbourhood of each of the finite number of the $y_{n+1}, y_{n+2} \dots$ that fail to lie in the correct handle. By abuse of notation let $\omega_i = w_i(y_j)\, dy_j$ be the representation of ω_i in these patches too. With respect to these coordinates the differential of $\mu^{(-n)}$ is

$$d\mu^{(-n)}(y) = [\vec{w}(y_{n+1})\vec{w}(y_{n+2}) \dots \vec{w}(y_j) \dots] \tag{8.4}$$

where $\vec{w}(y)$ is the column vector whose i^{th} row is $w_i(y)$. As we will use this derivative frequently, we state a more general lemma, only parts of which are needed in the proof of Proposition 8.1.

Let $\mathcal{L}(\mathcal{V}, \mathcal{V}')$ denote the set of bounded linear operators from the Banach space cv tothe Banach space $\mathcal{V}'$.

LEMMA 8.3. *Let $\mathcal{O}$ be an open subset of $\mathbb{C}^N$ and $\mathcal{O}'$ an open subset of $W^{(-n')}$ with $n' > n$. Let*

$$\vec{u}_j(x,y)\colon \mathcal{O}\times\mathcal{O}' \to B, \quad j = n+1,\dots,n'$$

be analytic maps. Define the analytic family of linear operators

$$D\colon \mathcal{O}\times\mathcal{O}' \to \mathcal{L}(B^{(-n)}, B)$$

by

$$D(x,y)\vec{\lambda} = \sum_{j=n+1}^{n'} \vec{u}_j(x,y)\lambda_i + \sum_{j=n'+1}^{\infty} \vec{w}(y_j)\lambda_j$$

where $\vec{w}(y_j)\,dy_j$ is the form $\vec{\omega}(y_j)$ represented in the local coordinates for $\widetilde{Y}_j$ defined just before (8.2). Then

(a) *For each $(x,y)\in\mathcal{O}\times\mathcal{O}'$, the operator $D(x,y)$ is Fredholm of index n. That is, $D(x,y)$ has finite-dimensional kernel and finite-codimensional range with*

$$\operatorname{codim}\bigl(\operatorname{range} D(x,y)\bigr) - \dim\bigl(\ker(x,y)\bigr) = n.$$

The range of $D(x,y)$ is closed.

(b) *Let $r = \min_{(x,y)\in\mathcal{O}\times\mathcal{O}'} \dim\bigl(\ker D(x,y)\bigr)$. The set on which $\dim\bigl(\ker D(x,y)\bigr) > r$ is an analytic variety of codimension least one.*

(c) *If $r \geq 1$ and $(x_0,y_0)\in\mathcal{O}\times\mathcal{O}'$ with*

$$\dim\bigl(\ker D(x_0,y_0)\bigr) = r = \min_{(x,y)\in\mathcal{O}\times\mathcal{O}'} \dim\bigl(\ker D(x,y)\bigr)$$

then there exist

$$\vec{\lambda}_1(x,y),\dots,\vec{\lambda}_r(x,y) \in \ker D(x,y)$$

that are independent and analytic in a neighbourhood of (x_0,y_0).

(d) *For any subset $\Sigma\subset\mathbb{N}$ let*

$$(P_{\mathbb{N}\setminus\Sigma}\vec{\lambda}) = \begin{cases}\lambda_j & j\notin\Sigma\\ 0 & j\in\Sigma\end{cases}$$

be viewed as a map from B to the Banach space $\{\vec{e}\in B \mid e_j = 0 \forall j\in\Sigma\}$. If $\dim\bigl(\ker D(x_0,y_0)\bigr) = 0$ then there exists $\Sigma\subset\mathbb{N}$ with $|\Sigma| = n$ such that $P_{\mathbb{N}\setminus\Sigma}D(x,y)$ is boundedly invertible for all (x,y) in a neighbourhood of (x_0,y_0).

(e) *If $r\geq 0$ and $(x_0,y_0)\in\mathcal{O}\times\mathcal{O}'$ with*

$$\dim\bigl(\ker D(x_0,y_0)\bigr) = r = \min_{(x,y)\in\mathcal{O}\times\mathcal{O}'} \dim\bigl(\ker D(x,y)\bigr)$$

then there exist

$$\ell_1(x,y),\dots,\ell_{n+r}(x,y)\in B^*$$

that are independent and analytic at each (x,y) in a neighbourhood of (x_0,y_0) such that

$$b\in\operatorname{range}\bigl(D(x,y)\bigr) \iff \langle\ell_i(x,y),b\rangle = 0 \quad \text{for } i = 1\dots n+r.$$

PROOF. (a) Pick a neighbourhood $\mathcal{N}$ of y sufficiently small and an integer $M > n'$ sufficiently large that $y_j'\in\widetilde{Y}_j(\eta)$ for all $j\geq M$ and $y'\in\mathcal{N}$ and such that

$$\sum_{j\geq M} t_j^{\eta}|\log t_j| \leq \frac{1}{192\sqrt{\pi}C} \tag{8.5}$$

where C is the constant in Theorem 6.4. Block

$$D = \begin{bmatrix} U & R \\ V & \mathbb{1} + S \end{bmatrix}$$

where the left half consists of columns $n+1 \ldots M$ and the upper half consists of rows $1 \ldots M$. First we show that

$$\|S\| \leq \tfrac{3}{4} \quad \text{for all } y' \in \mathcal{N}. \tag{8.6}$$

By Proposition 6.16

$$|w_i(y_j) - \delta_{ij}| \leq \begin{cases} \frac{1}{2} & \text{for } i = j \\ 48\sqrt{\pi} t_j^\eta \|\omega_i|_{Y_j'}\|_2 & \text{for } i \neq j \end{cases}.$$

Consequently, the operator norm of S is at most

$$\frac{1}{2} + 48\sqrt{\pi} \sup_i \left(\frac{1}{|\log t_i|} \sum_{j \geq M} t_j^\eta \|\omega_i|_{Y_j'}\|_2 |\log t_j| \right) \leq \frac{3}{4}.$$

Next we determine the range and kernel of $D(x,y)$ in terms of the $M \times (M-n)$ matrix $d(x,y) = U - R(\mathbb{1}+S)^{-1}V$. Note that the matrix elements of $d(x,y)$ are analytic functions of (x,y). We have

$$\begin{aligned} \ker D(x,y) &= \left\{ \begin{bmatrix} a \\ b \end{bmatrix} \middle| \begin{bmatrix} U & R \\ V & \mathbb{1}+S \end{bmatrix} \begin{bmatrix} a \\ b \end{bmatrix} = \begin{bmatrix} 0 \\ 0 \end{bmatrix} \right\} \\ &= \left\{ \begin{bmatrix} a \\ b \end{bmatrix} \middle| Ua + Rb = 0, Va + (\mathbb{1}+S)b = 0 \right\} \\ &= \left\{ \begin{bmatrix} a \\ b \end{bmatrix} \middle| b = -(\mathbb{1}+S)^{-1}Va, [U - R(\mathbb{1}+S)^{-1}V]a = 0 \right\}. \end{aligned} \tag{8.7}$$

In particular

$$\dim\bigl(\ker D(x,y)\bigr) = \dim\bigl(\ker d(x,y)\bigr).$$

Let $\Gamma(x,y)$ be a complementary subspace of the range of $d(x,y)$. The vector $\left[\begin{smallmatrix} \gamma \\ \delta \end{smallmatrix}\right]$ is the range of $D(x,y)$ if and only if

$$Ua + Rb = \gamma, \quad Va + (\mathbb{1}+S)b = \delta$$

has a solution, or equivalently if and only if

$$\begin{gathered} b = (\mathbb{1}+S)^{-1}(\delta - Va), \\ [U - R(\mathbb{1}+S)^{-1}V\ ]a = \gamma - R(\mathbb{1}+S)^{-1}\delta \end{gathered}$$

has a solution. But there always exists $\gamma' \in \Gamma$ and a such that

$$\gamma' + \ [U - R(\mathbb{1}+S)^{-1}V]a = \gamma - R(\mathbb{1}+S)^{-1}\delta.$$

Thus given any $\left[\begin{smallmatrix} \gamma \\ \delta \end{smallmatrix}\right] \in B$ there exists $\left[\begin{smallmatrix} a \\ b \end{smallmatrix}\right] \in B^{(-n)}$ and $\gamma' \in \Gamma(x,y)$ such that

$$\begin{bmatrix} \gamma \\ \delta \end{bmatrix} = \begin{bmatrix} \gamma' \\ 0 \end{bmatrix} + \begin{bmatrix} U & R \\ V & \mathbb{1}+S \end{bmatrix} \begin{bmatrix} a \\ b \end{bmatrix}.$$

Furthermore, if $\left[\begin{smallmatrix} \gamma \\ \delta \end{smallmatrix}\right] = \left[\begin{smallmatrix} 0 \\ 0 \end{smallmatrix}\right]$ then $\gamma' + [U - R(\mathbb{1}+S)^{-1}V]a = 0$ forces $\gamma' = 0$, since Γ is complementary to the range of $d(x,y)$. We have shown

$$B = \operatorname{range} D(x,y) \oplus \begin{bmatrix} \Gamma(x,y) \\ 0 \end{bmatrix}$$

and, in particular

$$\operatorname{codim}\big(\operatorname{range} D(x,y)\big) = \operatorname{codim}\big(\operatorname{range} d(x,y)\big).$$

From this the Fredholm condition

$$\begin{aligned}\operatorname{codim}\big(\operatorname{range} D(x,y)\big) - \dim\big(\ker D(x,y)\big)\\ &= \operatorname{codim}\big(\operatorname{range} d(x,y)\big) - \dim(\ker d(x,y))\\ &= M - \dim\big(\operatorname{range} d(x,y)\big) - \dim(\ker d(x,y))\\ &= M - (M-n)\\ &= n\end{aligned}$$

follows, since $d(x,y)$ is an $M \times (M-n)$ matrix.

We now verify that the range of $D(x,y)$ is closed. Let $\left[\begin{smallmatrix}\gamma_i\\ \delta_i\end{smallmatrix}\right]$ be a sequence of vectors in the range of $D(x,y)$ that converges to $\left[\begin{smallmatrix}\gamma\\ \delta\end{smallmatrix}\right]$. Then $\gamma_i - R(\mathbb{1}+S)^{-1}\delta_i$ is a sequence of vectors in the range of the finite matrix $[U - R(\mathbb{1}+S)^{-1}V]$ that converges to $\gamma - R(\mathbb{1}+S)^{-1}\delta$. Hence $\gamma - R(\mathbb{1}+S)^{-1}\delta$ is in the range of $[U - R(\mathbb{1}+S)^{-1}V]$ so that

$$\begin{aligned}b &= (\mathbb{1}+S)^{-1}(\delta - Va),\\ [U - R(\mathbb{1}+S)^{-1}V]a &= \gamma - R(\mathbb{1}+S)^{-1}\delta\end{aligned}$$

has a solution $\left[\begin{smallmatrix}a\\ b\end{smallmatrix}\right]$. As we have already observed, this implies that $\left[\begin{smallmatrix}\gamma\\ \delta\end{smallmatrix}\right]$ is in the range of $D(x,y)$.

(b) We have already observed that $\dim\big(\ker D(x,y)\big) = \dim\big(\ker d(x,y)\big)$ on a neighbourhood of any point $(x,y) \in \mathcal{O} \times \mathcal{O}'$ for a suitable finite matrix $d(x,y)$. The dimension of the range of $d(x,y)$ is strictly smaller than $M-n-r$ if and only if the determinant of every $(M-n-r)\times(M-n-r)$ minor of $d(x,y)$ is zero. As the matrix elements of $d(x,y)$ are analytic, we have shown that $\big\{(x,y) \mid \dim\big(\ker D(x,y)\big)\big\} > r$ is an analytic subvariety of $\mathcal{O} \times \mathcal{O}'$. By hypothesis this subvariety does not cover all $\mathcal{O} \times \mathcal{O}'$ and hence has codimension at least one.

(c) Suppose $\dim\big(\ker D(x_0,y_0)\big) = r \geq 1$. Then the corresponding $M \times (M-n)$ matrix $d(x_0,y_0)$ has an $(M-n-r)\times(M-n-r)$ minor with nonzero determinant. By renumbering rows and columns we may assume

$$d(x_0,y_0) = \begin{array}{c} \begin{array}{cc} M-n-r & \quad r \end{array} \\ \begin{bmatrix} v(x_0,y_0) & \sigma(x_0,y_0)\\ \tau(x_0,y_0) & u(x_0,y_0)\end{bmatrix} \end{array} \begin{array}{c} \\ M-n-r \\ n+r \end{array}$$

with $\det v(x_0,y_0) \neq 0$. By continuity, $\det v(x,y) \neq 0$ for (x,y) in a neighbourhood of (x_0,y_0). The vector $a = \left[\begin{smallmatrix}\vartheta\\ \psi\end{smallmatrix}\right]$ is in the kernel of $d(x,y)$ if and only if

$$\vartheta = -v(x,y)^{-1}\sigma(x,y)\psi, \quad [u(x,y) - \tau(x,y)v(x,y)^{-1}\sigma(x,y)]\psi = 0.$$

By hypothesis $d(x,y)$ has kernel of dimension at least r. As $v(x,y)$ is invertible its dimension must be exactly r. Thus all ψ's must satisfy $[u - \tau v^{-1}\sigma]\psi = 0$. Setting for $i = 1,\ldots,r$, $\psi_i = [0 \ldots 0\,1\,0 \ldots 0]^{t}$ with the one in the ℓth row, we have

$$\vartheta_i(x,y) = -v(x,y)^{-1}\sigma(x,y)\psi_i$$

and

$$\lambda_i(x,y) = \begin{bmatrix} -v(x,y)^{-1}\sigma(x,y)\psi_i \\ \psi_i \\ -(\mathbb{1} + S(x,y)^{-1}V(x,y)\begin{bmatrix} -v(x,y)^{-1}\sigma(x,y)\psi_i \\ \psi_i \end{bmatrix} \end{bmatrix}.$$

(d) Let $\dim\big(\ker D(x_0,y_0)\big) = 0$. Then by (8.7) the $M \times (M-n)$ matrix $U - R(\mathbb{1}+S)^{-1}V = d(x_0,y_0)$ has a trivial kernel. Consequently there exists $\Sigma \subset \{1\dots M\}$ with $|\Sigma| = n$ such that the matrix $\big(d(x,y)_{ij}\big)$ with indices $\{(i,j) \mid 1 \le i \le M, 1 \le j \le M-n, i \notin \Sigma\}$ is invertible for all (x,y) in a neighbourhood of (x_0,y_0). So given any $\left[\begin{smallmatrix}\gamma\\ \delta\end{smallmatrix}\right] \in \{\vec{e} \in B \mid e_j = 0\ \forall j \in \Sigma\}$ there exists a with

$$P_\Sigma d(x,y)a = \gamma - P_\Sigma R(\mathbb{1}+S)^{-1}\delta$$

and

$$\begin{aligned}\|a\| &\le C\|\gamma - P_\Sigma R(\mathbb{1}+S)^{-1}\delta\| \\ &\le C\big(\|\gamma\| + \|R\|\|(\mathbb{1}+S)^{-1}\|\|\delta\|\big) \\ &\le C\big(1 + \|R\|\|(\mathbb{1}+S)^{-1}\|\big)\,\|\left[\begin{smallmatrix}\gamma\\ \delta\end{smallmatrix}\right]\|.\end{aligned}$$

Define $b = (\mathbb{1}+S)^{-1}(\delta - Va)$ then

$$\begin{aligned}\|b\| &\le \|(\mathbb{1}+S)^{-1}\|(\|\delta\| + \|V\|\|a\|) \\ &\le \|(\mathbb{1}+S)^{-1}\|\big\{1 + C\|V\|\big(1 + \|R\|\|(\mathbb{1}+S)^{-1}\|\big)\big\}\,\|\left[\begin{smallmatrix}\gamma\\ \delta\end{smallmatrix}\right]\|\end{aligned}$$

and $\left[\begin{smallmatrix}a\\ b\end{smallmatrix}\right]$ obeys

$$P_\Sigma Ua + P_\Sigma Rb = \gamma, \quad Va + (\mathbb{1}+S)b = \delta$$

and

$$\|\left[\begin{smallmatrix}a\\ b\end{smallmatrix}\right]\| \le \text{const}\,\|\left[\begin{smallmatrix}\gamma\\ \delta\end{smallmatrix}\right]\|.$$

(e) By part (c) we may discard r columns of $D(x,y)$ in such a way that the resulting operator $\widetilde{D}(x,y)$ is injective and has the same range as $D(x,y)$ for all (x,y) in a neighbourhood of (x_0,y_0). (Discard the columns selected by the vectors $\left[\begin{smallmatrix}0 & \psi_i & 0\end{smallmatrix}\right]^t$, in the notation of part (c). The operator $\widetilde{D}(x,y)$ is still of the form specified in the statement of this Lemma and has Fredholm index $n+r$.

By part (d) there is a subset $\Sigma \subset \mathbb{N}$ with $|\Sigma| = n+r$ such that $P_{N\setminus\Sigma}\widetilde{D}(x,y)$ is boundedly invertible. Now

$$\begin{aligned} b \in \operatorname{range} D(x,y) &\implies b \in \operatorname{range}\widetilde{D}(x,y) \\ &\implies \exists\lambda \text{s.t.} b = \widetilde{D}(x,y)\lambda \\ &\implies \exists\lambda \text{s.t.} \begin{cases} P_{\mathbb{N}\setminus\Sigma}b &= P_{N\setminus\Sigma}\widetilde{D}(x,y)\lambda \\ P_\Sigma b &= P_\Sigma\widetilde{D}(x,y)\lambda \end{cases} \\ &\implies \exists\lambda \text{s.t.} \begin{cases} \lambda &= \big(P_{\mathbb{N}\setminus\Sigma}\widetilde{D}(x,y)\big)^{-1}P_{\mathbb{N}\setminus\Sigma}b \\ P_\Sigma b &= P_\Sigma\widetilde{D}(x,y)\lambda \end{cases} \\ &\implies P_\Sigma b - P_\Sigma\widetilde{D}(x,y)\big(P_{\mathbb{N}\setminus\Sigma}\widetilde{D}(x,y)\big)^{-1}P_{\mathbb{N}\setminus\Sigma}b = 0. \end{aligned}$$

The $n+r$ nonzero rows of $P_\Sigma\big(\mathbb{1} - \widetilde{D}(x,y)\big(\mathbb{P}_{\mathbb{N}\setminus\Sigma}\widetilde{D}(x,y)\big)^{-1}P_{\mathbb{N}\setminus\Sigma}\big)$ block

$$\begin{array}{ccc} & \Sigma & \mathbb{N}\setminus\Sigma \\ \Sigma & [\,\mathbb{1} & -\widetilde{D}(x,y)\big(P_{\mathbb{N}\setminus\Sigma}\widetilde{D}(x,y)\big)^{-1}\,] \end{array}$$

and give the dual vectors $\ell_1(x,y),\dots,\ell_{n+r}(x,y)$. □

PROOF OF PROPOSITION 8.1(c)–(d). Part (c) follows directly from Lemma 8.3(a) and (8.4). Finally, to prove part (d) observe that the derivative of $P_n \circ \mu^{(-n)}$ is the matrix of (8.4) with the first n rows deleted. As in (8.6), one shows that this matrix is of the form $\mathbb{1} + S$ with the operator norm of S bounded by $\frac{3}{4}$ whenever n is sufficiently large. □

Now fix lifts $\widetilde{P}_\nu$ of the paths P_ν under the covering map $\pi\colon \widetilde{X} \to X$ such that $\widetilde{P}_1(0) = \cdots = \widetilde{P}_m(0)$. For $x \in \widetilde{X}$ we define $\int_x^{\infty_\mu} \vec{\omega}$ as $\int_x^{\widetilde{P}_1(0)} \vec{\omega} + \int_{\widetilde{P}_\mu([0,\infty])} \vec{\omega}$. Again put $\int_x^{\infty} \vec{\omega} := \int_x^{\infty_1} \vec{\omega}$.

THEOREM 8.4. *Riemann's Vanishing Theorem For each* $(y_2, y_3, \dots) \in W^{(-1)}$

$$\theta\Big(\Big(\hat{e} - \int_{\tilde{x}_1}^{\infty} \vec{\omega}\Big) - \sum_{j=2}^{\infty} \int_{\tilde{x}_j}^{y_j} \vec{\omega}\Big) = 0.$$

Conversely, if $e \in B$ is a point such that $\theta(e) = 0$, but $\theta(e - \int_{\infty_\nu}^{x} \vec{\omega})$ is not identically zero for any $\nu = 1,\dots,m$, then there is $(y_2, y_3, \dots) \in W^{(-1)}$ such that

$$e = \Big(\hat{e} - \int_{\tilde{x}_1}^{\infty} \vec{\omega}\Big) - \sum_{j=2}^{\infty} \int_{\tilde{x}_j}^{y_j} \vec{\omega}. \tag{8.8}$$

Furthermore the sequence $\pi(y_2)$, $\pi(y_3), \dots$ is uniquely determined by (8.8) up to permutation. Finally, in this case for any $x \in X$ with $\theta(e - \int_{\infty_\nu}^{x} \vec{\omega}) \neq 0$ for $\nu = 1,\dots,m$ the points $x, \pi(y_2)$, $\pi(y_3)$, are just the zeroes (counted with multiplicity) of $\theta(e + \int_x^y \vec{\omega})$, considered as a multivalued function of y.

The rest of the chapter is devoted to the proof of this theorem. First we have

PROPOSITION 8.5. *Continuity of the roots of θ with respect to B*

(a) *Let $e \in B$ such that $\theta(e - \hat{e} + \hat{e}_\nu) \neq 0$ for $\nu = 1,\dots,m$. There is $y = (y_1, y_2, \dots) \in W^{(0)}$ such that*

$$\sum_{j=1}^{\infty} \int_{\tilde{x}_j}^{y_j} \vec{\omega} = \hat{e} - e \quad \textit{in } B$$

and $\pi(y_1)$, $\pi(y_2)\dots$ are the zeroes of $\theta(e + \int_\infty^x \vec{\omega})$.

(b) *Suppose that $y_i = y_j$ whenever $\pi(y_i) = \pi(y_j)$. Let $[y]$ be the image of y in $S^{(0)}$. Then there is a neighbourhood U of $[y]$ in $S^{(0)}$ and a neighbourhood V of $\hat{e} - e$ in B such that $j^{(0)}$ maps U biholomorphically onto V. For each $[y'] \in U$ the zeroes of $\theta(\hat{e} - j^{(0)}([y']) + \int_\infty^x \vec{\omega})$ are $\pi(y_1')$, $\pi(y_2'), \dots$.*

PROOF. (a) By Theorem 7.16 there are $\widetilde{\eta} > 0$, $y_1, y_2, \dots$ in $\widetilde{X}$ such that $y_j \in \widetilde{Y}_j(\widetilde{\eta})$ for all sufficiently big j and

$$\sum_{j=1}^{\infty} \int_{\tilde{x}_j}^{y_j} \vec{\omega} = \hat{e} - e \quad \text{in } B$$

and $\pi(y_1)$, $\pi(y_2)\dots$ are the zeroes of $\theta(e + \int_\infty^x \vec{\omega})$ and

$$\Big(\int_{\tilde{x}_1}^{y_1} \omega_1, \int_{\tilde{x}_1}^{y_2} \omega_2, \dots\Big) \in B.$$

(b) By Remark 7.13 and the first part of this proposition there are $E > 0$, $n > 0$, $\eta > 0$ such that for every $e' \in B$ with $\|e'\| < E$ there is $z(e') \in W^{(0)}$ such that $z_j(e') \in \widetilde{Y}_j(\eta)$ for $j \geq n$ and

$$\sum_{j=1}^{\infty} \int_{\widetilde{x}_j}^{z_j(e')} \vec{\omega} = \hat{e} - e - e' \tag{8.9}$$

$\pi\big(z_1(e')\big)$, $\pi\big(z_2(e')\big), \ldots$ are the roots of $\theta(e + e' + \int_{\infty}^{x} \vec{\omega})$. After possibly shrinking η and enlarging n we may assume that the conclusions of Proposition 8.1(d) hold. The point represented by $\big(z_1(e'), \ldots, z_n(e')\big) \in \widetilde{X}^n/\mathfrak{S}$ depends analytically on e'. Now (8.9) implies that

$$P_n\left(\sum_{j=n+1}^{\infty} \int_{\widetilde{x}_j}^{z_j(e')} \vec{\omega}\right) = P_n\left(\hat{e} - e - e' - \sum_{j=1}^{n} \int_{\widetilde{x}_j}^{z_j(e')} \vec{\omega}\right)$$

so by Proposition 8.1(d) $\big(z_{n+1}(e'), \ldots\big) \in W^{(-n)}$ also depends analytically on e'. So the map

$$\rho \colon \hat{e} - e - e' \mapsto \big(z_1(e'), z_2(e'), \ldots\big) \in S^{(0)}$$

is analytic. By (8.9)

$$j^{(0)} \circ \rho = id.$$

Since $dj^{(0)}$ is Fredholm of index 0 and $dj^{(0)} \circ d\rho = id$ this implies that $j^{(0)}$ is locally biholomorphic and has ρ as its inverse.

□

Proof of Theorem 8.4. Let U resp. V be neighbourhoods of $[\widetilde{x}]$ in $S^{(0)}$ resp. $\hat{e}$ in B such that $j^{(0)}$ maps U biholomorphically onto V. Then for each $[(y_1, y_2, \ldots)] \in U$ the points $\pi(y_1), \pi(y_2), \ldots)$ are zeroes of $\theta\big(\hat{e} - j^{(0)}([y]) + \int_{\infty}^{x} \vec{\omega}\big)$. In other words

$$\theta\left(\hat{e} - \sum_{j=1}^{\infty} \int_{\widetilde{x}_j}^{y_j} \vec{\omega} + \int_{\infty}^{y_k} \vec{\omega}\right) = 0 \quad \text{for all } k.$$

So

$$\theta\left(\left(\hat{e} - \int_{\widetilde{x}_k}^{\infty} \vec{\omega}\right) - \sum_{\substack{j=1 \\ j \neq k}}^{\infty} \int_{\widetilde{x}_j}^{y_j} \vec{\omega}\right) = 0$$

for all $y = (y_1, y_2, \ldots)$ in a neighbourhood of $\widetilde{x}$. Since $W^{(-1)}$ is a connected B-manifold it follows by analytic continuation that

$$\theta\left(\left(\hat{e} - \int_{\widetilde{x}_k}^{\infty} \vec{\omega}\right) - \sum_{\substack{j=1 \\ j \neq k}}^{\infty} \int_{\widetilde{x}_j}^{y_j} \vec{\omega}\right) = 0 \tag{8.10}$$

for all $y \in W^{(-1)}$. Putting $k = 1$ we get the first part of the theorem.

To prove the "Converse" part of Theorem 8.4, let $e \in B$ such that $\theta(e) = 0$, $\theta(e - \int_{\infty_\nu}^{x} \vec{\omega})$ is not identically zero for $\nu = 1, \ldots, m$. By assumption there exists $x_0 \in \widetilde{X}$ such that

$$\theta\left(e - \int_{\infty_\nu}^{x_0} \vec{\omega}\right) \neq 0 \quad \text{for } \nu = 1, \ldots, m.$$

Let $(y_1, y_2, \dots) \in W^{(0)}$ such that $\pi(y_1), \pi(y_2), \dots$ are the zeros of $\theta(e - \int_\infty^{x_0} \vec{\omega} + \int_\infty^{y} \vec{\omega})$ and

$$\sum_{j=1}^{\infty} \int_{\widetilde{x}_j}^{y_j} \vec{\omega} = \hat{e} - e + \int_\infty^{x_0} \vec{\omega}.$$

Without loss of generality we may assume that $y_1 = x_0$. Then

$$e = \hat{e} - \int_{\widetilde{x}_1}^{\infty} \vec{\omega} - \sum_{j=2}^{\infty} \int_{\widetilde{x}_j}^{y_j} \vec{\omega}.$$

Finally, to prove uniqueness let $(z_2, z_3, \dots) \in W^{(-1)}$ such that

$$e = \hat{e} - \int_{\widetilde{x}_1}^{\infty} \vec{\omega} - \sum_{j=2}^{\infty} \int_{\widetilde{x}_j}^{z_j} \vec{\omega}.$$

Then for $k = 2, 3, \dots$ and $u \in \widetilde{X}$

$$\begin{aligned}\theta\left(e + \int_{\widetilde{x}_1}^{u} \vec{\omega}\right) &= \theta\left(\left(\hat{e} - \int_{\widetilde{x}_k}^{\infty} \vec{\omega}\right) - \sum_{j \neq 1,k} \int_{\widetilde{x}_j}^{y_j} \vec{\omega} + \int_{y_k}^{u} \vec{\omega}\right) \\ &= \theta\left(\left(\hat{e} - \int_{\widetilde{x}_k}^{\infty} \vec{\omega}\right) - \sum_{j \neq 1,k} \int_{\widetilde{x}_j}^{z_j} \vec{\omega} + \int_{z_k}^{u} \vec{\omega}\right)\end{aligned}$$

Hence, by (8.10),

$$\theta\left(e + \int_{\widetilde{x}_1}^{u} \vec{\omega}\right) = \theta\left(\left(e - \int_{\infty}^{\widetilde{x}_1} \vec{\omega}\right) + \int_{\infty}^{u} \vec{\omega}\right)$$

is zero whenever $u \in \{y_2, y_3, \dots\}$ or $u \in \{z_2, z_3, \dots\}$ or $u = \widetilde{x}_1$. In the case that $\theta(e - \int_{\infty_\nu}^{\widetilde{x}_1} \vec{\omega}) \neq 0$ for $\nu = 1, \dots, m$ it follows from Theorem 7.11 that $\pi(y_2)$, $\pi(y_3), \dots$ and $\pi(z_2)$, $\pi(z_3), \dots$ coincide up to finite permutations. In the other case, choose $(\widetilde{x}_1', \widetilde{x}_2', \dots)$ in $W^{(0)}$ close to $(\widetilde{x}_1, \widetilde{x}_2, \dots)$ such that $\theta(e - \int_{\infty_\nu}^{\widetilde{x}_1'} \vec{\omega}) \neq 0$ for $\nu = 1, \dots, m$, and put $\hat{e}' := \mu^{(0)}(\widetilde{x}_1', \widetilde{x}_2', \dots)$. Then by Proposition 8.5(b)

$$\hat{e} = \hat{e}' - \sum_{j=1}^{\infty} \int_{\widetilde{x}_j'}^{\widetilde{x}_j} \vec{\omega}$$

and $\pi(\widetilde{x}_1')$, $\pi(\widetilde{x}_2'), \dots$ are the zeroes of $\theta(\hat{e}' + \int_\infty^{x} \vec{\omega})$. So

$$e = \hat{e}' - \int_{\widetilde{x}_1'}^{\infty} \vec{\omega} - \sum_{j=2}^{\infty} \int_{\widetilde{x}_j'}^{y_j} \vec{\omega} = \hat{e}' - \int_{\widetilde{x}_1'}^{\infty} \vec{\omega} - \sum_{j=2}^{\infty} \int_{\widetilde{x}_j}^{z_j} \vec{\omega}$$

and we can apply the previous argument with $\hat{e}'$ resp. $(\widetilde{x}_1', \widetilde{x}_2', \dots)$ as new base points. □

9. The Geometry of the Theta Divisor

We define the **Theta divisor** of a Riemann surface $X = X^{\text{com}} \cup X^{\text{reg}} \cup X^{\text{han}}$ fulfilling the hypotheses (GH1)–(GH6) of §5, as

$$\Theta := \{e \in B \mid \Theta(e) = 0\}.$$

Furthermore for a sequence $\nu_1, \nu_2, \dots, \nu_n$ of integers between 1 and m put

$$\Theta^{(\nu_1\nu_2\dots\nu_n)} := \Big\{ e \in \Theta \,\Big|\, \theta\Big(e - \int_{\infty_{\nu_1}}^{x_1} \vec{\omega} - \dots - \int_{\infty_{\nu_n}}^{x_n} \vec{\omega}\Big) = 0 \quad \text{for all } (x_1, \dots, x_n) \in \widetilde{X}^n \Big\},$$

$$\Theta_n := \bigcup_{1 \le \nu_1 \dots \nu_n \le m} \Theta^{(\nu_1 \dots \nu_n)}.$$

Taking the limit $x_j \to \infty_{\nu_j}$ for $j = k+1 \dots n$ we see that

$$\Theta^{(\nu_1 \dots \nu_k)} \subset \Theta^{(\nu_1 \dots \nu_{k+1} \dots \nu_n)}$$

for all $n \ge k \ge 0$.

Set

$$F_{(\nu_1 \dots \nu_{n-1})} \colon W^{(-n)} \to B$$

$$y \mapsto \hat{e} - \int_{\widetilde{x}_1}^{\infty} \vec{\omega} - \sum_{j=2}^{n} \int_{\widetilde{x}_j}^{\infty_{\nu_{j-1}}} \vec{\omega} - \sum_{j=n+1}^{\infty} \int_{\widetilde{x}_j}^{y_j} \vec{\omega}$$

$$= \hat{e} - \int_{\widetilde{x}_1}^{\infty} \vec{\omega} - \sum_{j=2}^{n} \int_{\widetilde{x}_j}^{\infty_{\nu_{j-1}}} \vec{\omega} - \mu^{(-n)}(y).$$

Since

$$\int_{\widetilde{x}_j}^{y_j} \vec{\omega} + \int_{\widetilde{x}_k}^{y_k} \vec{\omega} - \int_{\widetilde{x}_j}^{y_k} \vec{\omega} - \int_{\widetilde{x}_k}^{y_j} \vec{\omega} = \int_{y_k}^{y_j} \vec{\omega} + \int_{y_j}^{y_k} \vec{\omega} = 0$$

$F_{(\nu_1 \dots \nu_{n-1})}(y)$ is invariant under permutations of the y_j's. Let

$$f_{(\nu_1 \dots \nu_{n-1})} \colon S^{(-n)} \to B$$

be the map induced by $F_{(\nu_1 \dots \nu_{n-1})}$. In particular the maps $F = F_\varnothing \colon W^{(-1)} \to B$ and $f = f_\varnothing \colon S^{(-1)} \to B$ were the subject of Theorem 8.4, Riemann's vanishing theorem.

In the case of compact Riemann surfaces, F and f map $W^{(-1)}$ and $S^{(-1)}$ respectively onto Θ [**GH**, p. 546]. Furthermore by Riemann's singularity theorem $e \in \Theta_{\text{reg}}$ if and only if the fibre $F^{(-1)}(e)$ is discrete. Due to the noncompactness of X, these results do not carry over directly.

This is the reason why we introduced the subsets $\Theta^{(\nu_1, \dots, \nu_n)}$ of Θ and the maps $f_{(\nu_1 \dots \nu_{n-1})}$. First observe that, because we use the universal covering $\pi \colon \widetilde{X} \to X$ of X, the maps $f_{(\nu_1 \dots \nu_{n-1})}$ cannot possibly be injective. If

$$y_j = y'_j \quad \forall j \ge n+1, \ j \ne k, \ell,$$
$$\pi(y_k) = \pi(y'_k),$$
$$\pi(y_\ell) = \pi(y'_\ell)$$

with the projections $\pi(\gamma_k)$, $\pi(\gamma_\ell)$ to X of the paths

$$\gamma_k \colon [0,1] \to \widetilde{X} \quad \text{with } \gamma_k(0) = y_k, \ \gamma_k(1) = y'_k,$$
$$\gamma_\ell \colon [0,1] \to \widetilde{X} \quad \text{with } \gamma_\ell(0) = y_\ell, \ \gamma_\ell(1) = y'_\ell$$

obeying

$$\pi(\gamma_k) = -\pi(\gamma_\ell) \text{ with } H_1(X, \mathbb{Z})$$

then

$$f_{(\nu_1\dots\nu_{n-1})}([y]) = f_{(\nu_1\dots\nu_{n-1})}([y']). \tag{9.1}$$

Furthermore, if $y_j \neq y_k$, but $\pi(y_j) = \pi(y_k)$ for some $j \neq k$ then $df_{(\nu_1\dots\nu_{n-1})}([y])$ must have a nontrivial kernel since the jth and k^{th} columns $\vec{\omega}(y_j)$ and $\vec{\omega}(y_k)$ of $df_{(\nu_1\dots\nu_{n-1})}([y])$ are equal. On the other hand, given any $y \in S^{(-n)}$ we can find, using (9.1), a $z \in S^{(-n)}$ such that

$$z_j = z_k \text{ whenever } \pi(z_j) = \pi(z_k),$$
$$f_{(\nu_1\dots\nu_{n-1})}([y]) = f_{(\nu_1\dots\nu_{n-1})}([z]).$$

We shall see in Theorem 9.1, that the above discussion accounts for all the lack of injectiveness in $f_{(\nu_1\dots\nu_{n-1})}$ over $\Theta^{(\nu_1\dots\nu_{n-1})} \setminus \Theta_n$

In this chapter we discuss the geometry of the Theta divisor. We have already observed that

$$\Theta \supset \Theta_1 \supset \Theta_2 \supset \cdots .$$

We shall show in Theorem 9.1 that, for every $\nu_1 \dots \nu_{n-1}$, the map $f_{(\nu_1\dots\nu_{n-1})}$ almost provides a global parametrization of $\Theta^{(\nu_1\dots\nu_{n-1})} \setminus \Theta_n$. The map $f_{(\nu_1\dots\nu_{n-1})}$ suffers from the "trivial" lack of injectiveness discussed above. But, it is onto $\Theta^{(\nu_1\dots\nu_{n-1})} \setminus \Theta_n$ and does provide [Theorem 9.1(b)] local biholomorphisms to neighbourhoods of each $e \in \Theta^{(\nu_1\dots\nu_{n-1})} \setminus \Theta_n$.

In particular $\Theta^{(\nu_1\dots\nu_{n-1})} \setminus \Theta_n$ is connected and smooth of codimension n in B (despite the fact that $\theta(e - \int_{\infty_{\nu_1}}^{x_1} \vec{\omega} - \cdots - \int_{\infty_{\nu_{n-1}}}^{x_{n-1}} \vec{\omega}) \equiv 0$ contains, superficially infinitely many conditions) and the tangent space to $\Theta^{(\nu_1\dots\nu_{n-1})} \setminus \Theta_n$ at $e = f_{(\nu_1\dots\nu_{n-1})}([y])$ is spanned by the columns of $df_{(\nu_1\dots\nu_{n-1})}([y])$, provided we have choosen y so that $y_j = y_k$ whenever $\pi(y_j) = \pi(y_k)$. For example, if $\pi(y_j) \neq \pi(y_k)$ for all $j \neq k$, the tangent space is

$$\operatorname{span}\{\vec{\omega}(y_{n+1}), \vec{\omega}(y_{n+2}), \dots\}.$$

When the y_j's are not distinct the tangent space contains the derivatives $\vec{\omega}^{(n)}(y_j)$, $n < \#\{k \mid y_k = y_j\}$ too.

Since

$$f_{(\nu_1\dots\nu_{k-1})}([y]) = \lim_{\substack{y_j \to \infty_{\nu_{j-1}} \\ n+1 \leq j \leq k}} f_{(\nu_1\dots\nu_{n-1})}([y_{n+1}, \dots, y_k, y])$$

for all $k > n$, all points of $\Theta^{(\nu_1\dots\nu_{n-1})} \setminus \bigcap_{k=n+1}^{\infty} [\bigcup_{\nu_n=1}^{m} \cdots \bigcup_{\nu_{k-1}=1}^{m} \Theta^{(\nu_1\dots\nu_{k-1})}]$ are in the closure of $\Theta^{(\nu_1\dots\nu_{n-1})} \setminus \bigcup_{\nu_n=1}^{m} \Theta^{(\nu_1\dots\nu_n)}$. (While the statement of Theorem 9.1 only claims $\operatorname{range}(F_{(\nu_1\dots\nu_{n-1})}) \supset \Theta^{(\nu_1\dots\nu_{n-1})} \setminus \Theta_n$, the result

$$\operatorname{range}(F_{(\nu_1\dots\nu_{n-1})}) \supset \Theta^{(\nu_1\dots\nu_{n-1})} \setminus \bigcup_{\nu_n=1}^{m} \Theta^{(\nu_1\dots\nu_{n-1},\nu_n)}$$

is contained in the proof.)

The range of $f_{(\nu_1\dots\nu_{n-1})}$ can slop over into $\Theta^{(\nu_1\dots\nu_{n-1})} \cap \Theta_n$. By the discussion above, all singular points of $\Theta^{(\nu_1\dots\nu_{n-1})}$ lie in $\Theta^{(\nu_1\dots\nu_{n-1})} \cap \Theta_n$. In general, we are not able to give a description like the Riemann Singularity Theorem. See, however the discussion of the hyperelliptic case in the following section.

In Lemma 9.2 it is shown that, *even at nonsmooth points*, $\Theta^{(\nu_1\dots\nu_{n-1})}$ is of codimension at least n in B. For the nonhyperelliptic case, we show in Corollary 9.7

and Proposition 9.3 that Θ is smooth at all points of $\Theta \setminus (\Theta_1 \cap -\Theta_1)$ and that $\Theta_1 \cap (-\Theta_1)$ is of codimension 2 in Θ.

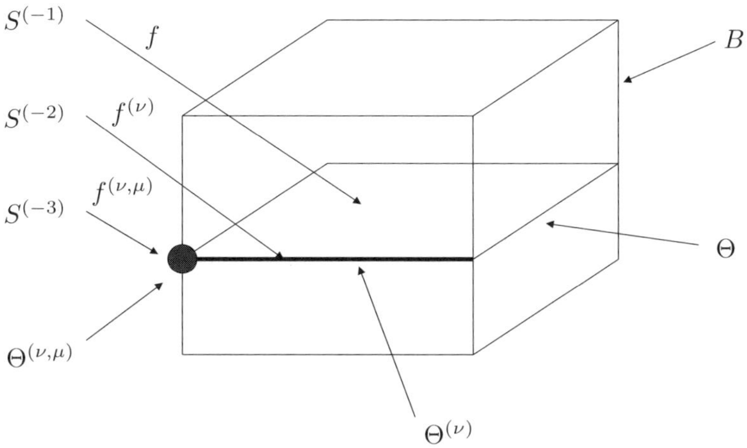

THEOREM 9.1. *The image of* $F_{(\nu_1\ldots\nu_{n-1})}$ *contains* $\Theta^{(\nu_1\ldots\nu_{n-1})} \setminus \Theta_n$ *and is contained in* $\Theta^{(\nu_1\ldots\nu_{n-1})}$. *If* $y \in W^{(-n)}$ *such that* $e = F_{(\nu_1\ldots\nu_{n-1})}(y) \in \Theta^{(\nu_1\ldots\nu_{n-1})} \setminus \Theta_n$ *then*

(a) *For any* $(x_1, \ldots, x_n) \in \widetilde{X}^n$ *such that*

$$\theta\left(e - \sum_{j=1}^{n-1} \int_{\infty_{\nu_j}}^{x_j} \vec{\omega} - \int_{\infty_\mu}^{x_n} \vec{\omega}\right) \neq 0 \quad \textit{for } \mu = 1, \ldots, m$$

the points $\pi(y_{n+1}), \pi(y_{n+2}), \ldots$ *are the zeroes of* $\theta\big(e - \sum_{j=1}^{n-1} \int_{\infty_{\nu_j}}^{x_j} \vec{\omega} + \int_{x_n}^{x} \vec{\omega}\big)$ *different from* $\pi(x_1), \ldots, \pi(x_n)$. *In particular* $\pi(y_{n+1}), \pi(y_{n+2}), \ldots$ *are uniquely determined by* e.

(b) *If in addition* $y_i = y_j$ *whenever* $\pi(y_i) = \pi(y_j)$ *then* $f_{(\nu_1\ldots\nu_{n-1})}$ *maps a neighbourhood of* $[y]$ *in* $S^{(-n)}$ *biholomorphically onto a neighbourhood of* e *in* $\Theta^{(\nu_1\ldots\nu_{n-1})}$

PROOF. If $y \in W^{(-n)}$ then for $(x_1, \ldots, x_{n-1}) \in \widetilde{X}^{(n-1)}$

$$\theta\left(F_{(\nu_1\ldots\nu_{n-1})}(y) - \int_{\infty_{\nu_1}}^{x_1} \vec{\omega} - \cdots - \int_{\infty_{\nu_{n-1}}}^{x_{n-1}} \vec{\omega}\right)$$
$$= \theta\left(\hat{e} - \int_{\tilde{x}_n}^{\infty} \vec{\omega} - \sum_{j=1}^{n-1} \int_{\tilde{x}_j}^{x_j} \vec{\omega} - \sum_{j=n+1}^{\infty} \int_{\tilde{x}_j}^{y_j} \vec{\omega}\right) = 0$$

by Riemann's Vanishing Theorem. This shows that the image of $F_{(\nu_1\ldots\nu_{n-1})}$ is contained in $\Theta^{(\nu_1\ldots\nu_{n-1})}$.

Now fix any $e \in \Theta_{(\nu_1\ldots\nu_{n-1})} \setminus \Theta_n$ and choose $(x_1, \ldots, x_n) \in \widetilde{X}^{(n)}$ such that $\Theta\big(e - \sum_{j=1}^{n-1} \int_{\infty_{\nu_j}}^{x_j} \vec{\omega} - \int_{\infty_\mu}^{x_n} \vec{\omega}\big) \neq 0$ for $\mu = 1, \ldots, m$. Then

$$e' = e - \sum_{j=1}^{n-1} \int_{\infty_{\nu_j}}^{x_j} \vec{\omega}$$

lies in $\Theta \setminus \Theta_1$ and we can apply Riemann's Vanishing theorem to it. So there is $y = (y_2, y_3, \dots) \in W^{(-1)}$ such that

$$e' = \hat{e} - \int_{\widetilde{x}_n}^{\infty} \vec{\omega} - \sum_{j=1}^{n-1} \int_{\widetilde{x}_j}^{y_{j+1}} \vec{\omega} - \sum_{j=n+1}^{\infty} \int_{\widetilde{x}_j}^{y_j} \vec{\omega}$$

and such that $\pi(x_n), \pi(y_2), \pi(y_3), \pi(y_4), \dots$ are the zeroes of $\theta(e' + \int_{x_n}^{x} \vec{\omega})$. As for $k = 1, \dots, n-1$

$$\theta\left(e' + \int_{x_n}^{x_k} \vec{\omega}\right) = \theta\left(e - \sum_{\substack{j=1 \\ j \neq k}}^{n-1} \int_{\infty_{\nu_j}}^{x_j} \vec{\omega} - \int_{\infty_{\nu_k}}^{x_n} \vec{\omega}\right) = 0$$

we may assume that $x_j = y_{j+1}$ for $j = 1, \dots, n-1$. So

$$e - \sum_{j=1}^{n-1} \int_{\infty_{\nu_j}}^{x_j} \vec{\omega} = e' = \hat{e} - \sum_{j=1}^{n-1} \int_{\widetilde{x}_j}^{x_j} \vec{\omega} - \int_{\widetilde{x}_n}^{\infty} \vec{\omega} - \sum_{j=n+1}^{\infty} \int_{\widetilde{x}_j}^{y_j} \vec{\omega}$$

or

$$e = F_{(\nu_1 \dots \nu_{n-1})}\big((y_{n+1}, y_{n+2}, \dots)\big)$$

If $e = F_{(\nu_1, \dots, \nu_{n-1})}\big((y'_{n+1}, y'_{n+2}, \dots)\big)$ then

$$e' = \hat{e} - \int_{\widetilde{x}_n}^{\infty} \vec{\omega} - \sum_{j=1}^{n-1} \int_{\widetilde{x}_j}^{x_j} \vec{\omega} - \sum_{j=n+1}^{\infty} \int_{\widetilde{x}_j}^{y'_j} \vec{\omega}.$$

Hence, by Riemann's Vanishing Theorem applied to e', the sequences $\pi(y_{n+1})$, $\pi(y_{n+2}), \dots$ and $\pi(y'_{n+1}), \pi(y'_{n+2}), \dots$ agree up to finite permutations. This proves part (a) of the Theorem. Part (b) follows from (a), as in Proposition 8.5(b), since the inverse map to $f_{(\nu_1 \dots \nu_{n-1})}$ is locally given by taking roots of $\Theta(e' + \int_{x_n}^{x} \vec{\omega})$. □

Next we point out that Θ_n has codimension at least n in Θ, that is codimension at least $n+1$ in B. A precise formulation of this fact is

LEMMA 9.2. *Let U be an open subset of $\mathbb{C}^N \times B^{(-n)}$ for some $N, n \in \mathbb{N}$, $g \colon U \to B$ a holomorphic map such that*

- *$dg(y)$ has bounded inverse and* codim range $dg(y) = k$ *for all $y \in U$.*
- *$g(U) \subset \Theta^{(\nu_1, \dots, \nu_n)}$ for some $\nu_j \in \{1, \dots, m\}$.*

Then $k \geq n+1$.

PROOF. Consider the map

$$\tilde{g} \colon U \times \widetilde{X}^n \to B,$$

$$(y; x_1, \dots, x_n) \mapsto g(y) - \int_{\infty_1}^{x_1} \vec{\omega} - \cdots - \int_{\infty_n}^{x_n} \vec{\omega}.$$

Since $g(U) \subset \Theta^{(\nu_1, \dots, \nu_n)}$ we have $\theta\big(\tilde{g}(y; x_1, \dots, x_n)\big) \equiv 0$. On the other hand the range of the differential $d\tilde{g}$ at $(y; x_1, \dots, x_n)$ is

$$\operatorname{range} dg(y) + \mathbb{C}\vec{\omega}(x_1) + \cdots + \mathbb{C}\vec{\omega}(x_n).$$

Since the image of X under the canonical map κ defined in §4 is not contained in a hyperplane one can choose $x_1, \dots, x_n$ so that

$$\operatorname{codim}\big(\operatorname{range} dg(y) + \mathbb{C}\vec{\omega}(x_1) + \cdots + \mathbb{C}\vec{\omega}(x_n)\big) \leq \max(0, k-n).$$

If $k \leq n$ then range $d\tilde{g}(y; x, \dots, x_n) = B$, so that the image of $\tilde{g}$ contains a neighbourhood of $\tilde{g}(y; x_1, \dots, x_n)$. Therefore Θ vanishes on an open subset of B. This implies that $\theta \equiv 0$, in contradiction to Theorem 7.1. $\square$

Since the Theta function is even, $\Theta = -\Theta$. We use this to show that for non hyperelliptic surfaces $\Theta_1 \cap (-\Theta_1)$ has codimension at least two in Θ.

PROPOSITION 9.3. *Assume that X is not hyperelliptic. Let U be an open nonempty subset of $\Theta^{(\nu)} \setminus \Theta_2, \nu = 1, \dots, m$. Then there is no $\mu \in \{1, \dots, m\}$ such that $(-U) \subset \Theta^{(\mu)} \setminus \Theta_2$.*

PROOF. Suppose that $(-U) \subset \Theta^{(\mu)} \setminus \Theta_2$. By Theorem 9.1 for every $e \in U$ there are $(y_3, y_4, y_5, y_6, \dots)$ and $(z_3, z_4, z_5, z_6 \dots)$ such that

$$\begin{aligned} e &= \hat{e} - \int_{\tilde{x}_1}^{\infty_\nu} \vec{\omega} - \int_{\tilde{x}_2}^{\infty} \vec{\omega} - \sum_{j=3}^{\infty} \int_{\tilde{x}_j}^{y_j} \vec{\omega} \\ &= -\left(\hat{e} - \int_{\tilde{x}_1}^{\infty_\mu} \vec{\omega} - \int_{\tilde{x}_2}^{z_2} \vec{\omega} - \int_{\tilde{x}_3}^{\infty} \vec{\omega} - \sum_{j=4}^{\infty} \int_{\tilde{x}_j}^{z_j} \vec{\omega}\right). \end{aligned}$$

By shrinking U we may assume without loss of generality that $(y_3, y_4, y_5, y_6, \dots)$ and $(z_3, z_4, z_5, x_6, \dots)$ run over open simply connected sets N resp. N' with the $\pi(y_j)$'s all distinct and the $\pi(z_j)$'s distinct. So, for each $y \in N$, there exists $z = z(y) \in N$ such that

$$\begin{aligned} F\big(z_2(y), y_3, y_4 \dots\big) &= \hat{e} - \int_{\tilde{x}_1}^{\infty_\nu} \vec{\omega} - \int_{\tilde{x}_2}^{z_2(y)} \vec{\omega} - \sum_{j=3}^{\infty} \int_{\tilde{x}_j}^{y_j} \vec{\omega} \\ &= -\left(\hat{e} - \int_{\tilde{x}_1}^{\infty_\mu} \vec{\omega} - \int_{\tilde{x}_2}^{\infty} \vec{\omega} - \int_{\tilde{x}_3}^{\infty} \vec{\omega} - \sum_{j=4}^{\infty} \int_{\tilde{x}_j}^{z_j(y)} \vec{\omega}\right) \in -\Theta^{(\mu,1)} \end{aligned}$$

by Theorem 9.1. The derivative of $f\big(z_2(y), y_3, y_4 \dots\big)$, as a map from the tangent space of $\mathcal{N}$ to B is Fredholm of index 2. If this derivative were injective its range would have codimension 2, violating Lemma 9.2, where it is shown that $\Theta^{(\mu,1)}$ has codimension 3. Therefore the derivative of the map $f\colon S^{(-1)} \to \Theta$ has a nontrivial kernel at every point $\big(z_2(y), y_3, y_4 \dots\big)$ with $(y_3, y_4 \dots) \in \mathcal{N}$.

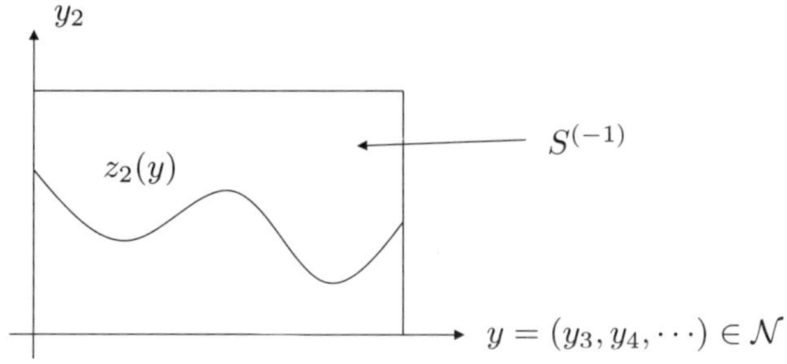

The first step in deriving a contradiction is to show that, on (a possibly shrunken) $\mathcal{N}$

$$(9.2) \qquad \begin{aligned} &\vec{\omega}\big(z_2(y)\big) \in \overline{\operatorname{span}\{\vec{\omega}(y_3), \vec{\omega}(y_4) \dots\}} && \text{if } z_2(y) \neq y_j \text{ for all } j \geq 3, \\ &\dot{\vec{\omega}}\big(z_2(y)\big) \in \overline{\operatorname{span}\{\vec{\omega}(y_3), \vec{\omega}(y_4) \dots\}} && \text{if } z_2(y) = y_j \text{ for some } j \geq 3. \end{aligned}$$

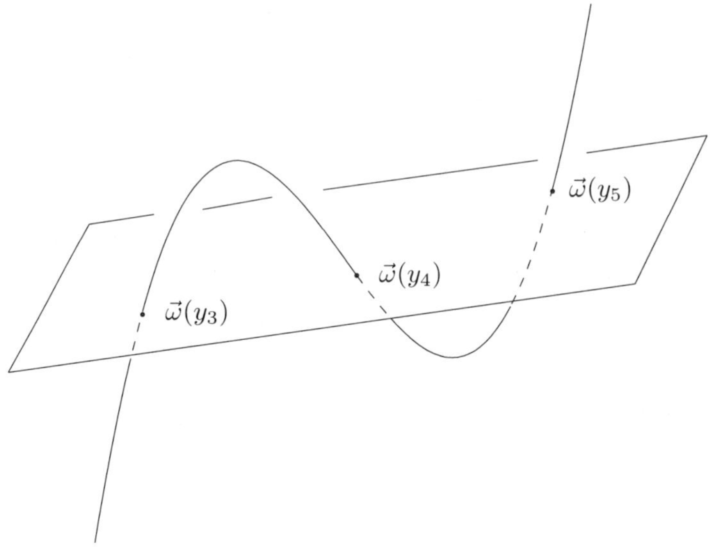

Pick $M \in \mathbb{Z}$ sufficiently large and $\mathcal{N}$ sufficiently small that $y_j \in \widetilde{Y}_j(\eta)$ for all $y \in \mathcal{N}$ and $j \geq M$, such that $z_2(y) \notin \widetilde{Y}_j$ for all $y \in \mathcal{N}$ and $j \geq M$ and such that M is larger than the n_0 of Proposition 10.1(b). Let

$$\omega_i = w_i(\xi)\, d\xi$$

using the local coordinates defined just before (8.2) for $\widetilde{Y}_j$, $j \geq M$ and some arbitrary coordinate patchs on $\{y_j \mid y \in \mathcal{N}\}$ for $j < M$. We may choose coordinates in $S^{(-1)}$ such that the tangent space to $S^{(-1)}$ at $[(y_2, y_3 \dots)]$ is $B^{(-1)}$ and such that $df(y_2, y_3 \dots)$ maps each such tangent vector $\lambda = (\lambda_2, \lambda_3, \dots)$ to $\sum_{j=2}^{\infty} \vec{w}(\xi(y_j))\lambda_j \in B$ when $y_2 \neq y_j$, $j \geq 3$ and to $\dot{\vec{w}}(\xi(y_2))\lambda_2 + \sum_{j=3}^{\infty} \vec{w}(\xi(y_j))\lambda_j \in B$ when $y_2 = y_j$ for some $j \geq 3$.

That df has a nontrivial kernel at $\big(z_2(y), y_3, y_4 \dots\big)$ then means that there exists $\big(\lambda_2(y), \lambda_3(y), \dots\big) \in B^{(-1)}$ such that

$$\text{(9.3)}\qquad \begin{aligned} &\vec{w}\big(\xi\big(z_2(y)\big)\big)\lambda_2(y) + \sum_{j=3}^{\infty} \vec{w}\big(\xi(y_j)\big)\lambda_j(y) = 0 \text{ if } z_2(y) \notin \{y_j \mid j \geq 3\},\\ &\dot{\vec{w}}\big(\xi\big(z_2(y)\big)\big)\lambda_2(y) + \sum_{j=3}^{\infty} \vec{w}\big(\xi(y_j)\big)\lambda_j(y) = 0 \text{ if } z_2(y) \in \{y_j \mid j \geq 3\}. \end{aligned}$$

To verify that $\vec{\omega}\big(z_2(y)\big)$, resp. $\dot{\vec{\omega}}\big(z_2(y)\big)$ is in $\overline{\operatorname{span}\{\vec{\omega}(y_3), \vec{\omega}(\vec{y}_4), \dots\}}$ it now suffices to verify that $\lambda_2(y) \neq 0$. By Proposition 10.1(c) we know that

$$\operatorname{codim} \overline{\operatorname{span}\{\vec{w}\big(\xi(y_M)\big), \vec{w}\big(\xi(y_{M+1})\big), \dots\}} = M - 1.$$

Since $\kappa(X)$ is not contained in a hyperplane one has, for generic, $y_3, \dots, y_{M-1}$

$$\operatorname{codim}\Big(\overline{\operatorname{span}\big\{\vec{w}\big(\xi(y_M)\big), \vec{w}\big(\xi(y_{M+1})\big), \dots\big\}} + \mathbb{C}\vec{w}\big(\xi(y_3)\big) + \dots + \mathbb{C}\vec{w}\big(\xi(y_{M-1})\big)\Big) = 2.$$

For these $y_3 \dots y_{M-1}$, $\lambda_2(y) \neq 0$. □

For nonhyperelliptic curves $x \mapsto [\vec{\omega}(x)]$ is injective. So the statement (9.2) is in contradiction to parts (b) and (a) respectively of

LEMMA 9.4. (a) *Let $n_1, n_2, \dots, n_p, n \in \mathbb{N}$ obey $n_1 + n_2 + \dots + n_p \leq n$. Then*

$$M = \Big\{ (x_1, \dots, x_p, y) \in \widetilde{X}^p \times W^{(-n)} \;\Big|\; \operatorname{codim} \operatorname{span}\big(\{\vec{\omega}^{(i)}(x_j) \mid 1 \leq j \leq p, 0 \leq i < n_j\} \cup \{\vec{\omega}(y_j) \mid j \geq n+1\}\big) = n \sum_{j=1}^{p} n_j \Big\}$$

is open and dense in $\widetilde{X}^p \times W^{(-n)}$.

(b) *Let $\mathcal{U} \subset W^{(-2)}$ be open and $v(y)\colon \mathcal{U} \to \widetilde{X}$ be analytic. If, for all $y \in \mathcal{U}$, $[\vec{\omega}(v(y))] \notin \{[\vec{\omega}(y_j)] \mid j \geq 3\}$ then*

$$\{ y \in \mathcal{U} \mid \vec{\omega}(v(y)) \notin \overline{\operatorname{span}\{\vec{\omega}(y_3), \vec{\omega}(y_4), \dots\}} \}$$

is open and dense.

(c) *Let $\mathcal{U} \subset W^{(-2)}$ be open and $v(y)\colon \mathcal{U} \to \widetilde{X}$ be analytic. If, for all $y \in \mathcal{U}$ one has $\kappa(v(y)) \neq \kappa(y_3)$ then*

$$\{ y \in \mathcal{U} \mid \{\vec{\omega}(v(y)), \dot{\vec{\omega}}(v(y))\} \not\subset \overline{\operatorname{span}\{\dot{\vec{\omega}}(y_3), \vec{\omega}(y_3), \vec{\omega}(y_4), \dots\}} \}$$

is open and dense.

REMARK. The analogs of Lemma 9.4 for compact Riemann surfaces are well-known facts about the canonical curve. Part (a) says that $g - n$ points in general position on the canonical curve span a linear subspace of codimension n. Part (b) corresponds to the fact that the subspace spanned by $g - 2$ points in general position on the canonical curve contains no further point of the canonical curve. Part (c) corresponds to the fact that for $g - 2$ points $y_3, \dots, y_g$ in general position on the canonical curve the hyperplane spanned by $y_4, \dots, y_g$ and the tangent line L to the canonical curve in the point y_3 contains no tangent line to the canonical curve apart from L.

PROOF OF LEMMA 9.4. (a) By Lemma 8.3(b) it suffices to show that M is nonempty. Suppose M is empty. To simplify the notation, suppose $p = n_1 = n_2 = 2$ and $n \geq 4$. The argument in general is identical to the one that follows. By Lemma 8.3(c) there exist (locally) analytic functions

$$\begin{aligned} &\mu_i(x_1, x_2, y), \quad 1 \leq i \leq 4, \\ &\lambda_i(x_1, x_2, y), \quad i \geq n+1 \end{aligned}$$

not all zero, such that, in local coordinates in some patch,

$$\mu_1 \vec{w}(x_1) + \mu_2 \dot{\vec{w}}(x_1) + \mu_3 \vec{w}(x_2) + \mu_4 \dot{\vec{w}}(x_2) + \sum_{i=n+1}^{\infty} \lambda_i \vec{w}(y_i) = 0.$$

If, for some $(\bar{x}_1, \bar{x}_2, \bar{y})$, there exists an $\bar{\imath} \geq n+1$ with $\lambda_{\bar{\imath}}(\bar{x}_1, \bar{x}_2, \bar{y}) \neq 0$ then for all y_i in a neighbourhood of $\bar{y}_{\bar{\imath}}$,

$$\vec{w}(y_{\bar{\imath}}) = \frac{1}{\lambda_{\bar{\imath}}} \Bigg[\mu_1 \vec{w}(\bar{x}_1) + \dots + \mu_4 \dot{\vec{w}}(\bar{x}_2) + \sum_{\substack{i \geq n+1 \\ i \neq \bar{\imath}}} \lambda_i \vec{w}(\bar{y}_i) \Bigg].$$

This contradicts the fact that $\vec{w}(y)$ is not contained in a hyperplane. Thus all λ_i, $i \geq n+1$ are identically zero.

Some μ_i is not identically zero. Let $\bar{\imath}$ be the index of the not identically zero μ_i associated to the $\vec{w}^{(j)}(x_i)$ of largest possible j. Suppose it is associated to $\vec{w}^{(j)}(x_1)$. If $\bar{\imath} = 1$ we have

$$\vec{w}(x_1) = -\frac{1}{\mu_1}[\mu_3\vec{w}(x_2) + \mu_4\dot{\vec{\omega}}(x_2)]$$

for (x_1, x_2) in some nonempty open ball. If $\bar{\imath} = 2$ we have

$$\dot{\vec{\omega}}(x_1) = -\frac{1}{\mu_2(x_1,x_2)}[\mu_1(x_1,x_2)\vec{w}(x_1) + \mu_3(x_1,x_2)\vec{w}(x_2) + \mu_4(x_1,x_2)\dot{\vec{\omega}}(x_2)].$$

In the first case we have $\vec{\omega}(x_1)$ contained, for all x_1, in the two-dimensional space spanned by $\vec{\omega}(x_2)$ and $\dot{\vec{\omega}}(x_2)$ for some fixed x_2. In the second case, by repeated differentiation, we have $\vec{\omega}(x_1)$ contained for all x_1 in $\operatorname{span}\{\vec{\omega}(\bar{x}_1), \vec{\omega}(x_2), \dot{\vec{\omega}}(x_2)\}$ for some fixed $(\bar{x}_1, x_2)$. In both cases we have reached a contradiction.

(b) By part (a) $\mathcal{U}' = \{y \in \mathcal{U} \mid \operatorname{codim} \overline{\operatorname{span}\{\vec{\omega}(y_3), \vec{\omega}(y_4), \dots\}} = 2\}$ is open and dense in $\mathcal{U}$. We now show that

$$\mathcal{V} = \{y \in \mathcal{U}' \mid \vec{\omega}(v(y)) \in \overline{\operatorname{span}\{\vec{\omega}(y_3), \vec{\omega}(y_4), \dots\}}\}$$

is not all of $\mathcal{U}'$. Then Lemma 8.3(b) implies that $\mathcal{V}$ is an analytic subvariety of $\mathcal{U}'$ of codimension at least one.

Suppose that $\mathcal{V} = \mathcal{U}'$. Then by Lemma 8.3(c) there exist analytic functions $\lambda_j(y)$, $j \geq 3$ such that, in local coordinates on a small ball in $\mathcal{U}'$

$$\vec{w}(v(y)) = \sum_{j\geq 3} \lambda_j(y)\vec{w}(y_j).$$

Suppose that $v(y)$ is not constant. Then there is a $k \geq 3$ such that $\partial v/\partial y_k$ is non zero on some nonempty neighbourhood. On this neighbourhood

$$\dot{\vec{w}}(v(y))\partial v/\partial y_k = \sum_{j\geq 3} \partial\lambda_j/\partial y_k \vec{w}(y_j) + \lambda_k(y)\dot{\vec{w}}(y_k)$$

implies

$$\dot{\vec{w}}(v(y)) \in \overline{\operatorname{span}\{\dot{\vec{w}}(y_k), \vec{w}(y_3), \vec{w}(y_4), \dots\}}.$$

Since $[\vec{\omega}(v(y))] \notin \{[\vec{\omega}(y_j)] \mid j \geq 3t\}$ there is a second $p \geq 3$, $p \neq k$ such that $\lambda_p(y) \neq 0$ on a non empty subneighbourhood. Then

$$\vec{w}(y_p) = -\frac{1}{\lambda_p(y)}\left(\sum_{\substack{j\geq 3\\ j\neq p}} \lambda_j(y)\vec{w}(y_j) + \vec{w}(v(y))\right).$$

By repeated differentiation, $\vec{w}(y_p)$ and all the derivatives $\vec{w}^{(n)}(y_p)$, $n \geq 1$ are in

$$H = \overline{\operatorname{span}\{\dot{\vec{w}}(y_k), \vec{w}(v), \vec{w}(y_j) \mid j \geq 3,\ j \neq p\}}.$$

By analyticity this forces $\vec{w}(y_p)$ to remain in the hyperplane H for all y_p in some nonempty open set, which is impossible. If $v(y)$ is constant we see directly that $\vec{w}(y_p)$ remains in $\overline{\operatorname{span}\{\vec{w}(v), \vec{w}(y_j) \mid j \geq 3,\ j \neq p\}}$.

(c) By part (a) $\mathcal{U}' = \{y \in \mathcal{U} \mid \operatorname{codim} \overline{\operatorname{span}\{\dot{\vec{\omega}}(y_3), \vec{\omega}(y_3), \vec{\omega}(y_4) \dots\}} = 1\}$ is open and dense. By continuity

$$\mathcal{V} = \{y \in \mathcal{U}' \mid \vec{\omega}(v(y)), \dot{\vec{\omega}}(v(y)) \in \overline{\operatorname{span}\{\dot{\vec{\omega}}(y_3), \vec{\omega}(y_3), \vec{\omega}(y_4) \dots\}}\}$$

is closed. So it only remains to show that $\mathcal{V}$ contains no nonempty open set $\mathcal{V}'$.

Suppose that $\mathcal{V}$ does contain a nonempty open set $\mathcal{V}'$. By Lemma 8.3(c), applied twice, there are analytic functions on some nonempty open subset of $\mathcal{V}'$ such that, in local coordinates,

$$\vec{w}\big(v(y)\big) = \lambda_2(y)\dot{\vec{w}}(y_3) + \sum_{j\geq 3} \lambda_j(y)\vec{w}(y_j), \quad \dot{\vec{w}}(v(y)) = \mu_2(y)\dot{\vec{w}}(y_3) + \sum_{j\geq 3} \mu_j(y)\vec{w}(y_j)$$

on the subset.

If there is some $\lambda_{j_0}(y)$ with $j_0 \geq 4$ which is not identically zero, then (shrinking $\mathcal{V}'$)

$$\vec{w}(y_{j_0}) = \frac{1}{\lambda_{j_0}(y)} \left\{ -\lambda_2(y)\dot{\vec{w}}(y_3) - \sum_{\substack{j\geq 3\\ j\neq j_0}} \lambda_j(y)\vec{w}(y_j) + \vec{w}\big(v(y)\big) \right\}.$$

Fix any $\overline{y} \in \mathcal{V}'$. Repeated differentiation with respect to y_{j_0}, followed by evaluation at $\overline{y}$ shows that $\vec{w}^{(n)}(\overline{y}_{j_0})$, $n \geq 0$, lies in $\overline{\operatorname{span}\big\{\vec{w}(v(\overline{y})), \dot{\vec{w}}(\overline{y}_3), \vec{w}(\overline{y}_j) \bigm| j \geq 3, j \neq j_0\big\}}$. But this implies that $\vec{w}(y_{j_0})$ lies in a hyperplane for y_{j_0} in a neighbourhood of $\overline{y}_{j_0}$ which is impossible.

On the other hand, if all $\lambda_j(y)$, $j \geq 4$ are identically zero, then

$$\vec{w}\big(v(y)\big) = \lambda_2(y)\dot{\vec{w}}(y_3) + \lambda_3(y)\vec{w}(y_3).$$

To complete the proof of the lemma it suffices to show that some $\partial v/\partial y_j$, $j \geq 4$, is non zero, because from this it would follow that $\vec{w}(z)$ lies in $\operatorname{span}\{\vec{w}(\overline{y}_3), \dot{\vec{w}}(\overline{y}_3)\}$ for all z in a neighbourhood of $v(\overline{y})$ which is impossible.

So suppose that $\lambda_j(y) = 0$ for all $j \geq 4$ and $v(y)$ is a function of y_3, only. By hypothesis $\lambda_2(y)$ is non zero. Differentiating with respect to y_3 gives

$$\dot{\vec{w}}\big(v(y)\big)\frac{\partial v}{\partial y_3} = \frac{\partial \lambda_3}{\partial y_3}\vec{w}(y_3) + \left(\lambda_3 + \frac{\partial \lambda_2}{\partial y_3}\right)\dot{\vec{w}}(y_3) + \lambda_2 \ddot{\vec{w}}(y_3).$$

This combines with

$$\dot{\vec{w}}\big(v(y)\big) = \mu_2(y)\dot{\vec{w}}(y_3) + \sum_{j\geq 3} \mu_j(y)\vec{w}(y_j)$$

to give that

$$\operatorname{codim} \overline{\operatorname{span}\{w(y_3), \dot{\vec{w}}(y_3), \ddot{\vec{w}}(y_3), \vec{w}(y_4), \vec{w}(y_5)\dots\}} \geq 1$$

on some nonempty open set of $W^{(-2)}$, in contradiction to part a. □

We now describe the tangent space of Θ at its regular points. The dual space B^* to B is

$$B^* = \left\{ \lambda = (\lambda_1, \lambda_2, \dots) \in \mathbb{C}^\infty \,\middle|\, \|\lambda\|_{B^*} = \sum_j |\lambda_j| |\log t_j| < \infty \right\}.$$

The derivative $\nabla\theta(e)$ of θ at e is a map from B to B^*.

Put

$$\Theta_{\mathrm{reg}} := \{e \in \theta \mid \nabla\theta(e) \neq 0\}, \quad \operatorname{Sing}\Theta := \Theta \setminus \Theta_{\mathrm{reg}}.$$

We have

LEMMA 9.5.

$$\Theta \setminus \big(\Theta_1 \cap (-\Theta_1)\big) \subset \Theta_{\mathrm{reg}}.$$

PROOF. Since $\theta(e)$ is even it suffices to show $\Theta \setminus \Theta_1 \subset \Theta_{\text{reg}}$. If $e \in \Theta \setminus \Theta_1$, then by Riemann's Vanishing theorem $\theta(e + \int_{x_0}^{x} \vec{\omega})$ has, for generic x_0, the point x_0 as a simple zero. Hence $\nabla\theta(e) \neq 0$. □

Furthermore put

$$\widetilde{\Omega} := \left\{ \sum_{j=1}^{\infty} \lambda_j \omega_j \,\middle|\, (\lambda_1, \lambda_2, \dots) \in B^* \right\}.$$

By Proposition 7.3(ii) every element of $\widetilde{\Omega}$ is a holomorphic differential form. In addition all elements of $\widetilde{\Omega}$ are square integrable since, by Corollary 6.17,

$$\begin{aligned} \left\| \sum_j \lambda_j \omega_j \right\|_2 &\leq \sum_j |\lambda_j| \|\omega_j\|_2 \leq \text{const} \sum_j |\lambda_j| \sqrt{|\log t_j|} \\ &\leq \text{const} \|\lambda\|_{B^*}. \end{aligned}$$

Let $\psi \colon B^* \to \widetilde{\Omega}$ be the map $(\lambda_1, \lambda_2, \dots) \mapsto \sum_{j=1}^{\infty} \lambda_j \omega_j$. It is bijective since $\int_{A_i} (\sum_{j=1}^{\infty} \lambda_j \omega_j) = \lambda_i$. For $y \in W^{(-n)}$ put

$$\widetilde{\Omega}(y) := \{\omega \in \widetilde{\Omega} \mid \omega \text{ vanishes at } y_j \text{ with multiplicity at least } \sharp\{k \mid \pi(y_k) = \pi(y_j)\}\}$$

and for a divisor $D = \sum m_p p$ on X put

$$\widetilde{\Omega}(D) := \{\omega \in \widetilde{\Omega} \mid \omega \text{ vanishes with muliplicity at least } m_p \text{ at } p\}.$$

Consider $e \in \Theta \setminus \Theta_1$. Recall that such an e has a representation $e = F(y)$ with $y = (y_2, y_3, \dots) \in W^{(-1)}$ obeying $y_i = y_j$ whenever $\pi(y_i) = \pi(y_j)$ and that $\{\pi(y_i) \mid i \geq 2\}$ is uniquely determined by e. We showed in Lemma 9.5 that $\nabla\theta(e) \neq 0$, i.e., Θ is smooth at e. In Corollary 9.9 we will show that $\mathbb{C}\nabla\theta(e)$, i.e., the "orthogonal complement" to the tangent space $T_e\Theta$, is uniquely determined by the condition that the holomorphic form $\sum_{k\geq 1} \nabla\theta(e)_k \omega_k(z)$ be zero for $z = y_2, y_3, \dots$.

Following Lemma 9.6 we will look at how $T_e\Theta$ varies as e moves in directions $v \in T_e\Theta$, i.e., study the Gauss map for θ. More precisely, we will look for directions v such that $T_e\Theta$ is stationary, in other words such that $\mathbb{C}\nabla\Theta_e$ is stationary. Such pairs (e, v) are given by the conditions

$$\nabla\theta(e) \neq 0, \quad \nabla\theta(e) \cdot v = 0, \quad \frac{d}{d\lambda} \nabla\theta(e + \lambda v)\big|_{z=0} \in \mathbb{C}\nabla\Theta_e. \tag{9.4}$$

In particular, when $e = F(y) \in \Theta \setminus \Theta_1$, we shall find in Proposition 9.8 necessary and sufficient conditions that the set of v's satisfying (9.4) is of dimension 1 and shall determine precisely what the set is. The conditions are that the form $\omega_e(z) = \sum_{k\geq 1} \nabla\theta(e)_k \omega_k(z)$ have a zero of precisely the right order, namely $\sharp\{i \mid \pi(y_i) = \pi(y_j)\}$, at each y_j $j \geq 2$ with one exception, say $y_j = x$. And that $\omega_e(z)$ have one excess zero, in other words a zero of order $\sharp\{i \mid \pi(y_i) = x\} + 1$, at $z = x$. Then the set of stationary directions $v \in T_e\Theta$ is precisely $\mathbb{C}\vec{\omega}(x)$.

Note that the conditions (9.4) are stated purely in terms of the function θ. They do not involve the Riemann surface that gave rise to θ. On the other hand the statement "the set of stationary directions $v \in T_e\Theta$ is precisely $\mathbb{C}\vec{\omega}(x)$" does involve the Riemann surface and indeed assigns, in the nonhyperelliptic case, a unique point $x \in X$ to the given $e \in \Theta$. In Proposition 9.10 we find, in the nonhyperelliptic case, a set $E \subset \Theta$ of such e's.

The set E is dense in a subset of codimension 1 in Θ. Such points are ramification points of the Gauss map. Furthermore, for x in a dense subset of X, the set $\{e \in E \mid e \text{ is paired with } x\}$ is, roughly speaking, of codimesion 2 in Θ. The pairing of points e in E with points $x \in X$ is the principal ingredient in the Proof of the Torelli Theorem 11.1 for the non hyperelliptic case. Of course since E and Θ_1 are both of codimension 1 in Θ, there is the danger that $E \cap \Theta \setminus \Theta_1 = \varnothing$. However, if $\big(e, v = \vec{\omega}(x)\big)$ obeys (9.4), then so does $\big(-e, v = \vec{\omega}(x)\big)$, since θ is an even function. As $\Theta_1 \cap (-\Theta_1)$ is of codimension 2 in Θ, most of E must be outside it.

On the other hand, we shall show in Proposition 9.11 and Proposition 10.5 that hyperelliptic curves are characterized by the existence, for each Weierstrass point $b \in X$, of a set $H^{(b)}$ which is dense in a subset of codimension 1 in Θ with every point $e \in H^{(b)}$ paired, as above, with b. We return to the hyperelliptic case in §10.

LEMMA 9.6. *Let $y \in W^{(-n)}$. Then*

$$\psi\big(\big(\operatorname{range} df_{(\nu_1,\dots,\nu_{n-1})}([y])\big)^{\perp}\big) = \widetilde{\Omega}(y).$$

PROOF. The range $df_{(\nu_1,\dots,\nu_{n-1})}([y])$ is generated by the vectors $\vec{\omega}(y_j)$, $\dot{\vec{\omega}}(y_j)$, $\ddot{\vec{\omega}}(y_j), \dots, \vec{\omega}^{(k_j-1)}(y_j)$ where $k_j = \sharp\{i \mid y_i = y_j\}$. □

For $e \in \Theta$ put

$$\omega_e := \psi\big(\nabla\theta(e)\big) = \frac{\partial\theta}{\partial e_1}(e)\omega_1 + \frac{\partial\theta}{\partial e_2}(e)\omega_2 + \cdots.$$

COROLLARY 9.7. *Let $e \in \Theta_{\mathrm{reg}}$, and let $y \in W^{(-1)}$ such that $F(y) = e$ and f maps a neighbourhood of $[y]$ biholomorphically onto a neighbourood of e in Θ. Then*

$$\widetilde{\Omega}(y) = \mathbb{C} \cdot \omega_e.$$

PROOF. $\nabla\Theta(e)$ clearly spans $\big(\operatorname{range} df([y])\big)^{\perp}$. □

For each fixed $e \in B$ denote by $\operatorname{Hess}(\theta)(e)\colon B \to B^*$ the second derivative of Θ. If $e \in \theta_{\mathrm{reg}}$ let

$$T_e\Theta := \ker \nabla\theta(e)$$

be the tangent space to Θ at e and let

$$H(e) := \big(q_e \circ \operatorname{Hess}(\theta)(e)\big|_{T_e\Theta}\big)\colon T_e\theta \to B^*/\mathbb{C}\nabla\theta(e)$$

be the restriction of $\operatorname{Hess}(\theta)(e)$ to $T_e(\theta)$, composed with the projection q_e from B^* to $B^*/\mathbb{C}\nabla\theta(e)$.

For $y = (y_{n+1}, y_{n+2}, \dots) \in W^{(-n)}$ and $x \in \widetilde{X}$ let

$$v(x;y) := \begin{cases} 0 & \text{if } x \neq y_{n+1}, y_{n+2}, \dots \\ \sharp\{i \mid \pi(y_i) = \pi(x)\} & \text{if } x = y_j \text{ for some } j \end{cases}$$

be the multiplicity of $\pi(x)$ in the sequence $\pi(y_2)$, $\pi(y_3), \dots$.

PROPOSITION 9.8. *Let $e \in \Theta_{\mathrm{reg}}$. Let $y \in W^{(-1)}$ such that $F(y) = e$ and such that f maps a neighbourhood of $[y]$ in $S^{(-1)}$ biholomorphically onto a neighbourhood of e in Θ. Let $k_j := \operatorname{mult}_{y_j} \omega_e - v(y_j, y)$ be the excess multiplicity of ω_e in y_j to the requirement that $\omega_e \in \widetilde{\Omega}(y)$. Then*

$$\ker H(e) = \operatorname{span}\big(\{\vec{\omega}(y_j), \dot{\vec{\omega}}(y_j), \dots, \vec{\omega}^{(\ell_j)}(y_j) \mid j = 2, 3, \dots\}\big)$$

where $\ell_j := \min(k_j, \operatorname{mult}_{y_i} \omega_e) - 1$.

COROLLARY 9.9. *In the situation above*

(a) $H(e)$ *is injective if and only if* $\operatorname{mult}_{y_j}\omega_e = v(y_j, y)$ *for* $j = 2, 3, \ldots$

(b) $\dim\ker H(e) = 1$ *if and only there is* $x \in \{y_2, y_3, \ldots\}$ *such that*

$$\operatorname{mult}_x \omega_e = v(x; y) + 1 \quad \textit{and} \quad \operatorname{mult}_{y_j}\omega_e = v(y_j; y) \textit{ whenever } \pi(y_j) \neq \pi(x).$$

In this case

$$\ker H(e) = \mathbb{C}\vec{\omega}(x).$$

PROOF OF PROPOSITION 9.8. After reordering the entries of y we may assume that there are integers $2 = j_p < j_{p+1} < j_{p+2} < \cdots$ such that $j_{i+1} - j_i = 1$ for sufficiently big i

$$\pi(y_{j_i}) \neq \pi(y_{j_{i'}}) \text{ if } i \neq i',$$
$$y_{j_i} = y_{j_{i+1}} = \cdots = y_{j_i+n_i-1} \text{ where } n_i = j_{i+1} - j_i = v(y_{j_i}; y).$$

Choose local coordinates ξ_i around y_{j_i} such that $\xi_i = 0$ corresponds to the point y_{j_i}. We use as local coordinates around the point $[(y_{j_i}, y_{j_i}, \ldots)] \in \widetilde{X}^{n_i}/\mathfrak{S}_{n_i}$ the coefficients of the polynomial $p_i(\xi_i) = \xi_i^{n_i} + a^{(i)}_{n_i-1}\xi_i^{n_i-1} + \cdots + a^{(i)}_0 = \prod_{\ell=0}^{n_i-1}\left(\xi_i - \xi_i(y'_{j_i+\ell})\right)$ whose zero set is a configuration of points $[(y'_{j_i}, y'_{j_i+1}, \ldots, y'_{j_i+n_i-1})]$ in $\widetilde{X}^{n_i}/\mathfrak{S}_{n_i}$ close to $[(y_{j_i}, \ldots, y_{j_i})]$.

For e' near e we write locally

$$\omega_{e'} = w(\xi_i, e')\, d\xi_i.$$

Then $w(\xi_i, e)$ vanishes at $\xi_i = 0$ with multiplicity $n_i + k_{j_i}$.

Now let $t \to e(t)$ be a holomorphic curve in Θ with $e(0) = e$, $\dot{e}(0) \neq 0$. The image of this curve under f^{-1} can coordinatewise be described by

$$t \mapsto P_i(\xi_i, t) = \xi_i^{n_i} + a^{(i)}_{n_i-1}(t)\xi_i^{n_i-1} + \cdots + a^{(i)}_0(t).$$

Observe that $p_i(\xi_i, 0) = \xi_i^{n_i}$.

Put

$$\dot{\omega}_e = \frac{d}{dt}\omega_{e(t)}\Big|_{t=0} = \psi([\operatorname{Hess}(\theta)(e)](\dot{e}(0))$$

and write locally

$$\dot{\vec{\omega}}_e = \dot{w}(\xi_i)\, d\xi_i = \frac{d}{dt}w\big(\xi_i, e(t)\big)\big|_{t=0}\, d\xi_i.$$

We can, by the local Nullstellensatz, write

$$w\big(\xi_i, e(t)\big) = g_i(\xi_i, t) \cdot p_i(\xi_i, t)$$

where g_i is an analytic function of ξ_i and t such that $g_i(\xi_i, 0)$ has a zero of order k_{j_i} at 0. By differentiating this identity we get

$$\dot{w}(\xi_i) = \left(\frac{d}{dt}g_i(\xi_i, t)\Big|_{t=0}\right)p_i(\xi_i, 0) + g_i(\xi_i, 0)\left(\frac{\partial p_i}{\partial t}(\xi_i, t)\right)_{t=0}.$$

The first term has a zero of order least n_i at 0, while the order of the zero of the second term at 0 is

$$k_{j_i} + \min\left\{\ell \,\Big|\, \frac{d}{dt}a^{(i)}_\ell(t)\Big|_{t=0} \neq 0\right\}$$

whenever the second term is not identically zero. Therefore

$$\begin{array}{c} \dot{w}(\xi_i) \text{ has a zero of multiplicity at least } n_i \text{ at } 0 \\ \Updownarrow \\ \left.\frac{d}{dt} a_\ell^{(i)}(t)\right|_{t=0} = 0 \quad \text{for } \ell = 0, \dots, n_i - k_{j_i} - 1. \end{array} \tag{9.5}$$

Now $\dot{e}(0)$ lies in the kernel of $H(e)$ if and only if $\dot{\omega}_e$ is proportional to ω_e. By the Corollary 9.7 this is the case if and only if $\dot{\omega}_e \in \widetilde{\Omega}(y)$, i.e., if $\dot{w}(\xi_i)$ has a zero of multiplicity at $\xi_i = 0$ of multiplicity at least n_i. By (9.5) this happens if and only if

$$\left.\frac{d}{dt} a_\ell^{(i)}(t)\right|_{t=0} = 0 \text{ for } \ell = 0, \dots, n_i - k_{j_i} - 1.$$

The formula for the kernel now follows from the fact that

$$\dot{e}(0) = \sum_i \sum_{\ell=1}^{n_i} \alpha_{\ell,n_i} \frac{\partial^{\ell-1}\vec{w}}{\partial \xi_h^{\ell-1}}(0) \dot{a}_{n_i-\ell}^{(i)}(0)$$

with some constants $\alpha_{\ell,n} \neq 0$. This proves the formula about the kernel. □

PROPOSITION 9.10. *Let X be non hyperelliptic. There is an open dense subset M of $\widetilde{X} \times W^{(-3)}$ such that each point of it has a neighbourhood U, for which the following holds: There is a holomorphic map*

$$\varphi \colon U \to \Theta$$

such that for all $(x, y) \in U$

$$\operatorname{codim}_B \operatorname{range} d\varphi(x; y) = 2$$

and for all (x, y) in a dense subset of U

$$\nabla\theta\big(\varphi(x; y)\big) \neq 0, \quad \ker H\big(\varphi(x; y)\big) = \mathbb{C} \cdot \vec{\omega}(x).$$

Loosely speaking, the proposition says that on a big part of the ramification set of the Gauss map the kernel of the derivative of the Gauss map is of the form $\mathbb{C} \cdot \vec{\omega}(x)$ with some $x \in X$.

PROOF OF PROPOSITION 9.10. We consider the map

$$\Phi \colon \widetilde{X} \times W^{(-3)} \to \Theta, \quad (x; y_4, y_5, \dots) \mapsto F(x, x, y_4, y_5, \dots).$$

By Lemma 8.3(b) and Lemma 9.4(a) the set M_1 of $(x; y) \in \widetilde{X} \times W^{(-3)}$ at which the differential $d\Phi(x; y)$ is injective and codim_B range $d\Phi(x; y) = 2$ is open and dense in $\widetilde{X} \times W^{(-3)}$. We have the commutative diagram

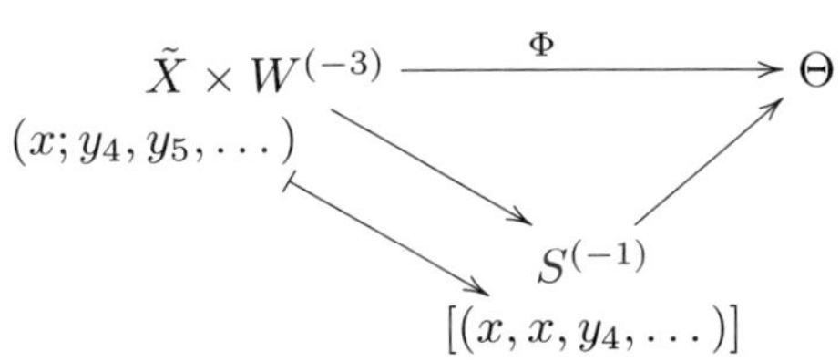

Therefore for $(x; y_4, y_5, \dots) \in M_1$ the points $\pi(x)$, $\pi(y_4)$, $\pi(y_5), \dots$ are pairwise different, for otherwise the projection $p \colon \widetilde{X} \times W^{(-3)} \to S^{(-1)}$ would not have injective differential. Put

$$M_2 := \{(x, y) \in M_1 \mid df \text{ has maximal rank at } p(x; y)\}.$$

As before one sees that M_2 is open at dense in $\widetilde{X} \times W^{(-3)}$.

Now put

$$M_3 := \{(x;y) \in M_2 \mid \Phi(x,y) \in \Theta \setminus (\Theta_1 \cap (-\Theta_1))\}.$$

Since codim_B range $d\Phi(x;y) = 2$ for all $(x,y) \in M_2$ it follows from Proposition 9.3 and Lemma 9.2 that $\Phi(M_2) \not\subset \Theta_1 \cap (-\Theta_1)$. As $\Theta_1 \cap (-\Theta_1)$ is an analytic subvariety of Θ it follows that M_3 is open and dense in M_2.

Corollary 9.7 applies to all the points $\Phi(x;y)$, $(x,y) \in M_3$. In particular $\nabla\theta(\Phi(x;y))$ and hence $\omega_{\Phi(x;y)}$ is non zero and generates $\widetilde{\Omega}(x,x;y_4,y_5,\dots)$.

Next we show that the set

$$M_4 := \{(x;y) \in M_3 \mid \Phi(x,y) \in \Theta \setminus (\Theta_1 \cup (-\Theta_1))\}$$

is open and dense in $\widetilde{X} \times W^{(-3)}$. Since Θ_1 and $(-\Theta_1)$ are closed analytic subvarieties of Θ is suffices to show that $\Phi(M_3)$ is not contained in any $\Theta^{(\nu)}$ or $(-\Theta^{(\nu)})$ for any $\nu = 1,\dots,m$. So assume that $\Phi(M_3) \subset \Theta^{(\nu)}$. Since codim range $d\Phi(x;y) = 2$ for $(x,y) \in M_1$ it follows from Lemma 9.2 that $\Phi(M_3)$ is not contained in Θ_2. So there would be an open subset O of M_3 such that $\Phi(O) \subset \Theta^{(\nu)} - \Theta_2$. By the condition defining M_2 we may, after posssibly shrinking O, assume that there is an open subset O' of $S^{(-1)}$ containing $p(O)$ such that f maps O' biholomorphically onto a neighbourhood of $\Phi(O)$ in Θ. After possibly shinking O again there is for every $z \in \widetilde{X}$ such that $\int_{\infty_\nu}^{z} \vec{\omega}$ is small and every $(x;y) \in O$ a unique point

$$[y'(x;y;z)] := [(y'_2(x;y;z), y'_3(x;y;z), y'_4(x;y;z), \dots)]$$

in O' such that

$$\Phi(x;y) - \int_{\infty_\nu}^{z} \vec{\omega} = f([y'(x;y;z)]). \tag{9.6}$$

The maps $(x,y,z) \mapsto y'_j(x;y;z)$ are holomorphic. For $j \geq 4$, $y_j(x;y;z) \to y_j$ as $\int_{\infty_\nu}^{z} \vec{\omega} \to 0$, and the points $\{y'_2(x;y;z), y'_3(x;y;z)\}$ go to x as $\int_{\infty_\nu}^{z} \vec{\omega} \to 0$.

By Lemma 9.6

$$\psi(\operatorname{range}(d\Phi(x;y))^{\perp}) = \widetilde{\Omega}(x,y_4,y_5,\dots).$$

For any fixed z the map

$$(x;y) \mapsto \Phi(x;y) - \int_{\infty_\nu}^{z} \vec{\omega}$$

has the same differential as $\Phi(x;y)$ at x, y. On the other hand

$$\nabla\theta\left(\Phi(x;y) - \int_{\infty_\nu}^{z} \vec{\omega}\right) = \nabla\theta(f(y'(x;y;z)))$$

annihilates the image of this differential. Therefore

$$\omega_{f(y'(x;y;z))} \in \widetilde{\Omega}(x,y_4,y_5,\dots).$$

Fix any $j \geq 4$. By Lemma 9.4(a) the set $M'_j \subset M_3$ on which $\omega_{\Phi(x,y)}$ has a zero of multiplicity one at y_j is open and dense in $\mathcal{O}$. Therefore, for every z such that $\int_{\infty_\nu}^{z} \vec{\omega}$ is small enough $\omega_{f(y'(x;y;z))} \approx \omega_{\Phi(x,y)}$ has only one zero near y_j. But y_j itself is such a zero as is $y'_j(x;y;z)$ by Lemma 9.6. Therefore $y'_j(x,y,z) = y_j$ for $(x;y) \in \mathcal{O} \cap M'_j$ and $\int_{\infty_\nu}^{z} \vec{\omega}$ small. By analyticity

$$y'_j(x;y;z) \equiv y_j \text{ for all } (x,y) \in \mathcal{O}, \quad z \in \widetilde{X},\ j \geq 4.$$

Similarly one sees that

$$x \in \{y_2'(x;y,z), \quad y_3'(x;y;z)\} \text{ for } (x,y) \in \mathcal{O}, \qquad z \in \widetilde{X}.$$

So may assume that $y_3'(x;y;z) \equiv x$. Hence (9.6) gives

$$\hat{e} - \int_{\widetilde{x}_1}^{\infty} \vec{\omega} - \int_{\widetilde{x}_2}^{x} \vec{\omega} - \int_{\widetilde{x}_3}^{x} \vec{\omega} - \sum_{j=4}^{\infty} \int_{\widetilde{x}_j}^{y_j} \vec{\omega} - \int_{\infty_\nu}^{z} \vec{\omega}$$
$$= \hat{e} - \int_{\widetilde{x}_1}^{\infty} \vec{\omega} - \int_{\widetilde{x}_2}^{y_2'(x;y;z)} \vec{\omega} - \int_{\widetilde{x}_3}^{x} \vec{\omega} - \sum_{j=4}^{\infty} \int_{\widetilde{x}_j}^{y_j} \vec{\omega}$$

so that

$$\int_{\widetilde{x}_2}^{y_2'(x,y,z)} \vec{\omega} = \int_{\widetilde{x}_2}^{x} \vec{\omega} + \int_{\infty_\nu}^{z} \vec{\omega}.$$

Differentiating with respect to z yields

$$\vec{\omega}\big(y_2'(x;y;z)\big) \cdot \frac{\partial y_2'}{\partial z} = \vec{\omega}(z).$$

Since the canonical map κ is injective

$$y_2'(x;y;z) = z.$$

This forces

$$\int_{\widetilde{x}_2}^{x} \vec{\omega} = \int_{\widetilde{x}_2}^{\infty_\nu} \vec{\omega} \text{ for all } x$$

which is imposible. The case $\Phi(M_3) \subset -\Theta^{(\nu)}$ is treated in the same way.

Let M_5 be the set of all points $(x,y) \in M_4$ such that for generic $u_0 \in \widetilde{X}$ the map $\widetilde{X} \to \mathbb{C}$, $u \mapsto \theta\big(\Phi(x;y) - \int_{u_0}^{u} \vec{\omega}\big)$ has $\pi(x)$ as at most a double zero, all the other zeroes are simple and one of the zeroes is diferent from $\pi(u_0), \pi(x), \pi(y_4), \pi(y_5) \ldots$. We claim that M_5 is open and dense in $\widetilde{X} \times W^{(-3)}$. In fact, for $(x,y) \in M_4$, $\Phi(x;y) \in \Theta \setminus (\Theta_1 \cup -\Theta_1)$ so that $-\Phi(x;y) \in \Theta \setminus (\Theta_1 \cup -\Theta_1)$ which implies, by Riemann's Vanishing Theorem,

$$-\Phi(x;y) = \hat{e} - \int_{\widetilde{x}_1}^{\infty} \vec{\omega} - \sum_{j=2}^{\infty} \int_{\widetilde{x}_j}^{z_\nu} \vec{\omega} \text{ with } (z_2, z_3, \dots) \in W^{(-1)}$$

with $\pi(u_0), \pi(z_2), \pi(z_3), \ldots$ being the zeroes of $\theta\big(\Phi(x,y) - \int_{u_0}^{u} \vec{\omega}\big)$. If U is any sufficiently small open subset of M_4 then, by Theorem 7.11, there is $N > 0$ such that for $j \geq N$ $\pi(z_i) \in \mathcal{Y}_j$ if and only if $i = j$. So far, we have shown

$$-\Phi(x;y) = \hat{e} - \int_{\widetilde{x}_1}^{\infty} \vec{\omega} - \sum_{j=2}^{\infty} \int_{\widetilde{x}_j}^{z_j} \vec{\omega};$$

$$\pi(u_0), \pi(z_2), \pi(z_3), \ldots \text{ are the roots of } u \mapsto \theta\left(\Phi(x;y) - \int_{u_0}^{u} \vec{\omega}\right); \tag{9.7}$$

$$\omega_{\Phi(x;y)} \in \widetilde{\Omega}(z);$$

$$\pi(z_N), \pi(z_{N+1}) \ldots \text{ are simple.}$$

We now show that the set U' of $(x,y) \in U$ for which x is at most a double zero and the remaining $\pi(z_2) \ldots \pi(z_{N-1})$ are simple zeroes is open and dense in U.

Openness is an immediate consequence of the continuity of the $z_i(x, y)$. If U' is not dense there is an open subset U' of U and a partition of $\{2, 3, \ldots, N-1\}$ such that for all $(x, y) \in U''$

- all $\pi(z_i)$'s with i in one element of the partition, are equal,
- $\pi(z_i) \neq \pi(z_j)$ if i and j are in different elements of the partition,
- at least one element of the partition is of cardinality at least two with $\pi(z_i) \neq \pi(x)$ for all i in that element and all $(x, y) \in U''$ or there is an element of cardinality at least three with $\pi(z_i) = \pi(x)$ for all i in that element and all $(x, y) \in U''$.

Note that if an analytic function $f(z)$ has a zero of order m at a point z_0 of an open disk D and if $f(z)$ has no other zeroes in D then

$$z_0 = \frac{1}{m} \int_{\partial D} z \frac{f'(z)}{f(z)}\, dz.$$

Consequently $z_2(x, y) \ldots z_{N-1}(x, y)$ are analytic on U''.

A triple zero of $\theta\big(\Phi(x, y) - \int_{u_0}^{u} \vec{\omega}\big)$ and hence, by Theorem 8.4, of $\omega_{-\Phi(x,y)} = -\omega_{\Phi(x,y)}$ at $\pi(x)$ for all $(x, y) \in U''$ is ruled out by Lemma 9.4(a). A double zero at $\pi(y_j)$, $j \geq 4$ is also ruled out by Lemma 9.4(a). A double zero $z_1(x, y)$, $z_2(x, y)$ with $\pi\big(z_1(x, y)\big) = \pi\big(z_2(x, y)\big) \notin \{\pi(x), \pi(y_4) \ldots\}$ and $z_i(x, y)$ analytic on U'' is ruled out by Lemma 9.4(c).

Clearly

$$\left\{ (x, y) \in U' \;\middle|\; \theta\left(\Phi(x, y) - \int_{u_0}^{u} \vec{\omega} \right) \text{ has a zero different from } \pi(u_0), \pi(x), \pi(y_4), \pi(y_5), \ldots \right\} = U' \cap M_5$$

is the complement of an analytic subvariety of U'. It suffices to show that it is not empty.

Otherwise the sequences $\pi(x), \pi(x), \pi(y_4), \pi(y_5), \ldots$ and $\pi(z_2), \pi(z_3), \pi(z_4), \ldots$ would agree up to finite permutation, so we could asume that $(x, x, y_4, \ldots) = z(x; y)$ for all $(x; y) \in U'$.

Now put

$$M := \left\{ (x, y) \in M_5 \;\middle|\; \begin{array}{l} \text{for generic } u_0 \in \widetilde{X} \text{ there is a zero } v \text{ of} \\ u \mapsto \theta\left(\Phi(x; y) - \int_{u_0}^{u} \vec{\omega} \right) \text{ such that} \\ \pi(v) \neq \pi(u_0) \text{ and } \vec{\omega}(v) \notin \overline{\operatorname{span}(\vec{\omega}(x), \vec{\omega}(y_4), \vec{\omega}(y_5), \ldots)} \end{array} \right\}.$$

On $U' \cap M_5$ the zero $v(x, y)$ can, locally, be chosen to depend analytically on (x, y), so Lemma 9.4(b) implies that M is open and dense in $\widetilde{X} \times W^{(-3)}$ whenever X is not hyperelliptic.

Let U be a neighbourhood of a point in M. Then there is a holomorphic map $U \to \widetilde{X}$ $(x; y) \mapsto z_2(x; y)$ such that for generic $u_0 \in \widetilde{X}$ the point $z_2(x; y)$ is a root of $\theta\big(\Phi(x; y) - \int_{u_0}^{u} \vec{\omega}\big)$ and $\vec{\omega}\big(z_2(x; y)\big) \notin \overline{\operatorname{span}\big(\vec{\omega}(x), \vec{\omega}(y_4), \ldots\big)}$.

After possibly shrinking U we can find $z_3(x; y), z_4(x, y), \ldots$ depdending holomorphically on $(x, y) \in \widetilde{U}$ such that

$$-\Phi(x; y) = \hat{e} - \int_{\widetilde{x}_1}^{\infty} \vec{\omega} - \sum_{j=2}^{\infty} \int_{\widetilde{x}_j}^{z_j(x;y)} \vec{\omega}.$$

Put

$$\begin{aligned}\varphi(x;y) &:= \hat{e} - \int_{\widetilde{x}_1}^{\infty} \vec{\omega} - \int_{\widetilde{x}_2}^{x} \vec{\omega} - \int_{\widetilde{x}_3}^{z_2(x;y)} \vec{\omega} - \sum_{j=4}^{\infty} \int_{\widetilde{x}_j}^{y_j} \vec{\omega} \\ &= F(x, z_2(x;y), y_4, y_5, \dots).\end{aligned}$$

We have the commutative diagram

$$\begin{array}{ccccc} & U & \xrightarrow{\quad\phi\quad} & \Theta \\ (x;y) & & \searrow \quad \nearrow & \\ & \searrow & W^{(-1)} & \\ & & (x, z_2(x,y), y_4, y_5, \dots) & \end{array}$$

The differential of the lower left arrow is injective and its range has codimension one. Furthermore for each $(x;y) \in U$

$$\operatorname{range} dF\big(x, z_2(x;y), y_4, \dots\big) = \overline{\operatorname{span}\big(\vec{\omega}(x), \vec{\omega}\big(z_2(x;y)\big), \vec{\omega}(y_4), \dots b\big)}$$

has codimension one in B. Therefore $dF\big(x, z_2(x;y), y_4, y_5, \dots\big)$ is injective, and

$$\operatorname{codim}_B \operatorname{range} d\varphi(x;y) = 2.$$

Furthermore by construction

$$\omega_{\varphi(x;y)}, \omega_{\Phi(x;y)} = -\omega_{-\Phi(x;y)} \in \widetilde{\Omega}\big(x, z_2(x;y), y_4, \dots\big).$$

By Lemma 9.4(a), 9.4(c) there is a dense subset U' of U on which the differential $\omega_{\Phi(x;y)}$ vanishes with multiplicity exactly two at x, and with multiplicity exactly one at $z_2(x;y), y_4, y_5, \dots$. For all $(x,y) \in U'$ with $\nabla\theta\big(\varphi(x;y)\big) \neq 0$ we have, by Corollary 9.7, that $\omega_{\varphi(x;y)} \propto \omega_{\Phi(x;y)}$ also vanishes with multiplicity exactly two at x, and with multiplicity exactly one at $z_2(x;y), y_4, y_5, \dots$.So by Corollary 9.9 we have

$$\ker H\big(\varphi(x,y)\big) = \mathbb{C}\vec{\omega}(x)$$

for all $(x,y) \in U'$ for which $\nabla\theta\big(\varphi(x;y)\big) \neq 0$. But $\nabla\theta$ does not vanish outside the analytic subvariety $\Theta_1 \cap (-\Theta_1)$ of codimension at least two in Θ (see Proposition 9.3 and Lemma 9.2) so $\big\{(x,y) \in U' \mid \nabla\theta\big(\varphi(x,y)\big) \neq 0\big\}$ is dense in U. □

PROPOSITION 9.11. *Assume that there is a vector $v \neq 0$ in B, an open subset U of $\mathbb{C}^N \times B^{(-n)}$ for some $n, N \in \mathbb{N}$ and a holomorphic map*

$$g\colon U \to \Theta_{\mathrm{reg}}$$

such that for all $e' \in U$

$$\operatorname{codim}_B \overline{\operatorname{range} dg(e')} = 2$$

and for all e' in a dense subset of U

$$\ker H\big(g(e')\big) = \mathbb{C}v.$$

Then X is hyperelliptic.

REMARK. A converse statement will be proven in Proposition 10.5. If X is hyperelliptic and $x_0 \in \widetilde{X}$ such that $\pi(x_0)$ is a Weierstrass point then one may take $g(y_3, y_4, \dots) = f(x_0, y_3, y_4, \dots)$ for $y \in W^{(-2)}$ and $\mathbb{C}v = \mathbb{C}v(x_0)$, independent of $y = (y_3, y_4, \dots)$.

PROOF. Assume that X is not hyperelliptic. By Proposition 9.3 and Lemma 9.2 we have $g(U) \not\subset \Theta_1$ or $g(U) \not\subset -\Theta_1$. We discuss the first case, the second being similar. Then we may shrink U so that $g(U) \subset \Theta \setminus \Theta_1$. By Corollary 9.7 and Corollary 9.9 we see that there is $x_0 \in \widetilde{X}$ and a map $\tilde{g}\colon U \to S^{(-2)}$ such that

$$\mathbb{C}v = \mathbb{C}\vec{\omega}(x_0)$$

and

$$\text{for all } e' \in U, \quad g(e') = f\big((x_0, \tilde{g}(e'))\big),$$
$$\widetilde{\Omega}(x_0, \tilde{g}(e')) = \widetilde{\Omega}(x_0, x_0, \tilde{g}(e')).$$

Since $\overline{\operatorname{codim}\operatorname{range} dg(e')} = 2$ it follows that $\tilde{g}$ is locally biholomorphic. Therefore there is $y \in W^{(-2)}$ such that $\pi(y_j) \neq \pi(y_k)$ for $j \neq k$ such that for all y' near y

$$\overline{\operatorname{span}\big(\vec{\omega}(x_0), \vec{\omega}(y_3'), \dots\big)} = \overline{\operatorname{span}\big(\vec{\omega}(x_0), \dot{\vec{\omega}}(x_0), \vec{\omega}(y_3'), \dots\big)},$$

that is

$$\dot{\vec{\omega}}(x_0) \in \overline{\operatorname{span}\big(\vec{\omega}(x_0), \vec{\omega}(y_3'), \vec{\omega}(y_4'), \dots\big)}$$

for all y' near y. We may assume that this space has codimension one in B for all such y'. If $\dot{\vec{\omega}}(x_0) \notin \mathbb{C}\vec{\omega}(x_0)$ then, by Lemma 8.3(c) there is $n \geq 0$ such that

$$\dot{\vec{\omega}}(x_0) \notin \overline{\operatorname{span}\big(\vec{\omega}(x_0), \vec{\omega}(y_n'), \vec{\omega}(y_{n+1}'), \dots\big)}.$$

Varying $y_3', \dots, y_{n-1}'$ slightly we may the assume that

$$\dot{\vec{\omega}}(x_0) \notin \overline{\operatorname{span}\big(\vec{\omega}(x_0), \vec{\omega}(y_3'), \dots, \vec{\omega}(y_{n-1}'), \vec{\omega}(y_n), \dots\big)}$$

which is a contradiction. So $\dot{\vec{\omega}}(x_0) \in \mathbb{C}\vec{\omega}(x_0)$, i.e., x_0 is a Weierstrass point. In particular, X is hyperelliptic by (A.5). □

10. Hyperelliptic Theta Divisors

In this section, we assume that the Riemann surface $X = X^{\mathrm{com}} \cup X^{\mathrm{reg}} \cup X^{\mathrm{han}}$ is hyperelliptic. Then there is a proper map $\tau\colon X \to \mathbb{P}^1 \setminus M$ of degree 2 onto the complement of a finite subset M of $\mathbb{P}^1$ which ramifies over a discrete subset S of $\mathbb{P}^1 \setminus M$. Without loss of generality, we may assume that $\infty \in M$. Let $i\colon X \to X$ be the hyperelliptic involution. Its fixed point set $\mathfrak{B} = \tau^{-1}(S)$ consists of the ordinary Weierstrass points on X. Since $i_*(\gamma) = -\gamma$ for all $\gamma \in \widetilde{H}_1(X, \mathbb{Z})$ one has

$$i^*(\omega_j) = -\omega_j \quad \text{for } j = 1, 2, \dots$$

and by (A.10)(ii),

$$\mathbb{C}\vec{\omega}(x) = \mathbb{C}\vec{\omega}(y) \iff x \in \{y, i(y)\}$$

for all $x, y \in X$.

We proved in §9 that $f\colon S^{(-1)} \to B$ almost provides a global paramaterization of $\Theta \setminus \Theta_1$. The "almost" is needed, first, because it is possible to have $f([y]) = f([z])$ with $y \neq z$ when $\pi(y_i) = \pi(z_i)$ for all i, second, because $df([y])$ trivially fails to be invertible when $\pi(y_i) = \pi(y_j)$ but $y_i \neq y_j$ for some $i, j \geq 2$ and, third, because the image of f can slop over into Θ_1. We also showed that Θ was smooth at all points of $\Theta \setminus \Theta_1$.

Now, in the hyperelliptic case, we improve these results. We show in Proposition 10.2 that $e \in \Theta \setminus \bigcap_{k=1}^{\infty} \Theta_k$ is a singular point of Θ if and only if $e \in \Theta^{(1,i(1))} = \dots = \Theta^{(m,i(m))}$. We show that any slop over of the range of f must be in Θ_2 and not in $\Theta_1 \setminus \Theta_2$. Indeed, we show in (10.4) that f maps all points of the form

$\big(x, i(x), y_4, y_5, \dots\big)$ into $\Theta^{(1,i(1))}$ and, provided $y_k = y_j$ whenever $\pi(y_k) = \pi(y_j)$, $\dim\ker df([y]) = 0$ if and only if $\pi(y_k) \neq i\pi(y_j)$ for all $k \neq j$.

Proposition 9.10, that relates the ramification set of the Gauss map to the canonical image of X, does not hold for hyperelliptic curves. In fact, we show in the remark after Lemma 10.4 that the Gauss-map is unramified at all points $e = F(y_2, y_3, \dots)$ of Θ_{reg} where $y \in W^{(-1)}$ such that none of the points y_j lies over a Weierstrass point of X. Consequently, the strategy for the proof of Torelli's theorem will be different in the hyperelliptic case. It will use the special structure of the sets

$$H^{(b)} = \{F(b, y_3, y_4, \dots) \mid y \in W^{(-2)}\} \cap \Theta_{\text{reg}}.$$

We will see in Proposition 10.5 that the Gauss map is ramified at all points of $H^{(b)}$, and that at generic points of $H^{(b)}$ the kernel of H is just $\mathbb{C}\vec{\omega}(b)$. So, while in the nonhyperelliptic case one can recover almost the whole image of X under the canonical map from an analysis of the spaces $\ker H(e)$, $e \in \Theta_{\text{reg}}$, in the hyperelliptic case we can only identify the points $\mathbb{C}\vec{\omega}(b)$ where b runs through the Weierstrass points of X. The proof of the Torelli theorem in the hyperelliptic case uses the image of $\widetilde{X}$ under the map

$$g\colon \widetilde{X} \to \Theta,$$

$$x \mapsto F(b, y_3, y_4, \dots) - \int_b^x \vec{\omega} = F(x, y_3, y_4, \dots)$$

where $\pi(b)$ is a Weierstrass point of X and $(y_3, y_4, \dots) \in W^{(-2)}$ is generic. By Corollary 9.7

$$\psi\big(\nabla\theta\big(g(x)\big)\big) \in \widetilde{\Omega}(y) \quad \text{for all } x \in \widetilde{X}.$$

In Proposition 10.5(c) we show that the curve $g(\widetilde{X})$ is essentially characterized by this property. This observation is the key point of the proof of Torelli's theorem for hyperelliptic curves.

PROPOSITION 10.1. *Let X be hyperelliptic. Then there is $\eta > 0$, $N > 0$ and there are closed loops a_j inside $Y_j(\eta)$ for $j \geq N$ such that*

(i) *$i(a_j) = a_j$, and a_j represents A_j in $\widetilde{H}_1(X, \mathbb{Z})$.*
(ii) *i has exactly two fixed points on a_j, and these are the only points of $Y_j \cap \mathfrak{B}$.*
(iii) *$\mathfrak{B} \setminus (\mathfrak{B} \cap \bigcup_{j=N}^{\infty} Y_j)$ consists of finitely many points.*
(iv) *If $(z_j)_{j \geq N}$ is a sequence of points in X such that $z_j \in a_j$ for all j, then there is a sequence $(y_j)_{j \geq N}$ in $W^{(-N+1)}$ such that $\pi(y_j) = z_j$ for all $j \geq N$.*

PROOF. Choose $b_0 \in \widetilde{X}$ such that $\pi(b_0) \in \mathfrak{B}$ and $e \in B$ such that

$$\theta\Big(e - \hat{e}_1 + \hat{e}_\nu + \int_{b_0}^{\infty} \vec{\omega}\Big) \neq 0, \quad \theta\Big(-e - \hat{e}_1 + \hat{e}_\nu + \int_{b_0}^{\infty} \vec{\omega}\Big) \neq 0.$$

Then, by Theorems 8.4 and 7.11, there exist x resp. x' in $W^{(0)}$ and $\eta' > 0$ such that $\pi(x_1), \pi(x_2), \dots$ are the zeroes of

$$\theta\Big(e + \int_{b_0}^{x} \vec{\omega}\Big) = \theta\Big(e + \int_{b_0}^{\infty} \vec{\omega} + \int_{\infty}^{x} \vec{\omega}\Big)$$

$\pi(x'_1), \pi(x'_2), \dots$ are the zeroes of

$$\theta\Big(-e + \int_{b_0}^{x} \vec{\omega}\Big) = \theta\Big(-e + \int_{b_0}^{\infty} \vec{\omega} + \int_{\infty}^{x} \vec{\omega}\Big)$$

$x_j, x'_j \in Y_j(\eta')$ for all sufficiently big j and

$$\hat{e} - e - \int_{b_0}^{\infty} \vec{\omega} = \sum_{j=1}^{\infty} \int_{\tilde{x}_j}^{x_j} \vec{\omega}, \quad \hat{e} + e - \int_{b_0}^{\infty} \vec{\omega} = \sum_{j=1}^{\infty} \int_{\tilde{x}_j}^{x'_j} \vec{\omega}$$

In particular

$$2e = \sum_{j=1}^{\infty} \int_{x_j}^{x'_j} \vec{\omega}.$$

Since $i^*(\vec{\omega}) = -\vec{\omega}$ the zeroes of $\theta(-e + \int_b^x \vec{\omega})$ are the images, under the hyperelliptic involution i, of the zeroes of $\theta(e + \int_b^x \vec{\omega})$. So $\{\pi(x'_1), \pi(x'_2), \dots\}$ is a permutation of $\{i(\pi(x_1)), i(\pi(x_2)), \dots\}$. By Proposition 6.16 and Theorem 6.4 one has for all j big enough

$$\left| \frac{\omega_k(\pi(x_j))}{\omega_j(\pi(x_j))} \right| \le \frac{1}{2}, \quad \left| \frac{\omega_k(\pi(x'_j))}{\omega_j(\pi(x'_j))} \right| \le \frac{1}{2} \quad \text{for } k \neq j.$$

Since $|\omega_k(i(\pi(x_j)))/\omega_j(i(\pi(x_j)))| = |\omega_k(\pi(x_j))/\omega_j(\pi(x_j))|$ this implies that $i(\pi(x_j)) = \pi(x'_j)$ for all j big enough. After performing further finite permutations we may assume that

$$i(\pi(x_j)) = \pi(x'_j) \quad \text{for } j = 1, 2, \dots.$$

For all sufficiently big j the equations

$$\int_{x_j}^{b_j^+} \omega_j = \frac{1}{2} \int_{x_j}^{x'_j} \omega_j, \quad \int_{x_j}^{b_j^-} \omega_j = \frac{1}{2} \int_{x_j}^{x'_j} \omega_j + \frac{1}{2} \tag{10.1}$$

have unique solutions b_j^+ resp. b_j^- in $\widetilde{Y}_j(\frac{1}{2}\eta')$. We claim that $\pi(b_j^+)$ and $\pi(b_j^-)$ are fixed points of i. To verify this for $\pi(b_j^+)$ let γ be a path in $\widetilde{Y}_j(\frac{1}{2}\eta')$ joining b_j^+ to x'_j such that for all $x \in \gamma$

$$\left| \int_{b_j^+}^{x} \omega_j \right| \le \left| \int_{b_j^+}^{x'_j} \omega_j \right|.$$

Then $\pi(\gamma)$ is a path in $\widetilde{Y}_j(\frac{1}{2}\eta')$ starting at $\pi(b_j^+)$ and ending at $\pi(x'_j)$. Then $i(\pi(\gamma))$ starts at $i(\pi(b_j^+))$ and ends at $i(\pi(x'_j)) = \pi(x_j)$, and also lies in $\widetilde{Y}_j(\frac{1}{2}\eta')$ provided that j is big enough. Let γ' be the lift of the path $i(\pi(\gamma))$, ending at x_j. Then

$$\begin{aligned}\int_{\gamma'} \omega_j = -\int_{\pi(\gamma')} i^*(\omega_j) = -\int_{\pi(\gamma)} \omega_j &= -\int_{b_j^+}^{x'_j} \omega_j \\ &= -\int_{b_j^+}^{x_j} \omega_j + \int_{x'_j}^{x_j} \omega_j = \frac{1}{2} \int_{x'_j}^{x_j} \omega_j = \int_{b_j^+}^{x_j} \omega_j\end{aligned}$$

Since $\gamma' \subset \widetilde{Y}_j(\frac{1}{2}\eta')$ and has endpoint x_j its starting point is b_j^+ (recall that $x \mapsto \int_{x_j}^{x} \omega_j$ is a coordinate on $\widetilde{Y}_j$). This shows that $i(\pi(b_j^+)) = \pi(b_j^+)$. The fact that $i(\pi(b_j^-)) = \pi(b_j^-)$ is verified in the same way. Also observe that $\pi(b_j^+) \neq \pi(b_j^-)$ since $\int_{b_j^+}^{b_j^-} \omega_j = \frac{1}{2}$. Thus we have shown that for j big enough $\sharp(Y_j(\frac{1}{2}\eta') \cap \mathfrak{B}) \ge 2$.

Now let a_j be image of $\tilde{a}_j := \{x \in \widetilde{Y}_j \mid \int_{b_j^+}^{x} \omega_j \in [0, 1]\}$ under π. If j is bigger than some constant N it is a closed loop in $Y_j(\frac{1}{4}\eta')$ that represents $\pm A_j$ in

$\widetilde{H}_1(X,\mathbb{Z})$. It is obviously invariant under i and contains the points $\pi(b_j^+)$, $\pi(b_j^-)$. So condition (i) of the proposition is fulfilled.

If $(z_j)_{j\geq N}$ is a sequence with $z_j \in a_j$ let y_j be a lift of z_j in $\tilde{a}_j$. Since $(x_j)_{j\in\mathbb{N}}$ and $(x'_j)_{j\in\mathbb{N}}$ lie in $W^{(0)}$ it follows that $(b_j^+)_{j\geq N} \in W^{(-N+1)}$. Now $|\int_{b_j^+}^{y_j}\omega_j| \leq 1$, so that also $(y_j)_{j\geq N} \in W^{(-N+1)}$. This proves part (iv) of the proposition.

Consider the exhaustion $\ldots X^{(n)} \subset X^{(n+1)} \subset \cdots \subset X$ of (7.6). Since a_j is invariant under the involution i and $\partial X^{(n)}$ does not meet a_j then $i(\partial X^{(n)})$ also does not meet any of the loops a_j. Put

$$\check{X}^{(n)} := X^{(n)} \cup i(X^{(n)}).$$

Clearly $a_j \cap \check{X}^{(n)} = a_j$ whenever $a_j \cap \check{X}_j^{(n)} \neq \varnothing$, and $\partial\check{X}^{(n)}$ is piecewise smooth consisting of m closed loops. Furthermore

$$\mathrm{genus}(\check{X}^{(n)}) = (N-1) + \sharp\{j \geq N \mid a_j \subset \check{X}^{(n)}\} \quad \text{for } n \gg 0.$$

So for n sufficiently big the topological Euler characteristic $\chi(\check{X}^{(n)})$ of $\check{X}^{(n)}$ is

$$2 - 2\sharp\{j \geq N \mid a_j \subset \check{X}^{(n)}\} - 2(N-1) - m.$$

Put

$$h_n := \sharp(\mathfrak{B} \cap \check{X}^{(n)}).$$

Clearly $h_n \geq 2\sharp\{j \geq N \mid a_j \subset \check{X}^{(n)}\}$.

The topological Euler characteristic of the quotient $\check{X}^{(n)}/(i)$ fulfils

$$\chi\big(\check{X}^{(n)}/(i)\big) \geq 2 - \sharp\{j \geq N \mid a_j \subset \check{X}^{(n)}\} - N - m + \tfrac{1}{2}h_n.$$

Since $\chi\big(\check{X}^{(n)} \mid (i)\big) \leq 2$ we get

$$h_n \leq 2(N+m) + 2\sharp\{j \geq N \mid a_j \subset \check{X}^{(n)}\}.$$

Thus there is a constant such that

$$h_n = \mathrm{const} + 2\sharp\{j \geq N \mid a_j \subset \check{X}^{(n)}\}$$

for all sufficiently big n. This proves parts (ii) and (iii) of the proposition. □

It follows from the preceeding discussion that for each $\nu \in \{1,\ldots,m\}$ there is a unique $i(\nu) \in \{1,\ldots,m\}$ such that

$$i(X_{\mathrm{reg}}^{(\nu)} \setminus X^{(n)}) \cap X_{\mathrm{reg}}^{(i(\nu))} \neq \varnothing \quad \text{for } n \gg 0.$$

The map $\nu \to i(\nu)$ is an involution on $\{1,\ldots,m\}$. We may assume without loss of generality that the starting point $x^{(0)}$ of the paths P_ν of Lemma 7.5 is a fixed point of the hyperelliptic involution i. Then for each $\nu = 1,\ldots,m$ there is $\ell_\nu \in B^*$, $c_\nu \neq 0$ such that for all $e \in B$

$$\Theta(e + \hat{e}_\nu) = c_\nu e^{\ell_\nu(e)}\Theta(e - \hat{e}_{i(\nu)}) \tag{10.2}$$

To see this observe that, by Proposition 7.8, for all t sufficiently big there are paths γ_t joining $i\big(P_\nu(t)\big)$ to $P_{i(\nu)}(t)$ inside $\big(X_{\mathrm{reg}}^{i(\nu)} \cap i(X_{\mathrm{reg}}^{(\nu)})\big) \setminus X^{(n(t))}$ with $\lim_{t\to\infty}\int_{\gamma_t}\vec{\omega} = 0$, $\lim_{t\to\infty} n(t) = \infty$.

Therefore there is a finite linear combination $v_B^{(\nu)}$ of B-cycles and that for all $t \gg 0$ the closed path consisting of $i(P_\nu[0,t])$, γ_t and $P_{i(\nu)}[t,0]$ is homologous to $v_B^{(\nu)}$ modulo a (t-dependent) finite linear combination of A-cycles.

Therefore by the transformation properties of the theta function

$$\begin{aligned}\theta(e+\hat{e}_\nu) &= \lim_{t\to\infty} \theta\Big(e + \int_{P_\nu([0,t])} \vec{\omega}\Big)\\ &= \lim_{t\to\infty} \theta\Big(e - \int_{i(P_\nu[0,t])} \vec{\omega}\Big)\\ &= \lim_{t\to\infty} c'_\nu e^{\ell_\nu(e-\int_{i(P_\nu[0,t])}\vec{\omega})}\theta\Big(e - \int_{P_{i(\nu)}[0,t]} \vec{\omega} - \int_{\gamma_t}\vec{\omega}\Big)\end{aligned}$$

with some $\ell_\nu \in B^*$, $c'_\nu \in \mathbb{C}\setminus\{0\}$ depending on $v_B^{(\nu)}$ only. Taking the limit we get the result.

For $x \in X$, $\nu = 1,\dots,m$, $e \in B$ we say that $\theta(e+\int_x^{i(\infty_\nu)}\vec{\omega}) = 0$ if $\theta(e+\int_\gamma \vec{\omega}) = 0$ for some path γ joining x to a point of $i\big(P_\nu([0,\infty])\big)$ and then following $i\big(P_\nu([0,\infty])\big)$ out to infinity. Using (10.2) one easily sees that

$$\theta\Big(e + \int_x^{i(\infty_\nu)} \vec{\omega}\Big) = 0 \iff \theta\Big(e + \int_x^{\infty_{i(\nu)}} \vec{\omega}\Big) = 0. \tag{10.3}$$

To study $\operatorname{Sing}\Theta$ we use the prime form

$$E_e(x_1,x_2) := \theta\Big(e + \int_{x_1}^{x_2}\vec{\omega}\Big), \quad (e\in B, x_1, x_2 \in \widetilde{X}).$$

PROPOSITION 10.2. *Assume that* X *is hyperelliptic*

(a) $\Theta^{(\nu_1,\dots,\nu_m)} = -\Theta^{(i(\nu_1),\dots,i(\nu_n))}$ *for all* $\nu_1,\dots,\nu_n \in \{1,\dots,m\}$.

(b) $\{e \in B \mid E_e(x_1,x_2) = 0 \forall x_1,x_2\} = \Theta^{(1,i(1))} = \dots = \Theta^{(m,i(m))}$. *This set is contained in* $\operatorname{Sing}\Theta$.

(c) *If* $e \in \Theta \setminus \bigcap_{k=1}^\infty \Theta_\nu$ *then* $e \in \operatorname{Sing}\Theta$ *if and only if* $E_e(x_1,x_2)$ *is identically zero.*

REMARK. If $i(\nu) = \nu$ for $\nu = 1,\dots,m$ then, by the pigeonhole principle and (b) every point of Θ_{m+1} lies in $\operatorname{Sing}\Theta$. So in this case

$$\operatorname{Sing}\Theta = \{e \in B \mid E_e(x_1,x_2) \equiv 0\}.$$

PROOF. (a) Let $e \in B$. Since $i^*(\vec{\omega}) = -\vec{\omega}$ one has for $x_1,\dots,x_{n-1} \in X$

$$\theta\Big(e - \int_{\infty_{\nu_1}}^{x_1}\vec{\omega} - \dots - \int_{\infty_{\nu_n}}^{x_n} i\Big) = 0 \iff \theta\Big(e + \int_{i(\infty_{\nu_1})}^{i(x_1)}\vec{\omega} + \dots + \int_{i(\infty_{\nu_1})}^{i(x_n)}\vec{\omega}\Big) = 0.$$

By (10.3) and the evenness of the Theta function this is the case if and only if

$$\Theta\Big(-e - \int_{\infty_{i(\nu_1)}}^{i(x_1)}\vec{\omega} - \dots - \int_{\infty_{i(\nu_n)}}^{i(x_n)}\vec{\omega}\Big) = 0.$$

This shows that

$$e \in \Theta^{(\nu_1,\dots,\nu_n)} \implies -e \in \Theta^{(i(\nu_1),\dots,i(\nu_n))}.$$

(b) Let $e \in B$. By (10.3), for any $x_1, x_2 \in X$, $\nu = 1,\dots,m$

$$\begin{aligned}\theta\Big(e + \int_{x_1}^{x_2}\vec{\omega}\Big) = 0 &\iff \theta\Big(e + \int_{x_1}^{\infty_\nu}\vec{\omega} + \int_{\infty_\nu}^{x_2}\vec{\omega}\Big) = 0\\ &\iff \theta\Big(e - \int_{\infty_\nu}^{x_1}\vec{\omega} - \int_{\infty_{i(\nu)}}^{i(x_2)}\vec{\omega}\Big) = 0.\end{aligned}$$

This shows that

$$\{e \in B \mid E_e(x_1, x_2) \equiv 0\} = \theta^{(\nu, i(\nu))} \quad for \nu = 1, \dots, m.$$

For any e in this set the vectors $\vec{\omega}(x) = d/dy \int_x^y \vec{\omega}\big|_{y=x}$ all lie in $\ker \nabla\theta(e)$. As these vectors span B, $e \in \operatorname{Sing}\Theta$.

(c) To verify (c) choose $e \in \Theta \setminus \Theta_k$ for some k such that $E_e(x_1, x_2)$ is not identically zero. There is $n \leq k$, $\nu_1, \dots, \nu_{n-1} \in \{1, \dots, m\}$ such that

$$e \in \Theta^{(\nu_1, \dots, \nu_{n-1})} \setminus \Theta_n, \quad \nu_i \neq i(\nu_j) \quad \text{for } i \neq j.$$

By Theorem 9.1 there is $y \in W^{(-n)}$ such that

$$e = F_{(\nu_1, \dots, \nu_{n-1})}(y).$$

Choose $z \in \widetilde{X}$ such that $\pi(z) \neq \pi(y_j)$ for $j = n+1, n+2, \dots$ and such that $x \mapsto E_e(z, x)$ is not identically zero on $\widetilde{X}$. This is possible by the assumption that $E_e(x_1, x_2)$ is not identically zero. Let U be a bounded neighbourhood of $\pi(z)$ in X containing none of the points $\pi(y_{n+1}), \pi(y_{n+2}), \dots$. As x_j goes to ∞_{ν_j} the function

$$x \mapsto \theta\left(e - \sum_{j=1}^{n-1} \int_{\infty_{\nu_j}}^{x_j} \vec{\omega} + \int_z^x \vec{\omega}\right)$$

on U goes to $x \mapsto E_e(z, x)$. By Theorem 9.1, for generic $x_1, \dots, x_{n-1}$ the function above has only one zero in U, namely $\pi(z)$, and that zero is a simple zero. Therefore

$$x \mapsto E_e(x, y) = \theta\left(e + \int_z^x \vec{\omega}\right)$$

has z as a simple zero. So $\nabla\theta(e) \neq 0$, i.e., $e \notin \operatorname{Sing}\Theta$.

□

Observe that for any $x \in X$ one has

$$\int_{\widetilde{x}_2}^x \vec{\omega} + \int_{\widetilde{x}_3}^{i(x)} \vec{\omega} = \int_{\widetilde{x}_2}^{\infty_\nu} \vec{\omega} + \int_{\widetilde{x}_3}^{i(\infty_\nu)} \vec{\omega} \quad \text{for all } \nu = 1, \dots, m.$$

Therefore

$$F(x, i(x), y_4, y_5, \dots) \in \Theta^{(1, i(1))} \subset \operatorname{Sing}\Theta \quad \forall y \in W^{(-3)},\ x \in \widetilde{X}. \tag{10.4}$$

This is essentially the only way the image of F meets Θ_1. Precisely,

LEMMA 10.3. *Let $y \in W^{(-1)}$ such that $F(y) \notin \operatorname{Sing}\Theta$, $F(y) \notin \bigcap_{k=1}^{\infty} \Theta_k$. Then $F(y) \notin \Theta_1$.*

PROOF. Put $e := F(y)$. Assume that $e \in \Theta_1$. Then there is $n \geq 2$ and $\nu_1, \dots, \nu_{n-1} \in \{1, \dots, n\}$ such that

$$e \in \Theta^{(\nu_1, \dots, \nu_{n-1})} \setminus \Theta_n, \quad \nu_i \neq i(\nu_j) \quad \text{for } i \neq j.$$

Write

$$e = F^{(\nu_1, \dots, \nu_{n-1})}(z)$$

with $z \in W^{(-n)}$. For generic $u, u_1, \dots, u_{n-1}$ the zeroes of

$$\theta\left(e - \int_{\infty_{\nu_1}}^{u_1} \vec{\omega} - \dots - \int_{\infty_{\nu_{n-1}}}^{u_{n-1}} \vec{\omega} + \int_u^x \vec{\omega}\right)$$

are precisely $\pi(u), \pi(u_1), \ldots, \pi(u_{n-1}), \pi(z_{n+1}), \pi(z_{n+2}), \ldots$. By the previous proposition we can choose u such that

$$\theta\left(e + \int_u^x \vec{\omega}\right)$$

is not identically zero. Taking the limit as $u_{\nu_j} \to \infty_{\nu_j}$, $j = 1, \ldots, n-1$, we see that the zeroes of $\theta(e + \int_u^x \vec{\omega})$ are precisely $\pi(u), \pi(z_{n+1}), \ldots$.

On the other hand there is $y' \in W^{(-1)}$ arbitrarily close to y such that $F(y') \in \Theta \setminus \Theta_1$. For generic u and such y' the function $\Theta(F(y') + \int_u^x \vec{\omega})$ has $\pi(u), \pi(y_2'), \ldots$ as zeroes. As y' goes to y this function converges to $\theta(e + \int_u^x \vec{\omega})$. Consequently this function has all the points $\pi(u), \pi(y_2), \pi(y_3), \ldots$ as zeroes. Before we noticed that the zeroes of this function are precisely $\pi(u), \pi(z_{n+1}), \ldots$. Since $n \geq 2$ this is a contradiction. □

LEMMA 10.4. *Let $i\colon X \to X$ be the hyperelliptic involution. Let $y = (y_2, y_3, \ldots) \in W^{(-1)}$ be such that for all $k \neq j$ one has $\pi(y_k) \neq i(\pi(y_j))$ and such that $y_k = y_\ell$ whenever $\pi(y_k) = \pi(y_\ell)$. Then*

(a) *$df([y])$ has* corank *one and* $\dim \widetilde{\Omega}(y) = 1$.
(b) *The roots of a nonzero differential $\omega \in \widetilde{\Omega}(y)$ are precisely $\pi(y_2), i(\pi(y_2)), \pi(y_3), i(\pi(y_3)), \ldots$.*

REMARK. Let $y \in W^{(-1)}$ be as in Lemma 10.4, and put $e = F(y)$. If none of the points $\pi(y_j)$ is a Weierstrass point, Corollary 9.9 shows that $H(e)$ is injective. On the other hand, if exactly one of the points $\pi(y_j)$ is a Weierstrass point b then $H(e)$ has a one-dimensional kernel, namely $\mathbb{C}\vec{\omega}(b)$.

PROOF OF LEMMA 10.4. First let $y' = (y_2', y_3', \ldots)$ be any element of $W^{(-1)}$ such that $\pi(y_k') \neq i(\pi(y_j'))$ for $k \neq j$ and such that $df([y'])$ has corank one. By Lemma 9.6 one has $\big((\operatorname{range} df([y']))^\perp\big) = \widetilde{\Omega}(y')$, so $\dim \widetilde{\Omega}(y') = 1$.

Let ω be a generator of $\widetilde{\Omega}(y')$. Assume that ω has a zero z different from $\pi(y_2'), i(\pi(y_2')), i(y_3'), i(\pi(y_3')), \ldots$ (counted with multiplicity). Since $i^*(\omega) = -\omega$, $i(z)$ is also a zero of ω different from $\pi(y_2')$, $i(\pi(y_2'))$, $\pi(y_3'), \ldots$. So

$$\omega'(x) := \frac{1}{\tau(x) - \tau(z)} \omega(x)$$

is a holomorphic differential which vanishes at y_j' with multiplicity at least $\sharp\{k \mid y_k' = y_j'\}$. Put

$$\lambda_j' := \int_{A_j} \omega'.$$

Clearly there is a constant C such that

$$|\lambda_j'| \leq C \left| \int_{A_j} \omega \right|.$$

Therefore the sequence $(\lambda_j')_{j=1,2,\ldots}$ lies in B^* and $\psi(\lambda') = \sum_{j=1}^\infty \lambda_j' \omega_j \in \widetilde{\Omega}$. So

$$\widetilde{\omega} := \omega' - \sum_{j=1}^\infty \lambda_j' \omega_j$$

is square integrable. By construction $\int_{A_j} \widetilde{\omega} = 0$ for all j, so that, by (A.8), $\widetilde{\omega} = 0$ and

$$\omega' = \sum_{j=1}^{\infty} \lambda'_j \omega_j.$$

In particular $\omega' \in \widetilde{\Omega}$. Now clearly $\omega' \in \widetilde{\Omega}(y)$, so ω' and ω only differ by a multiplicative constant. But this is impossible.

We have just proven Lemma 10.4 under the additional hypothesis that the range of $df([y])$ has codimension one. We now prove that there is no $y = (y_2, y_3, y_4, \dots) \in W^{(-1)}$ obeying

$$\begin{gathered} y_j = y_k \text{ whenever } \pi(y_j) = \pi(y_k), \\ \pi(y_j) \neq i(\pi(y_k)) \quad \text{for all } j \neq k, \\ \operatorname{codim} \operatorname{range} df([y]) \geq 2. \end{gathered}$$

The proof is by induction on the excess multiplicity

$$M([y]) = \sum_{z \in \widetilde{X}} [\sharp\{j \geq 2 \mid y_j = z\} - 1].$$

First suppose that $M([y]) = 0$, in other words the y_j's on all distinct, and that

$$\operatorname{corank} df([y]) = \operatorname{codim} \operatorname{span}[\omega(y_2), \omega(y_3), \dots] = n \geq 2.$$

Then, by Lemma 8.3(a), $df([y])$ has a nontrivial kernel so that there is an $m \geq 2$ with $\omega(y_m) \in \operatorname{span}[\omega(y_{m+1}), \omega(y_{m+2}) \dots]$. There is a nonempty open subset $\mathcal{U}$ of $\widetilde{X}$ that contains no element of $\{\pi^{-1}(\pi(y_j)), \pi^{-1}(i(\pi(y_j))) \mid j \geq 2\}$. Since $\{\vec{\omega}(y) \mid y \in \mathcal{U}\}$ cannot be contained in a hyperplane we can choose $y'_m \in \mathcal{U}$ such that $y' = (y_2, y_3, \dots, \not{y}_m, y'_m, y_{m+1}, \dots)$ obeys

$$\begin{gathered} \pi(y'_k) \neq \pi(y_m) \qquad \text{for all } k \geq 2, \\ \pi(y'_k) \neq \pi(y'_j) \qquad \text{for all } k \neq j, \\ \pi(y'_k) \neq i\pi(y'_j) \qquad \text{for all } k \neq j, \\ \operatorname{range} df([y]) \subset \operatorname{range} df([y']), \\ \operatorname{corank} df([y']) = n - 1. \end{gathered}$$

Repeating this process another $n-2$ times we can find $y'' \in W^{(-1)}$ such that

$$\begin{gathered} \pi(y''_k) \neq \pi(y_m) \qquad \text{for all } k \geq 2, \\ \pi(y''_k) \neq \pi(y''_j) \qquad \text{for all } k \neq j, \\ \pi(y''_k) \neq i\pi(y''_j) \qquad \text{for all } k \neq j, \\ \operatorname{range} df([y]) \subset \operatorname{range} df([y'']), \\ \operatorname{corank} df([y'']) = 1 \end{gathered}$$

Then $\widetilde{\Omega}(y) \supset \widetilde{\Omega}(y'') \neq \{0\}$ and any non zero $\omega \in \widetilde{\Omega}(y'')$ is zero at $\pi(y_m)$ as well as at $\pi(y''_2), i(\pi(y''_2)), \pi(y''_3), i(\pi(y''_3)), \dots$. This contradicts the part of Lemma 10.4 proven earlier and the proof of Lemma 10.4 for the case $M([y]) = 0$ is complete.

Finally, suppose that Lemma 10.4 has been proven for all y with $M([y]) \leq m$ and consider y with $M([y]) = m + 1$ and $\operatorname{corank} df([y]) = n > 1$. We may permute the y's so that $y_2 = y_3 = \cdots = y_{m+3}$ and $y_2 \neq y_j$ for $j \geq m+4$. As above, we can move the $(m+3)^{\text{th}}$ component of y a bit to give a $y' \in W^{(-1)}$ obeying

$$\begin{gathered} M([y']) = m \\ \sharp\{j \geq 2 \mid y'_j = y_2\} = m, \\ y'_j = y'_k \text{ whenever } \pi(y'_j) = \pi(y'_k), \\ \pi(y'_j) \neq i\big(\pi(y'_k)\big) \quad \text{for all } j \neq k. \end{gathered}$$

There are three possibilities. Either $\operatorname{corank} df([y']) = \operatorname{corank} df([y]) > 1$, which violates the inductive hypothesis, or $\operatorname{corank} df([y']) = \operatorname{corank} df([y]) - 1 > 1$, which also violates the inductive hypothesis, or

$$\operatorname{corank} df([y']) = \operatorname{corank} df([y]) - 1 = 1.$$

The last case arises only when $n = 2$ and

$$\vec{\omega}^{(m+1)}(y_2) \in \operatorname{span}\big\{\vec{\omega}(y_2), \ldots, \vec{\omega}^{(m)}(y_2), \vec{\omega}(y_{m+4}) \ldots\big\}.$$

But then $\operatorname{range} df([y']) \supset \operatorname{range} df([y])$ implies $\widetilde{\Omega}(y') \subset \widetilde{\Omega}(y)$ and we get the same contradiction as in the case $M([y]) = 0$. □

For $b \in \widetilde{X}$ such that $\pi(b) \in \mathfrak{B}$ we consider the map

$$F^{(b)} : W^{(-2)} \to \Theta, \quad y \mapsto F(b, y_3, y_4, \ldots),$$

and the induced map $f^{(b)} \colon S^{(-2)} \to \Theta$. We denote by $H^{(b)}$ the intersection of the image of $F^{(b)}$ with Θ_{reg}. Observe that by Lemma 9.6

$$\big(\operatorname{range} df^{(b)}(y)\big)^{\perp} = \psi^{-1}\big(\widetilde{\Omega}(y)\big). \tag{10.5}$$

PROPOSITION 10.5. (a) *Let $b \in \widetilde{X}$ such that $\pi(b) \in \mathfrak{B}$. Then there is an open dense subset U_b of $W^{(-2)}$ such that $F^{(b)}(U_b) \subset \Theta \setminus \Theta_1$. Furthermore*

$$\mathbb{C}\vec{\omega}(b) \subset \ker H\big(F^{(b)}(y)\big)$$

for all $y \in W^{(-2)}$ with $F^{(b)}(y) \in \Theta_{\text{reg}}$, and equality holds for all y in a dense subset of full Baire category in $W^{(-2)}$. Finally for each $e \in H^{(b)}$ there is $[y] \in S^{(-2)}$ such that $f^{(b)}([y]) = e$ and $\operatorname{corank} df^{(b)}([y]) = 2$.

(b) *Let $\mathcal{O}$ be an open subset of $\mathbb{C}^N \times B^{(-n)}$ for some $N, n \in \mathbb{N}$ and $g \colon \mathcal{O} \to B$ a holomorphic map such that $g(\mathcal{O}) \subset \Theta_{\text{reg}}$, for all $e' \in \mathcal{O}$ the linear map $dg(e')$ has* corank *two and is boundedly invertible onto its range, and such that there is a non zero vector $v \in B$ obeying*

$$\mathbb{C}v \subset \ker H\big(g(e')\big)$$

for all $e' \in \mathcal{O}$, with equality for some e'. Then

either $g(\mathcal{O}) \subset \Theta^{(\nu)}$ for some $\nu = 1, \ldots, m$, and $v \in \ker H(e)$ for all $e \in (\Theta^{(\nu)} \setminus \Theta_2) \cap \Theta_{\text{reg}}$ or there is an open dense subset $\mathcal{O}' \subset \mathcal{O}$ and $b \in \widetilde{X}$ with $\pi(b) \in \mathfrak{B}$ such that

$$g(\mathcal{O}') \subset H^{(b)} \setminus \Theta_1, \quad \mathbb{C}v = \mathbb{C}\vec{\omega}(b).$$

(c) *Let $b \in \widetilde{X}$ such that $\pi(b) \in \mathfrak{B}$, and let $y = (y_3, y_4, \ldots) \in U_b$. Let Z be a connected Riemann surface, $z_0 \in Z$ and*

$$g \colon Z \to \Theta_{\text{reg}} \text{ with } g(z_0) = F^{(b)}(y)$$

a holomorphic map such that for all $z \in Z$

$$\psi\big(\nabla\theta\big(g(z)\big)\big) \in \widetilde{\Omega}(y).$$

Then there is a subset $Z' \subset Z$ which is the complement of a discrete set and a holomorphic map

$$\Phi' \colon Z' \to \widetilde{X}$$

such that

$$g(z) = F^{(b)}(y) - \int_b^{\Phi'(z)} \vec{\omega} \quad \textit{for all } z \in Z'.$$

PROOF. If $e = F(y_2, y_3, \dots)$ is a point of Θ_{reg} then by Lemma 10.4(b) and Proposition 9.8

$$\dim\ker H(e) = \sharp\{j \mid \pi(y_j) \in \mathfrak{B}\}$$

and

$$\ker H(e) = \operatorname{span}\big\{\vec{\omega}(b) \mid b \in \mathfrak{B} \cap \{\pi(y_2), \pi(y_3), \dots\}\big\}. \tag{10.6}$$

We will use this fact for the proof of the first two parts of the Proposition.

(a) By Lemma 9.2 there is an open dense subset U_b of $W^{(-2)}$ such that $F^{(b)}(y) \notin \Theta_2$ for $y \in U_b$. Then

$$F^{(b)}(U_b) \subset \Theta \setminus \Theta_1$$

by Lemma 10.3. The fact that

$$\mathbb{C}\vec{\omega}(b) \subset \ker H\big(F^{(b)}(y)\big)$$

for all $y \in W^{(-2)}$ with $F^{(b)}(y) \in \Theta_{\text{reg}}$ and that equality holds on a set of full Baire category follows directly from (10.6).

Finally let $e \in H^{(b)}$. Then there is $y \in W^{(-2)}$ such that $e = F^{(b)}(y)$, $y_k = y_j$ whenever $\pi(y_k) = \pi(y_j)$. By (10.4) and Lemma 10.4 $\operatorname{corank} df^{(b)}([y]) = 2$.

(b) Assume that $g(\mathcal{O}) \not\subset \Theta^{(\nu)}$ for all $\nu = 1, \dots, m$. Then there is an open dense $\mathcal{O}'$ of $\mathcal{O}$ such that $g(\mathcal{O}') \subset \Theta \setminus \Theta_1$. The claim now is immediate from (10.6) and (10.4) and the fact that for a point $e \in \Theta \setminus \Theta_1$ the representation $y = F(e)$ is essentially unique (Riemann's Vanishing theorem).

Now assume that $g(\mathcal{O}) \subset \Theta^{(\nu)}$ for some $\nu \in \{1, \dots, m\}$. The function $e \mapsto \lambda\langle\nabla\theta(e), v\rangle$ is holomorphic and zero on the open subset $g(\mathcal{O})$ of $\Theta^{(\nu)}$. Since dg has corank 2, $g(\mathcal{O}) \not\subset \Theta_2$. As $\Theta^{(\nu)} \setminus \Theta_2$ is connected, this function is identically zero on $\Theta^{(\nu)} \setminus \Theta_2$ so that $v \in T_e(\Theta)$ for all $e \in \Theta_{\text{reg}} \cap (\Theta^{(\nu)} \setminus \Theta_2)$. Similarly, for all $e \in g(\mathcal{O})$, $\operatorname{Hess}(\theta)(e)v \in \mathbb{C}\nabla\theta(e)$. Again, by analyticity, $\operatorname{Hess}(\theta)(e)v \in \mathbb{C}\nabla\theta(e)$ for all $e \in \Theta_{\text{reg}} \cap (\Theta^{(\nu)} \setminus \Theta_2)$.

(c) $$Z' := \{z \in Z \mid g(z) \notin \Theta_1\}$$

is the complement of a discrete subset of Z and contains the point z_0, since $y \in U_b$. Let $\pi' \colon \widetilde{Z}' \to Z'$ be the universal covering of Z' and choose some $z_0' \in (\pi')^{-1}(z_0)$. By Riemann's Vanishing theorem the range of f contains $g\big(\pi'(\widetilde{Z}')\big)$. Since f, by Lemma 10.4(b), is locally invertible on $(\Theta \setminus \Theta_1) \cap \Theta_{\text{reg}}$ and since $\widetilde{Z}'$ is simply connected there exists a holomorphic map

$$\tilde{g} \colon \widetilde{Z}' \to S^{(-1)}$$

such that

$$g\big(\pi'(z')\big) = f\big(\tilde{g}(z')\big) \quad \text{for all } z' \in \widetilde{Z}', \tag{10.7}$$
$$\tilde{g}(z'_0) = [(b, y_3, y_4, \dots)].$$

By hypothesis

$$\omega_{f(\tilde{g}(z'))} = \psi\big(\nabla\theta\big(g(\pi'(z'))\big)\big) \subset \widetilde{\Omega}(y)$$

for all $z' \in \widetilde{Z}'$. Since $f\big(\tilde{g}(z')\big) \in \Theta_{\text{reg}}$, (10.4) implies that $\tilde{g}(z') = [y'_2, y'_3 \dots]$ satisfies the hypotheses of Lemma 10.4. Hence $y_j \in \{y'_k, i(y'_k) \mid k \geq 2\}$ for all $j \geq 3$. Since $\tilde{g}(z')$ depends analytically on z' and $\tilde{g}(z'_0) = (b, y_3, y_4, \dots)$ we have, up to permutations, $y_j = y'_j$, $j \geq 3$. Call $y'_2 = \widetilde{\Phi}(z')$. Thus

$$\begin{aligned} g\big(\pi'(z')\big) &= f\big([\widetilde{\Phi}(z'), y_3, y_4, \dots]\big), \\ &= F^{(b)}(y) - \int_b^{\widetilde{\Phi}(z')} \vec{\omega}. \end{aligned}$$

Since $g\big(\pi'(z')\big)$ is holomorphic and $\vec{\omega} \neq 0$, the inverse function theorem implies that $\widetilde{\Phi}(z')$ is holomorphic.

If $z', z'' \in \widetilde{Z}'$ such that $\pi'(z') = \pi'(z'')$ then by the formula above

$$\int_b^{\widetilde{\Phi}(z')} \vec{\omega} = \int_b^{\widetilde{\Phi}(z'')} \vec{\omega}$$

so that $\widetilde{\Phi}(z') = \widetilde{\Phi}(z'')$ by (A.4)(ii).

Therefore there is a map $\Phi\colon Z' \to \widetilde{X}$ such that $\widetilde{\Phi} = \Phi \circ \pi'$. □

11. The Torelli Theorem

THEOREM 11.1. *Let $X = X^{\text{com}} \cup X^{\text{reg}} \cup X^{\text{han}}$ and $X' = X'^{\text{com}} \cup X'^{\text{reg}} \cup X'^{\text{han}}$ be Riemann surfaces that fulfil the hypothese* (GH1)–(GH6) *of* §5. *Denote their canonical homology bases by $A_1, B_1, A_2, B_2, \dots$ and $A'_1, B'_1, A'_2, B'_2, \dots$. Let $\mathcal{R}_{ij}$ resp. $\mathcal{R}'_{ij}$ be the associated period matrices. If $\mathcal{R}_{ij} = \mathcal{R}'_{ij}$ for all $i, j \in \mathbb{Z}$ then there is a biholomorphic map $\mathfrak{F}\colon X \to X'$ and $\varepsilon \in \{\pm 1\}$ such that for all $j \in \mathbb{N}$*

$$\mathfrak{F}_*(A_j) = \varepsilon A'_j, \quad \mathfrak{F}_*(B_j) = \varepsilon B'_j.$$

In the course of the proof we will mark objects belonging to the curve X' by a prime, e.g., $W'^{(-n)}$, Θ'_n etc.

By Corollary 6.17

$$B = B'$$

as sets of sequences, and the two norms $\|\cdot\|$ and $\|\cdot\|'$ are equivalent. Using this identity throughout it follows that

$$\theta = \theta', \quad \Theta = \Theta'.$$

To prepare for the proof of Torelli's Theorem we note

LEMMA 11.2. *Assume that $\mathcal{R} = \mathcal{R}'$, and that either X and X' are both hyperelliptic or both non hyperelliptic. Assume furthermore that there is a dense subset X_1 of X and a holomorphic map $\mathfrak{F}_1\colon X_1 \to X'$ such that*

$$\mathbb{C}\vec{\omega}(x) = \mathbb{C}\vec{\omega}\big(\mathfrak{F}_1(x)\big) \quad \textit{for all } x \in X_1,$$

and that $\mathfrak{F}_1(i(x)) = i'(\mathfrak{F}_1(x))$ *in the hyperelliptic case. Then* $\mathfrak{F}_1$ *can be extended to a biholomorphic map*

$$\mathfrak{F}\colon X \to X'$$

such that there is $\varepsilon \in \{\pm 1\}$ *with* $\mathfrak{F}_*(A_j) = \varepsilon A'_j$, $\mathfrak{F}_*(B_j) = \varepsilon B'_j$ *for* $j = 1, 2, 3, \dots$.

PROOF. Our first step is to extend $\mathfrak{F}_1$ to a neighbourhood of X_1. Without loss of generality we may assume that X_1 does not contain any Weierstrass points. Let $x \in X_1$. Consider the exhaustion of X by submanifolds $X^{(n)}$ constructed in (7.6). Choose $n > 0$ such that $x \in X^{(n)}$, $\mathfrak{F}_1(x) \in X'^{(n)}$, and choose $N > 0$ such that

$$\kappa_N\colon X^{(n)} \to \mathbb{P}^{N-1}, \quad x \mapsto [\omega_1(x), \dots, \omega_N(x)]$$

and

$$\kappa'_N\colon X'^{(n)} \to \mathbb{P}^{N-1}$$

are well-defined, and such that $d\kappa_N(x) \neq 0$, $d\kappa'_N(\mathfrak{F}_1(x)) \neq 0$. Then there are neighbourhoods U of x, U' of $\mathfrak{F}(x')$ such that $\kappa_N\big|_U$ and $\kappa'_N\big|_{U'}$ are embeddings. Furthermore $\kappa_N(x') = \kappa'_N(\mathfrak{F}_1(x'))$ for all $x' \in U$. Therefore we may assume that $\kappa_N(U) = \kappa'_N(U')$. We can then extend $\mathfrak{F}_1$ on U by $(\kappa'_N)^{-1} \circ \kappa_N$.

So we may assume without loss of generality that X_1 is open and dense in X. Put

$$E = \{x \in X \mid [\omega_{g'+1}(x), \omega_{g'+2}(x), \dots] = [\sigma'_\nu(g'+1), \sigma'_\nu(g'+2), \dots] \\ \text{for some } \nu = 1, \dots, m' \text{ or } \omega_{g'+1}(x) = \omega_{g'+2}(x) = \cdots = 0\}.$$

Here $\sigma'_\nu(j)$ describes the leading part of $\Phi^*_\nu(\omega'_j)$ as in Proposition 6.18, and $[\dots]$ stands for elements in projective space. The set E is discrete in X. Observe that $\limsup_{j\to\infty} |\sigma'_\nu(j)| = \infty$ if there are infinitely many handles joining X'^{reg}_ν to some X'^{reg}_μ with $\mu \neq \nu$. In this case, there is no $x \in X$ with $[\omega_{g'+1}(x), \omega_{g'+2}(x), \dots] = [\sigma'_\nu(g'+1), \sigma'_\nu(g'+2), \dots]$ by Remark 6.15.

We first show that $\mathfrak{F}_1$ can be extended holomorphically to $X \setminus E$. So let x be in $X \setminus X_1$ with $x \notin E$. We show that $\mathfrak{F}_1$ can be extended holomorphically to x. By the definition of E and Remark 6.15, there is $i_0 \geq g'+1$ and a neighbourhood U of x in X such that $\omega_{i_0}(y) \neq 0$ for all $y \in U$ and $\sup_{y\in U, i\in\mathbb{N}} |\omega_i(y)/\omega_{i_0}(y)| < \infty$. Choose a sequence $(x_n)_{n\in\mathbb{N}}$ of points in $X_1 \cap U$ with $\lim_{n\to\infty} x_n = x$. Then there is a constant C such that $|\omega_i(x_n)/\omega_{i_0}(x_n)| \leq C$ for all $i \geq g'+1$, $n \in \mathbb{N}$. We can now apply Proposition 6.18 to the sequence $(\mathfrak{F}_1(x_n))_{n\in\mathbb{N}}$ in X' and see that it has an accumulation point x' in X'.

Choose $N > 0$ such that

$$\kappa_N\colon X \to \mathbb{P}^{N-1}, \quad y \mapsto [\omega_1(y), \omega_2(y), \dots, \omega_N(y)],$$
$$\kappa'_N\colon X' \to \mathbb{P}^{N-1}, \quad y' \mapsto [\omega'_1(y'), \omega'_2(y'), \dots, \omega'_N(y')]$$

are well-defined on neighbourhoods of x and x' respectively, and that these maps are embeddings if the surfaces are nonhyperelliptic, resp. only factor over the hyperelliptic involution in the hyperelliptic case. One easily sees that $\kappa_N(x) = \kappa'_N(x')$, and as above one verifies that $\mathfrak{F}_1$ can be extended holomorphically to x with $\mathfrak{F}_1(x) = x'$. Thus $\mathfrak{F}_1$ is extended to a holomorphic map $\mathfrak{F}_2\colon X \setminus E \to X'$ such that

$$\mathbb{C}\vec{\omega}(x) = \mathbb{C}\vec{\omega}(\mathfrak{F}_2(x)) \quad \text{for all } x \in X \setminus E. \tag{11.1}$$

Now let $x \in E$. We want to extend $\mathfrak{F}_2$ holomorphically to x. By (A.4)(i) and Remark 6.15 there is $i_0 \in \mathbb{N}$ and a neighbourhood U of x in X such that $\omega_{i_0}(y) \neq 0$

for all $y \in U$ and $\sup_{y\in U,j} |\omega_j(y)/\omega_{i_0}(y)| < \infty$. As E is discrete we may assume that $U \cap E = \{x\}$. If there is a sequence $(x_n)_{n\in\mathbb{N}}$ in U with $\lim_{n\to\infty} x_n = x$ such that the sequence $\big(\mathfrak{F}_2(x_n)\big)_{n\in\mathbb{N}}$ has an accumulation point in X' then, as before, $\mathfrak{F}_2$ can be extended holomorphically to x. So we assume that this is not the case and will derive a contradiction. By Proposition 6.16 there is a finite set $J \subset \mathbb{N}$ such that $\mathfrak{F}_2(U \setminus \{x\}) \cap \phi'_j\big(\{(z_1, z_2) \in H(t'_j) \mid |z_1|, |z_2| \le \frac{1}{4}\}\big) = \varnothing$ for $j \notin J$. As, by assumption, there is no sequence $(x_n)_{n\in\mathbb{N}}$ in $U \setminus \{x\}$ with $\lim_{n\to\infty} x_n = x$ such that $\mathfrak{F}_2(x_n)$ lies in the compact set $\bigcup_{j\in J} Y'_j$ for all $n \in \mathbb{N}$, we may assume that

$$\mathfrak{F}_2(U \setminus \{x\}) \cap \phi'_j\big(\{(z_1, z_2) \in H(t'_j) \mid |z_1|, |z_2| \le \tfrac{1}{4}\}\big) = \varnothing \quad \text{for all } j \ge g' + 1.$$

Similarly we may assume that

$$\mathfrak{F}_2(U \setminus \{x\}) \cap X'^{\text{com}} = \varnothing.$$

So $\mathfrak{F}_2(U \setminus \{x\}) \subset X'^{\text{reg}}$. As $U \setminus \{x\}$ is connected there is $\nu \in \{1, \dots, m'\}$ such that

$$\mathfrak{F}_2(U \setminus \{x\}) \subset X'^{\text{reg}}_{\nu}.$$

We may assume that U has the form of a disc. That is, there is a biholomorphic map $\phi\colon \{z \in \mathbb{C} \mid |z| \le 1\} \to U$ with $\phi(0) = x$. Put

$$C_r = \phi(\{z \in \mathbb{C} \mid |z| = r\}) \quad \text{and} \quad C'_r = \Phi_\nu^{-1}\big(\mathfrak{F}_2(C_r)\big)$$

C'_r is a closed curve in $\mathbb{C} \setminus \bigcup_{s\in S'_\nu} D'_\nu(s)$. Choose $p' \in \mathbb{C} \setminus \Phi_\nu^{-1}\big(\mathfrak{F}_2(U \setminus \{x\})\big)$ such that C'_1 has nonzero winding number around p'. Since S'_ν is infinite and discrete, there is $s' \in S'_\nu$ such that C'_1 has winding number zero around s'. Let Σ be the line segment joining p' to s'.

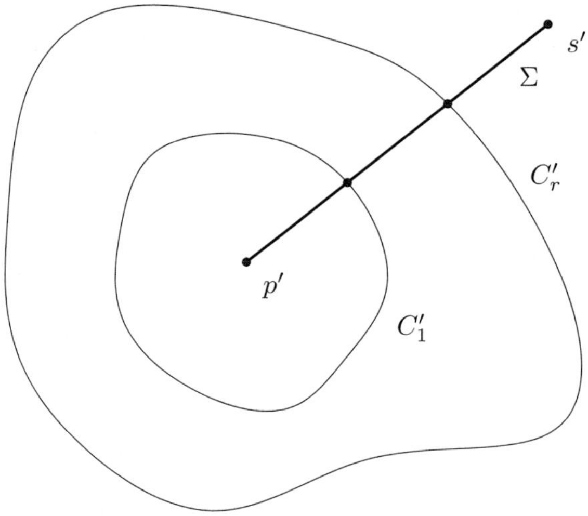

As $C'_r \subset \mathbb{C} \setminus D'_\nu(s')$, all the curves C'_r, $0 < r \le 1$ meet Σ. Therefore there is a sequence of points x_n in U with $\lim_{n\to\infty} x_n = x$ such that $\mathfrak{F}_2(x_n) \in \Sigma$ for all n. This is a contradiction, and so we have shown that $\mathfrak{F}_2$ extends holomorphically to all of X.

Next we show that $\mathfrak{F}$ is surjective. Since $\mathfrak{F}$ is holomorphic and not constant the set $\mathfrak{F}(X)$ is open in X'. We want to show that it is also closed. So let $x' \in \overline{\mathfrak{F}(X)} \setminus \mathfrak{F}(X)$. Choose a sequence $(x'_i)_{i\in\mathbb{N}}$ in $\mathfrak{F}(X)$ with $\lim_{i\to\infty} x'_i = x'$, and choose

$x_i \in X$ with $\mathfrak{F}(x_i) = x'_i$. As above one sees that the sequence $(x_i)_{i\in\mathbb{N}}$ has an accumulation point x'. Clearly $\mathfrak{F}(x) = x'$. So $x' \in \mathfrak{F}(X)$.

In the case that X' is not hyperelliptic κ' is an embedding (see (A.10)) and (11.1) implies that $\mathfrak{F}$ is injective. So in this case $\mathfrak{F}$ is biholomorphic. Now assume that X' is hyperelliptic. Then $\kappa'(X')$ is isomorphic to the complement of a finite set in $\mathbb{P}^1$. The same then holds for $\kappa(X)$. So X and X' are both hyperelliptic. Let $\tau\colon X \to \mathbb{P}^1 \setminus M$ resp. $\tau'\colon\ \to \mathbb{P}^1 \setminus M'$ be the hyperelliptic projections. Then by (11.1) there is an isomorphism $\varphi\colon \mathbb{P}^1 \setminus M \to \mathbb{P}^1 \setminus M'$ such that the diagram

$$\begin{array}{ccc} X & \xrightarrow{\ \Phi\ } & X' \\ \big\downarrow{\scriptstyle\tau} & & \big\downarrow{\scriptstyle\tau'} \\ \mathbb{P}^1 \setminus M & \xrightarrow[\simeq]{\ \phi\ } & \mathbb{P}^1 \setminus M' \end{array}$$

commutes. This implies that $\mathfrak{F}$ is an isomorphism.

Finally (11.1) implies that there is a nowhere vanishing holomorphic function $\varepsilon(x)$ on X such that $\mathfrak{F}^*(\omega'_j) = \varepsilon\omega_j$ for all $j = 1, 2, \dots$. For every finite linear combination $\omega = \sum_{j=1}^{\infty} \lambda_j\omega_j$ with $\lambda_j = 0$ for all but finitely many j we have

$$\begin{aligned} i\int_X (\varepsilon\omega) \wedge (\overline{\varepsilon\omega}) &= i\int_X \mathfrak{F}^*\left(\sum_{j=1}^{\infty} \lambda_j\omega'_j\right) \wedge \overline{\mathfrak{F}^*\left(\sum_{j=1}^{\infty} \lambda_j\omega'_j\right)} \\ &= i\int_{X'} \left(\sum_{j=1}^{\infty} \lambda_j\omega'_j\right) \wedge \overline{\left(\sum_{j=1}^{\infty} \lambda_j\omega'_j\right)} \\ &= \lambda^*(\operatorname{Im}\mathcal{R}')\lambda \\ &= \lambda^*(\operatorname{Im}\mathcal{R})\lambda \\ &= i\int_X \omega \wedge \overline{\omega} \end{aligned}$$

by Riemann's Bilinear Relations.

So $\omega \mapsto \varepsilon\omega$ is a bounded linear map T on the Hilbert space Ω of square integrable holomorphic differential forms such that $\|T\omega\|_\Omega = \|\omega\|_\Omega$ for all $\omega \in \Omega$. Here $\|\omega\|_\Omega = (i\int \omega \wedge \overline{\omega})^{1/2}$ denotes the norm on Ω. Since T is surjective, the adjoint of T

$$T^*: \Omega^* \to \Omega^*$$

then also fulfills $\|T^*\ell\| = \|\ell\|$ for all $\ell \in \Omega^*$. For each $x \in X$ let L_x be the linear subspace of Ω^* spanned by "the evaluation map"

$$\omega \mapsto \frac{\omega}{d\xi}(x)$$

where ξ is a local coordinate around x. Since

$$T^*\ell = \varepsilon(x)\ell \quad \text{for all } \ell \in L_x$$

we conclude that $|\varepsilon(x)| = 1$ for all $x \in X$. Since ε is holomorphic the function ε is constant and $|\varepsilon| = 1$.

As $\int_{A_i} \mathfrak{F}^*(\omega'_j) = \int_{\mathfrak{F}_*(A_i)} \omega'_j = \varepsilon\delta_{ij}$ it follows that $\varepsilon \in \{\pm 1\}$, and $\mathfrak{F}_*(A_i) = \varepsilon A'_i$ for all $i \in \mathbb{N}$. Since $A_i \cdot B_j = \delta_{ij} = A'_i \cdot B'_j$ it also follows that $\mathfrak{F}_*(B_i) = \varepsilon B'_i$ for $i \in \mathbb{N}$. $\square$

PROOF OF THEOREM 11.1. Assume first that X is not hyperelliptic. By Proposition 9.11 and Proposition 10.5(a), X' is also not hyperelliptic. Put

$$\widetilde{X}_0 := \{x \in \widetilde{X} \mid \{x\} \times W^{(-3)} \cap M \neq \varnothing\}$$

where M is the subset of $\widetilde{X} \times W^{(-3)}$ introduced in Proposition 9.10. Then $\widetilde{X}_0$ is open and dense in $\widetilde{X}$. Let $x_0 \in \widetilde{X}_0$, and choose a neighbourhood U_1 of $x_0 \in \widetilde{X}_0$ and an open subset U_2 in $W^{(-3)}$ such that $U := U_1 \times U_2 \subset M$ and the conclusions of Proposition 9.10 hold. Let $\varphi \colon U \to \Theta$ be a holomorphic map such that for all $(x; y) \in U$

$$\operatorname{codim}_B \overline{\operatorname{range} d\varphi(x; y)} = 2$$

and for all (x, y) in subset of full Baire category U_d in U

$$\nabla\Theta\big(\varphi(x; y)\big) \neq 0, \quad \ker H\big(\varphi(x; y)\big) = \mathbb{C}\vec{\omega}(x).$$

Put

$$U' := \{(x, y) \in U \mid \varphi(x, y) \notin \Theta_1' \cap (-\Theta_1')\}.$$

By Proposition 9.3 and Lemma 9.2 the set U' is open and dense in U.

Let $(x, y) \in U_d \cap U'$. If $\varphi(x, y) \in \Theta \setminus \Theta_1'$, there is, by Corollary 9.9(b) a point $x' \in X'$ such that

$$\mathbb{C}\vec{\omega}(x) = \mathbb{C}\vec{\omega}'(x').$$

If $\varphi(x, y) \in \Theta \setminus (-\Theta_1')$, then $-\varphi(x.y) \in \Theta \setminus \Theta_1'$ and we come to the same conclusion. As $\pi(U')$ is dense in U_1 we conclude that

> there is a dense subset X_1 of X such that for all $x \in X_1$ there is $x' \in X'$ such that $\mathbb{C}\vec{\omega}(x) = \mathbb{C}\vec{\omega}'(x')$.

As X' is non hyperelliptic, the point x' above is unique. So there is a map

$$\mathfrak{F}_1 \colon X_1 \to X'$$

such that $\mathbb{C}\vec{\omega}(x) = \mathbb{C}\vec{\omega}'\big(\mathfrak{F}_1(x)\big)$ for all $x \in X_1$. By Lemma 11.2 $(X; A_1, B_1, \dots)$ and $(X'; A_1', B_1', \dots)$ are isomorphic.

Now we consider the case that X is hyperelliptic. By Proposition 9.11 and Proposition 10.5(a) the surface X' is also hyperelliptic. Denote by $\tau \colon X \to \mathbb{P}^1 \setminus M$, resp. $\tau' \colon X' \to \mathbb{P}^1 \setminus M'$ the hyperelliptic projections, by $S \subset \mathbb{P}^1 \setminus M$ resp. $S' \subset \mathbb{P}^+ \setminus M'$ the set of branch points and put $\mathfrak{B} = \tau^{-1}(S)$, $\mathfrak{B}' = (\tau')^{-1}(S')$. By Proposition 10.5(a), for each $b \in \widetilde{X}$ with $\pi(b) \in \mathfrak{B}$ there is an open dense subset U_b of $W^{(-2)}$ such that

$$F^{(b)}(U_b) \subset \Theta \setminus \Theta_1 \subset \Theta_{\text{reg}}$$

and

$$\mathbb{C}\vec{\omega}(b) \subset \ker H\big(F^{(b)}(y)\big)$$

for all $y \in U_b$ with equality for y in a subset of full Baire category in U_b.

Now apply Proposition 10.5(b) to the hyperelliptic Riemann surfaces X' with $\mathcal{O} = U_b$ and $g = F^{(b)}$. For each b, there are two possible conclusions to Proposition 10.5(b). Let $\mathfrak{B}_0$ be the set of $b \in \mathfrak{B}$ for which $F^{(\tilde{b})}(U_{\tilde{b}}) \subset \Theta'^{(\nu(b))}$ for some $\nu(b) \in \{1, \dots, m'\}$, $\tilde{b} \in \widetilde{X}$ with $\pi(\tilde{b}) = b$.

Then $\mathbb{C}\vec{\omega}(b) \subset \ker H(e)$ for all $b \in \mathfrak{B}_0$, $e \in (\Theta'^{(\nu(b))} \setminus \Theta_2') \cap \Theta_{\text{reg}}$. By Proposition 10.5(a), $\ker H\big(F^{(c)}(y)\big) = \mathbb{C}\vec{\omega}(\tilde{c})$ for generic $y \in U_{\tilde{c}}$, $\tilde{c} \in \widetilde{X}$ with $\pi(\tilde{c}) \in \mathfrak{B}$. Hence $\mathbb{C}\vec{\omega}(c) = \mathbb{C}\vec{\omega}(b)$ where $b, c \in \mathfrak{B}_0$, $\nu(b) = \nu(c)$. Consequently $b \mapsto \nu(b)$ is injective and, in particular, $\mathfrak{B}_0$ contains at most m' elements. For each $b \in \widetilde{X}$ with $\pi(b) \in \mathfrak{B} \setminus \mathfrak{B}_0$, there is a $b' \in \widetilde{X}'$ with $\pi(b') \in \mathfrak{B}'$ such that

$$F^{(b)}(\widetilde{U}_b) \subset H^{(b')} \cap (\Theta \setminus \Theta'_1) \cap (\Theta \setminus \Theta_1)$$

for some open subset $\widetilde{U}_b$ of U_b.

Choose b and b' as in the last paragraph. By Proposition 10.5(a), for each $y \in \widetilde{U}_b$, there is $y' \in U'_b$ such that $F'^{(b')}(y') = F^{(b)}(y)$. Therefore the tangent spaces to the ranges of $F'^{(b)}(y')$ and $F^{(b)}(y)$ agree at corresponding points. Consequently the ranges of $df'^{(b')}(y')$ and $df^{(b)}(y)$ agree at corresponding points so that by (10.5)

$$\psi^{-1}\big(\widetilde{\Omega}(y)\big) = \psi'^{-1}\big(\widetilde{\Omega}'(y')\big).$$

Consider the map

$$g\colon \widetilde{X}' \to \Theta,$$
$$x' \mapsto F'^{(b')}(y') - \int_{b'}^{x'} \vec{\omega}' = F'(x'; y'_3, y'_4 \ldots).$$

Then $\widetilde{X}'_1 := \{x' \in \widetilde{X}' \mid g(x') \in \Theta_{\text{reg}}\}$ is the complement of a discrete set in $\widetilde{X}'$. Clearly for $x' \in \widetilde{X}'_1$

$$\psi'\big(\nabla\big(\theta\big(g(x')\big)\big) \in \widetilde{\Omega}'(x', y'_3, y'_4, \ldots) \subset \widetilde{\Omega}'(y'), \text{ so } \psi'\big(\nabla\big(\theta\big(g(x')\big)\big) \in \widetilde{\Omega}(y).$$

By Proposition 10.5(c), applied to $Z = \widetilde{X}_1$, there is a holomorphic map $\mathfrak{F}'\colon \widetilde{X}'_2 \to \widetilde{X}$, on the complement $\widetilde{X}'_2$ of a discrete set in $\widetilde{X}'_1$, such that for all $x' \in \widetilde{X}'_2$

$$\int_{b'}^{x'} \vec{\omega}' = \int_{b}^{\mathfrak{F}'(x')} \vec{\omega}. \tag{11.2}$$

We claim that $\mathfrak{F}'$ induces a map $\mathfrak{F}\colon X'_2 = \pi(\widetilde{X}'_2) \to X$. Indeed, if $\pi(x') = \pi(x'')$ then $\mathfrak{F}'(x')$ and $\mathfrak{F}'(x'')$ obey

$$\int_{\mathfrak{F}'(x')}^{\mathfrak{F}'(x'')} \vec{\omega} = \int_{x'}^{x''} \vec{\omega} = \int_{\Sigma n_i A'_i + \Sigma m_j B'_j} \vec{\omega}$$

for some finite integer linear combination of A- and B-cycles. Let $y' \in \widetilde{X}'$ be a point with the property that the π' projection of any curve from y' to $\mathfrak{F}'(x')$ is homologous to $-(\Sigma n_i A'_i + \Sigma m_j B'_j)$. Then $\pi'(y') = \pi'\big(\mathfrak{F}'(x'')\big)$ and $\int_{y'}^{\mathfrak{F}'(x'')} \vec{\omega} = 0$. So $\int_{y'}^{\mathfrak{F}'(x'')} \omega = 0$ for all $\omega \in \Omega'$ and $\pi(y') = \pi'\big(\mathfrak{F}'(x')\big)$ by (A.4)(ii), which is an application of the Riemann–Roch theorem.

Differentiating (11.2) we see that for all $x' \in X'_2$

$$\mathbb{C}\vec{\omega}'(x') = \mathbb{C}\vec{\omega}\big(\mathfrak{F}(x')\big).$$

Let γ' be a path in X' from b' to x' and let γ be a path in X from b to $\mathfrak{F}(x')$ such that

$$\int_{\gamma'} \vec{\omega}' = \int_{\gamma} \vec{\omega}.$$

Since $i \circ \gamma'$ is a path from b' to $i'(x')$, $i \circ \gamma$ is a path from b to $i\big(\mathfrak{F}(x')\big)$ and

$$\int_{i'\circ\gamma'} \vec{\omega}' = -\int_{i'\circ\gamma'} i'^{*}\vec{\omega}' = -\int_{\gamma'} \vec{\omega}' = -\int_{\gamma} \vec{\omega} = \int_{i\circ\gamma} \vec{\omega}$$

so that we have $i(\mathfrak{F}(x')) = \mathfrak{F}(ix')$. Lemma 11.2 implies that the marked Riemann surfaces $(X; A_1, B_1, \dots)$ and $(X'; A'_1, B'_1, \dots)$ are isomorphic. $\square$

CHAPTER 3

Examples

Introduction

There is a simple and elegant theory of $g \times g$ matrices, $g < \infty$, acting on $\mathbb{C}^g$. Indeed, from a few elementary concepts such as determinant, characteristic polynomial and spectrum one quickly derives many basic and useful facts. The arguments are typically algebraic and combinatorial.

The naive attempt to generalize the important concepts and results of linear algebra to all $\infty \times \infty$ matrices immediately fails. First, $\mathbb{C}^\infty$ must, at the very least, be replaced by a topological vector space of sequences. To obtain more than the softest structure it must be replaced by a Banach space or even a Hilbert space. Then, the class of matrices under consideration must be restricted.

The most rigid, but nevertheless useful, extension of linear algebra is to trace class perturbations of the identity acting on, for example, $\ell^2(\mathbb{N})$. There is a well-defined "infinite determinant" and "characteristic polynomial" and an analogue of most facts of matrix theory. The trace class condition puts strong analytic constraints on the matrix elements. It may be relaxed to Hilbert–Schmidt or to arbitrary von Neumann–Schatten without losing any structure. If we consider general compact perturbations of the identity, there is no longer a determinant but much of the structure survives.

By contrast, merely bounded operators can be extremely complicated. It is often an almost hopeless task to extract information about the spectrum of even a bounded, self adjoint operator.

There is an equally simple and elegant theory of compact Riemann surfaces X of genus g marked with a canonical homology basis $A_1, B_1, \ldots, A_g, B_g$. It begins with the observation that there exists a *unique* frame of holomorphic one forms $\omega_1, \ldots, \omega_g$ on X satisfying

$$\int_{A_j} \omega_i = \delta_{ij}.$$

Then, the associated Riemann matrix $R_X = (\int_{B_j} \omega_i)$, Jacobian variety and theta function are introduced. The elementary theory culminates in the Torelli theorem stating that two Riemann surfaces with the same Riemann matrix are biholomorphic.

To extend the geometry of finite genus surfaces to marked Riemann surfaces of infinite genus we introduce the Hilbert space $\Omega(X)$ of all square integrable holomorphic one forms. For any marked Riemann surface X one can prove that there is a sequence $\omega_1, \omega_2, \ldots,$ of forms in $\Omega(X)$ normalized by $\int_{A_j} \omega_i = \delta_{ij}$ for all $i, j \geq 1$. The naive attempt to generalize finite genus constructions to *all* marked surfaces of infinite genus now fails. There are noncompact surfaces X with non constant holomorphic functions satisfying

$$\int_X df \wedge \overline{*df} < \infty.$$

It follows that there are many normalized frames of square integrable one forms on a surface X of this kind. The class of surfaces under consideration must be restricted.

In Chapter 1 we considered surfaces $(X; A_1, B_1, \dots)$ with an exhaustion function h of finite charge. That is, marked surfaces with a proper, nonnegative Morse function h that satisfies

$$\int_X |d{*}dh| < \infty$$

and has the additional property that for each $t > 0$ large enough, there is an $n \geq 1$ such that the cycles

$$A_1, B_1, \dots, A_n, B_n$$

are homologous in X to cycles of a canonical homology basis for

$$X_t = h^{-1}([0, t]).$$

We showed that for such surfaces there is a *unique* normalized frame of square integrable holomorphic one forms and that the associated infinite Riemann matrix

$$R_X = \left(\int_{B_i} \omega_j \right)$$

is symmetric and $\operatorname{Im} R_X$ is positive definite. We also showed, among other things, that for every $x \in X$ there is an $\omega \in X$ with $\omega(x) \neq 0$, and that the canonical map

$$x \in X \to \delta_x(\omega) = \omega(x) \in \mathbb{P}\big(\Omega(X)^*\big)$$

is injective whenever X is not hyperelliptic. Roughly speaking, the elementary structure of compact Riemann surfaces, up to the existence of a theta function, extends to general surfaces with an exhaustion function of bounded charge. It appears that this is the "natural" class of surfaces with this property (see also [**Ac1, AS**]).

The theta function and the geometry of its divisor are fundamental for the theory of compact Riemann surfaces. Not surprisingly, there are marked Riemann surfaces with exhaustion functions with finite charge such that the associated theta series diverges. We must again restrict the class of surfaces under consideration.

In Chapter 2 we introduced a class of marked surfaces of infinite genus obtained by pasting plane domains and handles together subject to explicit geometric constraints. The theta series associated to surfaces in this class converge to entire functions on canonical Banach spaces. The Riemann vanishing theorem and the Torelli theorem hold. That is, there is an analogue of most facts of finite genus theory.

The class of surfaces introduced in Chapter 2 is a rigid extension of the classical theory. We think of it in the context of Riemann surfaces as the counterpart of trace class perturbations of the identity in operator theory. To justify this conviction we must show that the the geometric constraints arise "naturally".

To do this, we recall the relationship between Riemann surfaces of finite genus and solutions of the Kadomcev–Petviashvilii equation. For any marked Riemann surface X of genus g there exist vectors $U, V, W \in \mathbb{C}^g$ and a constant c such that for any $D \in \mathbb{C}^g$ the function

$$u(x_1, x_2, t) = 2\frac{\partial^2}{\partial x_2^2} \log \theta(x_1 U + x_2 V + tW + D) + 2c$$

is a solution of the Kadomcev–Petviashvilii equation

$$u_{x_1 x_1} = -\tfrac{2}{3}(u_t - 3uu_{x_2} + \tfrac{1}{2}u_{x_2 x_2 x_2})_{x_2}$$

(see [**K**]). This property characterizes theta functions of Riemann surfaces among all possible theta functions [**AC**].

This relationship can be made even more explicit when the initial data

$$u(x_1, x_2, 0) = 2\frac{\partial^2}{\partial x_1^2} \log \theta(x_1 U + x_2 V + D) + 2c$$

is periodic with respect to a lattice

$$\Gamma = (0, 2\pi)\mathbb{Z} \oplus (\omega_1, \omega_2)\mathbb{Z}.$$

In this case the Riemann surface X is the normalization of the "heat curve" $\mathcal{H}(u)$ consisting of the points $(\xi_1, \xi_2) \in \mathbb{C}^* \times \mathbb{C}^*$ for which there is a nontrivial distributional solution $\psi(x_1, x_2)$ in $L^\infty_{\text{loc}}(\mathbb{R}^2)$ of the "heat equation"

$$\left(\frac{\partial}{\partial x_1} - \frac{\partial^2}{\partial x_2^2}\right)\psi + u(x_1, x_2, 0)\psi = 0$$

satisfying

$$\psi(x_1 + \omega_1, x_2 + \omega_2) = \xi_1 \psi(x_1, x_2), \quad \psi(x_1, x_2 + 2\pi) = \xi_2 \psi(x_1, x_2).$$

In [**BEKT**] it is shown that the set of Riemann surfaces of genus g which have such a heat curve representation is dense in the moduli space. It can also be shown that for general $q \in L^2(\mathbb{R}^2/\Gamma)$ the corresponding $\mathcal{H}(q)$ is smooth and of infinite genus.

It follows from the discussion above that any reasonable theory of Riemann surfaces of infinite genus must treat all the curves $\mathcal{H}(q)$ for sufficiently regular $q \in L^2(\mathbb{R}^2/\Gamma)$, as well as "small" deformations of them. One of the main results of Chapter 3 (Theorem 15.2) is that for any smooth q the normalization of $\mathcal{H}(q)$ is either of finite genus or belongs to the class of infinite genus Riemann surfaces introduced in Chapter 2. The technique is to analyze the singular curve $\mathcal{H}(0)$ and to show that the influence of q is to open the singularities into handles that satisfy all the geometric constraints introduced in Chapter 2. In this context, we see a natural decomposition of $\mathcal{H}(q)$ into a regular piece and handles. Figuratively speaking, $\mathcal{H}(q)$ is a "trace class" perturbation of $\mathcal{H}(0)$. In Chapter 4 the theta function of $\mathcal{H}(q)$ is used to solve the initial value problem for the Kadomcev–Petviashvilii equation with inital data q.

The class of Riemann surfaces introduced in Chapter 2 is quite rich. It includes Fermi curves (Sections 16–18), rather general hyperelliptic curves (Section 12), and, we believe, the spectral curve associated to any ordinary differential operator with sufficiently regular peridic coefficients.

12. Hyperelliptic Surfaces

We start with a discrete subset $\mathcal{S} \subset \mathbb{C}$. Let $p\colon X \to \mathbb{C}$ be the double cover ramified at the points of $\mathcal{S}$. Not all such hyperelliptic Riemann surfaces need fulfill (GH1)–(GH6). We present cases where they do.

We assume that there is a finite subset $\mathcal{S}_0 \subset \mathcal{S}$ such that the complement $\mathcal{S}\setminus\mathcal{S}_0$ can be ordered $v_1, w_1, v_2, w_2, \ldots$ and that there are $\rho_j > |v_j - w_j|$ such that for all $j \neq j'$ the circle of radius $4\rho_j$ around

$$s(j) = \tfrac{1}{2}(v_j + w_j)$$

does not meet the circle of radius $4\rho_{j'}$ around $s(j')$ and does not contain any point of $\mathcal{S}_0$. We choose a canonical homology basis $A_1, B_1, A_2, B_2, \ldots$ of X such that there is n so that, for all $j \geq 1$, A_{j+n} is represented by the inverse image under p of the line segment joining v_j and w_j.

THEOREM 12.1. *Assume that for all* $\beta > 0$

$$\sum_j \frac{|v_j - w_j|^\beta}{\rho_j^\beta} < \infty. \tag{12.1}$$

Assume that there exist $0 < \Delta < D$ *such that*

$$\sum_j \frac{1}{|s(j)|^{D-4\Delta-3}} < \infty$$

and for all sufficiently big j

$$|v_j - w_j| < \frac{1}{3|s(j)|^D}, \quad \rho_j > \frac{3}{|s(j)|^\Delta}. \tag{12.2}$$

Assume that

$$\lim_{j\to\infty} \frac{\log|s(j)|}{\log|v_j - w_j|/2\rho_j} = 0, \tag{12.3a}$$

$$\lim_{j\to\infty} \frac{\rho_j \log|s(j)|}{\min_{j'\neq j}|s(j) - s(j')|} = 0. \tag{12.3b}$$

Then the hypotheses (GH1)–(GH6) *of Section* 5 *are fulfilled for the marked Riemann surface* $(X; A_1, B_1, A_2, B_2, \ldots)$.

EXAMPLE 1. $\mathcal{S}$ consists of real numbers γ_0 and $4\pi^2 n^2 + \gamma_n^\pm$ for $n \in \mathbb{N}$ with $\sum_n |\gamma_n^\pm|^\beta < \infty$ for all $\beta > 0$. Such surfaces occur as spectral curves for Hill's operators $-d^2/dx^2 + q(x)$ with $q \in C^\infty_{\mathbb{R}}(\mathbb{R}/\mathbb{Z})$, the space of smooth real-valued functions of period 1. See [**MT1**]. The hypotheses of Theorem 12.1 are verified if one chooses $v_j = 4\pi^2 j^2 + \gamma_j^+$, $w_j = 4\pi^2 j^2 + \gamma_j^-$ and $\rho_j = 1$. In particular, the Torelli theorem holds for Hill's surfaces. This extends the results of [**MT2**].

EXAMPLE 2. $\mathcal{S}$ consists of real numbers $n + \gamma_n^\pm$ for $n \in \mathbb{Z}$ with $\sum_n |\gamma_n^\pm|^\beta < \infty$ for all $\beta > 0$. The hypotheses of the Theorem are verified if one chooses $v_j = j + \gamma_j^+$, $w_j = j + \gamma_j^-$ and $\rho_j = 1/\log^2(2 + |j|)$.

The region lying over a connected open set that contains v_j and w_j will be used as a handle as in (GH2). In the following lemma we show that one can construct an isomorphism between such a region and our model handle

$$H(t) = \{(z_1, z_2) \in \mathbb{C}^2 \mid z_1 z_2 = t, |z_\mu| \leq 1\}.$$

The regular piece of X is parametrized by the projection p if $|\mathcal{S}_0|$ is even and by the square root of the projection if $|\mathcal{S}_0|$ is odd. In the lemma we also analyze the resulting glueing maps.

LEMMA 12.2. *Let v, w be different points of $\mathcal{S}$ and $\rho > |v - w|$ be such that the disk of radius ρ around*

$$s = \tfrac{1}{2}(v + w)$$

intersects $\mathcal{S}$ only in v and w. Put

$$a = \rho\left(1 + \sqrt{1 - \frac{|v - w|^2}{4\rho^2}}\right), \qquad t = \frac{|v - w|^2}{4a^2}.$$

Then there is an embedding

$$\phi\colon H(t) \to X$$

such that with

$$p'\colon H(t) \to \mathbb{C}, \quad (z_1, z_2) \mapsto s + \frac{a}{2}\frac{v - w}{|v - w|}(z_1 + z_2)$$

the diagram

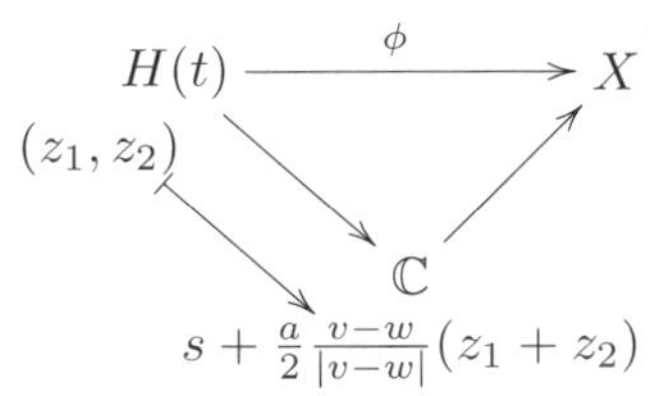

commutes.

(A) *If $\tau > \sqrt{t}$, $r > 0$, $R > 0$ are real numbers with*

$$\frac{a}{2}\left(4\tau + \frac{t}{4\tau}\right) < r < \frac{1}{4}R < \frac{a}{2}\left(\frac{1}{4} - 4t\right) < \frac{a}{2}\left(\frac{1}{2} + 2t\right) < R < \frac{a}{2}\left(1 - t\right)$$

then for $\mu = 1, 2$ and $(z_1, z_2) \in H(t)$

(12.4a) $\qquad |p'(z_1, z_2) - s| < r \quad$ *if* $|z_\mu| = 4\tau$,

(12.4b) $\qquad |p'(z_1, z_2) - s| > R \quad$ *if* $|z_\mu| = 1$,

(12.4c) $\qquad |p'(z_1, z_2) - s| < R \quad$ *if* $|z_\mu| = \dfrac{1}{2}$,

(12.4d) $\qquad |p'(z_1, z_2) - s| > \dfrac{R}{4} \quad$ *if* $|z_\mu| = \dfrac{1}{4}$.

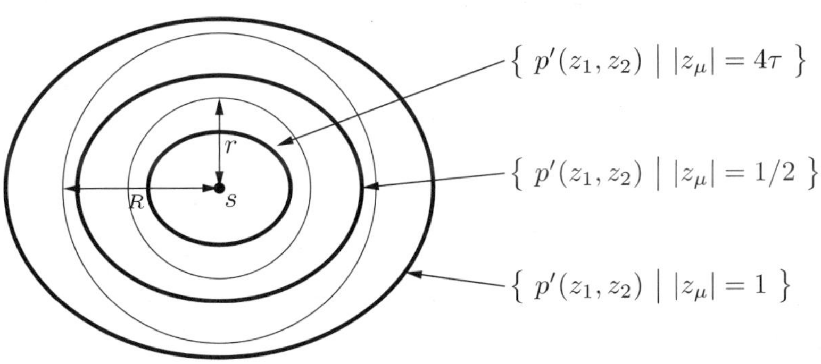

The map p' induces a biholomorphic map between

$$\{(z_1, z_2) \in H(t) \mid |z_\mu| \geq \tau\}$$

and the elliptical ring between

$$\left\{ z \in \mathbb{C} \;\middle|\; \frac{(\mathrm{Re}\,|v-w|/v-w(z-s))^2}{a^2(\tau+t/\tau)^2} + \frac{(\mathrm{Im}\,|v-w|/v-w(z-s))^2}{a^2(\tau-t/\tau)^2} = 4 \right\}$$

and

$$\left\{ z \in \mathbb{C} \;\middle|\; \frac{(\mathrm{Re}\,|v-w|/v-w(z-s))^2}{a^2(1+t)^2} + \frac{(\mathrm{Im}\,|v-w|/v-w(z-s))^2}{a^2(1-t)^2} = 4 \right\}.$$

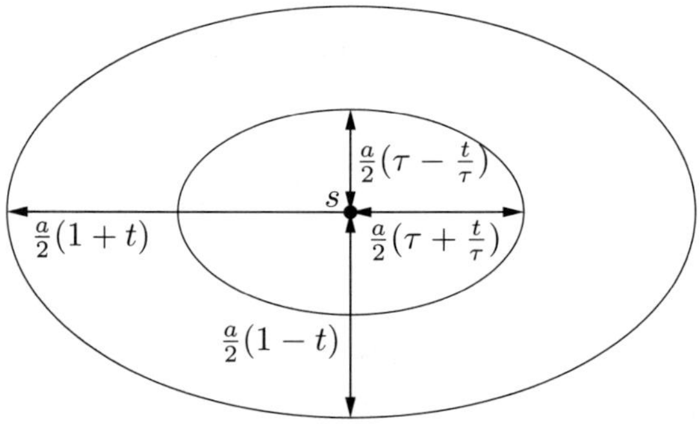

Denote this map by p'_μ and define $\alpha_\mu(z)$ by

$$\alpha_\mu(z) = (p'_\mu)_* \left(\frac{1}{2\pi i} \frac{dz_\mu}{z_\mu} \right) - \frac{1}{2\pi i} \frac{dz}{z-s}.$$

Then

$$\|\alpha_\mu(z)\, dz\big|_{\{z\in\mathbb{C} | r<|z-s|<R\}}\|_2 \leq 1$$

and

$$R \sup_{|z-s|=R} |\alpha_\mu(z)| \leq \frac{1}{\pi} \frac{4t}{1-4t}.$$

(B) *Assume that $\rho < |s|$. Choose a branch of $\sqrt{}$ outside of $\{\lambda s \mid \lambda \leq 0\}$. If $\tau > \sqrt{t}$, $R > 4r > 0$ are real numbers with*

$$\frac{a}{2}\left(4\tau + \frac{t}{4\tau}\right) < r(2\sqrt{|s|} - r), \quad \tfrac{1}{4}R(2\sqrt{|s|} + \tfrac{1}{4}R) < \frac{a}{2}(\tfrac{1}{4} - 4t),$$

$$\frac{a}{2}\left(\frac{1}{2} + 2t\right) < R(2\sqrt{|s|} - R) < R(2\sqrt{|s|} + R) < \frac{a}{2}(1-t)$$

then for $\mu = 1, 2$ and $(z_1, z_2) \in H(t)$

$$|\sqrt{p'(z_1,z_2)} - \sqrt{s}| < r \quad \text{if } |z_\mu| = 4\tau, \tag{12.5a}$$

$$|\sqrt{p'(z_1,z_2)} - \sqrt{s}| > R \quad \text{if } |z_\mu| = 1, \tag{12.5b}$$

$$|\sqrt{p'(z_1,z_2)} - \sqrt{s}| < R \quad \text{if } |z_\mu| = \tfrac{1}{2}, \tag{12.5c}$$

$$|\sqrt{p'(z_1,z_2)} - \sqrt{s}| > \tfrac{R}{4} \quad \text{if } |z_\mu| = \tfrac{1}{4}. \tag{12.5d}$$

The map $\sqrt{p'}$ *induces a biholomorphic map between*

$$\{(z_1, z_2) \in H(t) \mid |z_\mu| \geq \tau\}$$

and the preimage under the squareroot map of the elliptical ring between

$$\left\{ z \in \mathbb{C} \;\middle|\; \frac{(\operatorname{Re}|v-w|/v-w(z-s))^2}{a^2(\tau+t/\tau)^2} + \frac{(\operatorname{Im}|v-w|/v-w(z-s))^2}{a^2(\tau-t/\tau)^2} = 4 \right\},$$

and

$$\left\{ z \in \mathbb{C} \;\middle|\; \frac{(\operatorname{Re}|v-w|/v-w(z-s))^2}{a^2(1+t)^2} + \frac{(\operatorname{Im}|v-w|/v-w(z-s))^2}{a^2(1-t)^2} = 4 \right\}.$$

Denote this map by g_μ *and define* $\alpha_\mu(\zeta)$ *by*

$$\alpha_\mu(\zeta) = (g_\mu)_*\left(\frac{1}{2\pi i}\frac{dz_\mu}{z_\mu}\right) - \frac{1}{2\pi i}\frac{d\zeta}{\zeta-\sqrt{s}}.$$

Then

$$\|\alpha_\mu(\zeta)\, d\zeta|_{\{\zeta\in\mathbb{C}|r<|\zeta-\sqrt{s}|<R\}}\|_2 \leq 2\pi\frac{\sqrt{|s|}+R}{\sqrt{|s|}-R}$$

and

$$R \sup_{|\zeta-\sqrt{s}|=R} |\alpha_\mu(\zeta)| \leq \frac{1}{2\pi}\left(\frac{t}{1/4-t}\frac{4\sqrt{|s|}+3R}{2\sqrt{|s|}-R} + \frac{R}{(1/4-t)(2\sqrt{|s|}-R)}\right).$$

PROOF. The map

$$p'' \colon H(t) \to \mathbb{C}, \quad (z_1, z_2) \mapsto \frac{a}{2}(z_1+z_2)$$

maps the circle $|z_\mu| = \lambda$, $\sqrt{t} < \lambda \leq 1$ bijectively to the ellipse

$$\frac{(\operatorname{Re} z)^2}{a^2(\lambda+t/\lambda)^2} + \frac{(\operatorname{Im} z)^2}{a^2(\lambda-t/\lambda)^2} = 4$$

whose major semiaxis has length $a/2(\lambda+t/\lambda)$ and whose minor semiaxis has length $a/2(\lambda-t/\lambda)$. Since

$$\left(\frac{1}{2}(z_1+z_2)\right)^2 + \left(\frac{1}{2i}(z_1-z_2)\right)^2 = t$$

p'' ramifies precisely at the points with $1/2i(z_1-z_2) = 0$, in other words at the points $z_1 = z_2 = \pm\sqrt{t}$. The images of these points are

$$\pm a\sqrt{t} = \pm\frac{|v-w|}{2}$$

p' is the composition of p'' with the map $z \mapsto s + v - w/|v-w|z$. Therefore it is a double cover of the ellipse

$$E = \left\{ z \in \mathbb{C} \;\middle|\; \frac{(\operatorname{Re}|v-w|/v-w(z-s))^2}{a^2(1+t)^2} + \frac{(\operatorname{Im}|v-w|/v-w(z-s))^2}{a^2(1-t)^2} \leq 4 \right\}$$

ramified exactly over v and w. As $a/2(1+t) = \rho$ the ellipse E is contained in the disk of radius ρ around s and contains no other points of $\mathcal{S}$. Therefore we can lift p' to an embedding ϕ as in the statement of the lemma. The statements (12.4) are obvious. The statements (12.5) follow from the fact that the image of a circle of radius ε about $\sqrt{s}$ under the map $\zeta \mapsto z = \zeta^2$ lies between the circles around s of radii $\varepsilon(2\sqrt{s}-\varepsilon)$ and $\varepsilon(2\sqrt{s}+\varepsilon)$. It remains to prove the statements about α_μ.

(A) Since

$$z = s + \frac{a}{2}\frac{v-w}{|v-w|}\left(z_\mu + \frac{t}{z_\mu}\right)$$

we have

$$\frac{dz}{z-s} = \frac{z_\mu^2 - t}{z_\mu^2 + t}\frac{dz_\mu}{z_\mu}.$$

Therefore

$$\frac{dz_\mu}{z_\mu} - \frac{dz}{z-s} = \frac{2t}{z_\mu^2 + t}\frac{dz_\mu}{z_\mu} = \frac{2t}{z_\mu^2 - t}\frac{dz}{z-s}$$

and

$$\alpha_\mu(z) = \frac{1}{2\pi i(z-s)}\frac{2t}{z_\mu^2 - t}.$$

So

$$R \sup_{|z-s|=R} |\alpha_\mu(z)| \le \frac{1}{2\pi}\frac{2t}{1/4 - t}$$

by (12.4c). Also, by (12.4a) and (12.4b)

$$\begin{aligned}\left\| \alpha_\mu(z)\,dz\big|_{r\le|z-s|\le R}\right\|_2 &= \left\|(p_\mu')^*\left(\alpha_\mu(z)\,dz\right)\big|_{p_\mu'^{-1}(r\le|z-s|\le R)}\right\|_2 \\ &\le \frac{1}{2\pi}\left\|\frac{2t}{z_\mu^2+t}\frac{dz_\mu}{z_\mu}\bigg|_{4\sqrt{t}\le|z_\mu|\le 1}\right\|_2 \\ &\le 1\end{aligned}$$

for all $t > 0$.

(B) As before

$$\frac{dz}{z-s} = \frac{z_\mu^2 - t}{z_\mu^2 + t}\frac{dz_\mu}{z_\mu}.$$

Furthermore

$$\frac{dz}{z-s} = \frac{2\zeta}{\zeta+\sqrt{s}}\frac{d\zeta}{\zeta-\sqrt{s}}$$

so

$$\frac{dz_\mu}{z_\mu} = \frac{z_\mu^2 + t}{z_\mu^2 - t}\frac{2\zeta}{\zeta+\sqrt{s}}\frac{d\zeta}{\zeta-\sqrt{s}}.$$

Therefore

$$\alpha_\mu(\zeta)\,d\zeta = \frac{1}{2\pi i}\left(\frac{z_\mu^2+t}{z_\mu^2-t}\frac{2\zeta}{\zeta+\sqrt{s}} - 1\right)\frac{d\zeta}{\zeta-\sqrt{s}}$$

and

$$\alpha_\mu(\zeta) = \frac{1}{2\pi i}\left(\frac{z_\mu^2}{(z_\mu^2-t)(\zeta+\sqrt{s})} + \frac{t(3\zeta+\sqrt{s})}{(z_\mu^2-t)(\zeta+\sqrt{s})(\zeta-\sqrt{s})}\right).$$

So

$$R \sup_{|\zeta-\sqrt{s}|=R} |\alpha_\mu(\zeta)| \le \frac{1}{2\pi}\left(\frac{t}{1/4-t}\frac{4\sqrt{|s|}+3R}{2\sqrt{|s|}-R} + \frac{R}{(1/4-t)(2\sqrt{|s|}-R)}\right).$$

The L^2 norm of the first term in $2\pi\alpha_\mu$ is

$$\begin{aligned}
&\left\| \frac{z_\mu^2}{(z_\mu^2 - t)(\zeta + \sqrt{s})}\, d\zeta \Big|_{r \le |\zeta - \sqrt{s}| \le R} \right\|_2 \\
&\qquad \le \frac{1}{2\sqrt{|s|} - R} \sup_{4\sqrt{t} \le |z_\mu| \le 1} \left| \frac{z_\mu^2}{z_\mu^2 - t} \right| \left\| d\zeta \big|_{r \le |\zeta - \sqrt{s}| \le R} \right\|_2 \\
&\qquad \le \frac{4}{3} \frac{\sqrt{\pi} R}{2\sqrt{|s|} - R}
\end{aligned}$$

As in part (A), the second term is bounded by

$$\begin{aligned}
&\left\| g_\mu^* \left(\frac{t(3\zeta + \sqrt{s})}{(z_\mu^2 - t)(\zeta + \sqrt{s})} \frac{d\zeta}{\zeta - \sqrt{s}} \right) \Big|_{4\sqrt{t} \le |z_\mu| \le 1} \right\|_2 \\
&\qquad = \left\| \frac{t(3\zeta + \sqrt{s})}{2\zeta(z_\mu^2 + t)} \frac{dz_\mu}{z_\mu} \Big|_{4\sqrt{t} \le |z_\mu| \le 1} \right\|_2 \\
&\qquad \le \frac{4\sqrt{|s|} + 3R}{2\sqrt{|s|} - 2R} \left\| \frac{t}{z_\mu^2 + t} \frac{dz_\mu}{z_\mu} \Big|_{4\sqrt{t} \le |z_\mu| \le 1} \right\|_2 \\
&\qquad \le \pi \frac{4\sqrt{|s|} + 3R}{2\sqrt{|s|} - 2R}.
\end{aligned}$$

□

PROOF OF THE THEOREM. Fix g big enough and a simply connected subset K of $\mathbb{C}$ containing 0 and with smooth boundary ∂K such that

$$\begin{gathered}
\mathcal{S}_0 \subset K, \\
v_j, w_j \in K \quad \forall j \le g, \\
K \cap \{z \in \mathbb{C} \mid |z - s(j)| \le \rho_j\} = \varnothing \quad \forall j > g, \\
\{z \in \mathbb{C} \mid |z - s(j)| \le \rho_j\} \cap \{z \in \mathbb{C} \mid |z - s(j')| \le \rho_{j'}\} = \varnothing \quad \forall j, j' > g,\ j \ne j'.
\end{gathered}$$

Define

$$X^{\mathrm{com}} = p^{-1}(K)$$

and, for $j \ge g + 1$, set

$$a_j = \rho_j \left(1 + \sqrt{1 - \frac{|v_j - w_j|^2}{4\rho_j^2}} \right), \quad t_j = \frac{|v_j - w_j|^2}{4a_j^2}.$$

CASE (A) $\mathcal{S}_0$ *is even.* Then K contains an even number of points of $\mathcal{S}$. So the boundary $\partial X^{\mathrm{com}} = p^{-1}(\partial K)$ consists of two components. Define

$$\begin{aligned}
S &= \{s(j) \mid j \ge g + 1\}, \\
D(j) &= \left\{ z \in \mathbb{C} \;\middle|\; \frac{(\operatorname{Re} |v_j - w_j| / v_j - w_j (z - s(j)))^2}{a_j^2 (\tau_j + t_j/\tau_j)^2} \right. \\
&\qquad\qquad \left. + \frac{(\operatorname{Im} |v_j - w_j| / v_j - w_j (z - s(j)))^2}{a_j^2 (\tau_j - t_j/\tau_j)^2} \le 4 \right\}, \\
G &= \mathbb{C} \setminus \left(\operatorname{int} K \cup \bigcup_{j \ge g+1} \operatorname{int} D(j) \right), \\
X^{\mathrm{reg}} &= p^{-1}(G)
\end{aligned}$$

and for $\mu = 1, 2$

$$s_\mu(j) = s(j), \quad D_\mu\big(s(j)\big) = D(j)$$

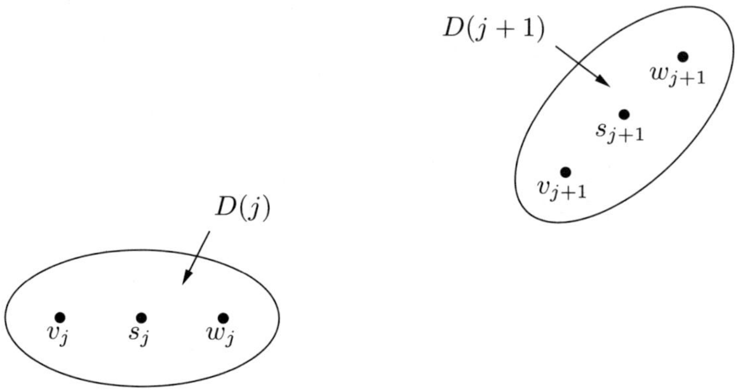

In this case X^{reg} consists of two components X_1^{reg}, X_2^{reg}. Let

$$\Phi_\nu \colon G \to X_\nu^{\text{reg}}$$

be the inverse of the biholomorphic map $p\big|_{X_\nu^{\text{reg}}}$. With these definitions hypothesis (GH1) is trivially satisfied.

By Lemma 12.2, there exist embeddings

$$\phi_j \colon H(t_j) \to X$$

such that with

$$p_j' \colon H(t_j) \to \mathbb{C}, \quad (z_1, z_2) \mapsto s(j) + \frac{a_j}{2} \frac{v_j - w_j}{|v_j - w_j|}(z_1 + z_2)$$

the diagram

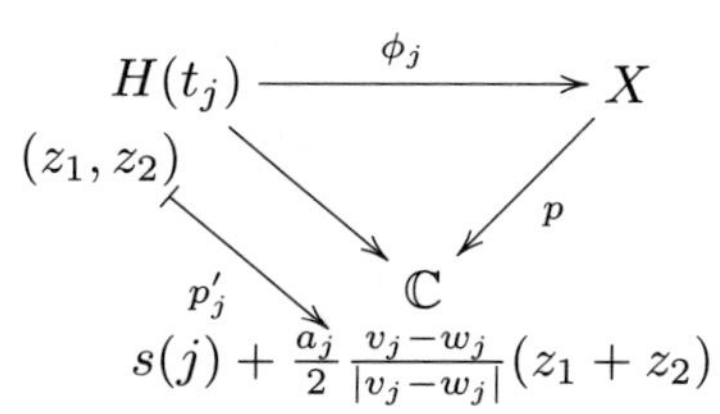

commutes. Hypothesis (GH2) is satisfied. The only nontrivial condition in this hypothesis, namely (GH2)(iv) follows from the assumption (12.1).

Define

$$\tau_\mu(j) = 2\sqrt{t_j}, \quad r_\mu(j) = 3|v_j - w_j|, \quad R_\mu(j) = \tfrac{1}{3}\rho_j.$$

If g was chosen big enough, then for all $j \geq g + 1$

$$\frac{a_j}{2}\left(4\tau_\mu(j) + \frac{t_j}{4\tau_\mu(j)}\right) < r_\mu(j)$$
$$< \tfrac{1}{4}R_\mu(j) < \frac{a_j}{2}\big(\tfrac{1}{4} - 4t_j\big) < \frac{a_j}{2}\big(\tfrac{1}{2} + 2t_j\big) < R_\mu(j) < \frac{a_j}{2}(1 - t_j).$$

By Lemma 12.2 hypothesis (GH3) is fulfilled. Hypothesis (GH4) is trivial and (GH6) is void.

We verify (GH5) part by part with $d = D$ and $\delta = \Delta$. Part (i) is automatic from the conditions on the ρ_j's. Part (ii) follows from (12.2). Observe that $\nu_1(j) \neq \nu_2(j)$

for all j so that the last condition of (GH5)(ii) is void. Part (iii) is trivial because $|s_1(j)| = |s_2(j)|$ for all j. Parts (iv) and (v) follow from (12.3a)–(12.3b). Part (vi) follows from Lemma 12.2. For $j \geq g+1$, let $s_1(j)$, $s_2(j)$ be the two square roots of $s(j)$. Define

$$S = \{s_\mu(j) \mid j \geq g+1, \mu = 1, 2\}, \quad G = \{\zeta \in \mathbb{C} \mid \zeta^2 \in G'\}$$

and for $\mu = 1, 2$ let $D_\mu\big(s_\mu(j)\big)$ be the component of $\{\zeta \in \mathbb{C} \mid \zeta^2 \in D(j)\}$ containing $s_\mu(j)$.

In this case X^{reg} is connected and $p\big|_{X^{\text{reg}}} \colon X^{\text{reg}} \to G'$ is an unramified double cover. Therefore there exists a biholomorphic map

$$\Phi \colon G \to X^{\text{reg}}$$

such that the diagram

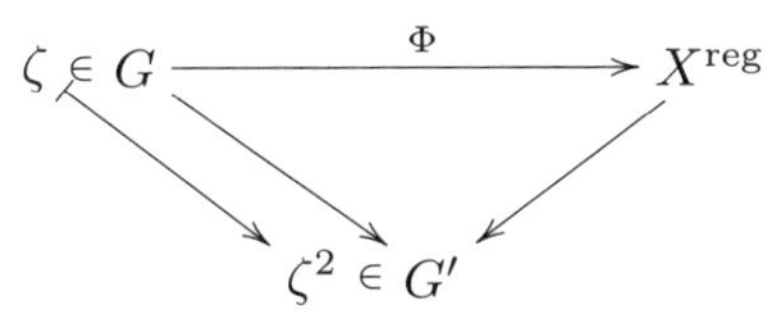

commutes. With these definitions hypothesis (GH1) is trivially satisfied.

Again by Lemma 12.2, there exist embeddings

$$\phi_j \colon H(t_j) \to X$$

such that with,

$$p_j' \colon H(t_j) \to \mathbb{C}, \quad (z_1, z_2) \mapsto s(j) + \frac{a_j}{2} \frac{v_j - w_j}{|v_j - w_j|}(z_1 + z_2)$$

the diagram

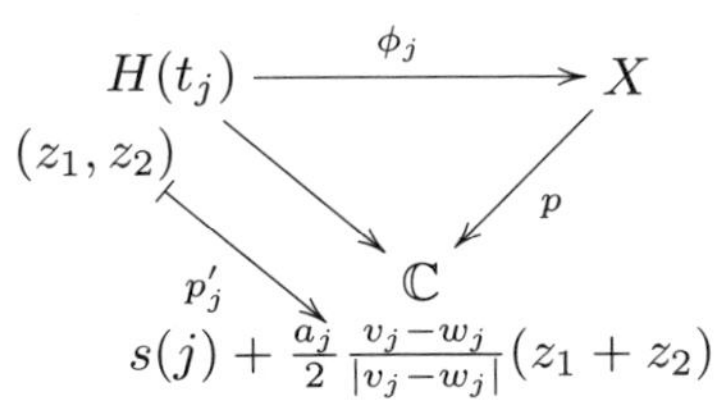

commutes and such that the curve

$$\Phi^{-1}\big(\phi_j(\{(z_1, z_2) \in H(t_j) \mid |z_\mu| = 1\})\big)$$

encloses $s_\mu(j)$. Again hypothesis (GH2) is satisfied.

Define

$$\tau_\mu(j) = 2\sqrt{t_j}, \quad r_\mu(j) = \frac{3|v_j - w_j|}{\sqrt{|s(j)|}}, \quad R_\mu(j) = \frac{\rho_j}{3\sqrt{|s(j)|}}.$$

Observe that, by (12.3b),

$$\lim_{j\to\infty} \frac{R_\mu(j)}{\sqrt{|s(j)|}} \lim_{j\to\infty} \frac{\rho_j}{|s(j)|} = 0.$$

If g was chosen big enough, then for all $j \geq g+1$ and $\mu = 1, 2$

$$\frac{a_j}{2}\left(4\tau_\mu(j) + \frac{t_j}{4\tau_\mu(j)}\right) < r_\mu(j)(2\sqrt{|s(j)|} - r_\mu(j)),$$

$$\frac{1}{4}R_\mu(j)(2\sqrt{|s(j)|} + \frac{1}{4}R_\mu(j)) < \frac{a_j}{2}\left(\frac{1}{4} - 4t_j\right),$$

$$\frac{a_j}{2}\left(\frac{1}{2} + 2t_j\right) < R_\mu(j)(2\sqrt{|s(j)|} - R_\mu(j))$$

$$< R_\mu(j)(2\sqrt{|s(j)|} + R_\mu(j)) < \frac{a_j}{2}(1 - t_j).$$

By Lemma 12.2 hypothesis (GH3) is fulfilled. Hypotheses (GH4) and (GH6) are trivial.

We verify (GH5) part by part with $d = 2D + 1$ and $\delta = 2\Delta + 1$. Part (i) is automatic from the conditions on the ρ_j's. Part (ii) follows from (12.2). Part (iii) is trivial because $|s_1(j)| = |s_2(j)|$ for all j. Parts (iv) and (v) follow from (12.3a)–(12.3b). Part (vi) follows from Lemma 12.2. □

13. Heat Curves: Basic Properties

Let

$$\Gamma = (0, 2\pi)\mathbb{Z} \oplus (\omega_1, \omega_2)\mathbb{Z}$$

where $\omega_1 > 0$, $\omega_2 \in \mathbb{R}$. The lattice dual to Γ is

$$\Gamma^\sharp = \{b \in \mathbb{R}^2 \mid < b, \gamma > \in 2\pi\mathbb{Z} \ \text{ forall } \gamma \in \Gamma\}$$
$$= \left(\frac{2\pi}{\omega_1}, 0\right)\mathbb{Z} \oplus \left(-\frac{\omega_2}{\omega_1}, 1\right)\mathbb{Z}.$$

Definition 13.1. The heat curve $\mathcal{H}(q)$ associated to $q \in L^2(\mathbb{R}^2/\Gamma)$ is the set of all points $(\xi_1, \xi_2) \in \mathbb{C}^* \times \mathbb{C}^*$ for which there is a nontrivial distributional solution $\psi(x_1, x_2)$ in $L^\infty_{\text{loc}}(\mathbb{R}^2)$ of the "heat equation"

$$\left(\frac{\partial}{\partial x_1} - \frac{\partial^2}{\partial x_2^2}\right)\psi + q(x_1, x_2)\psi = 0$$

satisfying

$$\psi(x_1 + \omega_1, x_2 + \omega_2) = \xi_1 \psi(x_1, x_2), \quad \psi(x_1, x_2 + 2\pi) = \xi_2 \psi(x_1, x_2).$$

The holomorphic map

$$k \in \mathbb{C}^2 \to \mathrm{E}(k) = (e^{i(\omega_1 k_1 + \omega_2 k_2)}, e^{i2\pi k_2}) \in \mathbb{C}^* \times \mathbb{C}^*$$

covers $\mathbb{C}^* \times \mathbb{C}^*$ with $\mathbb{C}^2$. An element b of the covering group $\Gamma^\sharp$ acts by translation $b \cdot k = kb$ on $\mathbb{C}^2$. Let

$$\widehat{\mathcal{H}}(q) = \mathrm{E}^{-1}\left(\mathcal{H}(q)\right). \tag{13.1}$$

Then, $\widehat{\mathcal{H}}(q)$ is the set of all points $k \in \mathbb{C}^2$ for which there is a nontrivial distribution $\phi(x_1, x_2)$ in $L^\infty(\mathbb{R}^2/\Gamma)$ satisfying

$$H_k \phi + q(x_1, x_2)\phi = 0.$$

Here,

$$\begin{aligned} H_k &= e^{-i\langle k,x\rangle}\left(\frac{\partial}{\partial x_1} - \frac{\partial^2}{\partial x_2^2}\right)e^{i\langle k,x\rangle} \\ &= \frac{\partial}{\partial x_1} - 2ik_2\frac{\partial}{\partial x_2} - \frac{\partial^2}{\partial x_2^2} + ik_1 + k_2^2. \end{aligned}$$

LEMMA 13.2.

(i) *For all* $p(x_1) \in L^2(\mathbb{R}/\omega_1\mathbb{Z})$ *with* $\int_0^{\omega_1} p(x_1)\,dx_1 = 0$,

$$\mathcal{H}\big(q(x_1,x_2) + p(x_1)\big) = \mathcal{H}(q).$$

(ii) *For all* $\lambda \in \mathbb{C}$,

$$\mathcal{H}(q-\lambda) = \{(k_1,k_2) \mid (k_1 + i\lambda, k_2) \in \widehat{\mathcal{H}}(q)\}/\Gamma^\sharp.$$

PROOF. (i) Suppose $\int_0^{\omega_1} p(x_1)\,dx_1 = 0$. Then, there is a solution f of

$$\frac{d}{dx_1}f(x_1) + p(x_1)f(x_1) = 0$$

belonging to $H^1(\mathbf{R}/\omega_1\mathbf{Z})$. In this case,

$$(H+q)\psi(x_1,x_2) = 0, \quad \psi \in H_k^2(\mathbf{R}^2)$$

is equivalent to

$$(H+q+p)f(x_1)\psi(x_1,x_2) = 0, \quad f\psi \in H_k^2(\mathbf{R}^2).$$

Therefore, $\mathcal{H}\big(q(x_1,x_2)+p(x_1)\big) = \mathcal{H}(q)$.

(ii) Note that $e^{-\lambda x_1}\psi(x_1,x_2)$ belongs to $H_k^2(\mathbf{R}^2)$ if $(k_1 i\lambda, k_2)$ is a point on $\widehat{\mathcal{H}}(q)$ and conversely. □

The Fourier coefficient $\widehat{q}(b)$, $b = (2\pi/\omega_1 m \omega_2/\omega_1 n, n) \in \Gamma^\sharp$, of $q \in L^2(\mathbf{R}^2/\Gamma)$ is defined by

$$\widehat{q}(b) = \langle e^{i\langle b,x\rangle}, q\rangle$$

where

$$\langle f,g\rangle = \frac{1}{|\mathbf{R}^2/\Gamma|}\int_{\mathbf{R}^2/\Gamma} \bar{f}(x)g(x)\,dx.$$

Concretely,

$$\widehat{q}(b) = \frac{1}{2\pi\omega_1}\int_0^{\omega_1} dx_1 \int_0^{2\pi} dx_2\, q(x_1,x_2)e^{-i(2\pi/\omega_1 m - \omega_2/\omega_1 n)x_1}e^{-inx_2}.$$

Write

$$q(x_1,x_2) = \widehat{q}(0) + p(x_1) + \sum_{b_2\neq 0}\widehat{q}(b)e^{i\langle b,x\rangle}$$

with

$$p(x_1) = \sum_{m\neq 0}\widehat{q}\Big(\frac{2\pi}{\omega_1}m, 0\Big)e^{i2\pi/\omega_1 m x_1}.$$

By the second part of Lemma 13.2, $\mathcal{H}(q)$ is biholomorphic to $\mathcal{H}\big(q - \widehat{q}(0)\big)$. By the first part,

$$\mathcal{H}\big(q - \widehat{q}(0)\big) = \mathcal{H}\Big(\sum_{b_2\neq 0}\widehat{q}(b)e^{i\langle b,x\rangle}\Big).$$

Thus, we may restrict our attention to q in

$$\begin{aligned}\mathcal{Q}(\Gamma) &= \left\{ q \in L^2(\mathbf{R}^2/\Gamma) \;\middle|\; \int_0^{2\pi} q(x_1, x_2)\, dx_2 = 0 \text{ for almost all } x_1 \right\} \\ &= \{ q \in L^2(\mathbf{R}^2/\Gamma) \mid \widehat{q}(b) = 0 \text{ for all } b \in \Gamma^\sharp \text{ with } b_2 = 0 \}.\end{aligned}$$

LEMMA 13.3. *The lift $\widehat{\mathcal{H}}(0)$ for $q = 0$ is the locally finite union $\bigcup_{b \in \Gamma^\sharp} \mathcal{P}_b$ of parabolas*

$$(13.2) \quad \mathcal{P}_b = \{(k_1, k_2) \in \mathbb{C}^2 \mid P_b(k_1, k_2) = 0\}, \quad P_b(k) = i(k_1 + b_1) + (k_2 + b_2)^2.$$

In particular, the heat curve $\mathcal{H}(0)$ is a complex analytic curve in $\mathbb{C}^2/\Gamma^\#$.

PROOF. For all $k \in \mathbb{C}^2$ the exponentials $e^{i\langle b, x\rangle}$, $b \in \Gamma^\sharp$ are a complete set of eigenfunctions for H_k in $L^2(\mathbf{R}^2/\Gamma)$ satisfying

$$H_k e^{i\langle b, x\rangle} = P_b(k) e^{i\langle b, x\rangle}.$$

Therefore,

$$\widehat{\mathcal{H}}(0) = \bigcup_{b \in \Gamma^\sharp} \mathcal{P}_b.$$

Observe that only a finite number of the parabolas $\mathcal{P}_b$ can intersect any bounded subset of $\mathbb{C}^2$. Thus, the union is locally finite. □

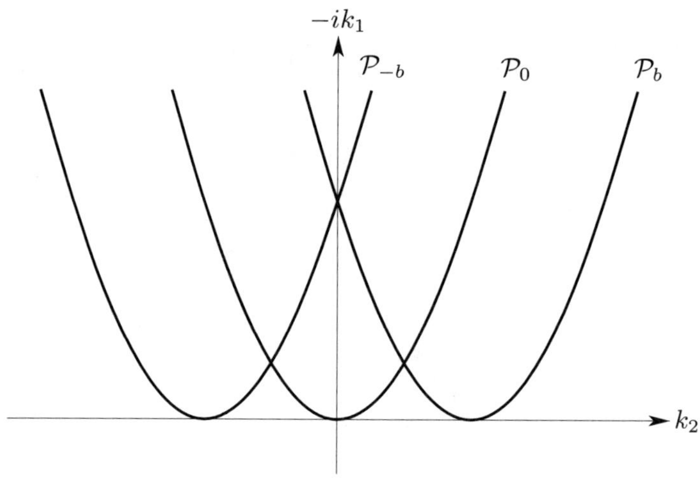

Reducing by the dual lattice, $\widehat{\mathcal{H}}(0)$ looks like

REMARK 13.4. The Weierstrass product construction yields

$$\widehat{\mathcal{H}}(0) = \left\{ (k_1, k_2) \in \mathbf{C}^2 \;\middle|\; P_0(k) \prod_{\substack{b \in \Gamma^\sharp \\ b \neq 0}} \frac{P_b(k)}{P_b(0)} R_b(k) = 0 \right\}$$

where

$$R_b(k) = e^{-(2k_2 b_2 + P_0(k)/P_b(0)) + 1/2(2k_2 b_2 + P_0(k)/P_b(0))^2}.$$

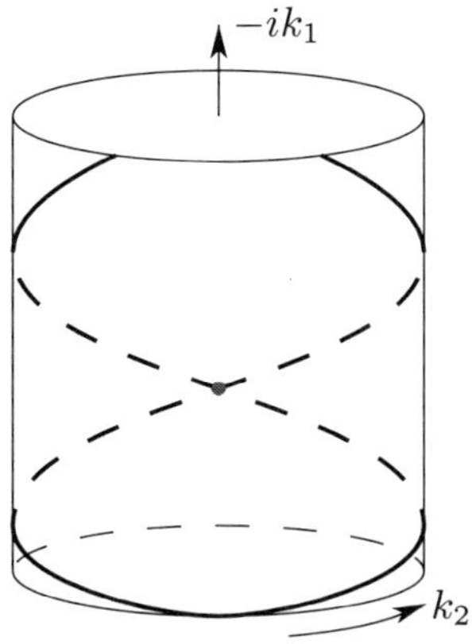

The parabolas $\mathcal{P}_b$ and $\mathcal{P}_c$, $b \neq c$, intersect transversely at the point with coordinates

$$k_1 = -b_1 + i(k_2 + b_2)^2 = -c_1 + i(k_2 + c_2)^2$$

and

$$k_2 = -\frac{b_2 + c_2}{2} - \frac{i}{2}\frac{b_1 c_1}{b_2 c_2}.$$

Note that, $b_2 = c_2$ forces $b = c$. In particular, $\mathcal{P}_b$ and $\mathcal{P}_0$, $b \neq 0$ intersect at

$$\begin{aligned} k_1 &= i\frac{1}{4}\left(b_2 - i\frac{b_1}{b_2}\right)^2 - b_1, \\ k_2 &= -\frac{1}{2}\left(b_2 + i\frac{b_1}{b_2}\right) \\ &= -\frac{1}{2}\left(n + i\frac{2\pi m \omega_2 n}{\omega_1 n}\right). \end{aligned}$$

The translate $c \cdot \mathcal{P}_b$ of $\mathcal{P}_b$ by $c \in \Gamma^\sharp$ is $\mathcal{P}_{b+c}$ and

$$c \cdot (\mathcal{P}_b \cap \mathcal{P}_d) = c \cdot \mathcal{P}_b \cap c \cdot \mathcal{P}_d = \mathcal{P}_{b+c} \cap \mathcal{P}_{d+c}.$$

We have

$$\mathrm{E}(\mathcal{P}_b \cap \mathcal{P}_0) = \mathrm{E}\big(-b \cdot (\mathcal{P}_b \cap \mathcal{P}_0)\big) = \mathrm{E}(\mathcal{P}_0 \cap \mathcal{P}_{-b}).$$

Conversely, $\mathrm{E}(\mathcal{P}_b \cap \mathcal{P}_0) = \mathrm{E}(\mathcal{P}_c \cap \mathcal{P}_0)$ implies that there is a $d \in \Gamma^\sharp$ with

$$\mathcal{P}_b \cap \mathcal{P}_0 = d \cdot (\mathcal{P}_c \cap \mathcal{P}_0) = \mathcal{P}_{c+d} \cap \mathcal{P}_d.$$

It follows that $\{c + d, d\} = \{b, 0\}$, since $\mathcal{P}_b \cap \mathcal{P}_c \cap \mathcal{P}_d = \varnothing$ when b, c and d are distinct. Hence, $c = b$ and $d = 0$, or $c = -b$ and $d = b$.

For each $b \in \Gamma^\sharp$, set

$$z_b = -\frac{1}{2}\left(b_2 + i\frac{b_1}{b_2}\right). \tag{13.3}$$

Note that

$$z_{-b} = -\frac{1}{2}\left(-b_2 + i\frac{b_1}{b_2}\right)$$

is the reflection of z_b across the real axis and for each $n \in \mathbb{Z}$, there is a "picket fence"

$$\left\{ z_{(2\pi/\omega_1 m\omega_2/\omega_1 n, n)} = -\frac{1}{2}n + i\left(\frac{\omega_2}{2\omega_1}\frac{\pi}{\omega_1}\frac{m}{n}\right) \middle| m \in \mathbb{Z} \right\}$$

of points lying on the line $\operatorname{Re} z = -\frac{1}{2}n$ at intervals of length $\pi/\omega_1 1/n$. In the special case of the square lattice $\Gamma = 2\pi\mathbb{Z}^2$ (that is, $\omega_1 = 2\pi$ and $\omega_2 = 0$) we have $\Gamma^\sharp = \mathbb{Z}^2$ and

$$z_{(m,n)} = -\frac{1}{2}\left(n + i\frac{m}{n}\right), \quad \mathcal{P}_{(m,n)} \cap \mathcal{P}_0 = \left(i\frac{1}{4}\left(ni\frac{m}{n}\right)^2 - m, z_{(m,n)}\right).$$

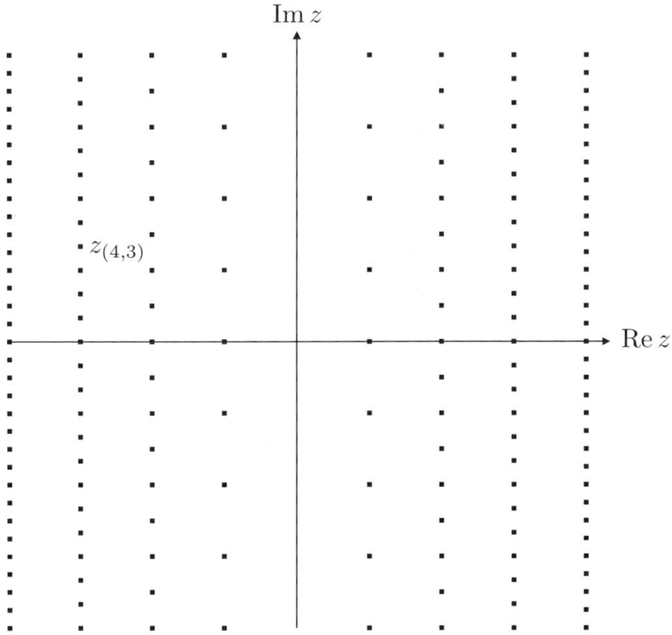

Also,

$$\mathrm{E}(\mathcal{P}_{(m,n)} \cap \mathcal{P}_0) = \mathrm{E}(iz^2_{(m,n)}, z_{(m,n)}) = \left((-1)^m e^{-\pi/2(n^2+m^2/n^2)}, (-1)^n e^{\pi m/n}\right).$$

The last three paragraphs are summarized in

LEMMA 13.5. *The composition*

$$z \in \mathbb{C} \to (iz^2, z) \in \widehat{\mathcal{H}}(0) \to \mathrm{E}(iz^2, z) \in \mathcal{H}(0)$$

is a biholomorphism between $\mathbb{C} \setminus \{z_b \mid b \in \Gamma^\sharp\}$ *and* $\mathcal{H}(0) \setminus \{\mathrm{E}(iz_b^2, z_b) \mid b \in \Gamma^\sharp\}$. *It collapses the pair* $\{z_b, z_{-b}\}$ *to the ordinary double point* $\mathrm{E}(iz_b^2, z_b)$ *on* $\mathcal{H}(0)$. *Thus, the curve* $\mathcal{H}(0)$ *is isomorphic to the variety obtained from* $\mathbb{C}$ *by identifying* z_b *with its reflection* z_{-b} *for all* $b \in \Gamma^\sharp$.

PROOF. If $\mathrm{E}(iz^2, z) = \mathrm{E}(iw^2, w)$, there is a $b \in \Gamma^\sharp$ with $(iw^2, w) = b \cdot (iz^2, z)$. That is, $(iz^2, z) \in \mathcal{P}_b \cap \mathcal{P}_0$ and consequently, $z = z_b$. Hence, the composition is a bijective holomorphic map between $\mathbb{C}\setminus\{z_b \mid b \in \Gamma^\sharp\}$ and $\mathcal{H}(0)\setminus\{\mathrm{E}(iz_b^2, z_b) \mid b \in \Gamma^\sharp\}$. The remaining statements follow from the discussion above. □

To show, for general $q \in \mathcal{Q}(\Gamma)$, that $\mathcal{H}(q)$ is a complex analytic curve in $\mathbb{C}^2$ we proceed as in [**KT**] and first obtain an analytic equation for $\widehat{\mathcal{H}}(q) \cap \left(\mathbb{C}^2 \setminus \mathcal{H}(0)\right)$.

LEMMA 13.6. *Let $q \in \mathcal{Q}(\Gamma)$. For all $k \in \mathbb{C}^2 \setminus \widehat{\mathcal{H}}(0)$*

$$(H_k + q)H_k^{-1} - \mathbf{1} = qH_k^{-1}$$

is a Hilbert–Schmidt operator on $L^2(\mathbf{R}^2/\Gamma)$ with matrix elements

$$\langle e^{i\langle b,x\rangle}, qH_k^{-1}e^{i\langle c,x\rangle}\rangle = \frac{\widehat{q}(bc)}{P_c(k)} = \frac{\widehat{q}(bc)}{i(k_1+c_1)+(k_2+c_2)^2}, \quad b,c \in \Gamma^\sharp$$

with respect to the orthormal basis of exponentials $e^{i\langle b,x\rangle}$, $b \in \Gamma^\sharp$. Furthermore, the regularized determinant

$$\det\nolimits_2\left((H_k+q)H_k^{-1}\right)$$

is analytic on $\left(\mathbb{C}^2 - \widehat{\mathcal{H}}(0)\right) \times \mathcal{Q}(\Gamma)$. In fact, the finite determinants

$$\det\left(\delta_{b,c} + \frac{\widehat{q}(bc)}{P_c(k)} : |b|, |c| \le r\right)$$

converge to $\det_2\left((H_k+q)H_k^{-1}\right)$ uniformly on closed bounded subsets of $\left(\mathbb{C}^2 \setminus \widehat{\mathcal{H}}(0)\right) \times \mathcal{Q}(\Gamma)$. Finally,

$$\widehat{\mathcal{H}}(q) \cap \left(\mathbb{C}^2 \setminus \mathcal{H}(0)\right) = \left\{(k_1,k_2) \mid \det\nolimits_2\left((H_k+q)H_k^{-1}\right) = 0\right\}.$$

Recall that the determinant of $\mathbf{1} + A$, when A is a trace class operator on a Hilbert space, is given by the convergent sum

$$\det(\mathbf{1}+A) = \sum_{n\ge 0} \operatorname{tr} \Lambda^n(A)$$

where $\Lambda^n(A)$, $n \ge 0$, is the nth exterior power of A. If A is Hilbert–Schmidt,

$$(\mathbf{1}+A)e^{-A} - \mathbf{1}$$

is trace class and one defines the regularized determinant

$$\det\nolimits_2(\mathbf{1}+A) = \det\left((\mathbf{1}+A)e^{-A}\right).$$

The regularized determinant has the property that $\det_2(\mathbf{1}+A) \ne 0$ if and only if $\mathbf{1}+A$ is invertible. Equivalently, by the Fredholm alternative, $\det_2(\mathbf{1}+A) = 0$ if and only if $Af = -f$ has a solution in the Hilbert space.

PROOF. The product qH_k^{-1} is Hilbert–Schmidt for all $k \in \mathbb{C}^2 \setminus \widehat{\mathcal{H}}(0)$ since

$$\sum_{b,c\in\Gamma^\sharp} |\langle e^{i\langle b,x\rangle}, qH_k^{-1}e^{i\langle c,x\rangle}\rangle|^2 = \sum_{b,c\in\Gamma^\sharp} \left|\frac{\widehat{q}(bc)}{i(k_1+c_1)+(k_2+c_2)^2}\right|^2$$

$$= \|q\|^2 \sum_{b\in\Gamma^\sharp} \left|\frac{1}{i(k_1+c_1)+(k_2+c_2)^2}\right|^2 < \infty.$$

Therefore, $\det_2\left((H_k+q)H_k^{-1}\right)$ is well-defined for $k \in \mathbb{C}^2 \setminus \widehat{\mathcal{H}}(0)$.

Suppose $k \in \mathbb{C}^2 \setminus \widehat{\mathcal{H}}(0)$ and

$$\det\nolimits_2\left((H_k+q)H_k^{-1}\right) = 0.$$

It follows from the remarks made directly before the proof of the lemma that there is a nontrivial $f \in L^2(\mathbb{R}^2/\Gamma)$ with

$$(H_k+q)H_k^{-1}f = 0.$$

Observe that the function $\phi = H_k^{-1} f$ belongs to $L^\infty(\mathbb{R}^2/\Gamma)$ since

$$\begin{aligned} |(H_k^{-1} f)(x)| &\leq \sum_{c \in \Gamma^\sharp} \left| \frac{\hat{f}(c)}{i(k_1+c_1)+(k_2+c_2)^2} \right| \\ &\leq \|f\| \Bigl(\sum_{c\in\Gamma^\sharp} |i(k_1+c_1)+(k_2+c_2)^2|^{-2} \Bigr)^{1/2} \\ &< \infty. \end{aligned}$$

Therefore, $k \in \widehat{\mathcal{H}}(q) \cap \bigl(\mathbb{C}^2 \setminus \widehat{\mathcal{H}}(0)\bigr)$.

Conversely, suppose $k \in \widehat{\mathcal{H}}(q) \cap \bigl(\mathbb{C}^2 \setminus \widehat{\mathcal{H}}(0)\bigr)$. By definition, there is a nontrivial distribution $\phi(x_1, x_2)$ in $L^\infty(\mathbb{R}^2/\Gamma)$ satisfying

$$H_k \phi + q(x_1, x_2)\phi = 0.$$

We have, $H_k\phi = -q\phi \in L^2(\mathbb{R}^2/\Gamma)$ and $(H_k+q)H_k^{-1}(H_k\phi) = 0$. This implies, again by the remarks above, that $\det_2\bigl((H_k+q)H_k^{-1}\bigr) = 0$. Thus,

$$\widehat{\mathcal{H}}(q) \cap \bigl(\mathbb{C}^2 \setminus \widehat{\mathcal{H}}(0)\bigr) = \bigl\{(k_1,k_2) \in \mathbb{C}^2 \setminus \widehat{\mathcal{H}}(0) \bigm| \det_2\bigl((H_k+q)H_k^{-1}\bigr) = 0\bigr\}.$$

Let π_r be the orthogonal projection onto the subspace of $L^2(\mathbf{R}^2/\Gamma)$ spanned by $e^{i\langle b,x\rangle}$, $|b| \leq r$. The truncated operator

$$\mathbf{1} + \pi_r q H_k^{-1} \pi_r$$

has matrix elements

$$\begin{cases} \delta_{b,c} + \frac{\widehat{q}(b-c)}{P_c(k)}, & |b|, |c| \leq r \\ \delta_{b,c}, & |b| \text{ or } |c| > r \end{cases}.$$

It is the direct sum of a principal minor of qH_k^{-1} and an identity matrix. We have

$$\begin{aligned} &\det_2(\mathbf{1} + \pi_r q H_k^{-1} \pi_r) \\ &\qquad = \det(\mathbf{1} + \pi_r q H_k^{-1}\pi_r) \exp\Bigl\{ -\widehat{q}(0) \sum_{|b|\leq r} \frac{1}{i(k_1+b_1)+(k_2+b_2)^2} \Bigr\} \\ &\qquad = \det\left(\delta_{b,c} + \frac{\widehat{q}(bc)}{P_c(k)} : |b|, |c| \leq r \right) \end{aligned}$$

since $\widehat{q}(0) = 0$ and

$$\det_2(\mathbf{1} + A) = \det(\mathbf{1} + A)\exp(-\operatorname{tr} A).$$

The right truncated operator

$$\mathbf{1} + qH_k^{-1}\pi_r$$

has matrix elements

$$\begin{cases} \delta_{b,c} + \frac{\widehat{q}(b-c)}{P_c(k)}, & |c| \leq r \\ \delta_{b,c}, & |c| > r \end{cases}.$$

The columns for $|c| > r$ have a single nonzero entry, namely $\delta_{cc} = 1$. By adding multiples of these columns to those with $|c| < r$, this matrix can be reduced to that of $\mathbf{1} + \pi_r q H_k^{-1} \pi_r$. In a *finite* number of dimensions, we would have

$$\det_2(\mathbf{1} + qH_k^{-1}\pi_r) = \det\left(\delta_{b,c} + \frac{\widehat{q}(bc)}{P_c(k)} : |b|, |c| \leq r \right).$$

To verify the last identity, first notice that for $s > r$

$$\det_2(\mathbf{1} + \pi_s q H_k^{-1} \pi_r) = \det\left(\delta_{b,c} + \frac{\widehat{q}(b-c)}{P_c(k)} : |b| \le s, |c| \le r : \delta_{b,c}; |b| < s, r < |c| \le s\right)$$
$$= \det\left(\delta_{b,c} + \frac{\widehat{q}(b-c)}{P_c(k)} : |b|, |c| \le r\right)$$

by column reduction for finite matrices. Next, we recall the estimate

$$|\det_2(\mathbf{1} + A) - \det_2(\mathbf{1} + B)| \le \|A - B\|_{\mathrm{HS}} \exp\{\alpha(\|A\|_{\mathrm{HS}} + \|B\|_{\mathrm{HS}} + 1)^2\}$$

where $\alpha > 0$ is a universal constant. In other words, $\det_2(\mathbf{1} + A)$ is Lipschitz with respect to the Hilbert–Schmidt norm. Our identity now follows from

$$\|q H_k^{-1} \pi_r - \pi_s q H_k^{-1} \pi_r\|_{\mathrm{HS}}^2 = \sum_{\substack{|b|>s \\ |c| \le r}} \left|\frac{\widehat{q}(bc)}{P_c(k)}\right|^2$$

since the right-hand side tends to zero as s goes to infinity.

Applying the Lipschitz estimate once again

$$\left|\det_2\left((H_k + q) H_k^{-1}\right) - \det\left(\delta_{b,c} + \frac{\widehat{q}(b-c)}{P_b(k)} : |b|, |c| \le r\right)\right|^2$$
$$= \left|\det_2\left((H_k + q) H_k^{-1}\right) - \det_2(\mathbf{1} + q H_k^{-1} \pi_r)\right|^2$$
$$\le \|q\|_2^2 \left(\sum_{|c|>r} \frac{1}{|P_c(k)|^2}\right) \exp\{\alpha(2\|q H_k^{-1}\|_{\mathrm{HS}} + 1)^2\}.$$

The remaining statements made in the lemma follow at once because

$$\det\left(\delta_{b,c} + \frac{\widehat{q}(b-c)}{P_b(k)} : |b|, |c| \le r\right)$$

is holomorphic on $\left(\mathbb{C}^2 - \widehat{\mathcal{H}}(0)\right) \times \mathcal{Q}(\Gamma)$, and

$$\|q\|_2^2 \sum_{|c|>r} \frac{1}{|P_c(k)|^2}$$

tends uniformly to zero on closed bounded subsets of $\left(\mathbb{C}^2 - \widehat{\mathcal{H}}(0)\right) \times \mathcal{Q}(\Gamma)$ while $\|q H_k^{-1}\|_{\mathrm{HS}}$ is uniformly bounded on them. □

For each finite subset B of $\Gamma^\sharp$ set

$$\mathbb{C}_B^2 = \mathbb{C}^2 \setminus \bigcup_{b \in \Gamma^\sharp \setminus B} \mathcal{P}_b.$$

For example,

$$\mathbb{C}_\varnothing^2 = \mathbb{C}^2 \setminus \widehat{\mathcal{H}}(0)$$

These sets are an open cover of $\mathbb{C}^2$. Also, let π_B be the orthogonal projection onto the subspace spanned by $e^{i\langle b,x\rangle}$, $b \in B$, and define a partial inverse $(H_k)_B^{-1}$ for $k \in \mathbb{C}_B^2$ by

$$(H_k)_B^{-1} = \pi_B + H_k^{-1}(\mathbf{1} - \pi_B).$$

Its matrix elements are

$$\langle e^{i\langle b,x\rangle}, (H_k)_B^{-1} e^{i\langle c,x\rangle}\rangle = \begin{cases} \delta_{b,c}, & c \in B \\ \delta_{b,c} \frac{1}{P_c(k)}, & c \notin B \end{cases}$$

and

$$(H_k+q)(H_k)_B^{-1} = \mathbf{1} + q(H_k)_B^{-1} + (H_k - \mathbf{1})\pi_B.$$

LEMMA 13.7. *Let $q \in \mathcal{Q}(\Gamma)$ and B a finite subset of $\Gamma^\sharp$. Then,*

$$e^{|B|}\det\left(\delta_{b,c} + \frac{\widehat{q}(bc)}{P_c(k)} : |b|,|c| \le r\right) \prod_{b\in B} P_b(k)e^{-P_b(k)}$$

for $r > \max\{|b| \mid b \in B\}$, converges uniformly on closed bounded subsets of $\mathbb{C}_B^2 \times \mathcal{Q}(\Gamma)$ to

$$\det_2\left((H_k+q)(H_k)_B^{-1}\right).$$

In particular,

$$\det_2\left((H_k+q)(H_k)_B^{-1}\right) = e^{|B|}\det_2\left((H_k+q)H_k^{-1}\right)\prod_{b\in B} P_b(k)e^{-P_b(k)}$$

is analytic on $\mathbb{C}_B^2 \times \mathcal{Q}(\Gamma)$. Furthermore,

$$\widehat{\mathcal{H}}(q) \cap \mathbb{C}_B^2 = \left\{(k_1,k_2) \in \mathbb{C}_B^2 \mid \det_2\left((H_k+q)(H_k)_B^{-1}\right) = 0\right\}.$$

PROOF. We have

$$\begin{aligned}\det_2&\left(\mathbf{1} + \pi_r\big(q(H_k)_B^{-1} + (H_k-\mathbf{1})\pi_B\big)\pi_r\right)\\ &= e^{|B|}\det\left(\delta_{b,c} + \frac{\widehat{q}(b-c)}{P_c(k)} : |b|,|c| \le r\right)\prod_{b\in B} P_b(k)e^{-P_b(k)}.\end{aligned}$$

All statements, but the last, now follow as in the proof of Lemma 13.6. For the last, observe that the partial inverse $(H_k)_B^{-1}$ is defined so that $k \in \widehat{\mathcal{H}}(q)\cap\mathbb{C}_B^2$ if and only if $(H_k+q)(H_k)_B^{-1}$ is not invertible. $\square$

Multiplying $\det_2\left((H_k+q)H_k^{-1}\right)$ by the function $P_0(k)\prod_{\substack{b\in\Gamma^\sharp\\ b\neq 0}} P_b(k)/P_b(0)R_b(k)$ of Remark 13.4 we get an entire function whose zero set is $\widehat{\mathcal{H}}(q)$. As in [**KT**, Theorem 2], one checks that it is of finite order.

We summarize the discussion above in

THEOREM 13.8. *For all $q \in \mathcal{Q}(\Gamma)$ the "lifted" heat curve $\widehat{\mathcal{H}}(q)$ is a one-dimensional complex analytic subvariety of $\mathbb{C}^2$. It is the zero set of an entire function of finite order. The intersection of $\widehat{\mathcal{H}}(q)$ with $\mathbb{C}_B^2$ is given by*

$$\widehat{\mathcal{H}}(q) \cap \mathbb{C}_B^2 = \left\{(k_1,k_2) \in \mathbb{C}_B^2 \mid \det_2\left((H_k+q)(H_k)_B^{-1}\right) = 0\right\}.$$

The heat curve $\mathcal{H}(q) = \widehat{\mathcal{H}}(q)/\Gamma^\sharp$ is an analytic subvariety of $\mathbb{C}^2/\Gamma^\sharp$.

LEMMA 13.9. *Suppose $k \in \widehat{\mathcal{H}}(q)$. Then there is a nontrivial distribution $\phi^\sharp$ in $L^\infty_{\mathrm{loc}}(\mathbb{R}^2/\Gamma)$ satisfying the adjoint equation*

$$-2i\bar{k}_2\phi^\sharp_{x_2} = \phi^\sharp_{x_1} + \phi^\sharp_{x_2x_2} - (-i\bar{k}_1 + \bar{k}_2^2 + \bar{q})\phi^\sharp.$$

PROOF. By hypothesis, there is a distribution $\phi \in L^\infty_{\mathrm{loc}}(\mathbb{R}^2/\Gamma)$ satisfying the equation

$$2ik_2\phi_{x_2} = \phi_{x_1} - \phi_{x_2x_2} + (ik_1 + k_2^2 + q)\phi.$$

Let B be a finite subset of $\Gamma^\sharp$ such that $k \in \widehat{\mathcal{H}}(q)\cap\mathbb{C}_B^2$. We have

$$(H_k^*)_B^{-1}(H_k^* + \bar{q}) = \left((H_k+q)(H_k)_B^{-1}\right)^*.$$

By Theorem 13.8,

$$\det_2\left((H_k^*)_B^{-1}(H_k^*+\overline{q})\right) = \overline{\det_2\left((H_k+q)(H_k)_B^{-1}\right)} = 0.$$

Hence there exists an $f \in L^2(\mathbb{R}^2/\Gamma)$ with

$$(H_k^*)_B^{-1}(H_k^*+\overline{q})f = 0$$

or equivalently,

$$\overline{P}_b(k)\hat{f}(b) = -\sum_{c\in\Gamma^\sharp}\widehat{\overline{q}}(cb)\hat{f}(c) \tag{13.4}$$

for all $b \in \Gamma^\sharp$. To complete the proof it suffices to show that $\hat{f} \in \ell^1$ since then $f \in L^\infty$.

By Lemma 13.6 the operator $\widehat{\overline{q}}(c-b)/\overline{P}_b(k)$ is Hilbert–Schmidt. Hence we can choose B finite but sufficiently large that

$$\sum_{b,c\in\Gamma^\#\setminus B}\left|\frac{\widehat{\overline{q}}(c-b)}{\overline{P}_b(k)}\right|^2 \le \frac{1}{2} \tag{13.5a}$$

$$\sup_{c\in\Gamma^\#}\sum_{b\in\Gamma^\#\setminus B}\left|\frac{\widehat{\overline{q}}(c-b)}{\overline{P}_b(k)}\right| \le \|\widehat{\overline{q}}\|_2\sqrt{\sum_{b\in\Gamma^\#\setminus B}\left|\frac{1}{\overline{P}_b(k)}\right|^2} \le \frac{1}{2}. \tag{13.5b}$$

Define the operators

$$R_{GB} = \left(\frac{\widehat{\overline{q}}(c-b)}{\overline{P}_b(k)}\right)_{b\notin B, c\in B}, \qquad R_{GG} = \left(\delta_{b,c} + \frac{\widehat{\overline{q}}(c-b)}{\overline{P}_b(k)}\right)_{b\notin B, c\notin B}$$

and the vectors

$$\hat{f}_B = \left(\hat{f}(c)\right)_{c\in B}, \quad \hat{f}_G = \left(\hat{f}(c)\right)_{c\notin B}.$$

Then, by (13.4)

$$R_{GB}\hat{f}_B + R_{GG}\hat{f}_G = 0$$

so that

$$\hat{f}_G = -R_{GG}^{-1}R_{GB}\hat{f}_B.$$

Now, $\widehat{f_B}$ only has finitely many components and so is trivially in $\ell^1(B)$. By (13.5b), R_{GB} maps $\ell^1(B)$ into $\ell^1(\Gamma^\#\setminus B)$. Also by (13.5b), $R_{GG} - \mathbb{1}$ has norm at most $\frac{1}{2}$ as an operator on $\ell^1(\Gamma^\#\setminus B)$ so that R_{GG}^{-1} is a bounded operator on $\ell^1(\Gamma^\#\setminus B)$. Hence $f_G \in \ell^1(\Gamma^\#\setminus B)$. □

There are horizontal and vertical projections of $\widehat{\mathcal{H}}(q)$ onto the k_1 and k_2 axes of $\mathbb{C}^2$. A vertical germ $\big(h(z,q), D\big)$ of the heat curve $\mathcal{H}(q)$ is a holomorphic function of z on the domain $D \subset \mathbb{C}$ such that

$$\big(h(z,q), z\big) \in \widehat{\mathcal{H}}(q)$$

for all $z \in D$. If $k \in \widehat{\mathcal{H}}(p) \cap \mathbb{C}_B^2$ and

$$\frac{\partial}{\partial k_1}\det_2\left((H_k+q)(H_k)_B^{-1}\right) \neq 0.$$

Then, by the implicit function theorem, there is a holomorphic function $h(z,q)$ on an open subset of $\mathbb{C}\times\mathcal{Q}(\Gamma)$ satisfying

$$\det_2\left((H_k+q)(H_k)_B^{-1}\right)\Big|_{k=(h(z,q),z)} = 0$$

and

$$k_1 = h(k_2, p).$$

Consequently, there is a holomorphic family of vertical germs.

LEMMA 13.10. *Let $\big(h(z,q), D\big)$ be a differentiable family of vertical germs and suppose that $\phi = \phi(x,k,q)$, where $k = \big(h(z,q), z\big)$, is the unique solution of*

$$(H_k + q)\phi = 0.$$

Then,

$$\frac{\partial}{\partial q(x)} h(z,q) = i \frac{\phi^\sharp(x,k)\overline{\phi(x,k)}}{\langle \phi, \phi^\sharp \rangle}\bigg|_{k=(h(z,q),z)}.$$

Here, $\phi^\sharp = \phi^\sharp(x,k,q)$ satisfies

$$(H_k + q)^* \phi^\sharp = 0.$$

Recall that the gradient is defined by the Riesz representation theorem through the identity

$$\frac{\partial}{\partial \varepsilon} F(q + \varepsilon v)\bigg|_{\varepsilon=0} = \left\langle \frac{\partial}{\partial q(\cdot)} F(q), v \right\rangle$$

for all $v \in L^2(\mathbb{R}^2/\Gamma)$.

PROOF. By hypothesis, $\phi = \phi(x,k,q+\varepsilon v)$, where $k = \big(h(z,q+\varepsilon v), z\big)$, is the unique solution of

$$(H_k + q + \varepsilon v)\phi = 0.$$

It follows from a familiar perturbative argument that $\phi(x,k,q+\varepsilon v)$ is a differentiable function of ε. Differentiating the equation for ϕ at $\varepsilon = 0$ and using the notation

$$\bullet = \frac{\partial}{\partial \varepsilon}\bigg|_{\varepsilon=0}$$

one obtains

$$(H_k^\bullet + v)\phi = -(H_k + q)\phi^\bullet.$$

Taking the inner product of both sides of the last equation with $\phi^\sharp$,

$$\begin{aligned}
\langle \phi^\sharp, (H_k^\bullet + v)\phi \rangle &= -\langle \phi^\sharp, (H_k + q)\phi^\bullet \rangle \\
&= -\langle (H_k + q)^* \phi^\sharp, \phi^\bullet \rangle \\
&= 0
\end{aligned}$$

or

$$\langle \phi^\sharp, H_k^\bullet \phi \rangle = -\langle \phi^\sharp, v\phi \rangle.$$

We have

$$H_k^\bullet = \left(\frac{\partial}{\partial x_1} - 2iz\frac{\partial}{\partial x_2} - \frac{\partial^2}{\partial x_2^2} + ih(z, q+\varepsilon v) + z^2 \right)^\bullet = i\frac{\partial}{\partial \varepsilon} h(z, q+\varepsilon v)\bigg|_{\varepsilon=0}.$$

Therefore,

$$i\langle \phi^\sharp, \phi \rangle \left\langle \frac{\partial}{\partial q(\cdot)} h(z,q), v \right\rangle = -\langle \phi^\sharp \overline{\phi}, v \rangle.$$

□

The rest of this section is devoted to the relationship between heat curves and the Kadomcev–Petviashvilli equation.

For each $u \in L^2(\mathbb{R}^2/\Gamma)$ define the function $I(u)$ by

$$I(u)(x_1,x_2) = \int_0^{x_2} u(x_1,s)\,ds - \frac{1}{2\pi}\int_0^{2\pi} dt \int_0^t u(x_1,s)\,ds.$$

The Kadomcev–Petviashvilli equation is

$$\text{(KP)} \qquad u_t = 3uu_{x_2} - \tfrac{1}{2}u_{x_2x_2x_2} - \tfrac{3}{2}I(u_{x_1x_1}).$$

If one differentiates both sides of (KP) with respect to x_2 one recovers the standard KPII equation (see, for example, [**K**])

$$(u_t - 3uu_{x_2} + \tfrac{1}{2}u_{x_2x_2x_2})_{x_2} + \tfrac{3}{2}u_{x_1x_1} = 0.$$

LEMMA 13.11. *Suppose, $u(x,t)$ is a classical solution of the* KP *equation with*

$$u(x_1, x_2+2\pi, t) = u(x_1,x_2,t)$$

for all x_1x_2 and t. Then,

$$\frac{\partial}{\partial t}\int_0^{2\pi} dx_2\, u(x_1,x_2,t) = 0$$

for all x_1 and t. In particular, if the initial data $u(x,0)$ satisfies $\int_0^{2\pi} dx_2\, u(x_1,x_2,0) = 0$ for all x_1, then $\int_0^{2\pi} dx_2\, u(x_1,x_2,t) = 0$ for all x_1 and $t \neq 0$.

PROOF. We have

$$\begin{aligned}\frac{\partial}{\partial t}\int_0^{2\pi} dx_2\, u(x_1,x_2,t) &= \int_0^{2\pi} dx_2\, 3uu_{x_2} - \frac{1}{2}u_{x_2x_2x_2} - \frac{3}{2}I(u_{x_1x_1})\\ &= \left(\frac{3}{2}u^2 - \frac{1}{2}u_{x_2x_2}\right)\Big|_0^{2\pi} - \int_0^{2\pi} dx_2\, I(u_{x_1x_1})\\ &= 0\end{aligned}$$

since,

$$\int_0^{2\pi} dx_2\, I(v)(x_1,x_2) = 0$$

for any function v. □

Let $\mathcal{U}(\Gamma) \subset \mathcal{Q}(\Gamma)$ be the space of all *real valued* functions $u(x_1,x_2)$ in $C^\infty(\mathbb{R}^2/\Gamma)$ satisfying

$$\int_0^{2\pi} u(x_1,x_2)\,dx_2 = 0$$

for all x_1. It follows from Lemma 13.9 that the initial value problem

$$u_t = 3uu_{x_2} - \tfrac{1}{2}u_{x_2x_2x_2} - \tfrac{3}{2}I(u_{x_1x_1}), \quad u(x,0) = u_0(x)$$

can be posed in $\mathcal{U}(\Gamma)$.

Suppose $u = u(x,t)$ is a solution of the initial value problem for the KP equation with intial data $u_0 \in \mathcal{U}(\Gamma)$. There is an associated family $\mathcal{H}(u(\cdot,t))$, $-\infty < t < \infty$, of heat curves. We will show that for all $-\infty < t < \infty$,

$$\mathcal{H}(u(\cdot,t)) = \mathcal{H}(u_0)$$

as subsets of $\mathbb{C}^* \times \mathbb{C}^*$.

Let $u \in \mathcal{U}(\Gamma)$. Define operators L_u and J_u by

$$L_u = u\frac{\partial}{\partial x_2} + \frac{\partial}{\partial x_2}u - \frac{1}{2}\frac{\partial^3}{\partial x_2^3} J_u = L_u - \frac{3}{2}\frac{\partial^2}{\partial x_1^2}I$$

and set

$$H(u) = \frac{1}{2}\int_{\mathbb{R}^2/\Gamma} u^2(x).$$

LEMMA 13.12. *Let u, v and w belong to $\mathcal{U}(\Gamma)$. Then,*

(i)
$$I(u)(x) = \sum_{\substack{b\in\Gamma^\sharp \\ b_2\neq 0}} \frac{1}{ib_2}\widehat{u}(b)e^{i\langle b,x\rangle}.$$

(ii) $\langle u, I(v)\rangle = -\langle I(u), v\rangle, \quad \langle v, L_u w\rangle = -\langle L_u v, w\rangle, \quad \langle v, J_u w\rangle = -\langle J_u v, w\rangle.$

(iii)
$$J_u\frac{\partial}{\partial u(x)}H(u) = J_u u = 3uu_{x_2} - \frac{1}{2}u_{x_2x_2x_2} - \frac{3}{2}I(u_{x_1x_1}).$$

PROOF. (i) In general,

$$\int_0^{x_2} u(x_1, s)\,ds = x_2\sum_{\substack{b\in\Gamma^\sharp \\ b_2=0}} \widehat{u}(b_1, 0)e^{ib_1x_1} + \sum_{\substack{b\in\Gamma^\sharp \\ b_2\neq 0}} \frac{1}{ib_2}\widehat{u}(b)(e^{i\langle b,x\rangle} - e^{ib_1x_1}).$$

Specializing to $\mathcal{U}(\Gamma)$ and integrating,

$$\int_0^{2\pi} dt \int_0^t u(x_1, s)\,ds = -\sum_{\substack{b\in\Gamma^\sharp \\ b_2\neq 0}} \frac{1}{ib_2}\widehat{u}(b)e^{ib_1x_1}.$$

Consequently,

$$I(u) = \sum_{\substack{b\in\Gamma^\sharp \\ b_2\neq 0}} \frac{1}{ib_2}\widehat{u}(b)(e^{i\langle b,x\rangle} - e^{ib_1x_1}) + \sum_{\substack{b\in\Gamma^\sharp \\ b_2\neq 0}} \frac{1}{ib_2}\widehat{u}(b)e^{ib_1x_1}$$

for $u \in \mathcal{U}(\Gamma)$.

Part (ii) is proved by using Parseval's identity to manipulate

$$\langle u, I(v)\rangle = \sum_{\substack{b\in\Gamma^\sharp \\ b_2\neq 0}} \overline{\widehat{u}(b)}\frac{1}{ib_2}\widehat{v}(b) = -\sum_{\substack{b\in\Gamma^\sharp \\ b_2\neq 0}} \overline{\frac{1}{ib_2}\widehat{u}(b)}\widehat{v}(b) = -\langle I(u), v\rangle.$$

For the remaining assertions, note that L_u is a skew symmetric ordinary differential operator.

For (iii), recall that the gradient of $H(u)$ is defined by

$$\left\langle \frac{\partial}{\partial u(x)}H(u), v\right\rangle = \frac{d}{d\varepsilon}H(u+\varepsilon v)\bigg|_{\varepsilon=0} = \int_{\mathbb{R}^2/\Gamma} dx\, u(x)v(x)$$

so that

$$\frac{\partial}{\partial u(x)}H(u) = u(x).$$

Now observe that

$$L_u u = 3uu_{x_2} - \tfrac{1}{2}u_{x_2x_2x_2}. \quad \square$$

LEMMA 13.13. *Let $u \in \mathcal{U}(\Gamma)$. Suppose that $\phi \in C^\infty(\mathbb{R}^2/\Gamma)$ satisfies $(H_k+u)\phi = 0$ and $\phi^\sharp \in C^\infty(\mathbb{R}^2/\Gamma)$ satisfies $(H_k+u)^*\phi^\sharp = 0$. Then,*

(i) $$L_u(\phi^\sharp\overline{\phi}) = 2i\overline{k}_1(\phi^\sharp\overline{\phi})_{x_2} - i\overline{k}_2(\phi^\sharp\overline{\phi})_{x_1} + \tfrac{1}{2}(\phi^\sharp_{x_1x_2}\overline{\phi} - \phi^\sharp\overline{\phi}_{x_1x_2}) + \tfrac{3}{2}(\phi^\sharp_{x_1}\overline{\phi}_{x_2} - \phi^\sharp_{x_2}\overline{\phi}_{x_1}).$$

(ii) $$(\phi^\sharp\overline{\phi})_{x_1} = -2i\overline{k}_2(\phi^\sharp\overline{\phi})_{x_2} + (\phi^\sharp\overline{\phi}_{x_2} - \phi^\sharp_{x_2}\overline{\phi})_{x_2}.$$

(iii) $$\begin{aligned}\langle(\phi^\sharp_{x_1x_2}\overline{\phi} - \phi^\sharp\overline{\phi}_{x_1x_2}), u\rangle &= 0,\\ \langle\phi^\sharp\overline{\phi}, u_{x_1}\rangle &= 0,\\ \langle\phi^\sharp\overline{\phi}, u_{x_2}\rangle &= 0.\end{aligned}$$

(iv) $$\langle\phi^\sharp\overline{\phi}, 3uu_{x_2} - \tfrac{1}{2}u_{x_2x_2x_2} - \tfrac{3}{2}I(u_{x_1x_1})\rangle = 0.$$

PROOF. The first two parts of the lemma are verified by direct calculation, using the equations

$$\begin{aligned}-2i\overline{k}_2\overline{\phi}_{x_2} &= \overline{\phi}_{x_1} - \overline{\phi}_{x_2x_2} + (-i\overline{k}_1 + \overline{k}_2^2 + u)\overline{\phi},\\ -2i\overline{k}_2\phi^\sharp_{x_2} &= \phi^\sharp_{x_1} + \phi^\sharp_{x_2x_2} - (-i\overline{k}_1 + \overline{k}_2^2 + u)\phi^\sharp\end{aligned}$$

to reduce derivatives.

We now derive the first statement of (iii). Multiply the equation for $\overline{\phi}$ by $\phi^\sharp_{x_1x_2}$ and the equation for $\phi^\sharp$ by $\overline{\phi}_{x_1x_2}$ and then add to obtain,

$$\begin{aligned}-2i\overline{k}_2(\phi^\sharp_{x_2}\overline{\phi}_{x_2})_{x_1} = {}&(-i\overline{k}_1 + \overline{k}_2^2 + u)\big(\phi^\sharp_{x_1x_2}\overline{\phi} - \phi^\sharp\overline{\phi}_{x_1x_2}\big)\\ &+ \phi^\sharp_{x_2x_2}\overline{\phi}_{x_1x_2} - \phi^\sharp_{x_1x_2}\overline{\phi}_{x_2x_2} + (\phi^\sharp_{x_1}\overline{\phi}_{x_1})_{x_2}.\end{aligned}$$

Integrating over $\mathbb{R}^2/\Gamma$,

$$\begin{aligned}\int_{\mathbb{R}^2/\Gamma} dx\, u(x)(\phi^\sharp_{x_1x_2}\overline{\phi} - \phi^\sharp\overline{\phi}_{x_1x_2}) &= -(-i\overline{k}_1 + \overline{k}_2^2)\int_{\mathbb{R}^2/\Gamma} dx(\phi^\sharp_{x_1x_2}\overline{\phi} - \phi^\sharp\overline{\phi}_{x_1x_2})\\ &\quad - \int_{\mathbb{R}^2/\Gamma} dx(\phi^\sharp_{x_2x_2}\overline{\phi}_{x_1x_2} - \phi^\sharp_{x_1x_2}\overline{\phi}_{x_2x_2})\\ &= 0\end{aligned}$$

by partial integration.

For the second statement multiply the equation for $\overline{\phi}$ by $\phi^\sharp_{x_1}$ and the equation for $\phi^\sharp$ by $\overline{\phi}_{x_1}$ and then subtract to obtain,

$$(-i\overline{k}_1 + \overline{k}_2^2 + u)(\phi^\sharp\overline{\phi})_{x_1} = 2i\overline{k}_2(\phi^\sharp_{x_2}\overline{\phi}_{x_1} - \phi^\sharp_{x_1}\overline{\phi}_{x_2}) - (\phi^\sharp_{x_1}\overline{\phi}_{x_2x_2} + \phi^\sharp_{x_2x_2}\overline{\phi}_{x_1}).$$

Again, integrate over $\mathbb{R}^2/\Gamma$ and then integrate by parts to complete the derivation.

To prove part (iv) we apply Lemma 13.12 and find

$$\begin{aligned}\langle\phi^\sharp\overline{\phi}, 3uu_{x_2} - \tfrac{1}{2}u_{x_2x_2x_2} - \tfrac{3}{2}I(u_{x_1x_1})\rangle &= \langle\phi^\sharp\overline{\phi}, J_u u\rangle\\ &= -\langle L_u(\phi^\sharp\overline{\phi}), u\rangle - \tfrac{3}{2}\langle\phi^\sharp\overline{\phi}, I(u_{x_1x_1})\rangle\\ &= -\langle L_u(\phi^\sharp\overline{\phi}), u\rangle + \tfrac{3}{2}\langle(\phi^\sharp\overline{\phi})_{x_1}, I(u_{x_1})\rangle\end{aligned}$$

By (i) and (iii),

$$-\langle L_u(\phi^\sharp\overline{\phi}), u\rangle = -\tfrac{3}{2}\langle(\phi^\sharp_{x_1}\overline{\phi}_{x_2} - \phi^\sharp_{x_2}\overline{\phi}_{x_1}), u\rangle$$

and by (ii),

$$\tfrac{3}{2}\big\langle(\phi^\sharp\overline{\phi})_{x_1}, I(u_{x_1})\big\rangle = \tfrac{3}{2}\big\langle -2i\overline{k}_2(\phi^\sharp\overline{\phi})_{x_2} + (\phi^\sharp\overline{\phi}_{x_2} - \phi^\sharp_{x_2}\overline{\phi})_{x_2}, I(u_{x_1})\big\rangle.$$

Applying Lemma 13.12(ii) again,

$$\begin{aligned}
\tfrac{3}{2}\langle(\phi^\sharp\overline{\phi})_{x_1}, I(u_{x_1})\rangle &= -\tfrac{3}{2}\big\langle I\big(-2i\overline{k}_2(\phi^\sharp\overline{\phi})_{x_2} + (\phi^\sharp\overline{\phi}_{x_2} - \phi^\sharp_{x_2}\overline{\phi})_{x_2}\big), u_{x_1}\big\rangle \\
&= -\tfrac{3}{2}\langle -2i\overline{k}_2(\phi^\sharp\overline{\phi}) + (\phi^\sharp\overline{\phi}_{x_2} - \phi^\sharp_{x_2}\overline{\phi}), u_{x_1}\rangle \\
&= -\tfrac{3}{2}\big\langle(\phi^\sharp\overline{\phi}_{x_2} - \phi^\sharp_{x_2}\overline{\phi}), u_{x_1}\big\rangle \\
&= \ \tfrac{3}{2}\big\langle(\phi^\sharp_{x_1}\overline{\phi}_{x_2} - \phi^\sharp_{x_2}\overline{\phi}_{x_1}), u\big\rangle + \tfrac{3}{2}\big\langle(\phi^\sharp\overline{\phi}_{x_1x_2} - \phi^\sharp_{x_1x_2}\overline{\phi}), u\big\rangle \\
&= \ \tfrac{3}{2}\big\langle(\phi^\sharp_{x_1}\overline{\phi}_{x_2} - \phi^\sharp_{x_2}\overline{\phi}_{x_1}), u\big\rangle.
\end{aligned}$$

To pass from the first to the second line, note that

$$\begin{aligned}
\langle I(f_{x_2}), g\rangle &= \left\langle f - \sum_{(b_1,0)\in\Gamma^\sharp} \hat{f}(b_1, 0)e^{ib_1x_1}, g\right\rangle \\
&= \langle f, g\rangle
\end{aligned}$$

for all $f \in C^\infty(\mathbb{R}^2/\Gamma)$ and $g \in \mathcal{Q}(\Gamma)$. Combining terms,

$$\langle \phi^\sharp\overline{\phi}, 3uu_{x_2} - \tfrac{1}{2}u_{x_2x_2x_2} - \tfrac{3}{2}I(u_{x_1x_1})\rangle = 0. \quad \square$$

THEOREM 13.14. *Suppose $u = u(x,t) \in \mathcal{U}(\Gamma)$ is a solution of the initial value problem for the Kadomcev–Petviashvilli equation*

$$u_t = 3uu_{x_2} - \tfrac{1}{2}u_{x_2x_2x_2} - \tfrac{3}{2}I(u_{x_1x_1})$$

with initial data

$$u(x,0) = u_0(x).$$

Let $\mathcal{H}\big(u(\cdot,t)\big)$, $-\infty < t < \infty$, be the associated family of heat curves. Then, for all $-\infty < t < \infty$,

$$\mathcal{H}\big(u(\cdot,t)\big) = \mathcal{H}(u_0)$$

as subsets of $\mathbb{C}^* \times \mathbb{C}^*$.

PROOF. Let $h(z,t)$, $s\varepsilon < t < s+\varepsilon$, be any smooth family of vertical germs for $\mathcal{H}\big(u(\cdot,t)\big)$. By Lemma 13.10, Lemma 13.13 and the chain rule,

$$\begin{aligned}
\frac{\partial}{\partial t}h(z,t) &= \left\langle \frac{\partial}{\partial q(\cdot)}h(z,t), \frac{\partial}{\partial t}u(\cdot,t)\right\rangle \\
&= \left\langle i\frac{\phi^\sharp(x,k)\overline{\phi(x,k)}}{\phi, \phi^\sharp\rangle}\bigg|_{k=(h(z,t),z)}, 3uu_{x_2} - \frac{1}{2}u_{x_2x_2x_2} - \frac{3}{2}I(u_{x_1x_1})\right\rangle \\
&= 0.
\end{aligned}$$

$\square$

14. Heat Curves: Asymptotics

In this section, we show that $\widehat{\mathcal{H}}(q)$ is close to $\widehat{\mathcal{H}}(0)$ when the imaginary parts of k_1 and k_2 are large. For $\varepsilon > 0$ and $b \in \Gamma^\sharp$ define the (ε-)tube about $\mathcal{P}_b$ by

(14.1)
$$\begin{aligned}
T_b &= \{k \in \mathbb{C}^2 \mid |P_b(k)| < \varepsilon\} \\
&= \{k \in \mathbb{C}^2 \mid |i(k_1 + b_1) + (k_2 + b_2)^2| < \varepsilon\}.
\end{aligned}$$

The pairwise intersection $\overline{T}_b \cap \overline{T}_{b'}$ is compact whenever $b \neq b'$. Indeed, if $k \in \overline{T}_b \cap \overline{T}_{b'}$ then

$$|i(b_1 - b_1') + 2k_2(b_2 - b_2') + b_2^2 - {b_2'}^2| = |P_b(k) - P_{b'}(k)| \leq 2\varepsilon.$$

Assuming the intersection is nonempty, $b \neq b'$ and ε is small enough, then b_2 and b_2' must be different and

$$\left| k_2 + \frac{b_2 + b_2'}{2} + \frac{i}{2} \frac{b_1 - b_1'}{b_2 - b_2'} \right| \leq \frac{\varepsilon}{|b_2 - b_2'|} \leq \varepsilon. \tag{14.2}$$

If ε is chosen small enough, we also have $\overline{T}_b \cap \overline{T}_{b'} \cap \overline{T}_{b''} = \varnothing$ for all distinct elements b, b', b'' of $\Gamma^\sharp$. We shall asymptotically confine $k \in \widehat{\mathcal{H}}(q)$ to the union of the tubes T_b, $b \in \Gamma^\sharp$.

For $\rho > 0$ define

$$\mathcal{K}_\rho = \{k \in \mathbb{C}^2 \mid |\operatorname{Im} k_1| + 2|\operatorname{Im} k_2|^2 \leq \rho\}.$$

Furthermore let $\mathrm{pr}\colon \mathbb{C}^2 \to \mathbb{C}$ be the projection $(k_1, k_2) \mapsto k_2$.

THEOREM 14.1. *Let $q \in L^2(\mathbb{R}^2/\Gamma)$ obey $\|b\widehat{q}(b)\|_1 := \sum_{b\in\Gamma^\#} |b\widehat{q}(b)| < \infty$ and $\widehat{q}(0) = 0$, and let $\varepsilon > 0$. Then there is a constant ρ, which depends only on $\|b\widehat{q}(b)\|_1$ and ε, such that*

(a) $$\{k \in \widehat{\mathcal{H}}(q) \mid k \notin \mathcal{K}_\rho\} \subset \bigcup_{b\in\Gamma^\#} T_b.$$

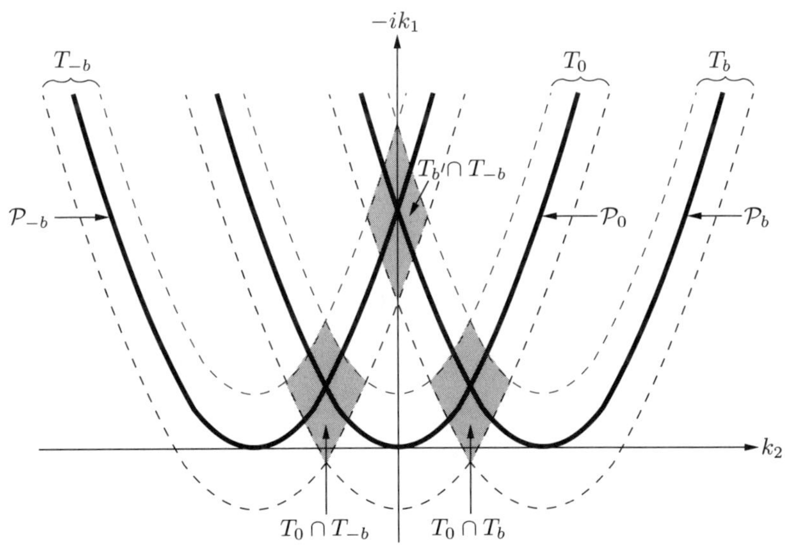

(b) *The projection* pr *induces a biholomorphic map between*

$$(\widehat{\mathcal{H}}(q) \cap T_0) \setminus \left(\mathcal{K}_\rho \cup \bigcup_{\substack{b\in\Gamma^\# \\ b_2 \neq 0}} T_b \right)$$

and its image in $\mathbb{C}$. This image contains

$$\left\{ z \in \mathbb{C} \,\middle|\, |z|^2 > 2\rho \text{ and } |z - z_b| > \frac{\varepsilon}{|b_2|} \text{ for all } b \in \Gamma^\# \text{ with } b_2 \neq 0 \right\}$$

and is contained in

$$\left\{ z \in \mathbb{C} \;\middle|\; |z - z_b| > \frac{\varepsilon}{4|b_2|} \text{ for all } b \in \Gamma^\# \text{ with } b_2 \neq 0 \right\}$$

where, as in §13, $z_b = -\frac{1}{2}(b_2 + ib_1/b_2)$

Clearly $\mathcal{K}_\rho$ is invariant under the $\Gamma^\#$-action and $\mathcal{K}_\rho/\Gamma^\#$ is compact. So the image of $\widehat{\mathcal{H}}(q) \cap \mathcal{K}_\rho$ under the exponential map $E\colon \widehat{\mathcal{H}}(q) \to \mathcal{H}(q)$ is compact in $\mathcal{H}(q)$. It will essentially play the role of X^{com} in the decomposition of $\mathcal{H}(q)$ that we need to apply the results of part II.

Since $c \cdot T_b = T_{b+c}$ for every $b, c \in \Gamma^\#$ the complement of $E\big(\widehat{\mathcal{H}}(q) \cap \mathcal{K}_\rho\big)$ in $\mathcal{H}(q)$ is the disjoint union of

$$E\left((\widehat{\mathcal{H}}(q) \cap T_0) \setminus \left(\mathcal{K}_\rho \cup \bigcup_{\substack{b \in \Gamma^\# \\ b_2 \neq 0}} T_b \right) \right)$$

and

$$\bigcup_{\substack{b \in \Gamma^\# \\ b_2 \neq 0}} E\big(\widehat{\mathcal{H}}(q) \cap T_0 \cap T_b\big).$$

Basicly the first of the two sets will be the regular piece of $\mathcal{H}(q)$, while the second sets will be the handles. The map Φ parametrizing the regular part will be the composition of E with the inverse of the map discussed in part (b) of Theorem 14.1. For the handles we will use

THEOREM 14.2. *Let* $\varepsilon > 0$ *be sufficiently small and let* $\beta \geq 4$. *Assume that* $q \in L^2(\mathbb{R}^2/\Gamma)$ *obeys* $\widehat{q}(0) = 0$ *and* $\| |b|^\beta \widehat{q}(b) \|_1 < \infty$. *Then there are constants such that for every sufficiently large* $d = (d_1, d_2) \in \Gamma^\# \setminus \{0\}$ *with* $d_2 \neq 0$ *there is a map*

$$\widehat{\phi}_d \colon \left\{ (z_1, z_2) \in \mathbb{C}^2 \;\middle|\; |z_1| \leq \frac{\varepsilon}{2}, |z_2| \leq \frac{\varepsilon}{2} \right\} \to T_0 \cap T_d$$

and a complex number $\hat{t}_d$ *with* $|\hat{t}_d| \leq \text{const}/|d|^{2\beta}$ *such that*

(i) $\widehat{\phi}_d$ *is biholomorphic to its image. The image contains*

$$\left\{ k \in \mathbb{C}^2 \;\middle|\; |P_0(k)| \leq \frac{\varepsilon}{8}, |P_d(k)| \leq \frac{\varepsilon}{8} \right\}.$$

Furthermore

$$D\widehat{\phi}_d = \frac{1}{2id_2} \begin{pmatrix} 2z_{-d} & -2z_d \\ -i & i \end{pmatrix} \left\{ \mathbb{1} + O\left(\frac{1}{|d_2 z_d|} \right) \right\}$$

and

$$|\widehat{\phi}_d(0) - (iz_d^2, z_d)| \leq \frac{\text{const}}{|d_2 z_d|} \left(1, \frac{1}{|z_d|} \right)$$

(ii) $$\widehat{\phi}_d^{-1}\big(T_0 \cap T_d \cap \widehat{\mathcal{H}}(q)\big) = \left\{ (z_1, z_2) \in \mathbb{C}^2 \;\middle|\; z_1 z_2 = \hat{t}_d, |z_1| \leq \frac{\varepsilon}{2}, |z_2| \leq \frac{\varepsilon}{2} \right\}.$$

(iii) $$\widehat{\phi}_d(z_1, z_2) = \widehat{\phi}_{-d}(z_2, z_1) - d.$$

As pointed out in Lemma 13.2 we may, for the proof of these theorems, assume that $\widehat{q}(b) = 0$ for all $b \in \Gamma^{\#}$ with $b_2 = 0$. To facilitate the discussion below write $k \in \mathbb{C}^2$ as

$$k_1 = u_1 + i(v_1 - v_2^2), k_2 = u_2 + iv_2 \tag{14.3a}$$

where $u_1, u_2, v_1 \mathrm{and} v_2$ are real. Then

$$P_b(k) = \big((u_2 + b_2)^2 - v_1\big) + i\big(u_1 + b_1 + 2(u_2 + b_2)v_2\big). \tag{14.3b}$$

In particular, $P_b(k) = 0$ if and only if

$$v_1 = (u_2 + b_2)^2, \quad v_2 = -\frac{1}{2}\frac{u_1 + b_1}{u_2 + b_2}. \tag{14.4}$$

By definition k is in $\widehat{\mathcal{H}}(q)$ if $H_k + q$ has a nontrivial kernel in $L^2(\mathbb{R}^2/\Gamma)$. To study the part of the curve in the intersection of $\cup_{d\in G} T_d$ and $\mathbb{C}^2 \setminus \cup_{b\notin G} T_b$ for some finite subset G of $\Gamma^\sharp$ it is natural to look for a nontrivial solution of

$$(H_k + q)\psi_G + (H_k + q)\psi_{G'} = 0$$

or equivalently of

$$(H_k + q)\phi_G + (\mathbb{1} + qH_k^{-1})\phi_{G'} = 0 \tag{14.5}$$

where

$$\begin{aligned}
\psi_G, \phi_G &\in L^2_G := \operatorname{span}\{e^{i\langle b,x\rangle} \mid b \in G\}^-, \\
\psi_{G'}, \phi_{G'} &\in L^2_{G'} := \operatorname{span}\{e^{i\langle b,x\rangle} \mid b \in \Gamma^\sharp \setminus G\}^-.
\end{aligned}$$

We shall shortly show that, for k in the region under consideration, $R_{G'G'}$, the restriction of $\mathbb{1} + qH_k^{-1}$ to $L^2_{G'}$, has a bounded inverse. Then the projection of (14.5) on $L^2_{G'}$ is equivalent to

$$\phi_{G'} = -R^{-1}_{G'G'} q\phi_G.$$

Substituting this into the projection on L^2_G yields

$$\pi_G(H_k + q - qH_k^{-1}R^{-1}_{G'G'}q)\phi_G = 0.$$

Here π_G is the obvious projection operator. This has a nontrivial solution if and only if the $|G| \times |G|$ determinant

$$\det[\pi_G(H_k + q - qH_k^{-1}R^{-1}_{G'G'}q)\pi_G] = 0,$$

or equivalently, expressing all operators as matrices in the basis $\{e^{i\langle b,x\rangle} \mid b \in \Gamma^\sharp\}$,

$$\det\left[P_d(k)\varepsilon_{d,d'} + \widehat{q}(d - d') - \sum_{b,c\in G'} \frac{\widehat{q}(d - b)}{P_b(k)}(R^{-1}_{G'G'})_{b,c}\widehat{q}(c - d')\right]_{d,d'\in G} = 0. \tag{14.6}$$

In general we define the operator R_{BC} to have matrix elements

$$(R_{BC})_{b,c} = \left[\delta_{b,c} + \frac{\widehat{q}(b - c)}{P_c(k)}\right]_{b\in B, c\in C}. \tag{14.7}$$

Our analysis of (14.6) is based on two lemmas. The first gives a collection of properties of $P_b(k)$. The second uses these to derive a number of properties of the operators R_{BC}.

LEMMA 14.3.

(a) *There is a constant such that, for all* $\alpha > 2$, $k \in \mathbb{C}$,

$$\sum_{\substack{b\in\Gamma^\sharp \\ |P_b(k)|>\alpha}} \frac{1}{|P_b(k)|^2} \le \text{const}\frac{\ln\alpha}{\sqrt{\alpha}}$$

(b)
$$D_\varepsilon = \sup_{B\subset\Gamma^\sharp} \sup_{k\in\mathbb{C}^2\setminus\bigcup_{b\in B}T_b} \left(\sum_{b\in B}\frac{1}{|P_b(k)|^2}\right)^{1/2} < \infty.$$

(c) *If* $b, c \in \Gamma^\sharp$ *obey* $b_2 \ne c_2$ *and*

$$|P_b(k)|,\ |P_c(k)| < \tfrac14[\sqrt{|v_1|} + \max\{|k_2+b_2|, |k_2+c_2|\}]$$

then

$$|b-c| > \tfrac18[\sqrt{|v_1|} + \max\{|k_2+b_2|, |k_2+c_2|\}].$$

(d) *Let* $b, d \in \Gamma^\sharp$ *obey* $b_2 \ne 0$, $|b| \le \frac18[\sqrt{|v_1|} + |k_2+d_2|]$ *and* $|P_d(k)| \le \varepsilon \le 1$. *Then*

$$\left|\frac{1}{P_{b+d}(k)} + \frac{1}{P_{-b+d}(k)}\right| \le 100\frac{b_2^2}{[1+\sqrt{|v_1|}+|k_2+d_2|]^2}.$$

(e) *Let* $d \in \Gamma^\sharp$ *obey* $|P_d(k)| < \frac14[\sqrt{|v_1|} + |k_2+d_2|]$. *Let* $\widehat{q}(b_1,0) = 0$ *for all* b_1. *Then*

$$\left[\sum_{\substack{b\in\Gamma^\sharp \\ |P_b(k)|\ge\varepsilon}} \left|\frac{\widehat{q}(d-b)}{P_b(k)}\right|^2\right]^{1/2} < \frac{13}{\varepsilon}\frac{\|b\widehat{q}(b)\|_2}{\sqrt{|v_1|}+|k_2+d_2|}.$$

(f) *Let* $|P_d(k)| < \varepsilon$ *and* $|P_{d+b}(k)| \ge \varepsilon$ *for all* $b \in \Gamma^\sharp \setminus \{0\}$. *Let* $\widehat{q}(b_1,0) = 0$ *for all* b_1. *Then*

$$\left|\sum_{b\in\Gamma^\sharp\setminus\{0\}} \frac{\widehat{q}(b)\widehat{q}(-b)}{P_{b+d}}\right| \le \frac{164}{\varepsilon}\frac{\|b\widehat{q}(b)\|_2^2}{[\sqrt{|v_1|}+|k_2+d_2|]^2}.$$

REMARK. We shall, for convenience, retain the terms $\sqrt{|v_1|}$ that appear in the above bounds. However, they do not strengthen the bounds. By (14.3b)

$$v_1 = (u_2+d_2)^2 - \mathrm{Re}P_d(k).$$

Consequently, whenever $|P_d(k)| \le \frac14[\sqrt{|v_1|} + |k_2+d_2|]$ we have

$$\begin{aligned} \text{const}(1+|k_2+d_2|) &\le 1+\sqrt{|v_1|}+|v_2| \\ &\le 1+\sqrt{|v_1|}+|k_2+d_2| \\ &\le \text{const}(1+|k_2+d_2|). \end{aligned} \tag{14.8}$$

To see the first inequality in (14.8), consider the cases $|u_2+d_2| \le |v_2|$ and $|u_2+d_2| > |v_2|$ separately. In the case $|u_2+d_2| > |v_2|$

$$|P_d(k)| \le \tfrac14\sqrt{|v_1|} + \tfrac12|u_2+d_2|$$

which implies that

$$|u_2+d_2|^2 \le |v_1| + \tfrac14\sqrt{|v_1|} + \tfrac12|u_2+d_2|$$

and hence

$$|u_2+d_2| \le \text{const}\sqrt{|v_1|}$$

when $|u_2+d_2|\ge 1$. To see the last inequality of (14.8) observe that

$$|v_1| \le |k_2+d_2|^2 + \tfrac{1}{4}[\sqrt{|v_1|} + |k_2+d_2|]$$

which implies

$$\tfrac{3}{4}|v_1| \le |k_2+d_2|^2 + \tfrac{1}{4}|k_2+d_2| \le \text{const}|k_2+d_2|^2$$

when $|v_1|\ge 1$.

PROOF OF (a). We are to bound the sum

$$\begin{aligned}
\sum_{\substack{b\in\Gamma^\sharp\\ |P_b(k)|>\alpha}} \frac{1}{|P_b(k)|^2} &= \sum_{\substack{b\in\Gamma^\sharp\\ |P_b(k)|>\alpha}} \frac{1}{(v_1-(u_2+b_2)^2)^2 + \big(u_1+b_1+2(u_2+b_2)v_2\big)^2}\\
&= \sum_{s\in\mathbf{Z}+u_2} \sum_{\substack{t\in 2\pi/\omega_1\mathbf{Z}-(s-u_2)\omega_2/\omega_1+u_1+2sv_2\\ (v_1-s^2)^2+t^2>\alpha^2}} \frac{1}{(v_1-s^2)^2+t^2}\\
&\le 2\sum_{s\in\mathbf{Z}+u_2} \sum_{\substack{t\in 2\pi/\omega_1\mathbf{Z}-(s-u_2)\omega_2/\omega_1+u_1+2sv_2\\ (v_1-s^2)^2+t^2>\alpha^2}} \frac{1}{(v_1-s^2)^2+t^2+\alpha^2}\\
&\le \text{const} \sum_{s\in\mathbf{Z}+u_2} \left[\int dt \frac{1}{(v_1-s^2)^2+\alpha^2+t^2} + \frac{1}{(v_1-s^2)^2+\alpha^2}\right]\\
&\le \text{const} \sum_{s\in\mathbf{Z}+u_2} \frac{1}{\sqrt{(v_1-s^2)^2+\alpha^2}}\\
&\le \text{const} \sum_{s\in\mathbf{Z}+u_2} \frac{1}{|s^2-v_1|+\alpha}.
\end{aligned}$$

In the event that $v_1\le 0$ we have

$$\sum_{s\in\mathbf{Z}+u_2} \frac{1}{|s^2-v_1|+\alpha} \le \sum_{s\in\mathbf{Z}+u_2} \frac{1}{s^2+\alpha} \le \frac{\text{const}}{\sqrt{\alpha}}$$

so it suffices to consider $v_1>0$. It also suffices to consider $s\ge 0$. The terms with $s\ge\sqrt{v_1}$ are bounded by

$$\begin{aligned}
\sum_{\substack{s\in\mathbf{Z}+u_2\\ s\ge\sqrt{v_1}}} \frac{1}{|s^2-v_1|+\alpha} &= \sum_{\substack{\sigma\in\mathbf{Z}+u_2-\sqrt{v_1}\\ \sigma\ge 0}} \frac{1}{(\sigma+\sqrt{v_1})^2-v_1+\alpha}\\
&= \sum_{\substack{\sigma\in\mathbf{Z}+u_2-\sqrt{v_1}\\ \sigma\ge 0}} \frac{1}{\sigma^2+2\sigma\sqrt{v_1}+\alpha}\\
&\le \sum_{\substack{\sigma\in\mathbf{Z}+u_2-\sqrt{v_1}\\ \sigma\ge 0}} \frac{1}{\sigma^2+\alpha} \le \frac{\text{const}}{\sqrt{\alpha}}.
\end{aligned}$$

The terms with $0 \le s \le \sqrt{v_1}$ are bounded by

$$\sum_{\substack{s\in\mathbf{Z}+u_2\\ 0\le s\le\sqrt{v_1}}} \frac{1}{v_1 - s^2 + \alpha} = \sum_{\substack{\sigma\in\mathbf{Z}-u_2+\sqrt{v_1}\\ 0\le\sigma\le\sqrt{v_1}}} \frac{1}{2\sigma\sqrt{v_1} - \sigma^2 + \alpha} \le \sum_{\substack{\sigma\in\mathbf{Z}-u_2+\sqrt{v_1}\\ 0\le\sigma\le\sqrt{v_1}}} \frac{1}{\sigma\sqrt{v_1} + \alpha}.$$

When $v_1 \le \alpha$ this last sum is bounded by $\text{const}\, 1 + \sqrt{v_1}/\alpha \le \text{const}/\sqrt{\alpha}$. Finally, when $v_1 > \alpha$ it is bounded by

$$\text{const}\left[\frac{1}{\alpha} + \frac{1}{\sqrt{v_1}}\int_1^{\sqrt{v_1}} d\sigma \frac{1}{\sigma}\right] = \text{const}\left[\frac{1}{\alpha} + \frac{\ln\sqrt{v_1}}{\sqrt{v_1}}\right] \le \text{const}\frac{\ln\alpha}{\sqrt{\alpha}}.$$

PROOF OF (b). It suffices to apply (a) with $\alpha = 3$ and observe that there are only finitely many points in $\Gamma^\sharp$ with $|P_b(k)| \le 3$ and hence with $|k_1 + b_1| \le 3$ and $|k_2 + b_2| \le \sqrt{3}$. The number of such points is bounded independent of $k \in \mathbb{C}^2$.

PROOF OF (c). First consider the case $\sqrt{|v_1|} \le \max\{|k_2 + b_2|, |k_2 + c_2|\}$. Assume, without loss of generality, that $|k_2 + b_2| \ge |k_2 + c_2|$. Then

$$|P_b(k) - P_c(k)| < |k_2 + b_2|.$$

Now suppose, contrary to the statement of the Lemma, that $|b - c| \le \frac{1}{2}|k_2 + b_2|$. Since

$$\begin{aligned} P_b(k) - P_c(k) &= 2(b_2 - c_2)k_2 + b_2^2 - c_2^2 + i(b_1 - c_1) \\ &= 2(b_2 - c_2)\left(k_2 + \frac{b_2 + c_2}{2}\right) + i(b_1 - c_1) \end{aligned} \tag{14.9}$$

and $b_2 - c_2 \in \mathbf{Z} - \{0\}$ we have

$$\begin{aligned} |P_b(k) - P_c(k)| &\ge 2\{|k_2 + b_2| - \tfrac{1}{2}|b_2 - c_2|\} - |b_1 - c_1| \\ &\ge 2 \times \tfrac{3}{4}|k_2 + b_2| - \tfrac{1}{2}|k_2 + b_2| \\ &= |k_2 + b_2| \end{aligned}$$

which is a contradiction.

Next consider the case $\sqrt{|v_1|} \ge \max\{|k_2 + b_2|,\ |k_2 + c_2|\}$, $\sqrt{|v_1|} \ge 4$. Then the real parts, $(u_2 + b_2)^2 - v_1, (u_2 + c_2)^2 - v_1$ of $P_b(k)$, $P_c(k)$ are both smaller in magnitude than $\frac{1}{2}\sqrt{|v_1|} \le \frac{1}{8}|v_1|$. This implies that v_1 is positive and

$$|(u_2 + b_2)^2 - (u_2 + c_2)^2| < \sqrt{|v_1|}, \quad (u_2 + b_2)^2 > \tfrac{3}{4}|v_1|, \quad (u_2 + c_2)^2 > \tfrac{3}{4}|v_1|$$

so that

$$\begin{aligned} \big||u_2 + b_2| - |u_2 + c_2|\big| &= \left|\frac{(u_2 + b_2)^2 - (u_2 + c_2)^2}{\sqrt{(u_2 + b_2)^2} + \sqrt{(u_2 + c_2)^2}}\right| \\ &\le \frac{\sqrt{|v_1|}}{\sqrt{3}/2\sqrt{|v_1|} + \sqrt{3}/2\sqrt{|v_1|}} < 1. \end{aligned}$$

Thus u_2+b_2 and u_2+c_2 must be nonzero and of opposite sign. Consequently

$$\begin{aligned}|b_2-c_2| &= |(u_2+b_2)-(u_2+c_2)|\\ &= |u_2+b_2|+|u_2+c_2|\\ &\geq \frac{\sqrt{3}}{2}\sqrt{|v_1|}+\frac{\sqrt{3}}{2}\sqrt{|v_1|}\\ &\geq \frac{\sqrt{3}}{2}[\sqrt{|v_1|}+\max\{|k_2+b_2|,|k_2+c_2|\}]\end{aligned}$$

as desired.

Finally, if $4>\sqrt{|v_1|}\geq \max\{|k_2+b_2|,|k_2+c_2|\}$

$$|b_2-c_2|\geq 1\geq \tfrac{1}{8}[\sqrt{|v_1|}+\max\{|k_2+b_2|,|k_2+c_2|\}].$$

Proof of (d). This is a simple consequence of part (c) and

$$\begin{aligned}(14.10)\quad \frac{1}{P_{b+d}(k)}+\frac{1}{P_{-b+d}(k)}\\ &= \frac{P_{b+d}(k)+P_{-b+d}(k)}{P_{b+d}(k)P_{-b+d}(k)}\\ &= \frac{2P_d(k)+(P_{b+d}(k)-P_d(k))+(P_{-b+d}(k)-P_d(k))}{P_{b+d}(k)P_{-b+d}(k)}\\ &= \frac{2P_d(k)+2b_2^2}{P_{b+d}(k)P_{-b+d}(k)}.\end{aligned}$$

The numerator is bounded by $4b_2^2$ and each factor in the denominator is at least

$$\tfrac{1}{4}[\sqrt{|v_1|}+|k_2+d_2|]\geq \tfrac{1}{5}[1+\sqrt{|v_1|}+|k_2+d_2|]$$

since

$$[\sqrt{|v_1|}+|k_2+d_2|]\geq 8|b|\geq 8.$$

Proof of (e). Define

$$B=\{b\in\Gamma^\sharp \mid |P_b(k)|\geq \tfrac{1}{4}[\sqrt{|v_1|}+|k_2+d_2|]\}$$

and

$$S=\{b\in\Gamma^\sharp \mid \varepsilon\leq |P_b(k)|<\tfrac{1}{4}[\sqrt{|v_1|}+|k_2+d_2|]\}.$$

Then, by part (c),

$$\begin{aligned}\sum_{\substack{b\in\Gamma^\sharp\\ |P_b(k)|\geq\varepsilon}}\left|\frac{\widehat{q}(d-b)}{P_b(k)}\right|^2 &= \sum_{b\in S}\left|\frac{\widehat{q}(d-b)}{P_b(k)}\right|^2+\sum_{b\in B}\left|\frac{\widehat{q}(d-b)}{P_b(k)}\right|^2\\ &\leq \sum_{|d-b|\geq 1/8[\sqrt{|v_1|}+|k_2+d_2|]}\left|\frac{\widehat{q}(d-b)}{\varepsilon}\right|^2+\sum_b\left|\frac{\widehat{q}(d-b)}{1/4[\sqrt{|v_1|}+|k_2+d_2|]}\right|^2\\ &\leq \left(\frac{64}{\varepsilon^2}+16\right)[\sqrt{|v_1|}+|k_2+d_2|]^{-2}\sum_b\left(b^2|\widehat{q}(b)|^2+|\widehat{q}(b)|^2\right).\end{aligned}$$

PROOF OF (f). Define

$$N = \{b \in \Gamma^\sharp \mid 0 < |b| \le 1/8[\sqrt{|v_1|} + |k_2 + d_2|]\}$$

and

$$F = \{b \in \Gamma^\sharp \| b| > 1/8[\sqrt{|v_1|} + |k_2 + d_2|]\}.$$

Then, by $b \to -b$ symmetry

$$\begin{aligned}\left|\sum_{b\in N} \frac{\widehat{q}(b)\widehat{q}(-b)}{P_{b+d}}\right| &= \frac{1}{2}\left|\sum_{b\in N} \widehat{q}(b)\widehat{q}(-b)\left[\frac{1}{P_{b+d}} + \frac{1}{P_{-b+d}}\right]\right| \\ &\le 100\frac{\|b\widehat{q}(b)\|^2}{[\sqrt{|v_1|} + |k_2 + d_2|]^2}\end{aligned}$$

by part (d). The sum over F is controlled as in part (e).

LEMMA 14.4. *Let $k \in C^2$. Let $\sqrt{|v_1|} + |v_2| > 8$ and*

$$S \subset \{b \in \Gamma^\sharp \mid \varepsilon \le |P_b(k)| < \tfrac{1}{4}[\sqrt{|v_1|} + |v_2|]\},$$
$$B \subset \{b \in \Gamma^\sharp \mid |P_b(k)| \ge \frac{1}{4}[\sqrt{|v_1|} + |v_2|]\}.$$

Then, if $\widehat{q}(b_1, 0) = 0$ for all b_1,

(a)
$$\begin{aligned}\|R_{SS} - \pi_S\| &\le \frac{8}{\varepsilon}\frac{\|b\widehat{q}(b)\|_1}{\sqrt{|v_1|} + |v_2|}, \\ \|R_{BB} - \pi_B\| &\le 4\frac{\|\widehat{q}(b)\|_1}{\sqrt{|v_1|} + |v_2|}, \\ \|R_{SB}\| &\le 4\frac{\|\widehat{q}(b)\|_1}{\sqrt{|v_1|} + |v_2|}, \\ \|R_{BS}\| &\le \frac{1}{\varepsilon}\|\widehat{q}(b)\|_1.\end{aligned}$$

(b)
$$\begin{aligned}\|R_{SS} - \pi_S\|_{HS} &\le \text{const}\frac{\|b\widehat{q}(b)\|_2}{\sqrt{|v_1|} + |v_2|}, \\ \|R_{BB} - \pi_B\|_{HS} &\le \text{const}\|\widehat{q}(b)\|_2\left(\frac{\ln(\sqrt{|v_1|} + |v_2|)}{\sqrt{\sqrt{|v_1|} + |v_2|}}\right)^{1/2}, \\ \|R_{SB}\|_{HS} &\le \text{const}\|\widehat{q}(b)\|_2\left(\frac{\ln(\sqrt{|v_1|} + |v_2|)}{\sqrt{\sqrt{|v_1|} + |v_2|}}\right)^{1/2}, \\ \|R_{BS}\|_{HS} &\le \text{const}\|\widehat{q}(b)\|_2.\end{aligned}$$

(c) *Let $\sqrt{|v_1|} + |v_2| \ge \max\{8, 16/\varepsilon(\|b\widehat{q}(b)\|_1 + \|\widehat{q}(b)\|_1^2)\}$. The operator*

$$\begin{pmatrix} R_{SS} & R_{SB} \\ R_{BS} & R_{BB}\end{pmatrix}$$

has a bounded inverse. The norm

$$\left\|\begin{pmatrix} R_{SS} & R_{SB} \\ R_{BS} & R_{BB}\end{pmatrix}^{-1} - \begin{pmatrix} \pi_S & 0 \\ R_{BS} & \pi_B\end{pmatrix}\right\| \le \text{const}\frac{1 + \|b\widehat{q}(b)\|_1^3}{\sqrt{|v_1|} + |v_2|}.$$

(d) *Let* $\sqrt{|v_1|}+|v_2| \geq \text{const}$. *Then*

$$\left|\det{}_2\begin{pmatrix} R_{SS} & R_{SB} \\ R_{BS} & R_{BB}\end{pmatrix} - 1\right| \leq \text{const}\big(1+\|\widehat{q}(b)\|_2\big)\|b\widehat{q}(b)\|_2\left(\frac{\ln(\sqrt{|v_1|}+|v_2|)}{\sqrt{\sqrt{|v_1|}+|v_2|}}\right)^{1/2}.$$

PROOF (a). In the case of R_{SB} and $R_{BB}-\pi_B$ it suffices to observe that, for $b \in B$,

$$|P_b(k)|^{-1} \leq 4[\sqrt{|v_1|}+|v_2|]^{-1}$$

and that the convolution operator $|\widehat{q}(b-c)|$ has operator norm bounded by $\|\widehat{q}\|_1$. In the case of $\|R_{BS}\|$

$$|P_b(k)|^{-1} \leq \frac{1}{\varepsilon}$$

is used instead. Finally

$$\begin{aligned} |(R_{SS}-\pi_S)_{b,c\in S}| &= \left|\frac{\widehat{q}(b-c)}{P_c(k)}\right| \\ &\leq 8\frac{|b-c|}{[\sqrt{|v_1|}+|v_2|]}\left|\frac{\widehat{q}(b-c)}{\varepsilon}\right| \end{aligned}$$

by Lemma 14.3(c). We may now continue as in the other cases.

PROOF OF (b). In the first case

$$\begin{aligned} \|R_{SS}-\pi_S\|_{HS}^2 &= \sum_{b,c\in S}\left|\frac{\widehat{q}(b-c)}{P_c(k)}\right|^2 \\ &\leq \sum_{b,c}\left(8\frac{|b-c|}{\sqrt{|v_1|}+|v_2|}\right)^2\left|\frac{\widehat{q}(b-c)}{P_c(k)}\right|^2 \\ &\leq \text{const}\left(\frac{\|b\widehat{q}\|_2}{\sqrt{|v_1|}+|v_2|}\right)^2. \end{aligned}$$

We have used Lemma 14.3(c) in the first inequality and Lemma 14.3(b) in the second. The next two cases may be treated at the same time.

$$\begin{aligned} \|R_{BB}-\pi_B\|_{HS}^2, \|R_{SB}\|_{HS}^2 &\leq \sum_{b\in S\cup B, c\in B}\left|\frac{\widehat{q}(b-c)}{P_c(k)}\right|^2 \\ &\leq \|\widehat{q}\|_2^2\sum_{c\in B}\frac{1}{|P_c(k)|^2} \\ &\leq \text{const}\|\widehat{q}\|_2^2\frac{\ln(\sqrt{|v_1|}+|v_2|)}{\sqrt{\sqrt{|v_1|}+|v_2|}} \end{aligned}$$

by Lemma 14.3(a) The final case is similar, but Lemma 14.3(b) must be used in place of Lemma 14.3(a)

PROOF OF (c). Define

$$\mathcal{R} = R_{SS} - R_{SB}R_{BB}^{-1}R_{BS}.$$

Then, by explicit calculation,

$$\begin{pmatrix} R_{SS} & R_{SB} \\ R_{BS} & R_{BB}\end{pmatrix}^{-1} = \begin{pmatrix} \mathcal{R}^{-1} & -\mathcal{R}^{-1}R_{SB}R_{BB}^{-1} \\ -R_{BB}^{-1}R_{BS}\mathcal{R}^{-1} & R_{BB}^{-1}+R_{BB}^{-1}R_{BS}\mathcal{R}^{-1}R_{SB}R_{BB}^{-1}\end{pmatrix} \tag{14.11}$$

By part (a)

$$\|\mathcal{R} - \pi_S\| \le \frac{8}{\varepsilon} \frac{\|b\widehat{q}(b)\|_1 + \|\widehat{q}(b)\|_1^2}{\sqrt{|v_1|} + |v_2|}$$

provided $4\|\widehat{q}\|_1[\sqrt{|v_1|} + |v_2|]^{-1} \le \frac{1}{2}$. This together with repeated applications of part (a) give the desired result.

PROOF OF (d). Write the matrix

$$\begin{aligned}\begin{pmatrix} R_{SS} & R_{SB} \\ R_{BS} & R_{BB} \end{pmatrix} &= \begin{pmatrix} \pi_S & 0 \\ R_{BS} & \pi_B \end{pmatrix} + \begin{pmatrix} R_{SS} - \pi_S & R_{SB} \\ 0 & R_{BB} - \pi_B \end{pmatrix} \\ &= \begin{pmatrix} \pi_S & 0 \\ R_{BS} & \pi_B \end{pmatrix} (\mathbb{1} + E)\end{aligned}$$

where

$$\begin{aligned}E &= \begin{pmatrix} \pi_S & 0 \\ R_{BS} & \pi_B \end{pmatrix}^{-1} \begin{pmatrix} R_{SS} - \pi_S & R_{SB} \\ 0 & R_{BB} - \pi_B \end{pmatrix} \\ &= \begin{pmatrix} \pi_S & 0 \\ -R_{BS} & \pi_B \end{pmatrix} \begin{pmatrix} R_{SS} - \pi_S & R_{SB} \\ 0 & R_{BB} - \pi_B \end{pmatrix} \\ &= \begin{pmatrix} R_{SS} - \pi_S & R_{SB} \\ R_{BS}(\pi_S - R_{SS}) & R_{BB} - \pi_B - R_{BS}R_{SB} \end{pmatrix}.\end{aligned}$$

First consider the case that B and S are finite sets. Then,

$$\det{}_2 \begin{pmatrix} R_{SS} & R_{SB} \\ R_{BS} & R_{BB} \end{pmatrix} = \det \begin{pmatrix} R_{SS} & R_{SB} \\ R_{BS} & R_{BB} \end{pmatrix}$$

since R_{SS} and R_{BB} both agree exactly with $\mathbb{1}$ on the diagonal. Consequently,

$$\begin{aligned}\det{}_2 \begin{pmatrix} R_{SS} & R_{SB} \\ R_{BS} & R_{BB} \end{pmatrix} &= \det(\mathbb{1} + E) = \det{}_2(\mathbb{1} + E) e^{\mathrm{tr}\, E} \\ &= 1 + O(\|E\|_{HS} + \mathrm{tr} E) \\ &= 1 + O(\|R_{SS} - \pi_S\|_{HS} + \|R_{SB}\|_{HS} + \|R_{BS}\| \, \|(\pi_S - R_{SS})\|_{HS}) \\ &\qquad + O(\|R_{BB} - \pi_B\|_{HS} + \|R_{BS}\|_{HS} \|R_{SB}\|_{HS}) \\ &= 1 + O\left((1 + \|\widehat{q}(b)\|_2) \|b\widehat{q}(b)\|_2 \left[\frac{\ln(\sqrt{|v_1|} + |v_2|)}{\sqrt{\sqrt{|v_1|} + |v_2|}} \right]^{1/2} \right)\end{aligned}$$

The bound when B or S are infinite is gotten by taking limits. □

It is now a simple matter to use the estimates of Lemmas 14.3, 14.4 to analyse the asymptotic behavior of the heat curve. Recall that $\widehat{q}(b_1, 0) = 0$ for all b_1. Define

$$R = \max \left\{ 8, \frac{16}{\varepsilon} \left(\|b\widehat{q}(b)\|_1 + \|\widehat{q}(b)\|_1^2 \right) \right\} \tag{14.12a}$$

and

$$\mathcal{K} = \{ k \in \mathbb{C}^2 \mid \sqrt{|v_1|} + |v_2| \le R \}. \tag{14.12b}$$

Then $\mathcal{K} \subset \mathcal{K}_\rho$ for some $\rho > 0$.

PROOF OF THEOREM 14.1(a). Let $k \in \mathbb{C}^2 \setminus (\mathcal{K} \cup \bigcup_{b\in\Gamma^\#} T_b)$. Use (14.5) with $G = \varnothing$ as a test for when $k \in \widehat{\mathcal{H}}(q)$. By Lemma 14.4(c) with

$$S = \{b \in \Gamma^\sharp \mid \varepsilon \le |P_b(k)| < \tfrac{1}{4}[\sqrt{|v_1|} + |v_2|]\},$$
$$B = \{b \in \Gamma^\sharp \mid |P_b(k)| \ge \tfrac{1}{4}[\sqrt{|v_1|} + |v_2|]\}$$

$\phi_{G'}$ must be zero. That is, there is no nontrivial solution of (14.5). □

To prove the rest of Theorem 14.1 and Theorem 14.2 we use

PROPOSITION 14.5. *Let $k \in \mathbb{C}^2 \setminus \mathcal{K}$.*

(a) *Let $k \in T_a \setminus \cup_{b\neq a} T_b$. Then $k \in \widehat{\mathcal{H}}(q)$ if and only if*

$$P_a(k) = \mathcal{A}(k)$$

where

$$\mathcal{A}(k) = \sum_{b,c\in A} \frac{\widehat{q}(a-b)}{P_b(k)} (R_{AA}^{-1})_{b,c} \widehat{q}(c-a),\ A = \Gamma^\sharp \setminus \{a\}$$

and obeys

$$|\mathcal{A}(k)| \le \frac{\mathrm{const}(\|b\widehat{q}(b)\|_1)}{1 + |k_2 + a_2|^2}.$$

Here $\mathrm{const}\big(\|b\widehat{q}(b)\|_1\big)$ *denotes that the constant* const *depends only on ε and the norm $\|b\widehat{q}(b)\|_1$.*

(b) *Let $k \in T_{d^{(1)}} \cap T_{d^{(2)}}$. Then $k \in \widehat{\mathcal{H}}(q)$ if and only if*

$$\big(P_{d^{(1)}}(k) - \mathcal{D}(k)_{1,1}\big)\big(P_{d^{(2)}}(k) - \mathcal{D}(k)_{2,2}\big)$$
$$= \big(\widehat{q}(d^{(1)} - d^{(2)}) - \mathcal{D}(k)_{1,2}\big)\big(\widehat{q}(d^{(2)} - d^{(1)}) - \mathcal{D}(k)_{2,1}\big)$$

where

$$\mathcal{D}(k)_{i,j} = \sum_{b,c\in D} \frac{\widehat{q}(d^{(i)} - b)}{P_b(k)} (R_{DD}^{-1})_{b,c} \widehat{q}(c - d^{(j)}), \quad D = \Gamma^\sharp \setminus \{d^{(1)}, d^{(2)}\}$$

and obeys

$$|\mathcal{D}(k)_{i,j}| \le \min_{i=1,2} \frac{\mathrm{const}(\|b\widehat{q}(b)\|_1)}{1 + |k_2 + d_2^{(i)}|^2}.$$

PROOF OF (a). For the region in question $\widehat{\mathcal{H}}(q)$ is given by (14.6) with $G = \{a\}$ and $G' = A$. This is precisely the desired equation. We now estimate $\mathcal{A}(k)$. Lemma 14.4(c) with

$$S = A \cap \{b \in \Gamma^\sharp \mid \varepsilon \le |P_b(k)| < \tfrac{1}{4}[\sqrt{|v_1|} + |v_2|]\},$$
$$B = A \cap \{b \in \Gamma^\sharp \mid |P_b(k)| \ge \tfrac{1}{4}[\sqrt{|v_1|} + |v_2|]\}$$

together with Lemma 14.3(e) and (14.8) gives the desired bound for the contribution from

$$\begin{pmatrix} R_{SS} & R_{SB} \\ R_{BS} & R_{BB} \end{pmatrix}^{-1} - \begin{pmatrix} \pi_S & 0 \\ R_{BS} & \pi_B \end{pmatrix}.$$

The two remaining contributions are

$$\sum_{b\in A} \frac{\widehat{q}(a-b)\widehat{q}(b-a)}{P_b(k)} + \sum_{b\in B, c\in S} \frac{\widehat{q}(a-b)}{P_b(k)} (R_{BS})_{b,c} \widehat{q}(c-a).$$

The first is estimated by Lemma 14.3(f). The second is estimated using Lemma 14.3(e), Lemma 14.4(a) and Lemma 14.3(c) (with the last used to show that $|c-a| \geq \sqrt{|v_1|}+|v_2|$). This discussion yields the bound

$$|\mathcal{A}(k)| \leq \frac{\text{const}(\|b\widehat{q}(b)\|_1)}{[\sqrt{|v_1|}+|v_2|]^2}.$$

But by (14.8)

$$\sqrt{|v_1|}+|v_2| \geq \text{const}(1+|k_2+a_2|).$$

PROOF OF (b). For the most part it suffices to repeat the above argument with $G=\{d^{(1)},d^{(2)}\}$, $G'=D$ and

$$\begin{aligned} S &= D\cap\{b\in\Gamma^\sharp \mid \varepsilon\leq|P_b(k)|<\tfrac14[\sqrt{|v_1|}+|v_2|]\},\\ B &= D\cap\{b\in\Gamma^\sharp \mid |P_b(k)|\geq\tfrac14[\sqrt{|v_1|}+|v_2|]\}.\end{aligned}$$

The one exception is the contribution

$$\sum_{b\in D}\frac{\widehat{q}(d^{(i)}-b)\widehat{q}(b-d^{(j)})}{P_b(k)}.$$

Note that, by (14.2),

$$k_2+d_2^{(i)} = \frac{d_2^{(i)}-d_2^{(j)}}{2} - \frac{i}{2}\frac{d_1^{(1)}-d_1^{(2)}}{d_2^{(1)}-d_2^{(2)}} + O(\varepsilon)$$

so that

$$|k_2+d_2^{(i)}| \leq \tfrac12|d^{(1)}-d^{(2)}|+O(\varepsilon).$$

First consider $i=j=1$. Then

$$\begin{aligned}\sum_{b\in D}\frac{\widehat{q}(d^{(1)}-b)\widehat{q}(b-d^{(1)})}{P_b(k)} &= \sum_{b\neq0,d^{(2)}-d^{(1)}}\frac{\widehat{q}(b)\widehat{q}(-b)}{P_{b+d^{(1)}}(k)}\\ &= \sum_{b\neq0,\pm(d^{(2)}-d^{(1)})}\frac{\widehat{q}(b)\widehat{q}(-b)}{P_{b+d^{(1)}}(k)} + \frac{\widehat{q}(d^{(1)}-d^{(2)})\widehat{q}(d^{(2)}-d^{(1)})}{P_{2d^{(1)}-d^{(2)}}(k)}.\end{aligned}$$

The sum is bounded as in Lemma 14.3(f). The last term

$$\begin{aligned}\left|\frac{\widehat{q}(d^{(1)}-d^{(2)})\widehat{q}(d^{(2)}-d^{(1)})}{P_{2d^{(1)}-d^{(2)}}(k)}\right| &\leq \frac{\|b\widehat{q}(b)\|_\infty^2}{\varepsilon|d^{(1)}-d^{(2)}|^2}\\ &\leq \text{const}\frac{\|b\widehat{q}(b)\|_\infty^2}{[1+|k_2+d_2^{(1)}|]^2}.\end{aligned}$$

The case $i=j=2$ is similar, so let $i\neq j$. The contribution from S is bounded by

$$\begin{aligned}\sum_{b\in S}\frac{\widehat{q}(d^{(i)}-b)\widehat{q}(b-d^{(j)})}{P_b(k)} &\leq \sum_{\substack{b\\ |b-d^{(1)}|,|b-d^{(2)}|>\frac18[\sqrt{|v_1|}+|v_2|]}}\frac{|\widehat{q}(d^{(i)}-b)\widehat{q}(b-d^{(j)})|)}{\varepsilon}\\ &\leq \text{const}\frac{\|b\widehat{q}(b)\|_2^2}{[\sqrt{|v_1|}+|v_2|]^2}\end{aligned}$$

while that from B is bounded by

$$\begin{aligned}\sum_{b\in B}\frac{\widehat{q}(d^{(i)}-b)\widehat{q}(b-d^{(j)})}{P_b(k)} &\le \sum_b 4\frac{|\widehat{q}(d^{(i)}-b)\widehat{q}(b-d^{(j)}|)}{\sqrt{|v_1|}+|v_2|}\\ &\le \text{const}\sum_b \frac{\{|d^{(i)}-b|+|b-d^{(j)}|\}\widehat{q}(d^{(i)}-b)\widehat{q}(b-d^{(j)})}{|d^{(1)}-d^{(2)}|[\sqrt{|v_1|}+|v_2|]}\\ &\le \text{const}\frac{2\|b\widehat{q}(b)\|_2\|\widehat{q}(b)\|_2}{[\sqrt{|v_1|}+|v_2|]^2}. \qquad\square\end{aligned}$$

To show that the curve does not wiggle too much we will also need bounds on the derivatives of $\mathcal{A}$ and $\mathcal{D}$. These are provided in

LEMMA 14.6. *Under the hypotheses of Proposition* 14.5, *if* $m+n=1$

$$\left|\frac{\partial^{n+m}}{\partial k_1^n\partial k_2^m}\mathcal{A}(k)\right| \le \frac{\text{const}(\|b\widehat{q}(b)\|_1)}{[1+|k_2+a_2|]^{2-m}},$$
$$\left|\frac{\partial^{n+m}}{\partial k_1^n\partial k_2^m}\mathcal{D}(k)_{i,j}\right| \le \min_{i=1,2}\frac{\text{const}(\|b\widehat{q}(b)\|_1)}{[1+|k_2+d_2^{(i)}|]^{2-m}}$$

and if $m+n\ge 2$

$$\left|\frac{\partial^{n+m}}{\partial k_1^n\partial k_2^m}\mathcal{A}(k)\right| \le \frac{\text{const}(\|b^2\widehat{q}(b)\|_1)}{[1+|k_2+a_2|]^{3-m}},$$
$$\left|\frac{\partial^{n+m}}{\partial k_1^n\partial k_2^m}\mathcal{D}(k)_{i,j}\right| \le \min_{i=1,2}\frac{\text{const}(\|b^2\widehat{q}(b)\|_1)}{[1+|k_2+d_2^{(i)}|]^{3-m}}$$

PROOF. Use $Q_{GG'}$ to denote the matrix $[\widehat{q}(b-c)]_{b\in G,c\in G'}$. Then $\mathcal{A}$ and $\mathcal{D}$ are given by $Q_{GG'}H_k^{-1}(R_{G'G'})^{-1}Q_{G'G}$ with $G=\{a\}$ and $G=\{d^{(1)},d^{(2)}\}$ respectively. Hence their first derivatives are given by

$$\begin{aligned}\frac{\partial}{\partial k_i}\mathcal{A}(k),\frac{\partial}{\partial k_i}\mathcal{D}(k) &= -Q_{GG'}H_k^{-1}\frac{\partial H_k}{\partial k_i}H_k^{-1}(R_{G'G'})^{-1}Q_{G'G}\\ &\quad + Q_{GG'}H_k^{-1}(R_{G'G'})^{-1}Q_{G'G'}H_k^{-1}\frac{\partial H_k}{\partial k_i}H_k^{-1}(R_{G'G'})^{-1}Q_{G'G}\\ &= -Q_{GG'}H_k^{-1}(R_{G'G'})^{-1}\frac{\partial H_k}{\partial k_i}H_k^{-1}(R_{G'G'})^{-1}Q_{G'G}\end{aligned}$$

since

$$Q_{G'G'}H_k^{-1} = R_{G'G'}-\mathbb{1}.$$

Hence the derivatives are bounded by

$$\|Q_{GG'}H_k^{-1}\|\|(R_{G'G'})^{-1}\|\left\|\frac{\partial H_k}{\partial k_i}H_k^{-1}(R_{G'G'})^{-1}Q_{G'G}\right\|.$$

By Lemma 14.3(e) and Lemma 14.4(c), 14.4(a)

$$\|Q_{GG'}H_k^{-1}\|\|(R_{G'G'})^{-1}\| \le \frac{\text{const}}{\sqrt{|v_1|}+|v_2|}.$$

When $i = 1$, $(\partial H_k/\partial k_1)_{b,c} = i\delta_{b,c}$ and

$$(14.13_1)\quad \left\| \frac{\partial H_k}{\partial k_1} H_k^{-1}(R_{G'G'})^{-1} Q_{G'G} \right\| \leq \|H_k^{-1}\pi_{G'}\| \, \|(R_{G'G'})^{-1} - \mathbb{1} - R_{BS}\| \, \|Q_{G'G}\| + \|H_k^{-1} Q_{G'G}\| + \|H_k^{-1}\pi_{G'}\| \|R_{BS}Q_{SG}\| \leq \frac{\text{const}}{\sqrt{|v_1|} + |v_2|}.$$

The first term was controlled by Lemma 14.4(c), the second by Lemma 14.3(e) and the third by Lemmas 14.4(a) and 14.3(c).

When $i = 2$, $(\partial H_k/\partial k_2)_{b,c} = 2(k_2 + b_2)\delta_{b,c}$ and

$$\left\| \frac{\partial H_k}{\partial k_2} H_k^{-1}\pi_{G'} \right\| \leq \text{const}[\sqrt{|v_1|} + |v_2|], \qquad \left\| \frac{\partial H_k}{\partial k_2} H_k^{-1} Q_{G'G} \right\| \leq \text{const}$$

because

$$(14.14a)\qquad P_b(k) \geq \tfrac{1}{4}[\sqrt{|v_1|} + |k_2 + b_2|] \implies \left| \frac{2(k_2 + b_2)}{P_b(k)} \right| \leq 8$$

and

$$(14.14b)\quad \varepsilon \leq P_b(k) < \tfrac{1}{4}[\sqrt{|v_1|} + |k_2 + b_2|] \implies \begin{cases} \left| \frac{2(k_2+b_2)}{P_b(k)} \right| \leq \frac{\text{const}[1+\sqrt{|v_1|}+|v_2|]}{\varepsilon} \\ \text{dist}(b, G) \geq \frac{1}{8}[\sqrt{|v_1|} + |v_2|] \end{cases}$$

by (14.8) and Lemma 14.3(c). Thus for $i = 2$ (16.13_1) is replaced by

$$(14.13_2)\quad \left\| \frac{\partial H_k}{\partial k_2} H_k^{-1}(R_{G'G'})^{-1} Q_{G'G} \right\| \leq \left\| \frac{\partial H_k}{\partial k_2} H_k^{-1}\pi_{G'} \right\| \|(R_{G'G'})^{-1} - \mathbb{1} - R_{BS}\| \, \|Q_{G'G}\| + \left\| \frac{\partial H_k}{\partial k_2} H_k^{-1} Q_{G'G} \right\| + \left\| \frac{\partial H_k}{\partial k_2} H_k^{-1}\pi_{G'} \right\| \|R_{BS}Q_{SG}\| \leq \text{const}$$

To complete the argument when one derivative is taken it suffices to use (14.8) to convert $\sqrt{|v_1|} + |v_2|$'s into $1 + |k_2 + a_2|$'s or $1 + |k_2 + d_2^{(i)}|$'s.

All higher derivatives are given by finite linear combinations of terms of the form

$$(14.15)\qquad Q_{GG'} \prod_j \left\{ H_k^{-1}(R_{G'G'})^{-1} \frac{\partial^{n_j} H_k}{\partial k_{i_j}^{n_j}} \right\} H_k^{-1}(R_{G'G'})^{-1} Q_{G'G}$$

where the sum of the n_j's for which i_j =1 (resp. 2) is n (resp. m). This is easily seen if $\mathcal{A}$ and $\mathcal{D}$ are written in the form

$$Q_{GG'}(H_k + Q)^{-1} Q_{G'G}$$

with the operator $H_k + Q$ being restricted to $L^2_{G'}$.

The proof will involve expanding each $R^{-1} = (R^{-1} - \mathbb{1} - R_{BS}) + \mathbb{1} + R_{BS}$ and applying

$$(14.16)\qquad \begin{aligned} \|Q_{GG'}H^{-1}\| &\le \frac{\text{const}}{\sqrt{|v_1|} + |v_2|}, \\ \left\| Q_{GG'}\frac{\partial H_k}{\partial k_i}H_k^{-1} \right\| &\le \text{const}[\sqrt{|v_1|} + |v_2|]^{-1+\delta_{i,2}}, \\ \left\| Q_{GG'}\left(\frac{\partial H_k}{\partial k_i}H_k^{-1}\right)^2 \right\| &\le \text{const}[\sqrt{|v_1|} + |v_2|]^{-2+2\delta_{i,2}}, \\ \left\| Q_{GG'}\frac{\partial H_k}{\partial k_i}H_k^{-2} \right\| &\le \text{const}[\sqrt{|v_1|} + |v_2|]^{-2+\delta_{i,2}}, \end{aligned}$$

the analogous bounds on R_{BS}

$$(14.17)\qquad \begin{aligned} \|H^{-1}R_{BS}\| &\le \frac{\text{const}}{\sqrt{|v_1|} + |v_2|}, \\ \left\| \frac{\partial H_k}{\partial k_i}H_k^{-1}R_{BS} \right\| &\le \text{const}[\sqrt{|v_1|} + |v_2|]^{-1+\delta_{i,2}}, \\ \left\| \left(\frac{\partial H_k}{\partial k_i}H_k^{-1}\right)^2 R_{BS} \right\| &\le \text{const}[\sqrt{|v_1|} + |v_2|]^{-2+2\delta_{i,2}}, \\ \left\| \frac{\partial H_k}{\partial k_i}H_k^{-2}R_{BS} \right\| &\le \text{const}[\sqrt{|v_1|} + |v_2|]^{-2+\delta_{i,2}}, \end{aligned}$$

and

$$(14.18)\qquad \|(R_{G'G'})^{-1} - \mathbb{1} - R_{BS}\| \le \text{const}[\sqrt{|v_1|} + |v_2|]^{-1},$$

$$(14.19)\qquad \left\| \frac{\partial H_k}{\partial k_2}H_k^{-1}\pi_{G'} \right\| \le \text{const}[\sqrt{|v_1|} + |v_2|]^{\delta_{i,2}}.$$

(The const's on the right-hand side of (14.16), (14.17)) depend on $\|b^2\widehat{q}\|_1$.) The moral of this long list of bounds is that we get a factor of $[\sqrt{|v_1|} + |v_2|]^{-1}$ for each H_k^{-1} that is

- a nearest or second nearest neighbor to a terminating $Q_{GG'}$ or $Q_{G'G}$(with intervening $\partial H_k/\partial k_i$'s but no intervening $(R_{G'G'})^{-1} - \mathbb{1} - R_{BS}$'s or R_{BS}'s allowed)
- a nearest or second nearest neighbor on the left of an R_{BS} (with the same intervention rules)

and for each $(R_{G'G'})^{-1} - \mathbb{1} - R_{BS}$ and that we get a $[\sqrt{|v_1|} + |v_2|]$ for each $\partial H_k/\partial k_2$. Thus it suffices to check that we always get at least three decay factors $[\sqrt{|v_1|} + |v_2|]^{-1}$.

Expand the leftmost R^{-1} in (14.15). Selecting either the $R^{-1} - \mathbb{1} - R_{BS}$ or the $\mathbb{1}$ yields two decay factors right away and we can always get a third one from the $H^{-1}R^{-1}Q$ on the far right-hand end.

That leaves $Q_{GG'}H^{-1}R_{BS}$ on the left end. Since

$$\text{dist}(S, G) \ge \tfrac{1}{8}[\sqrt{|v_1|} + |v_2|]$$

we can get two decay factors from the $\widehat{q}$'s in $Q_{GG'}H^{-1}R_{BS}$. As usual we can always get the third from the far right-hand end. $\square$

PROOF OF THEOREM 14.1(B). For $z \in \mathbb{C}$ put

$$F_z = \mathrm{pr}^{-1}(z) \cap \left(T_0 \setminus \bigcup_{\substack{b \in \Gamma^\# \\ b \neq 0}} T_b \right).$$

Observe that

$$F_z = \begin{cases} \{(k_1, z) \in \mathbb{C}^2 \mid |k_1 - iz^2| \leq \varepsilon\} \\ \qquad \text{if } |z - z_b| \geq \frac{\varepsilon}{b_2} \text{ for all } b \in \Gamma^\#,\ b_2 \neq 0 \\ \{(k_1, z) \in \mathbb{C}^2 \mid |k_1 - iz^2| \leq \varepsilon \text{ and } \left|k_1 - i\big(z^2 + 2d_2(z - z_d)\big)\right| > \varepsilon\} \\ \qquad \text{if } |z - z_d| < \frac{\varepsilon}{b_2} \end{cases}$$

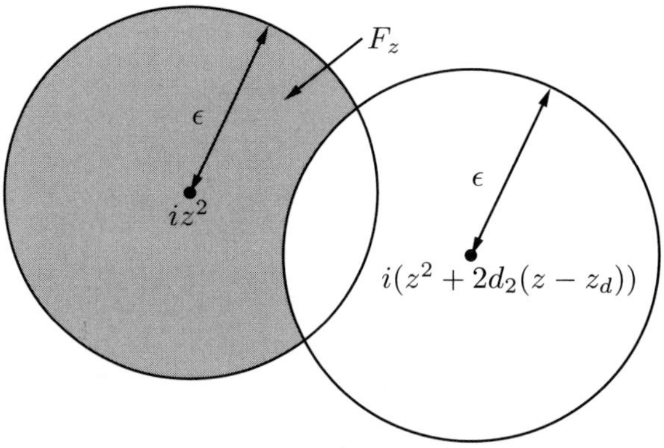

By Proposition 14.5 (k_1, z) lies in $\widehat{\mathcal{H}}(q)$ if and only if

$$ik_1 + z^2 = P_0(k_1, z) = \mathcal{A}(k_1, z) \tag{14.20}$$

for an analytic function $\mathcal{A}$ obeying

$$|\mathcal{A}(k_1, z)| \leq \frac{\text{const}}{1 + |z|^2}, \qquad \left| \frac{\partial}{\partial k_1} \mathcal{A}(k_1, z) \right| \leq \frac{\text{const}}{1 + |z|^2}$$

by Lemma 14.6. This shows, that for z big enough, the equation (14.20) has at most one solution in F_z, and that this solution is simple. Furthermore any such solution fulfils

$$|k_1 - iz^2| \leq \frac{\text{const}}{1 + |z|^2}.$$

So there is no solution in F_z, if

$$\left\{ k_1 \in \mathbb{C} \;\middle|\; |k_1 - iz^2| \leq \frac{\text{const}}{1 + |z|^2} \right\} \subset \left\{ k_1 \in \mathbb{C} \mid \left|k_1 - i\big(z^2 + 2d_2(z - z_d)\big)\right| \leq \varepsilon \right\}$$

that is, if

$$2d_2|z - z_d| + \frac{\text{const}}{1 + |z|^2} \leq \varepsilon$$

or equivalently,

$$|z - z_d| \leq \frac{\varepsilon}{2d_2} - \frac{\text{const}}{2d_2(1 + |z|^2)}.$$

Similarly equation (14.20) has a solution in F_z, if

$$|z - z_d| > \frac{\varepsilon}{2d_2} + \frac{\text{const}}{2d_2(1 + |z|^2)}. \qquad \square$$

PROOF OF THEOREM 14.2. First we perform the coordinate change

$$
\begin{aligned}
x_1 &= P_0(k) - \mathcal{D}_{1,1} = ik_1 + k_2^2 - \mathcal{D}_{1,1}, \\
x_2 &= P_d(k) - \mathcal{D}_{2,2} = i(k_1 + d_1) + (k_2 + d_2)^2 - \mathcal{D}_{2,2}, \\
k_2 &= \frac{x_2 - x_1}{2d_2} + z_d + \frac{\mathcal{D}_{2,2} - \mathcal{D}_{1,1}}{2d_2}, \\
k_1 &= -ix_1 - i\mathcal{D}_{1,1} + i\left(\frac{x_2 - x_1}{2d_2} + z_d + \frac{\mathcal{D}_{2,2} - \mathcal{D}_{1,1}}{2d_2}\right)^2
\end{aligned}
\tag{14.21}
$$

where $\mathcal{D}(k)_{i,j}$ are given by Proposition 14.5. The Jacobean of this map is

$$
\begin{aligned}
\begin{pmatrix} \frac{\partial x_1}{\partial k_1} & \frac{\partial x_1}{\partial k_2} \\ \frac{\partial x_2}{\partial k_1} & \frac{\partial x_2}{\partial k_2} \end{pmatrix}
&= \begin{pmatrix} i + O\big(\frac{1}{|z_d|^2}\big) & 2k_2 + O\big(\frac{1}{|z_d|}\big) \\ i + O\big(\frac{1}{|z_d|^2}\big) & 2(k_2 + d_2) + O\big(\frac{1}{|z_d|}\big) \end{pmatrix} \\
&= \begin{pmatrix} i + O\big(\frac{1}{|z_d|^2}\big) & 2z_d + O\big(\frac{1}{|d_2|}\big) \\ i + O\big(\frac{1}{|z_d|^2}\big) & 2z_{-d} + O(1|d_2|) \end{pmatrix} \\
&= \begin{pmatrix} i & 2z_d \\ i & 2z_{-d} \end{pmatrix} \left(\mathbb{1} + O\Big(\frac{1}{|d_2 z_d|}\Big)\right)
\end{aligned}
\tag{14.22a}
$$

since, by (14.2), $|k_2 - z_d| \le \varepsilon/|d_2|$. Its inverse is

$$
\begin{pmatrix} \frac{\partial k_1}{\partial x_1} & \frac{\partial k_1}{\partial x_2} \\ \frac{\partial k_2}{\partial x_1} & \frac{\partial k_2}{\partial x_2} \end{pmatrix}
= \frac{1}{2id_2} \begin{pmatrix} 2z_{-d} & -2z_d \\ -i & i \end{pmatrix} \left(\mathbb{1} + O\Big(\frac{1}{|d_2 z_d|}\Big)\right).
\tag{14.22b}
$$

The derivative of the inverse

$$
\begin{aligned}
\frac{\partial^2 k_m}{\partial x_i \partial x_j} &= -\sum_{\alpha,\beta} \frac{\partial k_m}{\partial x_\alpha} \frac{\partial}{\partial x_i}\left(\frac{\partial x_\alpha}{\partial k_\beta}\right) \frac{\partial k_\beta}{\partial x_j} = -\sum_{\alpha,\beta,\gamma} \frac{\partial k_m}{\partial x_\alpha} \frac{\partial^2}{\partial k_\gamma} \frac{x_\alpha}{\partial k_\beta} \frac{\partial k_\gamma}{\partial x_i} \frac{\partial k_\beta}{\partial x_j} \\
&= \sum_{\alpha,\beta,\gamma} O\left(\frac{1}{|d_2|^3} |z_d|^{\delta_{m,1} + \delta_{\gamma,1} + \delta_{\beta,1}} \left(2\delta_{\beta,2}\delta_{\gamma,2} + \frac{1}{|z_d|^{3 - \delta_{\beta,2} - \delta_{\gamma,2}}}\right)\right) \\
&= O\left(\frac{|z_d|^{\delta_{m,1}}}{|d_2|^3}\right).
\end{aligned}
$$

In these coordinates

$$
(P_0 - \mathcal{D}_{1,1})(P_d - \mathcal{D}_{2,2}) - (\widehat{q}(d) - \mathcal{D}_{1,2})(\widehat{q}(-d) - \mathcal{D}_{2,1}) = x_1 x_2 + h(x_1, x_2)
$$

where

$$
h(x_1, x_2) = -(\widehat{q}(d) - \mathcal{D}_{1,2})(\widehat{q}(-d) - \mathcal{D}_{2,1}).
$$

In Lemma 14.8 below we will improve the estimates of Lemmma 14.6 to

$$
\left|\frac{\partial^{n+m} l}{\partial k_1^n \partial k_2^m} \mathcal{D}(k)_{1,2}\right|, \quad \left|\frac{\partial^{n+m} l}{\partial k_1^n \partial k_2^m} \mathcal{D}(k)_{2,1}\right| \le \text{const} \frac{|d|^m}{(1 + |d|)^\beta}.
$$

In terms of the x-variables, when $r \neq s$

$$\left|\frac{\partial \mathcal{D}_{r,s}}{\partial x_i}(x_1,x_2)\right| \le \sum_{m=1,2}\left|\frac{\partial \mathcal{D}_{r,s}}{\partial k_m}\frac{\partial k_m}{\partial x_i}\right|$$
$$\le \sum_m |d|^{\delta_{m,2}}\frac{\text{const}}{(1+|d|)^\beta}\frac{\text{const}}{|d_2|}|z_d|^{\delta m,1}$$
$$\le \frac{\text{const}}{(1+|d|)^{\beta-1}}$$

and

$$\left|\frac{\partial^2 \mathcal{D}_{r,s}l}{\partial x_i \partial x_j}(x_1,x_2)\right| \le \sum_{m,n=1,2}\left|\frac{\partial^2 \mathcal{D}_{r,s}}{\partial k_m \partial k_n}\frac{\partial k_m}{\partial x_i}\frac{\partial k_n}{\partial x_j}\right| + \sum_{m=1,2}\left|\frac{\partial \mathcal{D}_{r,s}}{\partial k_m}\frac{\partial^2 k_m}{\partial x_i \partial x_j}\right|$$
$$\le \sum_{m,n}|d|^{\delta_{m,2}+\delta_{n,2}}\frac{\text{const}}{(1+|d|)^\beta}\frac{\text{const}}{d_2^2}|z_d|^{\delta m,1+\delta n,1}$$
$$+\sum_m |d|^{\delta_{m,2}}\frac{\text{const}}{(1+|d|)^\beta}\text{const}\frac{|z_d|^{\delta_{m,1}}}{|d_2|^3}$$
$$\le \frac{\text{const}}{(1+|d|)^{\beta-2}}$$

so that,

$$|h(0,0)| \le \frac{\text{const}}{(1+|d|)^{2\beta}}$$

and

$$\left|\frac{\partial h}{\partial x_i}(x_1,x_2)\right| \le \frac{\text{const}}{(1+|d|)^{\beta-1}}\left(|\widehat{q}(d)| + |\widehat{q}(-d)| + \frac{\text{const}}{(1+|d|)^\beta}\right) \le \frac{\text{const}}{(1+|d|)^{2\beta-1}},$$
$$\left|\frac{\partial^2 h}{\partial x_i \partial x_j}(x_1,x_2)\right| \le \frac{\text{const}}{(1+|d|)^{2\beta-2}}.$$

By the quantitative Morse lemma in the appendix, with $a = \text{const}/(1+|d|)^{2\beta-1}$ and $b = \text{const}/(1+|d|)^{2\beta-2}$, there is a biholomorphism ψ defined on $\{(z_1,z_2) \in \mathbb{C}^2 \mid |z_1| \le \varepsilon/2, |z_2| \le \varepsilon/2\}$ with range containing $\{(x_1,x_2) \in \mathbb{C}^2 \mid |x_1| \le \varepsilon/4, |x_2| \le \varepsilon/4\}$ and with

$$\|D\psi - \mathbb{1}\| \le \frac{\text{const}}{(1+|d|)^{2\beta-2}}, \quad (x_1x_2+h)\circ\psi = z_1z_2 - \hat{t}_d,$$
$$|\hat{t}_d| \le \frac{\text{const}}{(1+|d|)^{2\beta}}, \quad |\psi(0)| \le \frac{\text{const}}{(1+|d|)^{2\beta-1}}.$$

It now suffices to compose

$$\widehat{\phi}_d(z_1,z_2) = \big(k_1(\psi(z_1,z_2)), k_2(\psi(z_1,z_2))\big)$$

with $k(x)$ being the map of (14.21).

Conclusion (ii) of the theorem, as well as the first part of (i), is immediate. The Jacobean

$$D\widehat{\phi}_d = \frac{\partial k}{\partial x}D\psi$$
$$= \frac{1}{2id_2}\begin{pmatrix} 2z_{-d} & -2z_d \\ -i & i \end{pmatrix}\left\{\mathbb{1} + O\Big(\frac{1}{(1+|d|)^{2\beta-2}}\Big)\right\}\left\{\mathbb{1} + O\Big(\frac{1}{|d_2 z_d|}\Big)\right\}$$
$$= \frac{1}{2id_2}\begin{pmatrix} 2z_{-d} & -2z_d \\ -i & i \end{pmatrix}\left\{\mathbb{1} + O\Big(\frac{1}{|d_2 z_d|}\Big)\right\}.$$

The centre is

$$\begin{aligned}\widehat{\phi}_d(0) &= k\big(\psi(0)\big)\\ &= k\bigg(O\Big(\frac{1}{(1+|d|)^{2\beta-1}}\Big)\bigg)\end{aligned}$$

with

$$k_2\bigg(O\Big(\frac{1}{(1+|d|)^{2\beta-1}}\Big)\bigg) = z_d + O\Big(\frac{1}{|d_2 z_d^2|}\Big).$$

To prove part (iii) observe that $T_{-d} \cap T_0 \cap \widehat{\mathcal{H}}(q)$ is mapped to $T_0 \cap T_d \cap \widehat{\mathcal{H}}(q)$ by translation by $-d$. For $d_2 > 0$ define $\widehat{\phi}_d$ by the above construction. For $d_2 < 0$ define $\widehat{\phi}_d$ by

$$\widehat{\phi}_d(z_1, z_2) = \widehat{\phi}_{-d}(z_2, z_1) - d. \qquad \square$$

We now look more closely at the extent to which double points open up for various classes of potentials. Let f be a function on $\mathbb{R}^+$ satisfying

(i) $f \geq 1$, $f(0) = 1$.
(ii) $f(s)f(t) \geq f(s+t)$ for all $s, t \geq 0$.
(iii) f increases monotonically.

One may, for example , use $f(t) = e^{\beta t}$ or $f(t) = (1+t)^\beta$ for any $\beta \geq 0$. Define, for operators on $\ell^2(\Gamma^\#)$, the norm

$$\|A\|_f = \max\left\{ \sup_{b\in\Gamma^\#} \sum_{c\in\Gamma^\#} |A_{b,c}| f(|b-c|),\ \sup_{c\in\Gamma^\#} \sum_{b\in\Gamma^\#} |A_{b,c}| f(|b-c|) \right\}.$$

In particular for the convolution operator $\widehat{q}(b-c)$

$$\|\widehat{q}\|_f = \sum_{b\in\Gamma^\#} |\widehat{q}(b)| f(|b|).$$

By [**FKT**, (3.4)], the norm obeys

$$\|A\| \leq \|A\|_{f\equiv 1} \leq \|A\|_f, \quad \|AB\|_f \leq \|A\|_f \|B\|_f,$$

$$\|(\mathbb{1}+A)^{-1}\|_f \leq (1-\|A\|_f)^{-1} \text{ if } \|A\|_f < 1, \quad |A_{b,c}| \leq \frac{1}{f(|b-c|)}\|A\|_f. \tag{14.23}$$

The analogue of Lemma 14.4 for this norm is

LEMMA 14.7. *Let $k \in \mathbb{C}^2$ and*

$$\begin{aligned} S &\subset \{b \in \Gamma^\sharp \mid \varepsilon \leq |P_b(k)| < \tfrac{1}{4}[\sqrt{|v_1|} + |v_2|]\},\\ B &\subset \{b \in \Gamma^\sharp \mid |P_b(k)| \geq \tfrac{1}{4}[\sqrt{|v_1|} + |v_2|]\}.\end{aligned}$$

Then, if $\widehat{q}(b_1, 0) = 0$ for all b_1

(a)
$$\begin{aligned}\|R_{SS} - \pi_S\|_f &\leq \frac{8}{\varepsilon}\frac{\|b\widehat{q}(b)\|_f}{\sqrt{|v_1|} + |v_2|},\\ \|R_{BB} - \pi_B\|_f &\leq 4\frac{\|\widehat{q}(b)\|_f}{\sqrt{|v_1|} + |v_2|},\\ \|R_{SB}\|_f &\leq 4\frac{\|\widehat{q}(b)\|_f}{\sqrt{|v_1|} + |v_2|},\\ \|R_{BS}\|_f &\leq \frac{1}{\varepsilon}\|\widehat{q}(b)\|_f.\end{aligned}$$

(b) *Let* $\sqrt{|v_1|+|v_2|} \geq \max\{8, 16/\varepsilon(\|b\widehat{q}(b)\|_f + \|\widehat{q}(b)\|_f^2)\}$. *The operator*

$$\begin{pmatrix} R_{SS} & R_{SB} \\ R_{BS} & R_{BB} \end{pmatrix}$$

has a bounded inverse. The norm

$$\left\| \begin{pmatrix} R_{SS} & R_{SB} \\ R_{BS} & R_{BB} \end{pmatrix}^{-1} - \begin{pmatrix} \pi_S & 0 \\ R_{BS} & \pi_B \end{pmatrix} \right\|_f \leq \text{const} \frac{\|b\widehat{q}(b)\|_f(1+\|\widehat{q}\|_f^2)}{\sqrt{|v_1|+|v_2|}}.$$

PROOF OF (a). In the case of R_{SB} and $R_{BB} - \pi_B$ it suffices to observe that, for $b \in B$,

$$|P_b(k)|^{-1} \leq 4[\sqrt{|v_1|+|v_2|}]^{-1}$$

and that the convolution operator $|\widehat{q}(b-c)|$ has norm $\|\widehat{q}\|_f$. In the case of R_{BS}

$$|P_b(k)|^{-1} \leq \frac{1}{\varepsilon}$$

is used instead. Finally

$$|(R_{SS} - \pi_S)_{b,c \in S}| = \left| \frac{\widehat{q}(b-c)}{P_c(k)} \right| \leq 8 \frac{|b-c|}{[\sqrt{|v_1|+|v_2|}]} \left| \frac{\widehat{q}(b-c)}{\varepsilon} \right|$$

by Lemma 14.3(c). We may now continue as in the other cases.

PROOF OF (b). As in Lemma 14.3

$$\begin{pmatrix} R_{SS} & R_{SB} \\ R_{BS} & R_{BB} \end{pmatrix}^{-1} = \begin{pmatrix} \mathcal{R}^{-1} & -\mathcal{R}^{-1} R_{SB} R_{BB}^{-1} \\ -R_{BB}^{-1} R_{BS} \mathcal{R}^{-1} & R_{BB}^{-1} + R_{BB}^{-1} R_{BS} \mathcal{R}^{-1} R_{SB} R_{BB}^{-1} \end{pmatrix}$$

where

$$\mathcal{R} = R_{SS} - R_{SB} R_{BB}^{-1} R_{BS}.$$

By part (a)

$$\|\mathcal{R} - \pi_S\|_f \leq \frac{8}{\varepsilon} \frac{\|b\widehat{q}(b)\|_f + \|\widehat{q}(b)\|_f^2}{\sqrt{|v_1|+|v_2|}}$$

provided $4\|\widehat{q}\|_f[\sqrt{|v_1|+|v_2|}]^{-1} \leq \frac{1}{2}$. This together with repeated applications of part (a) give the desired result. ☐

The principal quantity that determines the degree of opening of the double point (iz_d^2, z_d) i.e. that determines the $\hat{t}_d$ of Theorem 14.2, is $(\widehat{q}(d) - \mathcal{D}_{1,2}) \times (\widehat{q}(-d) - \mathcal{D}_{2,1})$. The next lemma provides the estimates required to control it. Define

$$\mathcal{K}_f = \left\{ k \in \mathbb{C}^2 \;\middle|\; \sqrt{|v_1|+|v_2|} \leq \max\left\{8, \frac{16}{\varepsilon}(\|b\widehat{q}(b)\|_f + \|\widehat{q}(b)\|_f^2)\right\}\right\}.$$

LEMMA 14.8. *Let* $k \in T_{d^{(1)}} \cap T_{d^{(2)}} \cap \{\mathbb{C} \setminus \mathcal{K}_f\}$. *Then there is a constant, depending only on* m, n *and* $\|b\widehat{q}(b)\|_f$, *such that*

$$\left| \frac{\partial^{n+m} l}{\partial k_1^n \partial k_2^m} \mathcal{D}(k)_{1,2} \right|, \left| \frac{\partial^{n+m} l}{\partial k_1^n \partial k_2^m} \mathcal{D}(k)_{2,1} \right| \leq \text{const} \|\widehat{q}\|_f^2 d^{(1)} - d^{(2)}|^m \frac{1}{f(|d^{(1)} - d^{(2)}|)}.$$

REMARK. The constis bounded on compacts. The $\|\widehat{q}\|_f^2$ has significance only for small q.

PROOF. The bound for $m = n = 0$ is an immediate consequence of (14.23) and

$$\left\| \sum_{b,c\in D} \frac{\widehat{q}(d-b)}{P_b(k)} (R_{DD}^{-1})_{b,c} \widehat{q}(c-d') \right\|_f \le \frac{1}{\varepsilon} \|\widehat{q}\|_f \|R_{DD}^{-1}\|_f \|\widehat{q}\|_f$$
$$\le \text{const} \|\widehat{q}\|_f^2.$$

By (14.15) all derivatives are given by finite linear combinations of terms of the form

$$Q_{dD} \prod_j \left\{ H_k^{-1} (R_{DD})^{-1} \frac{\partial^{n_j} H_k}{\partial k_{i_j}^{n_j}} \right\} H_k^{-1} (R_{DD})^{-1} Q_{Dd'}$$

where the sum of the n_j's for which i_j =1 (resp. 2) is n (resp. m). Apply

$$\|Q\|_f = \|q\|_f, \quad \|R_{DD}^{-1}\|_f \le \text{const}, \quad \|H_k^{-1}\|_f \le \frac{1}{\varepsilon},$$

$$\left\| \frac{\partial^n H_k}{\partial\, k_i^n} H_k^{-1} \pi_D \right\|_f \le \begin{cases} \text{const}[\sqrt{|v_1|} + |v_2|] & i = 2,\ n = 1 \\ \frac{2}{\varepsilon} & \text{otherwise} \end{cases}.$$

Recall that H_k and its derivatives are diagonal operators and that, for diagonal operators, the operator norm and f-norm agree. So far we have

$$\left\| Q_{dD} \prod_j \left\{ H_k^{-1} (R_{DD})^{-1} \frac{\partial^{n_j} H_k}{\partial k_{i_j}^{n_j}} \right\} H_k^{-1} (R_{DD})^{-1} Q_{Dd'} \right\|_f \le \text{const} \|\widehat{q}\|_f^2 [\sqrt{|v_1|} + |v_2|]^m.$$

By (14.8)

$$\sqrt{|v_1|} + |v_2| \le \text{const}(1 + |k_2 + d_2^{(2)}|)$$

and, by (14.2)

$$\left| k_2 + \frac{d_2^{(1)} + d_2^{(2)}}{2} + \frac{i}{2} \frac{d_1^{(1)} - d_1^{(2)}}{d_2^{(1)} - d_2^{(2)}} \right| \le \varepsilon$$

for $k \in T_{d^{(1)}} \cap T_{d^{(2)}}$. Hence

$$\sqrt{|v_1|} + |v_2| \le \text{const} \left(1 + \left| \frac{d_2^{(1)} - d_2^{(2)}}{2} + \frac{i}{2} \frac{d_1^{(1)} - d_1^{(2)}}{d_2^{(1)} - d_2^{(2)}} \right| \right)$$
$$\le \text{const} |d^{(1)} - d^{(2)}|.$$

□

To conclude this section we note the following application of Theorem 14.1.

THEOREM 14.9. *Let $q \in L^2(\mathbb{R}^2/\Gamma)$ with $\|b\widehat{q}(b)\|_1 < \infty$. Then $\mathcal{H}(q)$ is a reduced and irreducible one-dimensional complex analytic variety.*

PROOF. Without loss of generality we may assume that $\widehat{q}(0) = 0$. Let $\varepsilon > 0$ be a small number, and choose ρ such that Theorem 14.1 holds. By part (ii) of this theorem there is a reduced component C of $\widehat{\mathcal{H}}(q)$ such that

$$(\widehat{\mathcal{H}}(q) \cap T_0) \setminus \left(\mathcal{K}_\rho \cup \bigcup_{\substack{b\in\Gamma^\# \\ b_2 \ne 0}} T_b \right) = (C \cap T_0) \setminus \left(\mathcal{K}_\rho \cup \bigcup_{\substack{b\in\Gamma^\# \\ b_2 \ne 0}} T_b \right).$$

Clearly $E(C)$ is a reduced component of $\mathcal{H}(q)$. Assume that $\mathcal{H}(q)$ has a component K different from $E(C)$. Then every component C' of $E^{-1}(K)$ lies in

$$\mathcal{K}_\rho \cup \bigcup_{b,c\in\Gamma^\#} T_b \cap T_c.$$

In particular the complement of $\mathrm{pr}(C')$ contains an open subset of $\mathbb{C}$.

On the other hand the indicator of growth ([**LG**, 3.6]) of $\widehat{\mathcal{H}}(q)$ is of finite order, since by Theorem 13.8 $\widehat{\mathcal{H}}(q)$ is the zero-set of an entire function of finite order. Therefore the indicator of growth of C' is also of finite order, and hence by the solution of the "Cousin problem with finite order" ([**LG**, 3.30]) C' is also the zero set of an entire function of finite order. Therefore by [**LG**, 3.44] the set $\{z \in \mathbb{C} \mid \mathrm{pr}^{-1}(z) \cap C' = \varnothing\}$ is either $\mathbb{C}$ itself or discrete. Since its complement contains an open set, it is in fact discrete. As C' is irreducible it follows that this set consists of one point z_0. So $C' \subset \mathbb{C} \times \{z_0\}$. If we now apply the same argument with the projection $(k_1, k_2) \mapsto k_1$ we conclude tha C' is a point, which is impossible. So $\mathcal{H}(q) = E(C)$ is irreducible and reduced. □

15. Heat Curves: Verification of the Geometric Hypotheses

Let $q \in C^\infty(\mathbb{R}^2/\Gamma)$. Fix $\varepsilon > 0$ sufficiently small. We construct a decomposition of $\mathcal{H}(q)$ into $\mathcal{H}(q)^{\mathrm{com}} \cup \mathcal{H}(q)^{\mathrm{reg}} \cup \mathcal{H}(q)^{\mathrm{han}}$ such that the geometric hypotheses of §5 hold.

First, we refine Theorem 14.2 to get control of the handles. For $d \in \Gamma^\#$ sufficiently large with $d_2 \neq 0$, let

$$\widehat{\phi}_d \colon \left\{ (z_1, z_2) \in \mathbb{C}^2 \;\middle|\; |z_1| \leq \frac{\varepsilon}{2}, |z_2| \leq \frac{\varepsilon}{2} \right\} \to T_0 \cap T_d$$

be the map of Theorem 14.2 and $\hat{t}_d$ the number such that

$$\widehat{\phi}_d^{-1}\big(T_0 \cap T_d \cap \widehat{\mathcal{H}}(q)\big) = \left\{ (z_1, z_2) \in \mathbb{C}^2 \;\middle|\; z_1 z_2 = \hat{t}_d, |z_1| \leq \frac{\varepsilon}{2}, |z_2| \leq \frac{\varepsilon}{2} \right\}$$

holds. Recall that

$$|\hat{t}_b| \leq \frac{\mathrm{const}(\beta')}{|b|^{2\beta'}} \tag{15.1a}$$

for all $\beta' > 0$ so that

$$\sum_{\beta} |\hat{t}_b|^\beta < \infty \tag{15.1b}$$

for all $\beta > 0$. Put

$$s_d = \mathrm{pr}\,\big(\widehat{\phi}_d(0)\big).$$

Then

$$|s_d - z_d| \leq \frac{\mathrm{const}}{|d_2 z_d^2|}. \tag{15.2}$$

Put

$$\tau_d = \frac{1}{|z_d|^{13}}, \quad r_d = \frac{2\varepsilon}{|d_2|}\frac{1}{|z_d|^{14}}, \quad R_d = \frac{\varepsilon}{6|d_2|}\frac{1}{|z_d|},$$

$$\widehat{g}_d \colon \left\{ \zeta \in \mathbb{C} \;\middle|\; \frac{\varepsilon}{2|z_d|}\tau_d \leq |\zeta| \leq \frac{\varepsilon}{2} \right\} \to \mathbb{C}, \quad \zeta \mapsto \mathrm{pr}\left(\widehat{\phi}_d\left(\zeta, \frac{\hat{t}_d}{\zeta}\right)\right).$$

LEMMA 15.1. *If $|d|$ is big enough, then*

(a) $\widehat{g}_d$ *is biholomorphic onto its image. Furthermore*

$$\left|\widehat{g}_d\left(\frac{2\varepsilon}{|z_d|}\tau_d e^{i\theta}\right) - s_d\right| < r_d$$

and

$$\left|\widehat{g}_d\left(\frac{\varepsilon}{2|z_d|}e^{i\theta}\right) - s_d\right| > R_d > \left|\widehat{g}_d\left(\frac{\varepsilon}{4|z_d|}e^{i\theta}\right) - s_d\right|,$$

$$\left|\widehat{g}_d\left(\frac{\varepsilon}{8|z_d|}e^{i\theta}\right) - s_d\right| > \frac{1}{4}R_d,$$

$$\left|\widehat{g}_d\left(\frac{\varepsilon}{2}e^{i\theta}\right) - s_d\right| > \frac{\varepsilon}{6|d_2|}$$

for all $0 \le \theta \le 2\pi$.

(b) *Define* $\alpha_d(z)$ *by*

$$\alpha_d(z)\,dz = (\widehat{g}_d)_*\left(\frac{1}{2\pi i}\frac{d\zeta}{\zeta}\right) + \frac{\operatorname{sgn} d_2}{2\pi i}\frac{1}{z - s_d}\,dz.$$

Then

$$\sup_d \|\alpha_d(z)\,dz|_{\{z\in\mathbb{C}|r_d<|z-s_d|<R_d\}}\|_2 < \infty$$

and

$$\lim_{d\to\infty} R_d \sup_{|z-s_d|=R_d} |\alpha_d(z)| = 0.$$

PROOF. Write

$$\widehat{\phi}_d(z_1,z_2) = \big(k_1(z_1,z_2), k_2(z_1,z_2)\big).$$

By the estimates in Theorem 14.2(i), we have, for all ζ with $2/\varepsilon|\hat{t}_d| \le |\zeta| \le \varepsilon/2$

$$\begin{aligned} k_2\left(\zeta, \frac{\hat{t}_d}{\zeta}\right) - s_d &= \big(k_2(\zeta,0) - s_d\big) + \left(k_2\left(\zeta,\frac{\hat{t}_d}{\zeta}\right) - k_2(\zeta,0)\right) \\ &= \int_0^\zeta \frac{\partial k_2}{\partial z_1}(\xi,0)\,d\xi + \int_0^{\hat{t}_d/\zeta} \frac{\partial k_2}{\partial z_2}(\zeta,\xi)\,d\xi \\ &= \frac{-\zeta}{2d_2} + \frac{\hat{t}_d}{2d_2\zeta} + \left(|\zeta| + \frac{|\hat{t}_d|}{|\zeta|}\right)O\left(\frac{1}{|d_2^2 z_d|}\right). \end{aligned}$$

Therefore

$$\left|\widehat{g}_d(\zeta) - s_d + \frac{1}{2d_2}\left(\zeta - \frac{\hat{t}_d}{\zeta}\right)\right| \le \frac{\text{const}}{|d_2^2 z_d|}\left(|\zeta| + \frac{|\hat{t}_d|}{|\zeta|}\right) \tag{15.3}$$

and the estimates of part a) are obeyed.

To see that $\widehat{g}_d$ is biholomorphic onto its range, we first estimate its derivative. Again by part (i) of Theorem 14.2

$$\begin{aligned} \frac{d\widehat{g}_d}{d\zeta}(\zeta) &= \frac{\partial k_2}{\partial z_1}\left(\zeta,\frac{\hat{t}_d}{\zeta}\right) - \frac{\hat{t}_d}{\zeta^2}\frac{\partial k_2}{\partial z_2}\left(\zeta,\frac{\hat{t}_d}{\zeta}\right) \\ &= -\frac{1}{2d_2}\left(1 + \frac{\hat{t}_d}{\zeta^2}\right)\left(1 + O\left(\frac{1}{|d_2 z_d|}\right)\right). \end{aligned}$$

In particular,

$$\left|\frac{d\widehat{g}_d}{d\zeta} + \frac{1}{d_2}\right| \le \frac{\text{const}}{|d_2^2 z_d|}. \tag{15.4}$$

Therefore, if d is big enough, its derivative vanishes nowhere. It remains to show that $\widehat{g}_d$ is injective. Let ζ and ζ' be two distinct points in the annulus $\{\xi \in \mathbb{C} \mid \varepsilon/2\tau_d \le |\xi| \le \varepsilon/2$. Connect ζ and ζ' by a path γ in this annulus with length at most $\pi/2|\zeta - \zeta'|$. By (15.4)

$$|\widehat{g}_d(\zeta') - \widehat{g}_d(\zeta)| = \left| \int_\gamma \frac{d\widehat{g}_d}{d\zeta} d\xi \right| \ge \left(\frac{1}{|d_2|} - \frac{\text{const}}{|d_2^2 z_d|} \frac{\pi}{2} \right) |\zeta - \zeta'|$$

so that $\widehat{g}_d(\zeta)$ and $\widehat{g}_d(\zeta')$ are different if d is sufficiently large. This concludes the proof of part (i) of the lemma.

To prove part (ii) we observe that, by (15.3)–(15.4),

$$\begin{aligned}
&\frac{d\widehat{g}_d}{\widehat{g}_d(\zeta) - s_d} - \frac{d\zeta}{\zeta} \\
&\qquad = \frac{1 + \hat{t}_d/\zeta^2}{\zeta - \hat{t}_d/\zeta} d\zeta \left(1 + O\left(\frac{1}{|d_2 z_d|} \right) \right) - \frac{d\zeta}{\zeta} \\
&\qquad = \frac{2\hat{t}_d/\zeta^2}{\zeta - \hat{t}_d/\zeta} d\zeta + \frac{1 + \hat{t}_d/\zeta^2}{\zeta - \hat{t}_d/\zeta} d\zeta \, O\left(\frac{1}{|d_2 z_d|} \right) \\
&\qquad = \frac{2\hat{t}_d}{\zeta^2(1 + \hat{t}_d/\zeta^2)} \frac{d\widehat{g}_d}{\widehat{g}_d(\zeta) - s_d} \left(1 + O\left(\frac{1}{|d_2 z_d|} \right) \right) + \frac{d\widehat{g}_d}{\widehat{g}_d(\zeta) - s_d} O\left(\frac{1}{|d_2 z_d|} \right).
\end{aligned}$$

Applying $(\widehat{g}_d)_*$

$$\frac{dz}{z - s_d} - (\widehat{g}_d)_* \left(\frac{d\zeta}{\zeta} \right) = \frac{dz}{z - s_d} \left(\frac{2\hat{t}_d}{\zeta^2 + \hat{t}_d} + O\left(\frac{1}{|d_2 z_d|} \right) \right).$$

Therefore

$$|\alpha_d(z)| \le \text{const} \frac{1}{|z - s_d|} \frac{1}{|d_2 z_d|}$$

since $|\zeta|^2 \ge \varepsilon^2/4|z_d|^{26}$. Now part (ii) of the lemma follows easily. $\square$

We now define the data appearing in the geometric hypotheses and verify (GH1)–(GH6). Fix $\rho > 0$ such that the conclusions of Theorem 14.1 hold with ε replaced by $\varepsilon/8$. Let K be a simply connected subset of $\mathbb{C}$ with smooth boundary that is symmetric with respect to the imaginary axis, such that

$$\{z \in \mathbb{C} \mid |z| < \rho\} \subset K$$

such that for all $d \in \Gamma^\#$ with $d_2 \ne 0$

$$\text{either} \quad \left\{ z \in \mathbb{C} \,\middle|\, |z - z_d| < \frac{\varepsilon}{|d_2|} \right\} \subset K \quad \text{or} \quad \left\{ z \in \mathbb{C} \,\middle|\, |z - z_d| < \frac{\varepsilon}{|d_2|} \right\} \cap K = \varnothing$$

and such that Lemma 15.1 holds for all d obeying the second alternative. Put

$$\widehat{\mathcal{H}}(q)^{\text{com}} = \widehat{\mathcal{H}}(q) \cap \left(\mathcal{K}_\rho \cup \left(\text{pr}^{-1}(K) \cap T_0 \right) \right).$$

Then

$$\mathcal{H}(q)^{\text{com}} = E\left(\widehat{\mathcal{H}}(q)^{\text{com}} \right)$$

is a compact subset of $\mathcal{H}(q)$ whose boundary is diffeomorphic to ∂K.

Let

$$\Gamma_K^\# = \{ b \in \Gamma^\# \mid b_2 \ne 0, z_b \in K \}.$$

For each $b \in \Gamma_K^\#$, let D_b be the region enclosed by the curve $\widehat{g}_b(\{(\varepsilon/2|z_b|)\tau_b e^{i\theta} \mid 0 \le \theta < 2\pi\})$. Define

$$G = \mathbb{C} \setminus \Big(K \cup \bigcup_{b \in \Gamma_K^\#} D_b \Big)$$

and

$$\widehat{\mathcal{H}}(q)^{\mathrm{reg}} = \Big\{ k \in \widehat{\mathcal{H}}(q) \cap \Big(T_0 \setminus \bigcup_{b \neq 0} T_b \Big) \Bigm| \mathrm{pr}(k) \in G \Big\}$$
$$\cup \bigcup_{b \in \Gamma_K^\#} \widehat{\phi}_b \Big(\Big\{ (z_1, z_2) \in \mathbb{C}^2 \Bigm| z_1 z_2 = \hat{t}_b, \frac{\varepsilon}{2|z_b|}\tau_b \le |z_1| \le \frac{\varepsilon}{2} \Big\} \Big).$$

By Theorem 14.1 and Lemma 15.1(a), pr induces a biholomorphic map from $\widehat{\mathcal{H}}(q)^{\mathrm{reg}}$ onto G. Since two points k and k' of T_0 are identified by the map E if and only if there is a $b \in \Gamma^\#$ with $b_2 \neq 0$ such that $k \in T_{-b} \cap T_0$, $k' \in T_0 \cap T_b$ and $k' = k - b$, the map E induces a biholomorphism between $\widehat{\mathcal{H}}(q)^{\mathrm{reg}}$ and

$$\mathcal{H}(q)^{\mathrm{reg}} = E\big(\widehat{\mathcal{H}}(q)^{\mathrm{reg}}\big).$$

Define

$$\Phi \colon G \to \mathcal{H}(q)^{\mathrm{reg}}$$

as the composition of $(\mathrm{pr}\,|_{\widehat{\mathcal{H}}(q)^{\mathrm{reg}}})^{-1}$ and E. If we put

$$S = \{ s_b \mid b \in \Gamma_K^\# \}, \quad D(s_b) = D_b$$

then (GH1) and (GH4) are fulfilled.

Next put, for $b \in \Gamma_K^\#$, $b_2 > 0$

$$Y_b = E\Big(\widehat{\phi}_b \Big(\Big\{ (z_1, z_2) \in \mathbb{C}^2 \Bigm| z_1 z_2 = \hat{t}_b, |z_1|, |z_2| \le \frac{\varepsilon}{2|z_b|} \Big\} \Big) \Big), \quad t_b = \Big(\frac{2|z_b|}{\varepsilon} \Big)^2 |\hat{t}_b|.$$

Then

$$\phi_b \colon H(t_b) \to Y_b, \quad (z_1, z_2) \mapsto E\Big(\widehat{\phi}_b \Big(\frac{2|z_b|}{\varepsilon} \frac{\hat{t}_b}{|\hat{t}_b|} z_1, \frac{2|z_b|}{\varepsilon} z_2 \Big) \Big)$$

is a biholomorphic map. Observe that $t_b \neq 0$ for all $b \in \Gamma^\#$ if $\mathcal{H}(q)$ is smooth. In this case choose a homology basis of $\mathcal{H}(q)$ such that Y_b represents an A cycle. By construction and (15.1) hypothesis (GH2) holds.

THEOREM 15.2. *Let $q \in C^\infty(\mathbb{R}^2/\Gamma)$ be such that $\mathcal{H}(q)$ is smooth. Then the marked Riemann surface $\mathcal{H}(q) = \mathcal{H}(q)^{\mathrm{com}} \cup \mathcal{H}(q)^{\mathrm{reg}} \cup \mathcal{H}(q)^{\mathrm{han}}$ obeys the geometric hypotheses* (GH1)–(GH6) *of* §5. *One pair of handles looks like*

PROOF. We have already verified (GH1), (GH2) and (GH4). Define, for all $b \in \Gamma_K^\#$ with $b_2 > 0$

$$\tau(b) = \tau_b,$$
$$s_1(b) = s_b, \quad s_2(b) = s_{-b},$$
$$R_\mu(b) = R_b, \quad r_\mu(b) = r_b.$$

Then (GH3) follows from Lemma 15.1(a).

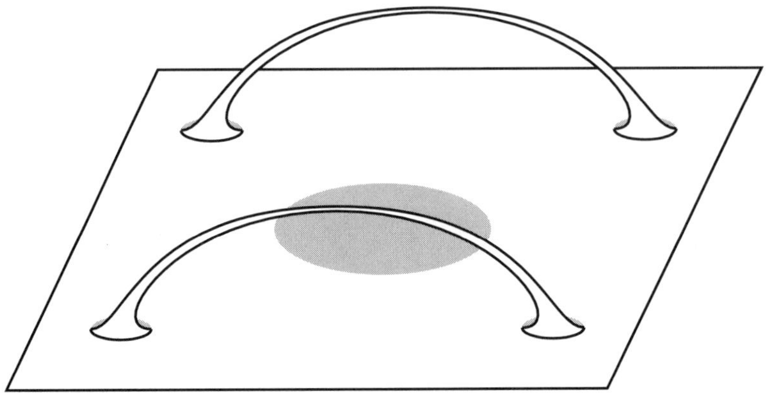

Part (i) of (GH5) is trivial. With $\delta = 2$ and $d = 14$ hypothesis (GH5)(ii) follows from the summability of $1/|z_b|^4$. The summability of $1/|b_2 z_b^3|$ together with (15.2) yield (GH5)(iii). Part (iv) of (GH5) follows from (15.1).

Finally (GH6) follows from the fact that $|s_b - s_{-b}| = O(|b_2|)$. □

REMARK 15.3. If $t_b = 0$ for some b then Φ can be extended to a map from $G \cup D_b \cup D_{-b}$ to the normalization of $\mathcal{H}(q)$. In this way one sees that the normalization of $\mathcal{H}(q)$ always fulfills the geometric hypotheses whenever $t_b \neq 0$ for infinitely many b. If, on the other hand, $t_b = 0$ for all but finitely many b, then the normalization of $\mathcal{H}(q)$ has finite genus and q is a "finite gap" potential [**K**].

REMARK 15.4. Denote by $\Gamma_+^\# = \{b \in \Gamma^\# \mid b_2 > 0\}$ and by $\{\omega_b \mid b \in \Gamma_+^\#\}$ the basis of L^2 holomorphic differential forms dual to the A cycles

$$\int_{A_b} \omega_c = \delta_{b,c}.$$

This crucial set of forms can be constructed as residues. First define, for each $b \in \Gamma_+^\#$

$$\widehat{\rho}_b = \frac{b_2}{\pi} \operatorname{res}_{\widehat{\mathcal{H}}(q)} \left(\sum_{c \in \Gamma^\sharp} \frac{1}{P_c P_{c+b}} \right) \frac{dk_1 \wedge dk_2}{\det_2(\mathbb{1} + qH_k^{-1})}$$

and let ρ_b be the corresponding form on $\mathcal{H}(q)$. One can show that for any vector in the space $\ell_w^2 = \{(\lambda_b)_{b\in\Gamma_+^\#} \mid \sum_{b\in\Gamma_+^\#} |b_2 z_b^2 \lambda_b|^2 < \infty\}$ the linear combination $\langle \lambda, \rho \rangle = \sum_{b\in\Gamma_+^\#} \lambda_b \rho_b$ is an L^2 holomorphic differential form on $\mathcal{H}(q)$, at least when $\langle \lambda, \rho \rangle$ has only finitely many nonzero A periods. One can also show that the matrix

$$\mathcal{M}_{b,c} = \int_{A_c} \rho_b$$

is boundedly invertible on ℓ_w^2 and that

$$\omega_c = \sum_{b\in\Gamma_+^\#} \mathcal{M}_{c,b}^{-1} \rho_b.$$

The technical details are available from the authors.

16. Fermi Curves: Basic Properties

Let Γ be a lattice in $\mathbb{R}^2$ and q a real valued function in $L^2(\mathbb{R}^2/\Gamma)$. For each k in $\mathbb{R}^2$ the self-adjoint boundary value problem

$$(-\Delta + q)\psi = \lambda\psi, \quad \psi(x+\gamma) = e^{i\langle k,\gamma\rangle} \text{ for all } \gamma \in \Gamma$$

has a discrete spectrum customarily denoted by

$$E_1(k) \le E_2(k) \le E_3(k) \le \cdots .$$

The eigenvalue $E_n(k)$, $n \ge 1$, defines a function of k called the nth band function. It is continuous and periodic with respect to the lattice

$$\Gamma^{\#} = \{b \in \mathbb{R}^2 \mid \langle b, \gamma\rangle \in 2\pi\mathbb{Z} \text{ for all } \gamma \in \Gamma\}$$

dual to Γ.

The (real "lifted") Fermi curve for energy λ is defined as

$$\widehat{\mathcal{F}}_{\lambda,\mathbb{R}}(q) = \{k \in \mathbb{R}^2 \mid E_n(k) = \lambda \text{ for some } n \in \mathbb{N}\}. \tag{16.1}$$

If we define

$$H_k = -(\Delta + 2ik\cdot\nabla - k^2)$$

then

$$\widehat{\mathcal{F}}_{\lambda,\mathbb{R}}(q) = \{k \in \mathbb{R}^2 \mid (H_k + q - \lambda)\phi = 0 \text{ for some } 0 \ne \phi \in H^2(\mathbb{R}^2/\Gamma)\}. \tag{16.2}$$

As we may replace q by $q - \lambda$, we only discuss the case $\lambda = 0$ and write $\widehat{\mathcal{F}}_{\mathbb{R}}(q)$ in place of $\widehat{\mathcal{F}}_{0,\mathbb{R}}(q)$. Clearly $\Gamma^{\#}$ acts on $\widehat{\mathcal{F}}_{\mathbb{R}}(q)$. We put

$$\mathcal{F}_{\mathbb{R}}(q) = \widehat{\mathcal{F}}_{\mathbb{R}}(q)/\Gamma^{\#},$$

$\mathcal{F}_{\mathbb{R}}(q)$ is a curve in the torus $\mathbb{R}^2/\Gamma^{\#}$.

We consider the complexifications of $\widehat{\mathcal{F}}_{\mathbb{R}}(q)$ and $\mathcal{F}_{\mathbb{R}}(q)$

$$\widehat{\mathcal{F}}(q) = \{k \in \mathbb{C}^2 \mid (H_k + q)\phi = 0 \text{ for some } 0 \ne \phi \in H^2(\mathbb{R}^2/\Gamma)\}, \quad \mathcal{F}(q) = \widehat{\mathcal{F}}(q)/\Gamma^{\#}.$$

We call $\mathcal{F}(q)$ the Fermi curve of q. It is the image of $\widehat{\mathcal{F}}(q)$ under

$$E\colon \mathbb{C}^2 \to \mathbb{C}^* \times \mathbb{C}^*, \quad k \mapsto (e^{i\langle k,\gamma_1\rangle}, e^{i\langle k,\gamma_2\rangle})$$

where γ_1, γ_2 is a basis of Γ. These definitions make sense for any complex valued q in $L^2(\mathbb{R}^2/\Gamma)$.

LEMMA 16.1. *The curve $\widehat{\mathcal{F}}(0)$ for $q = 0$ is the locally finite union $\bigcup_{\substack{b\in\Gamma^\sharp\\ \nu=1,2}} \mathcal{N}_\nu(b)$ of lines*

$$\begin{aligned} \mathcal{N}_\nu(b) &= \{(k_1,k_2) \in \mathbb{C}^2 \mid N_{b,\nu}(k_1,k_2) = 0\}, \\ N_{b,\nu}(k) &= (k_1 + b_1) + i(-1)^\nu(k_2 + b_2)\}. \end{aligned} \tag{16.3}$$

In particular, the Fermi curve $\mathcal{F}(0)$ is a complex analytic curve in $\mathbb{C}^2/\Gamma^{\#}$.

PROOF. Put

$$N_b(k) = N_{b,1}(k)N_{b,2}(k) = (k_1 + b_1)^2 + (k_2 + b_2)^2, \quad \mathcal{N}_b = \mathcal{N}_1(b) \cup \mathcal{N}_2(b).$$

For all $k \in \mathbb{C}^2$ the exponentials $e^{i\langle b,x\rangle}, b \in \Gamma^\sharp$ are a complete set of eigenfunctions for H_k in $L^2(\mathbb{R}^2/\Gamma)$ satisfying

$$H_k e^{i\langle b,x\rangle} = N_b(k)e^{i\langle b,x\rangle}.$$

Therefore,

$$\widehat{\mathcal{F}}(0) = \bigcup_{b \in \Gamma^\sharp} \mathcal{N}_b.$$

Observe that only a finite number of the line pairs $\mathcal{N}_b$ can intersect any bounded subset of $\mathbb{C}^2$. Thus, the union is locally finite. $\square$

Observe that

$$\mathcal{N}_\nu(b) \cap \mathcal{N}_\nu(c) = \varnothing \quad \text{if } b \neq c,$$

$$\mathcal{N}_1(0) \cap \mathcal{N}_2(b) = \{(iz_1(b), z_1(b))\}, \quad \mathcal{N}_1(-b) \cap \mathcal{N}_2(0) = \{(-iz_2(b), z_2(b))\}$$

where

$$z_\nu(b) = \tfrac{1}{2}\big((-1)^\nu b_2 + ib_1\big)$$

and that the map $k \mapsto k + b$ maps $\mathcal{N}_1(0) \cap \mathcal{N}_2(b)$ to $\mathcal{N}_1(-b) \cap \mathcal{N}_2(0)$.

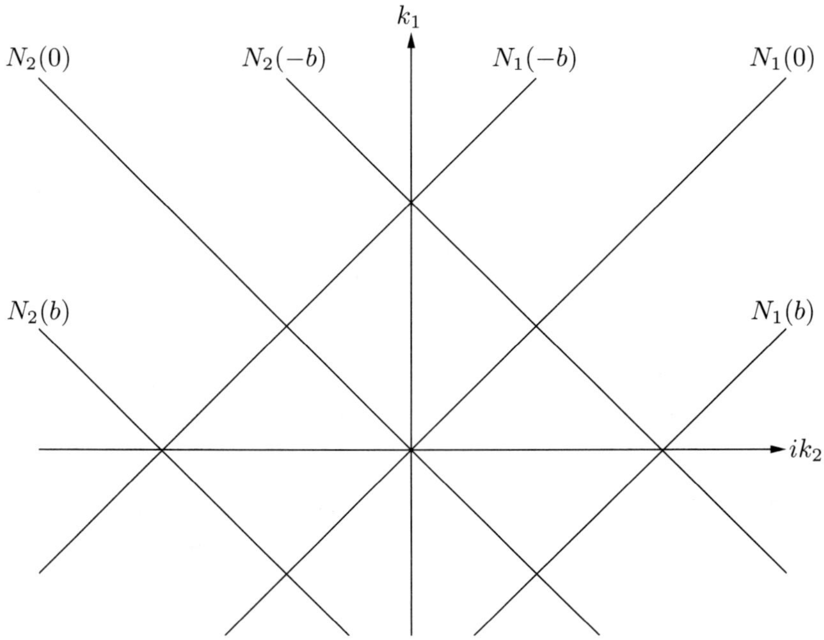

For each finite subset B of $\Gamma^\sharp$ set

$$\mathbb{C}^2_B = \mathbb{C}^2 \setminus \bigcup_{b \in \Gamma^\sharp \setminus B} \mathcal{N}_b.$$

Also, let π_B be the orthogonal projection onto the subspace spanned by $e^{i\langle b,x\rangle}$, $b \in B$, and define a partial inverse $(H_k)_B^{-1}$ for $k \in \mathbb{C}^2_B$ by

$$(H_k)_B^{-1} = \pi_B + H_k^{-1}(\mathbf{1} - \pi_B).$$

Its matrix elements are

$$\langle e^{i\langle b,x\rangle}, (H_k)_B^{-1} e^{i\langle c,x\rangle}\rangle = \begin{cases} \delta_{b,c} & \text{if } c \in B \\ \delta_{b,c} \frac{1}{N_c(k)} & \text{if } c \notin B \end{cases}$$

and

$$(H_k + q)(H_k)_B^{-1} = \mathbb{1} + q(H_k)_B^{-1} + (H_k - \mathbf{1})\pi_B.$$

In [**KT**] it is shown that $q(H_k)_B^{-1}$ is a Hilbert–Schmidt operator and that

THEOREM 16.2. *For all q in $L^2(\mathbb{R}^2/\Gamma)$ the "lifted" Fermi curve $\widehat{\mathcal{F}}(q)$ is a one-dimensional complex analytic subvariety of $\mathbb{C}^2$. It is the zero set of an entire function of finite order. The intersection of $\widehat{\mathcal{F}}(q)$ with $\mathbb{C}^2_B$ is given by*

$$\widehat{\mathcal{F}}(q) \cap \mathbb{C}^2_B = \big\{(k_1, k_2) \in \mathbb{C}^2_B \;\big|\; \det{}_2\big((H_k + q)(H_k)_B^{-1}\big) = 0\big\}.$$

The Fermi curve $\mathcal{F}(q) = \widehat{\mathcal{F}}(q)/\Gamma^\sharp$ is an analytic subvariety of $\mathbb{C}^2/\Gamma^\sharp$.

17. Fermi Curves: Asymptotics

In this section, we show that $\widehat{\mathcal{F}}(q)$ is close to $\widehat{\mathcal{F}}(0)$ when the imaginary parts of k_1 and k_2 are large. To facilitate the discussion write $k \in \mathbb{C}^2$ as

$$k_1 = u_1 + iv_1, \quad k_2 = u_2 + iv_2$$

where u_1, u_2, v_1 and v_2 are real. Then

$$\begin{aligned} N_{b,\nu}(k) &= (k_1 + b_1) + i(-1)^\nu (k_2 + b_2) \\ &= i\big(v_1 + (-1)^\nu (u_2 + b_2)\big) - \big((-1)^\nu v_2 - (u_1 + b_1)\big) \end{aligned} \tag{17.1}$$

so that

$$|N_{b,\nu}(k)| = |v + (-1)^\nu (u + b)^\perp|$$

where

$$(w_1, w_2)^\perp = (w_2, -w_1).$$

Recall that $N_b(k) = N_{b,1}(k) N_{b,2}(k)$. Hence $N_b(k) = 0$ if and only if

$$v = (u + b)^\perp \quad \text{or} \quad v = -(u + b)^\perp. \tag{17.2}$$

Let 2Λ be the length of the shortest nonzero vector in $\Gamma^\#$. Then there is at most one $b \in \Gamma^\#$ with $|v + (u+b)^\perp| < \Lambda$ and at most one $b \in \Gamma^\#$ with $|v - (u+b)^\perp| < \Lambda$.

For $\min\{1, \Lambda/6\} > \varepsilon > 0$ and $b \in \Gamma^\sharp$ define the (ε-)tube about $\mathcal{N}_b = \{k \in \mathbb{C}^2 \mid N_b(k) = 0\}$ by

$$T_b = T_1(b) \cup T_2(b),$$

$$T_\nu(b) = \left\{ k \in \mathbb{C}^2 \;\middle|\; |N_\nu(b)| = |v + (-1)^\nu (u + b)^\perp| < \frac{\varepsilon}{1 + |v|^{1-\varepsilon}} \right\}. \tag{17.3a}$$

Since $[v + (u + b)^\perp] + [v - (u + b)^\perp] = 2v$ at least one of the factors $|v + (u + b)^\perp|$, $|v - (u + b)^\perp|$ in $|N_b(k)|$ must always be at least $|v|$. If $k \in T_b$ one of the factors is bounded by $\varepsilon/(1 + |v|^{1-\varepsilon})$ and the other must lie within $\varepsilon/(1 + |v|^{1-\varepsilon})$ of $|2v|$. Thus

$$k \notin T_b \implies |N_b(k)| \geq \frac{\varepsilon |v|}{1 + |v|^{1-\varepsilon}}, \tag{17.3b}$$

$$k \in T_b \implies |N_b(k)| \leq \frac{\varepsilon(2|v| + \varepsilon)}{1 + |v|^{1-\varepsilon}}. \tag{17.3c}$$

The pairwise intersection $\overline{T}_b \cap \overline{T}_{b'}$ is compact whenever $b \neq b'$. Indeed, $\overline{T}_\nu(b) \cap \overline{T}_\nu(b') = \varnothing$ if $b \neq b'$ since $|(u + b)^\perp - (u + b')^\perp| \geq 2\Lambda$. If $k \in \overline{T}_1(b) \cap \overline{T}_2(b')$ then, we have

$$|u + \tfrac{1}{2}(b + b')| = \tfrac{1}{2}|v - (u + b)^\perp - v - (u + b')^\perp| \leq \frac{\varepsilon}{1 + |v|^{1-\varepsilon}}$$

and

$$\left|v - \tfrac{1}{2}(b-b')^{\perp}\right| \le \frac{\varepsilon}{1+|v|^{1-\varepsilon}}.$$

We also have $\overline{T}_b \cap \overline{T}_{b'} \cap \overline{T}_{b''} = \varnothing$ for all distinct elements b, b', b'' of $\Gamma^\sharp$. We shall asymptotically confine $k \in \widehat{\mathcal{F}}(q)$ to the union of the tubes $T_b, b \in \Gamma^\sharp$.

For $\rho > 0$ define

$$\mathcal{K}_\rho = \{k \in \mathbb{C}^2 \mid |v| \le \rho\}.$$

Furthermore let $\mathrm{pr}\colon \mathbb{C}^2 \to \mathbb{C}$ be the projection $(k_1,k_2) \mapsto k_2$.

THEOREM 17.1. *Let $q \in L^2(\mathbb{R}^2/\Gamma)$ obey $\|\widehat{q}(b)\|_1 := \sum_{b\in\Gamma^\#} |b\widehat{q}(b)| < \infty$ and let $\min\{1,\Lambda\} > \varepsilon > 0$. Then there is a constant ρ, which depends only on $\|b\widehat{q}(b)\|_1$, Λ and ε, such that*

(a) $$\{k \in \widehat{\mathcal{F}}(q) \mid k \notin \mathcal{K}_\rho\} \subset \bigcup_{b\in\Gamma^\#} T_b.$$

(b) *For $\nu = 1,2$ the projection* pr *induces a biholomorphic map between*

$$\left(\widehat{\mathcal{F}}(q) \cap T_\nu(0)\right) \setminus \left(\mathcal{K}_\rho \cup \bigcup_{\substack{b\in\Gamma^\# \\ b\neq 0}} T_b\right)$$

and its image in $\mathbb{C}$. This image component contains

$$\left\{ z \in \mathbb{C} \;\middle|\; |z| > 2\rho \text{ and } |z - z_\nu(b)| > \frac{\varepsilon}{|b|^{1-\varepsilon}} \text{ for all } b \in \Gamma^\# \text{ with } b \neq 0 \right\}$$

and is contained in

$$\left\{ z \in \mathbb{C} \;\middle|\; |z - z_\nu(b)| > \frac{\varepsilon}{2|b|^{1-\varepsilon}} \text{ for all } b \in \Gamma^\# \text{ with } b \neq 0 \right\}$$

where $z_\nu(b) = \frac{1}{2}\big((-1)^\nu b_2 + ib_1\big)$.

Clearly $\mathcal{K}_\rho$ is invariant under the $\Gamma^\#$-action and $\mathcal{K}_\rho/\Gamma^\#$ is compact. So the image of $\widehat{\mathcal{F}}(q) \cap \mathcal{K}_\rho$ under the exponential map $E\colon \widehat{\mathcal{F}}(q) \to \mathcal{F}(q)$ is compact in $\mathcal{F}(q)$. It will essentially play the role of X^{com} in the decomposition of $\mathcal{F}(q)$ that we need to apply the results of Chapter 2.

Since $c \cdot T_b = T_{b+c}$ for every $b, c \in \Gamma^\#$ the complement of $E\big(\widehat{\mathcal{F}}(q) \cap \mathcal{K}_\rho\big)$ in $\mathcal{F}(q)$ is the disjoint union of

$$E\left(\left(\widehat{\mathcal{F}}(q) \cap T_0\right) \setminus \left(\mathcal{K}_\rho \cup \bigcup_{\substack{b\in\Gamma^\# \\ b_2\neq 0}} T_b\right)\right)$$

and

$$\bigcup_{\substack{b\in\Gamma^\# \\ b_2\neq 0}} E\big(\widehat{\mathcal{F}}(q) \cap T_0 \cap T_b\big).$$

Basicly the first of the two sets will be the regular piece of $\mathcal{F}(q)$, while the second sets will be the handles. The map Φ parametrizing the regular part will be the composition of E with the inverse of the map discussed in part (b) of Theorem 17.1. For the handles we will use

THEOREM 17.2. *Let $\varepsilon > 0$ be sufficiently small and let $\beta \geq 1$. Assume that $q \in L^2(\mathbb{R}^2/\Gamma)$ obeys $\||b|^\beta \widehat{q}(b)\|_1 < \infty$. There are constants such that for every sufficiently large $d \in \Gamma^\# \setminus \{0\}$ there are maps*

$$\widehat{\phi}_{d,1}\colon \left\{(z_1,z_2)\in\mathbb{C}^2 \,\middle|\, |z_1| \leq \frac{\varepsilon}{2|d|^{1-\varepsilon}}, |z_2| \leq \frac{\varepsilon}{2|d|^{1-\varepsilon}}\right\} \to T_1(0)\cap T_2(d),$$
$$\widehat{\phi}_{d,2}\colon \left\{(z_1,z_2)\in\mathbb{C}^2 \,\middle|\, |z_1| \leq \frac{\varepsilon}{2|d|^{1-\varepsilon}}, |z_2| \leq \frac{\varepsilon}{2|d|^{1-\varepsilon}}\right\} \to T_1(-d)\cap T_2(0)$$

and a complex number $\hat{t}_d$ with $|\hat{t}_d| \leq \mathrm{const}/|d|^{2\beta+2}$ such that

(i) *$\widehat{\phi}_{d,\nu}$ is biholomorphic to its image. The image contains*

$$\left\{k\in\mathbb{C}^2 \,\middle|\, |k_1 + i(-1)^\nu k_2| \leq \frac{\varepsilon}{8|d|^{1-\varepsilon}},\right.$$
$$\left. |k_1 + (-1)^{\nu+1}d_1 - i(-1)^\nu(k_2 + (-1)^{\nu+1}d_2)| \leq \frac{\varepsilon}{8|d|^{1-\varepsilon}}\right\}.$$

Furthermore

$$D\widehat{\phi}_{d,\nu} = \frac{1}{2}\begin{pmatrix} 1 & 1 \\ -i(-1)^\nu & i(-1)^\nu \end{pmatrix}\left\{\mathbb{1} + O\left(\frac{1}{|d|^2}\right)\right\}$$

and

$$\phi_{d,\nu}(0) = \big((-1)^{\nu+1} i z_\nu(d), z_\nu(d)\big) + \widehat{q}(0)\left(\frac{-id_2}{d_1^2+d_2^2}, \frac{id_1}{d_1^2+d_2^2}\right) + \frac{\mathrm{const}}{|d|^{2-\varepsilon}}.$$

(ii)
$$\widehat{\phi}_{d,1}^{-1}\big(T_1(0)\cap T_2(d)\cap \widehat{\mathcal{F}}(q)\big) = \left\{(z_1,z_2)\in\mathbb{C}^2 \middle| z_1z_2 = \hat{t}_d, |z_1| \leq \frac{\varepsilon}{2|d|^{1-\varepsilon}}, |z_2| \leq \frac{\varepsilon}{2|d|^{1-\varepsilon}}\right\},$$
$$\widehat{\phi}_{d,2}^{-1}\big(T_1(-d)\cap T_2(0)\cap \widehat{\mathcal{F}}(q)\big) = \left\{(z_1,z_2)\in\mathbb{C}^2 \middle| z_1z_2 = \hat{t}_d, |z_1| \leq \frac{\varepsilon}{2|d|^{1-\varepsilon}}, |z_2| \leq \frac{\varepsilon}{2|d|^{1-\varepsilon}}\right\}.$$

(iii)
$$\widehat{\phi}_{d,1}(z_1,z_2) = \widehat{\phi}_{d,2}(z_2,z_1) - d.$$

By definition k is in $\widehat{\mathcal{F}}(q)$ if $H_k + q$ has a nontrivial kernel in $L^2(\mathbb{R}^2/\Gamma)$. To study the part of the curve in the intersection of $\cup_{d\in G} T_d$ and $\mathbb{C}^2 \setminus \cup_{b\notin G} T_b$ for some finite subset G of $\Gamma^\sharp$ it is natural to look for a nontrivial solution of

$$(H_k+q)\psi_G + (H_k+q)\psi_{G'} = 0$$

or equivalently of

$$(H_k+q)\phi_G + (\mathbb{1} + qH_k^{-1})\phi_{G'} = 0. \tag{17.4}$$

where

$$\psi_G, \phi_G \in L^2_G := \mathrm{span}\{e^{i\langle b,x\rangle} \mid b\in G\}^-,$$
$$\psi_{G'}, \phi_{G'} \in L^2_{G'} := \mathrm{span}\{e^{i\langle b,x\rangle} \mid b\in \Gamma^\sharp\setminus G\}^-.$$

We shall shortly show that, for k in the region under consideration, $R_{G'G'}$, the restriction of $\mathbb{1} + qH_k^{-1}$ to $L^2_{G'}$, has a bounded inverse. Then the projection of (17.4) on $L^2_{G'}$ is equivalent to

$$\phi_{G'} = -R_{G'G'}^{-1} q\phi_G.$$

Substituting this into the projection on L^2_G yields

$$\pi_G(H_k + q - qH_k^{-1}R_{G'G'}^{-1}q)\phi_G = 0.$$

Here π_G is the obvious projection operator. This has a nontrivial solution if and only if the $|G| \times |G|$ determinant

$$\det[\pi_G(H_k + q - qH_k^{-1}R_{G'G'}^{-1}q)\pi_G] = 0,$$

or equivalently, expressing all operators as matrices in the basis $\{e^{i\langle b,x\rangle} \mid b \in \Gamma^\sharp\}$,

$$\det\left[N_d(k)\delta_{d,d'} + \widehat{q}(d-d') - \sum_{b,c\in G'} \frac{\widehat{q}(d-b)}{N_b(k)}(R_{G'G'}^{-1})_{b,c}\widehat{q}(c-d')\right]_{d,d'\in G} = 0. \tag{17.5}$$

In general we define the operator R_{BC} to have matrix elements

$$(R_{BC})_{b,c} = \left[\delta_{b,c} + \frac{\widehat{q}(b-c)}{N_c(k)}\right]_{b\in B, c\in C}. \tag{17.6}$$

Our analysis of (17.5) is based on two Lemmas. The first gives of properties of $N_b(k)$. The second uses these to derive a number of properties of the operators R_{BC}.

LEMMA 17.3.

(a) *If* $|b + (u + v^\perp)| \geq \Lambda$ *and* $|b + (u - v^\perp)| \geq \Lambda$ *then*

$$(|v| + |u+b|)^2 \geq |N_b(k)| \geq \frac{\Lambda}{2}(|v| + |u+b|).$$

If $|v| > 2\Lambda$ *and* $k \in T_0$, *then* $|N_b(k)| \geq \Lambda/2(|v| + |u+b|)$ *for all but at most one* $b \neq 0$. *This exceptional* $\tilde{b}$ *obeys* $|\tilde{b}| \geq |v|$ *and* $\big||u+\tilde{b}| - |v|\big| < \Lambda$. *If* $|v| > 2\Lambda$ *and* $k \in T_0 \cap T_d$ *with* $d \neq 0$, *then* $|N_b(k)| \geq \Lambda/2(|v| + |u+b|)$ *for all* $b \neq 0, d$. *Furthermore* $|d| \geq |v|$.

(b) *There is a constant such that, for all* $k \in \mathbb{C}$,

$$\sum_{\substack{b\in\Gamma^\sharp \\ |b-(u\pm v^\perp)|\geq\Lambda}} \frac{1}{|N_b(k)|^2} \leq \text{const}\frac{\ln|v|}{|v|^2}.$$

(c) *If* $k \in T_0 \backslash (\bigcup_{b\in\Gamma^\#\backslash\{0\}})$ *with* $|v| > \max\{1, 2\Lambda\}$ *then there is a constant depending only on* ε *and* Λ *such that*

$$\left|\sum_{b\neq 0} \frac{\widehat{q}(-b)\widehat{q}(b)}{N_b(k)}\right| \leq \frac{\text{const}}{|v|^2}\big(|N_0(k)|\,\|\widehat{q}\|_2^2 + \|b\widehat{q}(b)\|_2^2\big).$$

If $k \in T_0 \cap T_d$ *with* $d \neq 0$ *and* $|v| > 2\Lambda$ *then there is a constant depending only on* Λ *such that for* $d' \in \{0, d\}$

$$\left|\sum_{b\neq 0,d} \frac{\widehat{q}(d'-b)\widehat{q}(b-d')}{N_b(k)}\right| \leq \frac{\text{const}}{|v|^2}\big(|N_{d'}(k)|\,\|\widehat{q}\|_2^2 + \|b\widehat{q}(b)\|_2^2\big).$$

PROOF OF (a). By hypothesis both factors in

$$|N_b(k)| = |v + (u+b)^\perp|\,|v - (u+b)^\perp|$$

are at least Λ. Suppose $|v| \geq |u+b|$. The other case is similar. Then, since $[v + (u+b)^\perp] + [v - (u+b)^\perp] = 2v$ at least one of the factors must be at least $|v| \geq \frac{1}{2}[|v| + |u+b|]$.

If $k \in T_0$, then by (17.3)(c)

$$|N_0(k)| \le \varepsilon(2|v| + \varepsilon) < 3\varepsilon|v| < \frac{\Lambda}{2}|v|$$

for $|v| > \varepsilon$ and $\varepsilon < \frac{1}{6}\Lambda$. Thus one of $|u \pm v^\perp| < \Lambda$. Suppose $|u + v^\perp| < \Lambda$. The other case is similar. There is no $b \in \Gamma^\# \setminus \{0\}$ obeying $|b + (u + v^\perp)| < \Lambda$ and there is at most one $\tilde{b} \in \Gamma^\# \setminus \{0\}$ obeying $|\tilde{b} + (u - v^\perp)| < \Lambda$. For this $\tilde{b}$

$$|\tilde{b}| = |2v^\perp - (u + v^\perp) + (\tilde{b} + u - v^\perp)| \ge 2|v| - 2\Lambda \ge |v|$$

for $|v| \ge 2\Lambda$.

If $k \in T_0 \cap T_d$, then d must be the exceptional $\tilde{b}$ of the last paragraph.

PROOF OF (b). The sum is bounded by

$$\begin{aligned}
\sum_{\substack{b \in \Gamma^\sharp \\ |N_b(k)| > \Lambda}} \frac{1}{|N_b(k)|^2} &\le \text{const} \int_{\substack{|x + (u+v^\perp)| \ge \Lambda \\ |x + (u - v^\perp)|^2| \ge \Lambda}} \frac{dx_1\, dx_2}{|x + (u + v^\perp)|^2 |x + (u - v^\perp)|^2} \\
&= \text{const} \int_{\substack{|x| \ge \Lambda \\ |x + (u - v^\perp)| \ge \Lambda}} \frac{dx_1\, dx_2}{|x|^2 |x + 2v^\perp|^2} \\
&\le \text{const} \frac{\ln |v|}{|v|^2}.
\end{aligned}$$

□

PROOF OF (c). By part (a) we have that, for all but one exceptional value of b, $|N_b(k)| \ge \Lambda/2|v|$. The exceptional $\tilde{b}$ obeys $|\tilde{b}| \ge |v|$ and $|N_{\tilde{b}}(k)| \ge \varepsilon/2$ by (17.3)(b). Hence

$$\begin{aligned}
\left| \sum_{b \ne 0} \frac{\widehat{q}(-b)\widehat{q}(b)}{N_b(k)} \right| &\le \left| \frac{\widehat{q}(-\tilde{b})\widehat{q}(\tilde{b})}{N_{\tilde{b}}(k)} \right| + \left| \frac{\widehat{q}(-\tilde{b})\widehat{q}(\tilde{b})}{N_{-\tilde{b}}(k)} \right| + \left| \sum_{b \ne 0, \pm\tilde{b}} \frac{\widehat{q}(-b)\widehat{q}(b)}{N_b(k)} \right| \\
&\le 2 \frac{2}{\varepsilon} \frac{1}{|v|^2} \|b\widehat{q}(b)\|_2^2 + \left| \frac{1}{2} \sum_{b \ne 0, \pm\tilde{b}} \widehat{q}(-b)\widehat{q}(b) \left(\frac{1}{N_b(k)} + \frac{1}{N_{-b}(k)} \right) \right| \\
&= \frac{4}{\varepsilon |v|^2} \|b\widehat{q}(b)\|_2^2 + \left| \sum_{b \ne 0, \pm\tilde{b}} \widehat{q}(-b)\widehat{q}(b) \frac{N_0(k) + b^2}{N_b(k) N_{-b}(k)} \right| \\
&\le \frac{4}{\varepsilon |v|^2} \|b\widehat{q}(b)\|_2^2 + \frac{4}{\Lambda^2 |v|^2} \big(|N_0(k)|\, \|\widehat{q}\|_2^2 + \|b\widehat{q}(b)\|_2^2 \big).
\end{aligned}$$

When $k \in T_0 \cap T_d$ and $d' \in \{0, d\}$

$$\begin{aligned}
&\left| \sum_{b \ne 0, d} \frac{\widehat{q}(d' - b)\widehat{q}(b - d')}{N_b(k)} \right| \\
&\qquad = \left| \sum_{b \ne -d', d - d'} \frac{\widehat{q}(-b)\widehat{q}(b)}{N_{b+d'}(k)} \right| \\
&\qquad \le \left| \frac{\widehat{q}(-d)\widehat{q}(d)}{N_{-d}(k)\delta_{d',0} + N_{2d}(k)\delta_{d',d}} \right| + \left| \sum_{b \ne 0, \pm d} \frac{\widehat{q}(-b)\widehat{q}(b)}{N_{b+d'}(k)} \right| \\
&\qquad \le \frac{2}{\Lambda} \frac{1}{|v|^3} \|b\widehat{q}(b)\|_2^2 + \left| \frac{1}{2} \sum_{b \ne 0, \pm d} \widehat{q}(-b)\widehat{q}(b) \left(\frac{1}{N_{b+d'}(k)} + \frac{1}{N_{-b+d'}(k)} \right) \right|
\end{aligned}$$

$$= \frac{2}{\Lambda|v|^3}\|b\widehat{q}(b)\|_2^2 + \left|\sum_{b\neq 0,\pm d} \widehat{q}(-b)\widehat{q}(b)\frac{N_{d'}(k)+b^2}{N_{b+d'}(k)N_{-b+d'}(k)}\right|$$

$$\leq \frac{2}{\Lambda|v|^3}\|b\widehat{q}(b)\|_2^2 + \frac{4}{\Lambda^2|v|^2}\left(|N_{d'}(k)|\,\|\widehat{q}\|_2^2 + \|b\widehat{q}(b)\|_2^2\right). \qquad \square$$

REMARK. Note that if $k \in T_d$ then

$$\big||v| - |u+d|\big| < \frac{\varepsilon}{1+|v|^{1-\varepsilon}}$$

so that

$$|k_2 + d_2| \leq |u+d| + |v| \leq \varepsilon + 2|v| \tag{17.7}$$

LEMMA 17.4. *Let $k \in \mathbb{C}^2$. Let $|v| > 4\Lambda$ and*

$$S \subset \left\{ b \in \Gamma^\sharp \;\middle|\; \frac{\varepsilon|v|}{1+|v|^{1-\varepsilon}} \leq |N_b(k)| < \frac{\Lambda}{2}|v| \right\}, \quad B \subset \left\{ b \in \Gamma^\sharp \;\middle|\; |N_b(k)| \geq \frac{\Lambda}{2}|v| \right\}$$

(a)
$$\begin{aligned}
\|R_{SS} - \pi_S\| &\leq \frac{1}{\varepsilon|v|^\varepsilon}\|\widehat{q}(b)\|_1, \\
\|R_{BB} - \pi_B\| &\leq \frac{2\|\widehat{q}(b)\|_1}{\Lambda|v|}, \\
\|R_{SB}\| &\leq \frac{2\|\widehat{q}(b)\|_1}{\Lambda|v|}, \\
\|R_{BS}\| &\leq \frac{1}{\varepsilon|v|^\varepsilon}\|\widehat{q}(b)\|_1.
\end{aligned}$$

(b)
$$\begin{aligned}
\|R_{SS} - \pi_S\|_{HS} &\leq \text{const}\frac{1}{\varepsilon|v|^\varepsilon}\|\widehat{q}(b)\|_2, \\
\|R_{BB} - \pi_B\|_{HS} &\leq \text{const}\|\widehat{q}(b)\|_2\frac{\sqrt{\ln|v|}}{|v|}, \\
\|R_{SB}\|_{HS} &\leq \text{const}\|\widehat{q}(b)\|_2\frac{\sqrt{\ln|v|}}{|v|}, \\
\|R_{BS}\|_{HS} &\leq \text{const}\frac{1}{\varepsilon|v|^\varepsilon}\|\widehat{q}(b)\|_2.
\end{aligned}$$

(c) *Let $|v| \geq \max\{4\|\widehat{q}(b)\|_1/\Lambda, (4\|\widehat{q}(b)\|_1/\varepsilon)^{1/\varepsilon}\}$. The operator*

$$\begin{pmatrix} R_{SS} & R_{SB} \\ R_{BS} & R_{BB} \end{pmatrix}$$

has a bounded inverse. The norm

$$\left\| \begin{pmatrix} R_{SS} & R_{SB} \\ R_{BS} & R_{BB} \end{pmatrix}^{-1} - \begin{pmatrix} \pi_S & 0 \\ 0 & \pi_B \end{pmatrix} \right\| \leq \text{const}\frac{\|\widehat{q}(b)\|_1}{\varepsilon|v|^\varepsilon}.$$

(d) *Let $|v| \geq$ const. Then*

$$\left|\det_2 \begin{pmatrix} R_{SS} & R_{SB} \\ R_{BS} & R_{BB} \end{pmatrix} - 1\right| \leq \text{const}\frac{\|\widehat{q}(b)\|_2}{\varepsilon|v|^\varepsilon}.$$

PROOF OF (a). It suffices to observe that,

$$b \in S \implies |N_b(k)|^{-1} \le \frac{1}{\varepsilon |v|^\varepsilon}, \qquad b \in B \implies |N_b(k)|^{-1} \le \frac{2}{\Lambda |v|}$$

and that the convolution operator $|\widehat{q}(b-c)|$ has operator norm bounded by $\|\widehat{q}\|_1$.

PROOF OF (b). The two cases with "second argument B" may be treated at the same time.

$$\begin{aligned}
\|R_{BB} - \pi_B\|_{HS}^2, \|R_{SB}\|_{HS}^2 &\le \sum_{b\in S\cup B, c\in B} \left| \frac{\widehat{q}(b-c)}{N_c(k)} \right|^2 \\
&\le \|\widehat{q}\|_2^2 \sum_{c\in B} \frac{1}{|N_c(k)|^2}, \\
&\le \text{const} \|\widehat{q}\|_2^2 \frac{\ln |v|}{|v|^2}
\end{aligned}$$

by Lemma 17.3(b). The other two cases are similar, but

$$\sum_{c\in S} \frac{1}{|N_c(k)|^2} \le \frac{2}{\varepsilon^2 |v|^{2\varepsilon}}$$

is used in place of Lemma 17.3(b).

PROOF OF (c). Write the matrix

$$\begin{pmatrix} R_{SS} & R_{SB} \\ R_{BS} & R_{BB} \end{pmatrix} = \mathbb{1} + E,$$

where

$$E = \begin{pmatrix} R_{SS} - \pi_S & R_{SB} \\ R_{BS} & R_{BB} - \pi_B \end{pmatrix}$$

and expand $(\mathbb{1} + E)^{-1}$ as a geometric series in E. □

PROOF OF (d). Again write the matrix

$$\begin{pmatrix} R_{SS} & R_{SB} \\ R_{BS} & R_{BB} \end{pmatrix} = \mathbb{1} + E.$$

Then,

$$\begin{aligned}
\det{}_2 \begin{pmatrix} R_{SS} & R_{SB} \\ R_{BS} & R_{BB} \end{pmatrix} &= 1 + O(\|E\|_{HS}) \\
&= 1 + O(\|R_{SS} - \pi_S\|_{HS} + \|R_{SB}\|_{HS} + \|R_{BS}\|_{HS} \\
&\qquad\qquad + \|R_{BB} - \pi_B\|_{HS}) \\
&= 1 + O\left(\frac{\|\widehat{q}(b)\|_2}{\varepsilon |v|^\varepsilon} \right).
\end{aligned}$$

□

It is now a simple matter to use the estimates of Lemmas 17.3, 17.4 to analyse the asymptotic behavior of the Fermi curve. Define

(17.8a) $$R = \max \left\{ 1, 2\Lambda, \frac{4\|\widehat{q}(b)\|_1}{\Lambda}, \left(\frac{4\|\widehat{q}(b)\|_1}{\varepsilon} \right)^{1/\varepsilon} \right\}$$

and recall that

(17.8b) $$\mathcal{K}_R = \{ k \in \mathbb{C}^2 \mid |v| \le R \}.$$

PROOF OF THEOREM 17.1(a). Let $k \in \mathbb{C}^2 \setminus (\mathcal{K}_R \cup \bigcup_{b \in \Gamma^\#} T_b)$. Use (17.4) with $G = \varnothing$ as a test for when $k \in \widehat{\mathcal{F}}(q)$. By Lemma 17.4(c) with

$$S = \left\{ b \in \Gamma^\sharp \;\middle|\; \frac{\varepsilon|v|}{1+|v|^{1-\varepsilon}} \le |N_b(k)| < \frac{\Lambda}{2}|v| \right\}, \quad B = \left\{ b \in \Gamma^\sharp \;\middle|\; |N_b(k)| \ge \frac{\Lambda}{2}|v| \right\}$$

$\phi_{G'}$ must be zero. That is, there is no nontrivial solution of (17.4). □

To prove the rest of Theorem 17.1 and Theorem 17.2 we use

PROPOSITION 17.5. *Let $k \in \mathbb{C}^2 \setminus \mathcal{K}_R$.*

(a) *Let $k \in T_0 \setminus \cup_{b \ne 0} T_b$. Then $k \in \widehat{\mathcal{F}}(q)$ if and only if*

$$N_0(k) = \mathcal{A}(k)$$

where

$$\mathcal{A}(k) = -\widehat{q}(0) + \sum_{b,c \in A} \frac{\widehat{q}(-b)}{N_b(k)} (R_{AA}^{-1})_{b,c} \widehat{q}(c), \quad A = \Gamma^\sharp \setminus \{0\}$$

and obeys

$$|\mathcal{A}(k) + \widehat{q}(0)| \le \frac{\mathrm{const}(\|\hat{b}q(b)\|_1)}{1+|k_2|^2} \big(1 + |N_0(k)|\big).$$

Here const$\big(\|b\widehat{q}(b)\|_1\big)$ *denotes that the constant* const *depends only on ε and the norm $\|b\widehat{q}(b)\|_1$.*

(b) *Let $k \in T_0 \cap T_d$. Then $k \in \widehat{\mathcal{F}}(q)$ if and only if*

$$\big(N_0(k) + \widehat{q}(0)\mathcal{D}(k)_{1,1}\big)\big(N_d(k) + \widehat{q}(0)\mathcal{D}(k)_{2,2}\big) = \big(\widehat{q}(-d)\mathcal{D}(k)_{1,2}\big)\big(\widehat{q}(d)\mathcal{D}(k)_{2,1}\big)$$

where $d^{(1)} = 0$, $d^{(2)} = d$ and

$$\mathcal{D}(k)_{i,j} = \sum_{b,c \in D} \frac{\widehat{q}(d^{(i)} - b)}{N_b(k)} (R_{DD}^{-1})_{b,c} \widehat{q}(c - d^{(j)}), \quad D = \Gamma^\sharp \setminus \{0, d\}$$

and obeys

$$|\mathcal{D}(k)_{i,j}| \le \min_{\ell=1,2} \frac{\mathrm{const}(\|b\widehat{q}(b)\|_1)}{1 + |k_2 - d_2^{(\ell)}|^2} \big(1 + |N_{d^{(j)}}(k)|\delta_{i,j}\big).$$

PROOF OF (a). For the region in question $\widehat{\mathcal{F}}(q)$ is given by (17.5) with $G = \{0\}$ and $G' = A$. This is precisely the desired equation. We now estimate $\mathcal{A}(k) = \widehat{q}(0)$ using Lemma 17.4 with

$$S = A \cap \left\{ b \in \Gamma^\sharp \;\middle|\; \frac{\varepsilon|v|}{1+|v|^{1-\varepsilon}} \le |N_b(k)| < \frac{\Lambda}{2}|v| \right\},$$

$$B = A \cap \left\{ b \in \Gamma^\sharp \;\middle|\; |N_b(k)| \ge \frac{\Lambda}{2}|v| \right\}.$$

By Lemma 17.3(a) S contains at most one element $\tilde{b}$ and it must obey $|\tilde{b}| \ge |v|$. Write

$$R_{AA}^{-1} = \mathbb{1} + R_{AA}^{-1}(\mathbb{1} - R_{AA}).$$

The contribution from $\mathbb{1}$ is

$$\sum_{b \in A} \frac{\widehat{q}(a-b)\widehat{q}(b-a)}{N_b(k)}$$

and is estimated by Lemma 17.3(c). The other contribution is

$$\begin{aligned}
&\left| \sum_{b,c,c' \in A} \frac{\widehat{q}(-b)}{N_b(k)} (R_{AA}^{-1})_{b,c} (\mathbb{1} - R_{AA})_{c,c'} \widehat{q}(c') \right| \\
&\quad \le \left\| \frac{\widehat{q}(-b)}{N_b(k)} \right\| \left\| R_{AA}^{-1} \right\| \|(\mathbb{1} - R_{AA})\widehat{q}\| \\
&\quad \le \text{const} \left\| \frac{\widehat{q}(-b)}{N_b(k)} \right\|_2 \left(\|(\pi_B - R_{BB} - R_{SB})\| \, \|\widehat{q}\|_2 + \|(\pi_S - R_{SS} - R_{BS})\| \, |\widehat{q}(\tilde{b})|_2 \right) \\
&\quad \le \text{const} \left\| \frac{\widehat{q}(-b)}{N_b(k)} \right\|_2 \left(\frac{4\|\widehat{q}\|_1}{\Lambda|v|} \|\widehat{q}\|_2 + \frac{1}{2} |\widehat{q}(\tilde{b})| \right) \qquad \text{by Lemma 17.4(a)} \\
&\quad \le \text{const} \left(\frac{2}{\Lambda|v|} \|\widehat{q}\|_2 + \left| \frac{\widehat{q}(-\tilde{b})}{N_{\tilde{b}}(k)} \right| \right) \left(\frac{4\|\widehat{q}\|_1}{\Lambda|v|} \|\widehat{q}\|_2 + \frac{1}{2} |\widehat{q}(\tilde{b})| \right) \\
&\quad \le \text{const} \left(\frac{2}{\Lambda|v|} \|\widehat{q}\|_2 + \frac{2}{\varepsilon|v|} \|b\widehat{q}(b)\|_1 \right) \left(\frac{4\|\widehat{q}\|_1}{\Lambda|v|} \|\widehat{q}\|_2 + \frac{1}{2|v|} \|b\widehat{q}(b)\|_1 \right).
\end{aligned}$$

It now suffices ot apply (17.7).

Proof of (b). In this case the two exceptional b's of Lemma 17.3(a) must be 0 and d so that

$$S = D \cap \left\{ b \in \Gamma^\# \;\middle|\; \frac{\varepsilon|v|}{1 + |v|^{1-\varepsilon}} \le |N_b(k)| < \frac{\Lambda}{2}|v| \right\} = \varnothing,$$

$$B = D \cap \left\{ b \in \Gamma^\# \;\middle|\; |N_b(k)| \ge \frac{\Lambda}{2}|v| \right\} = D.$$

As $|N_b(k)|^{-1} \le 2/\Lambda|v|$ for all $b \in D$ and, by Lemma 17.4(a), $\|R_{BB}^{-1} - \mathbb{1}\| \le \text{const}/|v|$ it suffices to bound

$$\sum_{b \in D} \frac{\widehat{q}(d^{(i)} - b)\widehat{q}(b - d^{(j)})}{N_b(k)}.$$

The cases

$$\sum_{b \in D} \frac{\widehat{q}(-b)\widehat{q}(b)}{N_b(k)} \quad \text{and} \quad \sum_{b \in D} \frac{\widehat{q}(d - b)\widehat{q}(b - d)}{N_b(k)}$$

are bounded in Lemma 17.3(c). Finally, let $i \ne j$. We have

$$\begin{aligned}
\sum_{b \in D} \frac{\widehat{q}(d^{(i)} - b)\widehat{q}(b - d^{(j)})}{N_b(k)} &\le \sum_b 2 \frac{|\widehat{q}(d^{(i)} - b)\widehat{q}(b - d^{(j)}|)}{\Lambda|v|} \\
&\le \text{const} \sum_b \frac{\{|d^{(i)} - b| + |b - d^{(j)}|\} \widehat{q}(d^{(i)} - b)\widehat{q}(b - d^{(j)})}{|d^{(1)} - d^{(2)}||v|} \\
&\le \text{const} \frac{2\|b\widehat{q}(b)\|_2 \|\widehat{q}(b)\|_2}{|v|^2}. \qquad \square
\end{aligned}$$

To show that the curve does not wiggle too much we will also need bounds on the derivatives of $\mathcal{A}$ and $\mathcal{D}$. These are provided in

LEMMA 17.6. *Under the hypotheses of Proposition* 17.5,

$$\left|\frac{\partial^{n+m}}{\partial k_1^n \partial k_2^m}\mathcal{A}(k)\right| \leq \frac{\operatorname{const}(\|b\widehat{q}(b)\|_1)}{[1+|k_2|]} \qquad \textit{if } m+n=1,$$

$$\left|\frac{\partial^{n+m}}{\partial k_1^n \partial k_2^m}\mathcal{D}(k)_{i,j}\right| \leq \min_{i=1,2} \frac{\operatorname{const}(\|\widehat{q}(b)\|_1)}{[1+|k_2+d_2^{(i)}|]} \qquad \textit{if } m+n\geq 1.$$

PROOF. Use $Q_{GG'}$ to denote the matrix $[\widehat{q}(b-c)]_{b\in G, c\in G'}$. Then $\mathcal{A}+\widehat{q}(0)$ and $\mathcal{D}$ are given by $Q_{GG'}H_k^{-1}(R_{G'G'})^{-1}Q_{G'G}$ with $G=\{0\}$ and $G=\{0,d\}$ respectively. Hence their first derivatives are given by

$$\begin{aligned}\frac{\partial}{\partial k_i}\mathcal{A}(k), \frac{\partial}{\partial k_i}\mathcal{D}(k) &= -Q_{GG'}H_k^{-1}\frac{\partial H_k}{\partial k_i}H_k^{-1}(R_{G'G'})^{-1}Q_{G'G} \\ &\quad + Q_{GG'}H_k^{-1}(R_{G'G'})^{-1}Q_{G'G'}H_k^{-1}\frac{\partial H_k}{\partial k_i}H_k^{-1}(R_{G'G'})^{-1}Q_{G'G} \\ &= -Q_{GG'}H_k^{-1}(R_{G'G'})^{-1}\frac{\partial H_k}{\partial k_i}H_k^{-1}(R_{G'G'})^{-1}Q_{G'G}\end{aligned}$$

since

$$Q_{G'G'}H_k^{-1} = R_{G'G'} - \mathbb{1}.$$

All higher derivatives are given by finite linear combinations of terms of the form

$$Q_{GG'}\prod_j\left\{H_k^{-1}(R_{G'G'})^{-1}\frac{\partial^{n_j}H_k}{\partial k_{i_j}^{n_j}}\right\}H_k^{-1}(R_{G'G'})^{-1}Q_{G'G} \tag{17.9}$$

where the sum of the n_j's for which i_j =1 (resp. 2) is n (resp. m). This is easily seen if $\mathcal{A}$ and $\mathcal{D}$ are written in the form

$$Q_{GG'}(H_k+Q)^{-1}Q_{G'G}$$

with the operator H_k+Q being restricted to $L^2_{G'}$.

The first step in bounding these derivatives is to compute

$$\left(\frac{\partial^{n_j}H_k}{\partial k_{i_j}^{n_j}}H_k^{-1}\right)_{b,c} = \delta_{b,c}\frac{1}{N_b(k)}\begin{cases}2(k_{i_j}+b_{i_j}) & \text{if } n_j=1\\ 2 & \text{if } n_j=2\\ 0 & \text{if } n_j\geq 3\end{cases}.$$

Since $|k_i+b_i|\leq|u_i+b_i|+|v_i|\leq|v|+|u+b|$, Lemma 17.3(a) implies

$$\left\|\frac{\partial^{n_j}H_k}{\partial k_{i_j}^{n_j}}H_k^{-1}\pi_{A\setminus\{\tilde{b}\}}\right\|, \quad \left\|\frac{\partial^{n_j}H_k}{\partial k_{i_j}^{n_j}}H_k^{-1}\pi_D\right\| \leq \frac{4}{\Lambda}.$$

As

$$\|(R_{G'G'})^{-1}\|\leq \operatorname{const}, \quad \|Q_{GG'}\|,\ \|Q_{G'G}\|\leq\|\widehat{q}(b)\|_1,$$
$$\|H_k^{-1}\pi_{A\setminus\{\tilde{b}\}}\|,\ \|H_k^{-1}\pi_D\|\leq\frac{2}{\Lambda|v|}$$

the bound on $\mathcal{D}$ follows and the bound on $\mathcal{A}$ is reduced to terms containing π_S.

We now return to $\mathcal{A}$ with $m+n=1$. To bound

$$Q_{GG'}H_k^{-1}(R_{G'G'})^{-1}\frac{\partial H_k}{\partial k_i}H_k^{-1}(R_{G'G'})^{-1}Q_{G'G}$$

we observe that

$$\begin{aligned}
\|Q_{GG'}H_k^{-1}\| &\le \|Q_{GS}\pi_S H_k^{-1}\| + \|Q_{GB}\pi_B H_k^{-1}\| \\
&\le \frac{\|b\widehat{q}(b)\|_1}{|v|}\frac{1}{\varepsilon|v|^\varepsilon} + \|\widehat{q}(b)\|_1\frac{2}{\Lambda|v|} \\
\|(R_{G'G'})^{-1}\| &\le \text{const} \\
\left\|\frac{\partial H_k}{\partial k_i}H_k^{-1}\pi_B(R_{G'G'})^{-1}Q_{G'G}\right\| &\le \frac{2}{\Lambda}\text{const}\|\widehat{q}(b)\|_1 \\
\left\|\frac{\partial H_k}{\partial k_i}H_k^{-1}\pi_S(R_{G'G'})^{-1}Q_{G'G}\right\| &\le \left\|\frac{\partial H_k}{\partial k_i}\pi_S\right\| \|H_k^{-1}\pi_S\| \|\pi_S(R_{G'G'})^{-1}Q_{G'G}\| \\
&\le 2|v|\frac{1}{\varepsilon|v|^\varepsilon}\|\pi_S(R_{G'G'})^{-1}Q_{G'G}\| \\
\|\pi_S(R_{G'G'})^{-1}Q_{G'G}\| &\le \|\pi_S(R_{G'G'})^{-1}\pi_B\| \|\widehat{q}(b)\|_1 \\
&\qquad + \|\pi_S(R_{G'G'})^{-1}\pi_S\|\frac{\|b\widehat{q}(b)\|_1}{|v|} \\
\|\pi_S(R_{G'G'})^{-1}\pi_B\| &\le \sum_{\ell=1}^{\infty}\|\pi_S(\mathbb{1}-R_{G'G'})^\ell\pi_B\| \\
&\le \sum_{\ell=1}^{\infty}\frac{1}{2^{\ell-1}}\frac{4\|\widehat{q}(b)\|_1}{\Lambda|v|} \qquad \text{by Lemma 17.4(a)} \\
&\le \frac{4\|\widehat{q}(b)\|_1}{\Lambda|v|}.
\end{aligned}$$

□

PROOF OF THEOREM 17.1(b). We give the proof for the case $\nu = 1$. Recall that $k \in T_1(0)$ if and only if

$$|k_1 - ik_2| = |u + v^\perp| < \frac{\varepsilon}{1+|v|^{1-\varepsilon}}.$$

Since $k_2 = u_2 + iv_2 = v_1 + iv_2 + u_2 - v_1 = v_1 + iv_2 + (u+v^\perp)_2$ we have that in $T_1(0)$

$$\big||v| - |k_2|\big| \le \frac{\varepsilon}{1+|v|^{1-\varepsilon}} \le \varepsilon$$

so that

$$\frac{\varepsilon}{1+(|k_2|+\varepsilon)^{1-\varepsilon}} \le \frac{\varepsilon}{1+|v|^{1-\varepsilon}} \le \begin{cases} \frac{\varepsilon}{1+(|k_2|-\varepsilon)^{1-\varepsilon}} & \text{if } |k_2| \ge \varepsilon \\ \varepsilon & \text{if } |k_2| \le \varepsilon \end{cases}.$$

If $k \in T_1(0) \cap T_2(d)$ then

$$|k_1 - ik_2| < \frac{\varepsilon}{1+|v|^{1-\varepsilon}} \quad \text{and} \quad |k_2 + \tfrac{1}{2}(d_2 - id_1)| < \frac{2\varepsilon}{1+|v|^{1-\varepsilon}}.$$

Conversely, if

$$|k_1 - ik_2| < \frac{\varepsilon/2}{1+|v|^{1-\varepsilon}} \quad \text{and} \quad |k_2 + \tfrac{1}{2}(d_2 - id_1)| < \frac{\varepsilon/2}{1+|v|^{1-\varepsilon}}$$

then $k \in T_1(0) \cap T_2(d)$.

For $z \in \mathbb{C}$ put

$$F(z) = \mathrm{pr}^{-1}(z) \cap \left(T_1(0) \setminus \bigcup_{\substack{b\in\Gamma^\# \\ b\neq 0}} T_b \right) \cap \{(k_1, z) \in \mathbb{C}^2 \mid |k_1 - iz| < \varepsilon\}.$$

Observe that if $|z| > 2\varepsilon$ and $|z + \frac{1}{2}(b_2 - ib_1)| \geq 2\varepsilon/(1+|v|^{1-\varepsilon})$ for all $b \in \Gamma^\#$, $b \neq 0$ then

$$F(z) = \left\{ (k_1, z) \in \mathbb{C}^2 \;\middle|\; |k_1 - iz| < \frac{\varepsilon}{1+|v|^{1-\varepsilon}} \right\}$$

while if $|z| > 2\varepsilon$ and $|z + \frac{1}{2}(d_2 - id_1)| < 2\varepsilon/(1+|v|^{1-\varepsilon})$ for some $d \in \Gamma^\# \setminus \{0\}$ then

$$F(z) = \left\{ (k_1, z) \in \mathbb{C}^2 \;\middle|\; |k_1 - iz| < \frac{\varepsilon}{1+|v|^{1-\varepsilon}} \text{ and } |k_1 + d_1 + i(z+d_2)| \geq \frac{\varepsilon}{1+|v|^{1-\varepsilon}} \right\}.$$

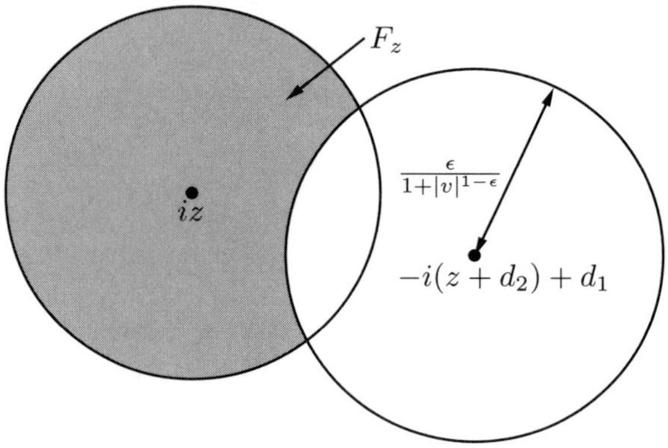

By Proposition 17.5 (k_1, z) lies in $\widehat{\mathcal{F}}(q)$ if and only if

$$k_1^2 + z^2 = N_0(k_1, z) = \mathcal{A}(k_1, z)$$

for an analytic function $\mathcal{A}$ obeying

$$|\mathcal{A}(k_1, z) + \widehat{q}(0)| \leq \frac{\text{const}}{1+|z|^2}\big(1 + |N_0(k_1, z)|\big) \leq \frac{\text{const}}{1+|z|^2}|z|^\varepsilon,$$

$$\left| \frac{\partial}{\partial k_1} \mathcal{A}(k_1, z) \right| \leq \frac{\text{const}}{1+|z|}$$

by (17.3c) and Lemma 17.6. For $(k_1, z) \in F(z)$ with $|z| > 2\varepsilon$

$$k_1 - iz = \frac{1}{k_1 + iz} \mathcal{A}(k_1, z) \tag{17.10}$$

with

$$\left| \frac{1}{k_1 + iz} \mathcal{A}(k_1, z) \right| \leq \text{const} \frac{\widehat{q}(0)}{1+|z|} + \frac{\text{const}}{1+|z|^2}(1 + |k_1 - iz|),$$

$$\left| \frac{\partial}{\partial k_1} \left(\frac{1}{k_1 + iz} \mathcal{A}(k_1, z) \right) \right| \leq \frac{\text{const}}{1+|z|^2}.$$

This shows, that for z big enough, the equation (17.10) has at most one solution in $F(z)$, and that this solution is simple. Furthermore any such solution fulfils

$$|k_1 - iz| \le \frac{\text{const}}{1+|z|}.$$

So there is no solution in $F(z)$, if for some $d \in \Gamma^\# \setminus \{0\}$

$$\left\{k_1 \in \mathbb{C} \;\middle|\; |k_1 - iz| \le \frac{\text{const}}{1+|z|}\right\} \subset \left\{k_1 \in \mathbb{C} \;\middle|\; \left|k_1 - iz + 2i\big(z - z_1(d)\big)\right| \le \frac{\varepsilon}{1+|v|^{1-\varepsilon}}\right\}$$

that is, if

$$2|z - z_1(d)| + \frac{\text{const}}{1+|z|} \le \frac{\varepsilon}{1+|v|^{1-\varepsilon}}$$

which is certainly the case if

$$|z - z_1(d)| \le \frac{1}{2}\left[\frac{\varepsilon}{1+(|z|+\varepsilon)^{1-\varepsilon}} - \frac{\text{const}}{1+|z|}\right] \approx \frac{1}{2}\frac{\varepsilon}{(1/2|d|)^{1-\varepsilon}}.$$

Similarly equation (17.10) has a solution in $F(z)$, if for all $d \in \Gamma^\# \setminus \{0\}$

$$2|z - z_1(d)| > \frac{\varepsilon}{1+|v|^{1-\varepsilon}} + \frac{\text{const}}{1+|z|}.$$

□

PROOF OF THEOREM 17.2. We first construct the maps $\widehat{\phi}_{d,1}$. For notational simplicity define

$$\begin{aligned} X_1 &= k_1 - ik_2, & Y_1 &= k_1 + ik_2, \\ X_2 &= k_1 + d_1 + i(k_2 + d_2), & Y_2 &= k_1 + d_1 - i(k_2 + d_2) \end{aligned}$$

and observe that, for $|k_1 - ik_2| \le \varepsilon/1 + |v|^{1-\varepsilon}$, $|k_1 + d_1 + i(k_2 + d_2)| \le \varepsilon/1 + |v|^{1-\varepsilon}$

$$\begin{aligned} X_1 &= O\Big(\frac{1}{|d|^{1-\varepsilon}}\Big), \quad Y_1 = 2iz_1(d) + O\Big(\frac{1}{|d|^{1-\varepsilon}}\Big), \\ X_2 &= O\Big(\frac{1}{|d|^{1-\varepsilon}}\Big), \quad Y_2 = -2iz_2(d) + O\Big(\frac{1}{|d|^{1-\varepsilon}}\Big). \end{aligned}$$

First we perform the coordinate change

$$\begin{aligned} x_1 &= X_1 + \frac{\widehat{q}(0)}{Y_1} - \frac{1}{Y_1}\mathcal{D}_{1,1}, \\ x_2 &= X_2 + \frac{\widehat{q}(0)}{Y_2} - \frac{1}{Y_2}\mathcal{D}_{2,2}, \\ k_1 &= \tfrac{1}{2}(X_1 + X_2) + iz_1(d) \\ &= \frac{x_1 + x_2}{2} + iz_1(d) - \widehat{q}(0)\Big(\frac{1}{2Y_1} + \frac{1}{2Y_2}\Big) + \frac{\mathcal{D}_{1,1}}{2Y_1} + \frac{\mathcal{D}_{2,2}}{2Y_2}, \\ k_2 &= \frac{i}{2}(X_1 - X_2) + z_1(d) \\ &= \frac{i}{2}(x_1 - x_2) + z_1(d) - \frac{i}{2}\Big(\frac{\widehat{q}(0)}{Y_1} - \frac{\widehat{q}(0)}{Y_2} - \frac{\mathcal{D}_{1,1}}{Y_1} + \frac{\mathcal{D}_{2,2}}{Y_2}\Big) \end{aligned} \tag{17.11}$$

where $\mathcal{D}(k)_{i,j}$ are given by Proposition 17.5. The Jacobean of this map is

$$(17.12\text{a})\qquad \begin{pmatrix} \frac{\partial x_1}{\partial k_1} & \frac{\partial x_1}{\partial k_2} \\ \frac{\partial x_2}{\partial k_1} & \frac{\partial x_2}{\partial k_2} \end{pmatrix} = \begin{pmatrix} 1+O(\frac{1}{|d|^2}) & -i+O(\frac{1}{|d|^2}) \\ 1+O(\frac{1}{|d|^2}) & i+O(\frac{1}{|d|^2}) \end{pmatrix} = \begin{pmatrix} 1 & -i \\ 1 & i \end{pmatrix} \left(\mathbb{1}+O\left(\frac{1}{|d|^2}\right)\right).$$

Its inverse is

$$(17.12\text{b})\qquad \begin{pmatrix} \frac{\partial k_1}{\partial x_1} & \frac{\partial k_1}{\partial x_2} \\ \frac{\partial k_2}{\partial x_1} & \frac{\partial k_2}{\partial x_2} \end{pmatrix} = \frac{1}{2}\begin{pmatrix} 1 & 1 \\ i & -i \end{pmatrix} \left(\mathbb{1}+O\left(\frac{1}{|d|^2}\right)\right).$$

The derivative of the inverse

$$\frac{\partial^2 k_m}{\partial x_i \partial x_j} = -\sum_{\alpha,\beta} \frac{\partial k_m}{\partial x_\alpha}\frac{\partial}{\partial x_i}\left(\frac{\partial x_\alpha}{\partial k_\beta}\right)\frac{\partial k_\beta}{\partial x_j} = -\sum_{\alpha,\beta,\gamma} \frac{\partial k_m}{\partial x_\alpha}\frac{\partial^2 x_\alpha}{\partial k_\gamma \partial k_\beta}\frac{\partial k_\gamma}{\partial x_i}\frac{\partial k_\beta}{\partial x_j} = O\left(\frac{1}{|d|^2}\right).$$

In these coordinates

$$\frac{1}{Y_1Y_2}\left[\left(N_0+\widehat{q}(0)\mathcal{D}_{1,1}\right)\left(N_d+\widehat{q}(0)\mathcal{D}_{2,2}\right)-\left(\widehat{q}(d)\mathcal{D}_{1,2}\right)\left(\widehat{q}(-d)\mathcal{D}_{2,1}\right)\right] = x_1x_2+h(x_1,x_2)$$

where

$$h(x_1,x_2) = -\frac{1}{Y_1Y_2}\left(\widehat{q}(d)-\mathcal{D}_{1,2}\right)\left(\widehat{q}(-d)-\mathcal{D}_{2,1}\right).$$

In Lemma 17.8 below we will improve the estimates of Lemma 17.6 to

$$\left|\frac{\partial^{n+m}}{\partial k_1^n \partial k_2^m}\mathcal{D}(k)_{1,2}\right|,\quad \left|\frac{\partial^{n+m}}{\partial k_1^n \partial k_2^m}\mathcal{D}(k)_{2,1}\right| \le \frac{\text{const}}{|d|^{\beta+1}}.$$

In terms of the x-variables, when $r \neq s$

$$\begin{aligned}\left|\frac{\partial \mathcal{D}_{r,s}}{\partial x_i}(x_1,x_2)\right| &\le \sum_{m=1,2}\left|\frac{\partial \mathcal{D}_{r,s}}{\partial k_m}\frac{\partial k_m}{\partial x_i}\right| \\ &\le \frac{\text{const}}{|d|^{\beta+1}}\end{aligned}$$

and

$$\begin{aligned}\left|\frac{\partial^2 \mathcal{D}_{r,s}}{\partial x_i \partial x_j}(x_1,x_2)\right| &\le \sum_{m,n=1,2}\left|\frac{\partial^2 \mathcal{D}_{r,s}}{\partial k_m \partial k_n}\frac{\partial k_m}{\partial x_i}\frac{\partial k_n}{\partial x_j}\right| + \sum_{m=1,2}\left|\frac{\partial \mathcal{D}_{r,s}}{\partial k_m}\frac{\partial^2 k_m}{\partial x_i \partial x_j}\right| \\ &\le \frac{\text{const}}{|d|^{\beta+1}}\end{aligned}$$

so that,

$$\begin{aligned}|h(0,0)| &\le \text{const}\frac{1}{|d|^2}\left(|\widehat{q}(d)|+\frac{\text{const}}{|d|^{\beta+1}}\right)\left(|\widehat{q}(-d)|+\frac{\text{const}}{|d|^{\beta+1}}\right) \\ &\le \frac{\text{const}}{|d|^{2\beta+2}}\end{aligned}$$

and

$$\begin{aligned}\left|\frac{\partial h}{\partial x_i}(x_1,x_2)\right| &\le \text{const}\frac{1}{|d|^3}\left(\frac{1}{|d|^\beta}+\frac{1}{|d|^{\beta+1}}\right)^2 + \text{const}\frac{1}{|d|^2}\frac{1}{|d|^{\beta+1}}\left(\frac{1}{|d|^\beta}+\frac{1}{|d|^{\beta+1}}\right)\\ &\le \frac{\text{const}}{|d|^{2\beta+3}}\end{aligned}$$

$$\left|\frac{\partial^2 h}{\partial x_i \partial x_j}(x_1,x_2)\right| \le \frac{\text{const}}{|d|^{2\beta+3}}.$$

By the quantitative Morse lemma in the appendix, with $a=b=\text{const}/|d|^{2\beta+3}$, and $\delta = 2\varepsilon/3|d|^{1-\varepsilon}$ there is a biholomorphism ψ defined on $\{(z_1,z_2)\in\mathbb{C}^2 \mid |z_1|\le \varepsilon/2|d|^{1-\varepsilon}, |z_2|\le\varepsilon/2|d|^{1-\varepsilon}\}$ with range containing $\{(x_1,x_2)\in\mathbb{C}^2 \mid |x_1|\le \varepsilon/4|d|^{1-\varepsilon}, |x_2|\le\varepsilon/4|d|^{1-\varepsilon}\}$ and with

$$\|D\psi - \mathbb{1}\| \le \frac{\text{const}}{|d|^{2\beta+3}}, \qquad (x_1x_2+h)\circ\psi = z_1z_2 - \hat{t}_d,$$

$$|\hat{t}_d| \le \frac{\text{const}}{|d|^{2\beta+2}}, \qquad \left|\psi(0)\right| \le \frac{\text{const}}{(1+|d|)^{2\beta+3}}.$$

It now suffices to compose

$$\widehat{\phi}_{d,1}(z_1,z_2) = \big(k_1\big(\psi(z_1,z_2)\big), k_2\big(\psi(z_1,z_2)\big)\big)$$

with $k(x)$ being the map of (17.11).

Conclusion (ii) of the theorem, as well as the first part of (i), is immediate. The Jacobean

$$\begin{aligned}D\widehat{\phi}_{d,1} &= \frac{\partial k}{\partial x}D\psi\\ &= \frac{1}{2}\begin{pmatrix}1 & 1\\ i & -i\end{pmatrix}\left\{\mathbb{1}+O\Big(\frac{1}{|d|^{2\beta+3}}\Big)\right\}\left\{\mathbb{1}+O\Big(\frac{1}{|d|^2}\Big)\right\}\\ &= \frac{1}{2}\begin{pmatrix}1 & 1\\ i & -i\end{pmatrix}\left\{\mathbb{1}+O\Big(\frac{1}{|d|^2}\Big)\right\}.\end{aligned}$$

The centre is

$$\begin{aligned}\widehat{\phi}_{d,1}(0) &= k\big(\psi(0)\big)\\ &= k\bigg(O\Big(\frac{1}{|d|^{2\beta+3}}\Big)\bigg)\end{aligned}$$

with

$$\begin{aligned}k_2\bigg(O\Big(\frac{1}{(1+|d|)^{2\beta+3}}\Big)\bigg) &= z_1(d) - \frac{i}{2}\left(\frac{\widehat{q}(0)}{2iz_1(d)} - \frac{\widehat{q}(0)}{-2iz_2(d)}\right) + O\Big(\frac{1}{|d|^{2-\varepsilon}}\Big)\\ &= z_1(d) - \frac{\widehat{q}(0)}{4}\left(\frac{1}{z_1(d)}+\frac{1}{z_2(d)}\right) + O\Big(\frac{1}{|d|^{2-\varepsilon}}\Big)\\ &= z_1(d) + i\widehat{q}(0)\frac{d_1}{d_1^2+d_2^2} + O\Big(\frac{1}{|d|^{2-\varepsilon}}\Big).\end{aligned}$$

The computation of $k_1\big(O(1/(1+|d|)^{2\beta+3})\big)$ is similar.

To prove part (iii) observe that $T_1(0)\cap T_2(d)\cap\widehat{\mathcal{F}}(q)$ is mapped to $T_1(-d)\cap T_2(0)\cap\widehat{\mathcal{F}}(q)$ by translation by d. If we define $\widehat{\phi}_{d,2}$ by

$$\widehat{\phi}_{d,2}(z_1,z_2) = \widehat{\phi}_{d,1}(z_2,z_1) + d$$

then Theorem 17.2 holds. □

We now look more closely at the extent to which double points open up for various classes of potentials. As in §14 let f be a function on $\mathbb{R}^+$ satisfying

(i) $f \geq 1$, $f(0) = 1$.
(ii) $f(s)f(t) \geq f(s+t)$ for all $s, t \geq 0$.
(iii) f increases monotonically

and define, for operators on $\ell^2(\Gamma^\#)$, the norm

$$\|A\|_f = \max\left\{ \sup_{b\in\Gamma^\#} \sum_{c\in\Gamma^\#} |A_{b,c}| f(|b-c|),\ \sup_{c\in\Gamma^\#} \sum_{b\in\Gamma^\#} |A_{b,c}| f(|b-c|) \right\}.$$

In particular for the convolution operator $\widehat{q}(b-c)$

$$\|\widehat{q}\|_f = \sum_{b\in\Gamma^\#} |\widehat{q}(b)| f(|b|).$$

The part of the analogue of Lemma 17.4 for this norm that we need for our analysis of $\mathcal{D}$ is

LEMMA 17.7. *Let $k \in \mathbb{C}^2$ and $B \subset \{b \in \Gamma^\sharp \mid |N_b(k)| \geq \Lambda/2|v|\}$. Then*

$$\|R_{BB} - \pi_B\|_f \leq 2\frac{\|\widehat{q}(b)\|_f}{\Lambda|v|}$$

PROOF. It suffices to observe that, for $b \in B$,

$$|N_b(k)|^{-1} \leq \frac{2}{\Lambda|v|}$$

and that the convolution operator $|\widehat{q}(b-c)|$ has norm $\|\widehat{q}\|_f$. □

The principal quantity that determines the degree of opening of the double point $\big(-i(-1)^\nu z_\nu(d), z_\nu(d)\big)$ in other words that determines the $\hat{t}_d$ of Theorem 17.2, is $\big(\widehat{q}(d) - \mathcal{D}_{1,2}\big)\big(\widehat{q}(-d) - \mathcal{D}_{2,1}\big)$. The next lemma provides the estimates required to control it. Define

$$\mathcal{K}_f = \left\{ k \in \mathbb{C}^2 \,\middle|\, v| \leq \max\left\{1\ 2\Lambda\,,\ \frac{4\|\widehat{q}(b)\|_f}{\Lambda}\right\}\right\}.$$

LEMMA 17.8. *Let $k \in T_0 \cap T_d \cap \{\mathbb{C} \setminus \mathcal{K}_f\}$. Then there is a constant, depending only on $m+n$ and Λ, such that for all $m, n \in \mathbb{N}$*

$$\left|\frac{\partial^{n+m}}{\partial k_1^n \partial k_2^m}\mathcal{D}(k)_{1,2}\right|,\quad \left|\frac{\partial^{n+m}}{\partial k_1^n \partial k_2^m}\mathcal{D}(k)_{2,1}\right| \leq \mathrm{const}\|\widehat{q}\|_f^2 \frac{1}{|d| f(|d|)}.$$

PROOF. The bound for $m = n = 0$ is an immediate consequence of (14.23) and

$$\begin{aligned}\left\| \sum_{b,c\in D} \frac{\widehat{q}(d-b)}{N_b(k)} (R_{DD}^{-1})_{b,c} \widehat{q}(c-d') \right\|_f &\leq \frac{2}{\Lambda|v|} \|\widehat{q}\|_f \|R_{DD}^{-1}\|_f \|\widehat{q}\|_f \\ &\leq \frac{4\|\widehat{q}\|_f^2}{\Lambda|v|}.\end{aligned}$$

By (17.9) all derivatives are given by finite linear combinations of terms of the form

$$Q_{dD} \prod_j \left\{ H_k^{-1} (R_{DD})^{-1} \frac{\partial^{n_j} H_k}{\partial k_{i_j}^{n_j}} \right\} H_k^{-1} (R_{DD})^{-1} Q_{Dd'}$$

where the sum of the n_j's for which i_j =1 (resp. 2) is n (resp. m). Apply

$$\|Q\|_f = \|q\|_f, \quad \|R_{DD}^{-1}\|_f \le 2,$$

$$\|H_k^{-1}\pi_D\|_f \le \frac{2}{\Lambda|v|}, \quad \left\|\frac{\partial^n H_k}{\partial k_i^n} H_k^{-1}\pi_D\right\|_f \le \frac{4}{\Lambda}\begin{cases} 1 & \text{if } n=1 \\ \frac{1}{|v|} & \text{if } n=2 \\ 0 & otherwise \end{cases}.$$

Recall that H_k and its derivatives are diagonal operators and that, for diagonal operators, the operator norm and f-norm agree. So far we have

$$\left\| Q_{dD} \prod_j \left\{ H_k^{-1}(R_{DD})^{-1} \frac{\partial^{n_j} H_k}{\partial k_{i_j}^{n_j}} \right\} H_k^{-1}(R_{DD})^{-1} Q_{Dd'} \right\|_f \le \text{const}\|\widehat{q}\|_f^2 \frac{1}{|v|}.$$

Finally, as we observed at the beginning of the proof of Theorem 17.1(b), on $T_0 \cap T_d$ we have

$$\text{const}|d| \le |v| \le \text{const}|d|. \qquad \square$$

To conclude this section we note the following application of Theorem 17.1

THEOREM 17.9. *Let $q \in L^2(\mathbb{R}^2/\Gamma)$ with $\|b\widehat{q}(b)\|_1 < \infty$. Then $\mathcal{F}(q)$ is a reduced one-dimensional complex analytic variety, which consists of at most two components. If $\mathcal{F}(q)$ is smooth then it is irreducible.*

PROOF. Let $\varepsilon > 0$ be a small number and choose ρ such that Theorem 17.1 holds. By part (ii) of this Theorem and part (iii) of Theorem 17.2 there are reduced components C_1, C_2 of $\widehat{\mathcal{F}}(q)$ such that

$$(\widehat{\mathcal{F}}(q) \cap T_0) \setminus \left(\mathcal{K}_\rho \cup \bigcup_{\substack{b\in\Gamma^\# \\ b_2 \neq 0}} T_b \right) = ((C_1 \cup C_2) \cap T_0) \setminus \left(\mathcal{K}_\rho \cup \bigcup_{\substack{b\in\Gamma^\# \\ b_2 \neq 0}} T_b \right).$$

Clearly $E(C_1)$ and $E(C_2)$ are reduced components of $\mathcal{F}(q)$. Assume that $\mathcal{F}(q)$ has a component K not contained in $E(C_1) \cup E(C_2)$. Then every component C' of $E^{-1}(K)$ lies in

$$\mathcal{K}_\rho \cup \bigcup_{b,c\in\Gamma^\#} T_b \cap T_c.$$

In particular the complement of $\text{pr}(C')$ contains an open subset of $\mathbb{C}$.

On the other hand the indicator of growth ([**LG**, 3.6]) of $\widehat{\mathcal{F}}(q)$ is of finite order, since by Theorem 13.8 $\widehat{\mathcal{F}}(q)$ is the zero-set of an entire function of finite order. Therefore the indicator of growth of C' is also of finite order, and hence by the solution of the "Cousin problem with finite order" ([**LG**, 3.30]) C' is also the zero set of an entire function of finite order. Therefore by [**LG**, 3.44] the set $\{z \in \mathbb{C} \mid \text{pr}^{-1}(z) \cap C' = \varnothing\}$ is either $\mathbb{C}$ itself or discrete. Since its complement contains an open set, it is in fact discrete. As C' is irreducible it follows that this set consists of one point z_0. So $C' \subset \mathbb{C} \times \{z_0\}$. If we now apply the same argument with the projection $(k_1, k_2) \mapsto k_1$ we conclude tha C' is a point, which is impossible. So $\mathcal{F}(q) = E(C_1) \cup E(C_2)$ consists of at most two components and is reduced.

If $\mathcal{F}(q)$ is smooth then the constants $\hat{t}_d$ of Theorem 17.2 are all different from zero. Therefore C_1 and C_2 can be connected by an arc inside the set of smooth points of $\widehat{\mathcal{F}}(q)$. Thus $C_1 = C_2$ and $\mathcal{F}(q)$ is irreducible. $\square$

COROLLARY 17.10. *Let q be a real valued function in $L^2(\mathbb{R}^2/\Gamma)$ with $\|b\widehat{q}(b)\|_1 < \infty$. Then the maxima and minima of the band functions $E_n(k)$ are all isolated.*

PROOF. Assume that the nth band function $E_n(k)$ has a nonisolated extremum with extremal value μ. After replacing q by $q-\mu$ we may assume that $\mu = 0$.

In [**KT**] there was constructed, for each finite subset B of $\Gamma^\#$, an analytic function F_B on

$$\{(k,\lambda) \in \mathbb{C}^2 \times \mathbb{C} \mid N_b(k) - \lambda \neq 0 \text{ for all } b \in \Gamma^\# \setminus B\}$$

whose zero set is the Bloch-variety

$$B(q) = \{(k,\lambda) \in \mathbb{C}^2 \times \mathbb{C} \mid \exists \psi \in H^2_{\text{loc}}(\mathbb{R}^2) \text{ such that } \psi \neq 0, \\ (-\Delta + q)\psi = \lambda\psi \text{ and } \psi(x+\gamma) = e^{i\langle k,\gamma\rangle}\psi(x) \forall \gamma \in \Gamma\}$$

and such that $F_B(k,0) = 0$ is the equation of $\widehat{\mathcal{F}}(q)$ discussed in §16. Since 0 is the value of E_n at a nonisolated extremum there is a curve γ in $\mathbb{R}^2$ such that the plane $\mathbb{R}^2 \times \{0\}$ is tangent to the real Bloch variety

$$B(q) \cap \mathbb{R}^2 \times \mathbb{R} = \cup_{m \in \mathbb{N}} \{(k,\lambda) \in \mathbb{R}^2 \times \mathbb{R}\} E_m(k) = \lambda$$

along $\gamma \times \{0\}$. After possibly shrinking Γ we can find a finite subset B of $\Gamma^\#$ with $\gamma \cap \mathbb{C}^2_B = \varnothing$. Clearly $\gamma \subset \widehat{\mathcal{F}}(q)$ and all partial derivatives of $F_B(k,0)$ vanish at all points of γ. By analytic continuation all partial derivatives of $F_B(k,0)$ vanish along every component of $\widehat{\mathcal{F}}(q)$ that contains an open subset of γ. Therefore there is at least one component of $\widehat{\mathcal{F}}(q)$ that is not reduced, in contradiction to Theorem 17.9. □

18. Fermi Curves: Verification of the Geometric Hypotheses

Let $q \in C^\infty(\mathbb{R}^2/\Gamma)$. Fix $\varepsilon > 0$ sufficiently small. We construct a decomposition of $\mathcal{F}(q)$ into $\mathcal{F}(q)^{\text{com}} \cup \mathcal{F}(q)^{\text{reg}} \cup \mathcal{F}(q)^{\text{han}}$ such that the geometric hypotheses of §5 hold.

First, we refine Theorem 17.2 to get control of the handles. For $d \in \Gamma^\# \setminus \{0\}$ sufficiently large, let

$$\widehat{\phi}_{d,1} \colon \left\{ (z_1,z_2) \in \mathbb{C}^2 \;\middle|\; |z_1| \le \frac{\varepsilon}{2|d|^{1-\varepsilon}}, |z_2| \le \frac{\varepsilon}{2|d|^{1-\varepsilon}} \right\} \to T_1(0) \cap T_2(d),$$
$$\widehat{\phi}_{d,2} \colon \left\{ (z_1,z_2) \in \mathbb{C}^2 \;\middle|\; |z_1| \le \frac{\varepsilon}{2|d|^{1-\varepsilon}}, |z_2| \le \frac{\varepsilon}{2|d|^{1-\varepsilon}} \right\} \to T_1(-d) \cap T_2(0)$$

be the maps of Theorem 17.2 and $\hat{t}_d$ the number such that

$$\widehat{\phi}^{-1}_{d,1}(T_1(0) \cap T_2(d) \cap \widehat{\mathcal{F}}(q)) \\ = \left\{ (z_1,z_2) \in \mathbb{C}^2 \;\middle|\; z_1 z_2 = \hat{t}_d, |z_1| \le \frac{\varepsilon}{2|d|^{1-\varepsilon}}, |z_2| \le \frac{\varepsilon}{2|d|^{1-\varepsilon}} \right\}$$
$$\widehat{\phi}^{-1}_{d,2}(T_1(-d) \cap T_2(0) \cap \widehat{\mathcal{F}}(q)) \\ = \left\{ (z_1,z_2) \in \mathbb{C}^2 \;\middle|\; z_1 z_2 = \hat{t}_d, |z_1| \le \frac{\varepsilon}{2|d|^{1-\varepsilon}}, |z_2| \le \frac{\varepsilon}{2|d|^{1-\varepsilon}} \right\}$$

holds. Recall that

$$|\hat{t}_b| \le \frac{\text{const}(\beta)}{|b|^\beta} \tag{18.1a}$$

for all $\beta > 0$ so that

$$\sum_{\mathbf{b}} |\hat{t}_b|^\beta < \infty \tag{18.1b}$$

for all $\beta > 0$. Put

$$s_\nu(d) = \mathrm{pr}\left(\widehat{\phi}_{d,\nu}(0)\right).$$

Then

$$\left| s_\nu(d) - z_\nu(d) - \widehat{q}(0)\frac{id_1}{d_1^2 + d_2^2} \right| \le \frac{\mathrm{const}}{|d|^{2-\ varepsilon}}. \tag{18.2}$$

Put

$$\tau_d = \frac{1}{|d|^{5-\varepsilon}}, \qquad r_d = \frac{2\varepsilon}{|d|^5}, \qquad R_d = \frac{\varepsilon}{6|d|^\varepsilon}$$

$$\widehat{g}_{d,\nu} : \left\{ \zeta \in \mathbb{C} \;\middle|\; \frac{\varepsilon}{2|d|^\varepsilon}\tau_d \le |\zeta| \le \frac{\varepsilon}{2|d|^{1-\varepsilon}} \right\} \to \mathbb{C}$$

$$\zeta \mapsto \mathrm{pr}\left(\widehat{\phi}_{d,\nu}\left(\zeta, \frac{\hat{t}_d}{\zeta}\right)\right)$$

LEMMA 18.1. *If $|d|$ is big enough, then*

(a) *$\widehat{g}_d$ is biholomorphic onto its image. Furthermore*

$$\left| \widehat{g}_{d,\nu}\left(\frac{2\varepsilon}{|d|^\varepsilon}\tau_d e^{i\theta}\right) - s_d \right| < r_d$$

and

$$\left| \widehat{g}_d\left(\frac{\varepsilon}{2|d|^\varepsilon}e^{i\theta}\right) - s_d \right| > R_d > \left| \widehat{g}_d\left(\frac{\varepsilon}{4|d|^\varepsilon}e^{i\theta}\right) - s_d \right|$$

$$\left| \widehat{g}_d\left(\frac{\varepsilon}{8|d^\varepsilon|}e^{i\theta}\right) - s_d \right| > \frac{1}{4}R_d, \qquad \left| \widehat{g}_d\left(\frac{\varepsilon}{2|d|^{1-\varepsilon}}e^{i\theta}\right) - s_d \right| > \frac{\varepsilon}{6|d|^{1-\varepsilon}}$$

for all $0 \le \theta \le 2\pi$.

(b) *Define $\alpha_d(z)$ by*

$$\alpha_d(z)\,dz = (\widehat{g}_d)_*\left(\frac{1}{2\pi i}\frac{d\zeta}{\zeta}\right) + \frac{\mathrm{sgn}\, d_2}{2\pi i}\frac{1}{z - s_d}\,dz.$$

Then

$$\sup_d \left\| \alpha_d(z)\,dz \big|_{\{z\in\mathbb{C} | r_d < |z-s_d| < R_d\}} \right\|_2 < \infty$$

and

$$\lim_{d\to\infty} R_d \sup_{|z-s_d|=R_d} |\alpha_d(z)| = 0.$$

PROOF. Write

$$\widehat{\phi}_{d,\nu}(z_1, z_2) = \left(k_1(z_1,z_2), k_2(z_1,z_2)\right).$$

By the estimates in Theorem 17.2(i), we have, for all ζ with $2/\varepsilon|\hat{t}_d| \leq |\zeta| \leq \varepsilon/2$

$$\begin{aligned}
k_2\Big(\zeta, \frac{\hat{t}_d}{\zeta}\Big) - s_\nu(d) &= \big(k_2(\zeta,0) - s_\nu(d)\big) + \Big(k_2\Big(\zeta, \frac{\hat{t}_d}{\zeta}\Big) - k_2(\zeta,0)\Big) \\
&= \int_0^\zeta \frac{\partial k_2}{\partial z_1}(\xi,0)\,d\xi + \int_0^{\hat{t}_d/\zeta} \frac{\partial k_2}{\partial z_2}(\zeta,\xi)\,d\xi \\
&= \frac{(-1)^\nu i}{2}\Big(-\zeta + \frac{\hat{t}_d}{\zeta}\Big) + \Big(|\zeta| + \Big|\frac{\hat{t}_d}{\zeta}\Big|\Big) O\Big(\frac{1}{|d|^2}\Big).
\end{aligned}$$

Therefore

$$\Big|\widehat{g}_{d,\nu}(\zeta) - s_\nu(d) + \frac{(-1)^\nu i}{2}\Big(\zeta - \frac{\hat{t}_d}{\zeta}\Big)\Big| \leq \frac{\text{const}}{|d|^2|}\Big(|\zeta| + \frac{|\hat{t}_d|}{|\zeta|}\Big) \tag{18.3}$$

and the estimates of part (a) are obeyed.

To see that $\widehat{g}_{d,\nu}$ is biholomorphic onto its range, we first estimate its derivative. Again by part (i) of Theorem 17.2

$$\begin{aligned}
\frac{d\widehat{g}_{d,\nu}}{d\zeta}(\zeta) &= \frac{\partial k_2}{\partial z_1}\Big(\zeta, \frac{\hat{t}_d}{\zeta}\Big) - \frac{\hat{t}_d}{\zeta^2}\frac{\partial k_2}{\partial z_2}\Big(\zeta, \frac{\hat{t}_d}{\zeta}\Big) \\
&= -\frac{(-1)^\nu i}{2}\Big(1 + \frac{\hat{t}_d}{\zeta^2}\Big)\Big(1 + O\Big(\frac{1}{|d|^2}\Big)\Big).
\end{aligned}$$

In particular,

$$\Big|\frac{d\widehat{g}_{d,\nu}}{d\zeta} + \frac{(-1)^\nu i}{2}\Big| \leq \frac{\text{const}}{|d|^2}. \tag{18.4}$$

Therefore, if d is big enough, its derivative vanishes nowhere. The fact that $\widehat{g}_{d,\nu}$ is injective is proven as in §15.

To prove part (ii) we observe that, by (18.3)–(18.4),

$$\begin{aligned}
\frac{d\widehat{g}_{d,\nu}}{\widehat{g}_{d,\nu}(\zeta) - s_\nu(d)} - \frac{d\zeta}{\zeta} &= \frac{1 + \hat{t}_d/\zeta^2}{\zeta - \hat{t}_d/\zeta}\,d\zeta\Big(1 + O\Big(\frac{1}{|d|^2}\Big)\Big) - \frac{d\zeta}{\zeta} \\
&= \frac{2\hat{t}_d/\zeta^2}{\zeta - \hat{t}_d/\zeta}\,d\zeta + \frac{1 + \hat{t}_d/\zeta^2}{\zeta - \hat{t}_d/\zeta}\,d\zeta O\Big(\frac{1}{|d|^2}\Big) \\
&= \frac{2\hat{t}_d}{\zeta^2(1 + \hat{t}_d/\zeta^2)}\frac{d\widehat{g}_{d,\nu}}{\widehat{g}_{d,\nu}(\zeta) - s_\nu(d)}\Big(1 + O\Big(\frac{1}{|d|^2}\Big)\Big) \\
&\qquad + \frac{d\widehat{g}_{d,\nu}}{\widehat{g}_{d,\nu}(\zeta) - s_\nu(d)} O\Big(\frac{1}{|d|^2}\Big).
\end{aligned}$$

Applying $(\widehat{g}_{d,\nu})_*$

$$\frac{dz}{z - s_\nu(d)} - (\widehat{g}_{d,\nu})_*\Big(\frac{d\zeta}{\zeta}\Big) = \frac{dz}{z - s_\nu(d)}\Big(\frac{2\hat{t}_d}{\zeta^2 + \hat{t}_d} + O\Big(\frac{1}{|d|^2}\Big)\Big).$$

Therefore

$$|\alpha_d(z)| \leq \text{const}\frac{1}{|z - s_d|}\frac{1}{|d|^2}.$$

Now part (ii) of the Lemma follows easily. □

We now define the data appearing in the geometric hypotheses and verify (GH1)–(GH6). Fix $\rho > 0$ such that the conclusions of Theorem 17.1 hold with ε replaced by $\varepsilon/8$. For $\nu = 1, 2$ let K_ν be simply connected subsets of $\mathbb{C}$ with smooth boundary such that

$$\{z \in \mathbb{C} \mid |z| < \rho\} \subset K_\nu$$

and such that for all $d \in \Gamma^\# \setminus \{0\}$

$$\text{either } \left\{ z \in \mathbb{C} \;\middle|\; |z - z_\nu(d)| < \frac{\varepsilon}{|d|^{1-\varepsilon}} \right\} \subset K_\nu \qquad \text{for both } \nu = 1, 2$$

$$\text{or } \left\{ z \in \mathbb{C} \;\middle|\; |z - z_\nu(d)| < \frac{\varepsilon}{|d|^{1-\varepsilon}} \right\} \cap K\nu = \varnothing \qquad \text{for both } \nu = 1, 2$$

and such that Lemma 18.1 holds for all d obeying the second alternative. Put

$$\widehat{\mathcal{F}}(q)^{\text{com}} = \widehat{\mathcal{F}}(q) \cap \big(\mathcal{K}_\rho \cup \big(\operatorname{pr}^{-1}(K_1) \cap T_1(0)\big) \cup \big(\operatorname{pr}^{-1}(K_2) \cap T_2(0)\big)\big).$$

Then

$$\mathcal{F}(q)^{\text{com}} = E\big(\widehat{\mathcal{F}}(q)^{\text{com}}\big)$$

is a compact subset of $\mathcal{F}(q)$ whose boundary is diffeomorphic to $\partial K_1 \cup \partial K_2$.

Let

$$\Gamma_K^\# = \{b \in \Gamma^\# \mid z_1(b) \in K_1\} = \{b \in \Gamma^\# \mid z_2(b) \in K_2\}.$$

For each $b \in \Gamma_K^\#$, let $D_\nu(b)$ be the region enclosed by the curve

$$\widehat{g}_{b,\nu}\left(\left\{ \frac{\varepsilon}{2|b|^\varepsilon} \tau_b e^{i\theta} \;\middle|\; 0 \le \theta < 2\pi \right\} \right).$$

Define

$$G_\nu = \mathbb{C} \setminus \left(K_\nu \cup \bigcup_{b \in \Gamma_K^\#} D_\nu(b) \right)$$

and

$$\widehat{\mathcal{F}}(q)_\nu^{\text{reg}} = \left\{ k \in \widehat{\mathcal{F}}(q) \cap \left(T_\nu(0) \setminus \bigcup_{b \neq 0} T_b \right) \;\middle|\; \operatorname{pr}(k) \in G_\nu \right\}$$
$$\cup \bigcup_{b \in \Gamma_K^\#} \widehat{\phi}_{b,\nu}\left(\left\{ (z_1, z_2) \in \mathbb{C}^2 \;\middle|\; z_1 z_2 = \hat{t}_b,\ \frac{\varepsilon}{2|b|^\varepsilon} \tau_b \le |z_1| \le \frac{\varepsilon}{2|b|^{1-\varepsilon}} \right\} \right).$$

By Theorem 17.1 and Lemma 18.1(a), pr induces a biholomorphic map from $\widehat{\mathcal{F}}(q)_\nu^{\text{reg}}$ onto G_ν. Since two points k and k' of T_0 are identified by the map E if and only if there is a $b \in \Gamma^\#$ such that $k \in T_1(0) \cap T_2(b)$, $k' \in T_1(-b) \cap T_2(0)$ and $k = k' - b$, or conversely, the map E induces a biholomorphism between $\widehat{\mathcal{F}}(q)_\nu^{\text{reg}}$ and

$$\mathcal{F}(q)_\nu^{\text{reg}} = E\big(\widehat{\mathcal{F}}(q)_\nu^{\text{reg}}\big).$$

Define

$$\Phi_\nu \colon G_\nu \to \mathcal{F}(q)_\nu^{\text{reg}}$$

as the composition of $(\operatorname{pr}|_{\widehat{\mathcal{F}}(q)_\nu^{\text{reg}}})^{-1}$ and E. If we put

$$S_\nu = \{s_\nu(b) \mid b \in \Gamma_K^\#\}$$

then (GH1) and (GH4) are fulfilled.

Next put, for $b \in \Gamma_K^{\#}$

$$Y_b = E\Big(\widehat{\phi}_{b,1}\Big(\Big\{(z_1,z_2)\in\mathbb{C}^2 \,\Big|\, z_1z_2 = \hat{t}_b, |z_1|,|z_2| \le \frac{\varepsilon}{2|b|^{\varepsilon}}\Big\}\Big)\Big)$$
$$= E\Big(\widehat{\phi}_{b,2}\Big(\Big\{(z_1,z_2)\in\mathbb{C}^2 \,\Big|\, z_1z_2 = \hat{t}_b, |z_1|,|z_2| \le \frac{\varepsilon}{2|b|^{\varepsilon}}\Big\}\Big)\Big)$$
$$t_b = \left(\frac{2|b|^{\varepsilon}}{\varepsilon}\right)^2 |\hat{t}_b|.$$

Then

$$\phi_b \colon H(t_b) \to Y_b$$
$$(z_1,z_2) \to E\Big(\widehat{\phi}_{b,1}\Big(\frac{2|b|^{\varepsilon}}{\varepsilon}\frac{\hat{t}_b}{|\hat{t}_b|}z_1, \frac{2|b|^{\varepsilon}}{\varepsilon}z_2\Big)\Big) = E\Big(\widehat{\phi}_{b,2}\Big(\frac{2|b|^{\varepsilon}}{\varepsilon}z_2, \frac{2|b|^{\varepsilon}}{\varepsilon}\frac{\hat{t}_b}{|\hat{t}_b|}z_1\Big)\Big)$$

is a biholomorphic map. Observe that $t_b \neq 0$ for all $b \in \Gamma^{\#}$ if $\mathcal{F}(q)$ is smooth. In this case choose a homology basis of $\mathcal{F}(q)$ such that Y_b represents an A cycle. By construction and (18.1) hypothesis (GH2) holds.

THEOREM 18.2. *Let $q \in C^{\infty}_{\mathbb{R}}(\mathbb{R}^2/\Gamma)$ be such that $\mathcal{F}(q)$ is smooth. Then the marked Riemann surface $\mathcal{F}(q) = \mathcal{F}(q)^{\mathrm{com}} \cup \mathcal{F}(q)^{\mathrm{reg}} \cup \mathcal{F}(q)^{\mathrm{han}}$ obeys the geometric hypotheses* (GH1)–(GH6) *of* §5. *It looks like*

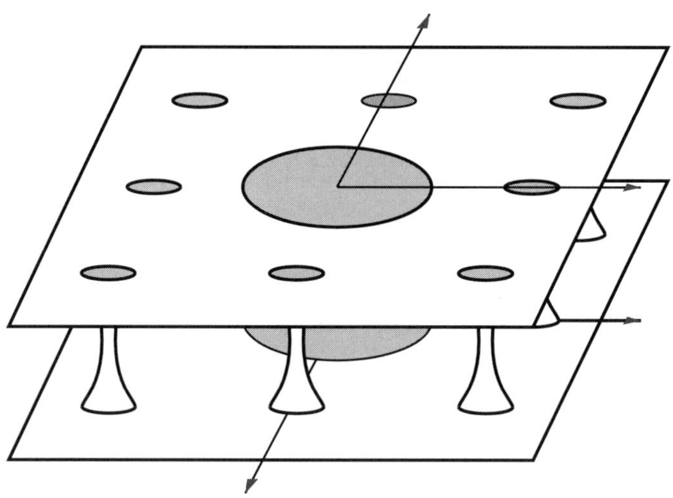

PROOF. We have already verified (GH1), (GH2) and (GH4). Define, for all $b \in \Gamma_K^{\#}$

$$\tau_\mu(b) = \tau_b, \quad \nu_\mu(b) = \mu,$$
$$R_\mu(b) = R_b, \quad r_\mu(b) = r_b.$$

Then (GH3) follows from Lemma 18.1(a).

Part (i) of (GH5) is trivial. With $\delta = 2\varepsilon$ and $d = 5$ hypothesis (GH5)(ii) follows from the summability of $1/|b|^{3-8\varepsilon}$. Since $\widehat{q}(0)$ is real

$$\left|z_1(b) + \widehat{q}(0)\frac{ib_1}{b_1^2+b_2^2}\right| = \left|z_2(b) + \widehat{q}(0)\frac{ib_1}{b_1^2+b_2^2}\right|.$$

By (18.2)

$$\big||s_1(b)| - |s_2(b)|\big| = O\Big(\frac{1}{|b|^{2-\varepsilon}}\Big).$$

This yields (GH5)(iii). Part (iv) of (GH5) follows from (18.1). Part (v) follows from the definition of R_b and the fact that $\min_{\substack{s\in S\\ s\neq s_b}} |s - s_b| = O(1)$. Lemma 18.1(b) implies (GH5)(vi).

Finally (GH6) is void. □

REMARK 18.3. If $t_b = 0$ for some b then Φ_ν can be extended to a map from $G_\nu \cup D_\nu(b)$ to the normalization of $\mathcal{F}(q)$. In this way one sees that the normalization of $\mathcal{F}(q)$ always fulfills the geometric hypotheses whenever $t_b \neq 0$ for infinitely many b. If, on the other hand, $t_b = 0$ for all but finitely many b, then the normalization of $\mathcal{F}(q)$ has finite genus and q is a "finite gap" potential [**K**].

REMARK 18.4. If $q \in C^\infty(\mathbb{R}^2/\Gamma)$ and $\widehat{q}(0) \notin \mathbb{R}$ then hypothesis (GH5)(iii) fails, but all the other geometric hypotheses of §5 hold. One can modify the construction of the exhaustion function of bounded charge given in Lemma 5.3 replacing $\log|z|$ by $\log|z| - \mathrm{Im}(\mathrm{Im}\,\widehat{q}(0)/2z^2)$ on one of the sheets. Using this construction one can show that all the results of Chapter 2 hold for $q \in C^\infty(\mathbb{R}^2/\Gamma)$.

B. A Quantitative Morse Lemma

LEMMA B.1. *Let*

$$f(x_1, x_2) = x_1 x_2 + h(x_1, x_2)$$

be a holomorphic function on

$$D_\delta = \{(x_1, x_2) \in \mathbb{C}^2 \mid |x_1| \le \delta \text{ and } |x_2| \le \delta\}, \quad \delta < 1$$

where h is a function that fulfils the estimates

$$\left|\frac{\partial h}{\partial x_i}(x)\right| \le a, \quad \left\|\frac{\partial^2 h}{\partial x_i \partial x_j}(x)\right\| \le b \text{ for } x \in D_\delta$$

with constants $a, b > 0$ such that

$$a < \delta, \quad b < \frac{1}{30}$$

Then f has unique critical point $\xi = (\xi_1, \xi_2)$ in D_δ, and

$$|\xi_1| \le a, \quad |\xi_2| \le a.$$

Put $s = \max(|\xi_1|, |\xi_2|)$. Then there is a biholomorphic map Φ from $D_{(\delta-s)(1-10b)}$ to a neighbourhood of ξ in D_δ that contains $\{(z_1, z_2) \in \mathbb{C}^2 \mid |z_i - \xi_i| < (\delta - s)(1 - 30b)\}$ such that

$$f \circ \Phi(z_1, z_2) = z_1 z_2 + c$$

with a constant $c \in \mathbb{C}$ fulfilling $|c - h(0,0)| \le a^2$. The differential $D\Phi$ fulfils

$$\|D\Phi - \mathbb{1}\| \le 12b.$$

In the case that $\partial h/\partial x_1(0,0) = \partial h_2/\partial x_2(0,0) = 0$ one has $\xi = 0$ and $s = 0$.

PROOF. Without loss of generality we may assume that $h(0,0)=0$. Put

$$C_i = \left\{ (x_1,x_2) \in D_\delta \;\middle|\; \frac{\partial f}{\partial x_i}(x) = 0 \right\}.$$

To prove the first claim, we show that C_1 and C_2 have a unique point of intersection. Observe that

$$\frac{\partial f}{\partial x_1} = x_2 + \frac{\partial h}{\partial x_1}(x_1,x_2).$$

By Rouché's theorem for each x_1 with $|x_1| < \delta$ the equation $\partial f/\partial x_1(x_1,x_2) = 0$ has a unique solution $\widetilde{x}_2(x_1)$. By the estimate on $\partial h/\partial x_1$ and on $\partial^2 h/\partial x_1 \partial x_2$ this solution fulfils

$$|\widetilde{x}_2(x_1)| \le a, \qquad \left|\frac{\partial \widetilde{x}_2}{\partial x_1}\right| \le \frac{b}{1-b}.$$

Similarly one can parametrise the curve C_2 by a map $x_2 \mapsto (\widetilde{x}_1(x_2), x_2)$ fulfilling

$$|\widetilde{x}_1(x_2)| \le a, \qquad \left|\frac{\partial \widetilde{x}_1}{\partial x_2}\right| \le \frac{b}{1-b}.$$

Therefore the curves C_1 and C_2 intersect in a unique point (ξ_1, ξ_2) , and this point fulfils $\widetilde{x}_1(\xi_2) = \xi_1$, $\widetilde{x}_2(\xi_1) = \xi_2$. So $|\xi_1| \le a$ and $|\xi_2| \le a$. Now write

$$f(x_1,x_2) = (x_1 - \xi_1)(x_2 - \xi_2) + \tilde{h}(x_1 - \xi_1, x_2 - \xi_2)$$

where $\tilde{h}(x_1', x_2')$ is a function defined in $D_{\delta - s}$ with $\tilde{h}(0,0) = \xi_1\xi_2$, $\partial\tilde{h}/\partial x_1(0,0) = \partial\tilde{h}/\partial x_2(0,0) = 0$, and one still has the bound

$$\left\| \left(\frac{\partial^2 \tilde{h}}{\partial x_i \partial x_j} \right) \right\| \le b.$$

This shows that it suffices to prove the lemma in the special case that $h(0,0) = \partial h/\partial x_1(0,0) = \partial h/\partial x_2(0,0) = 0$ and then replace δ by $\delta - s$. In this case put

$$f_t(x_1,x_2) = x_1x_2 + th(x_1,x_2).$$

We construct a t-dependent vector field X^t on $D_{\delta(1-4b)}$ such that

$$h(x_1,x_2) + \nabla f_t \cdot X^t = 0. \tag{B.1}$$

$$\|X^t(x)\| \le 5b(|x_1| + |x_2|) \le 10b\delta. \tag{B.2}$$

$$\left\| \frac{\partial}{\partial x_i} X^t(x) \right\| \le 8b \quad \text{for } i = 1, 2. \tag{B.3}$$

Integrating X^τ from 0 to t $(0 \le \tau \le 1)$ gives a map

$$\Phi_\tau \colon D_{\delta(1-10b)} \to \mathbb{C}^2, \quad x \mapsto x + \int_0^\tau X^t(x)\,dt$$

which is biholomorphic into its image, fulfils

$$\Phi_0(x) = x, \quad \frac{d}{d\tau} f_\tau(\Phi_\tau(x)) = 0,$$
$$\|\Phi_\tau(x) - x\| \le 5b(|x_1| + |x_2|) \le 10b\delta,$$
$$\|D_x\Phi_t - \mathbb{1}\| \le \sqrt{2}\int_0^t \sup_x \left(\left\| \frac{\partial}{\partial x_1} X^\tau(x) \right\|, \left\| \frac{\partial}{\partial x_2} X^\tau(x) \right\| \right) d\tau \le 12b$$

$\Phi = \Phi_1$ then has the desired properties, since $f_1 \circ \Phi_1 = f_0$ and the image of Φ contains $D_{\delta(1-30b)}$.

To construct X^t observe that equation (B.1) is

$$h(x_1,x_2)+\left\langle \begin{pmatrix} x_2+t\frac{\partial h}{\partial x_1}(x) \\ x_1+t\frac{\partial h}{\partial x_2}(x)\end{pmatrix}, \begin{pmatrix} X_1^t(x) \\ X_2^t(x)\end{pmatrix}\right\rangle = 0. \tag{B.4}$$

By the assumptions on h

$$\left|\frac{\partial h}{\partial x_i}(x)\right| \le b(|x_1|+|x_2|) \tag{B.5}$$

and hence

$$|h(x)| \le b(|x_1|+|x_2|)^2.$$

Therefore, for $0 \le t \le 1$, the map

$$P_t \colon D_\delta \to \mathbb{C}^2, \quad (x_1,x_2) \mapsto \left(x_2+t\frac{\partial h}{\partial x_1}(x), x_1+t\frac{\partial h}{\partial x_2}(x)\right)$$

is biholomorphic into its image. Furthermore

$$\left\| D_xP_t - \begin{pmatrix} 0 & 1 \\ 1 & 0\end{pmatrix}\right\| = \left\| t\frac{\partial^2 h}{\partial x_i \partial x_j}\right\| \le tb, \tag{B.6}$$

so that the image contains $D_{\delta(1-2b)}$ and

$$\left\| DP_t^{-1} - \begin{pmatrix} 0 & 1 \\ 1 & 0\end{pmatrix}\right\| \le \frac{tb}{1-tb}. \tag{B.7}$$

To solve (B.4) we first solve the equation

$$g(y_1,y_2) = \left\langle \begin{pmatrix} y_1 \\ y_2 \end{pmatrix}, \begin{pmatrix} Y_1^t(y) \\ Y_2^t(y)\end{pmatrix}\right\rangle$$

on $D_{\delta(1-2b)}$, where $g(y) = -h \circ P_t^{-1}(y)$. This is done by putting

$$Y_1^t(y) = \frac{1}{y_1}g(y_1,0), \quad Y_2^t(y) = \frac{1}{y_2}\left(g(y_1,y_2)-g(y_1,0)\right)$$

Observe that by (B.5) the change of variables

$$y = P_t(x) = \left(x_2+t\frac{\partial h}{\partial x_1}(x), x_1+t\frac{\partial h}{\partial x_2}(x)\right)$$

fullfils

$$(1-2b)(|x_1|+|x_2|) \le |y_1|+|y_2| \le (1+2b)(|x_1|+|x_2|). \tag{B.8}$$

By the chain rule and (B.5), (B.7)

$$\left|\frac{\partial g}{\partial y_i}\right| \le b(|x_1|+|x_2|)\left(1+\frac{\sqrt{2}b}{1-b}\right) \le b\frac{1+2b}{1-2b}(|y_1|+|y_2|).$$

Therefore

$$|g(y_1,y_2)| \le \frac{b}{2}\frac{1+2b}{1-2b}(|y_1|+|y_2|)^2$$

which shows that Y_1^t, Y_2^t are holomorphic. Furthermore we see that

$$\left|Y_1^t(y)\right| \le \frac{b}{2}\frac{1+2b}{1-2b}|y_1| \left|\frac{\partial}{\partial y_1}Y_1(y)\right| \le \frac{3b}{2}\frac{1+2b}{1-2b}, \quad \frac{\partial}{\partial y_2}Y_1(y) = 0.$$

To get estimates for Y_2 we discuss regions $|y_2| \ge |y_1|$ and $|y_2| \le |y_1|$ separately.

If $|y_2| \geq |y_1|$ then

$$|Y_2^t(y)| \leq \frac{1}{|y_2|}(|g(y_1,y_2)| + |g(y_1,0)|) \leq b\frac{1+2b}{1-2b}\frac{(|y_1|+|y_2|)^2}{|y_2|},$$
$$\leq 2b\frac{1+2b}{1-2b}(|y_1|+|y_2|)$$

and similarly

$$\left|\frac{\partial}{\partial y_1}Y_2^t(y)\right| \leq \frac{1}{|y_2|}\left(\left|\frac{\partial g}{\partial y_1}(y_1,y_2)\right| + \left|\frac{\partial g}{\partial y_1}(y_1,0)\right|\right) \leq 4b\frac{1+2b}{1-2b},$$
$$\left|\frac{\partial}{\partial y_2}Y_2^t(y)\right| \leq \frac{1}{|y_2|}\left|Y_2^t(y)\right| + \frac{1}{|y_2|}\left|\frac{\partial g}{\partial y_2}(y_1,y_2)\right| \leq 6b\frac{1+2b}{1-2b}.$$

This estimate holds in particular for $|y_1| = |y_2|$. For fixed y_1 we can now apply the maximum principle on the functions $Y_2(y_1,\cdot)$ and $\partial/\partial y_i Y_2(y_1,\cdot)$ in the disc $|z| \leq |y_1|$ to get

$$|Y_2^t(y_1,z)| \leq 2b\frac{1+2b}{1-2b}(|y_1|+|y_2|) \leq 4b\frac{1+2b}{1-2b}(|y_1|+|z|)$$

and similarly

$$\left|\frac{\partial}{\partial y_1}Y_2(y_1,z)\right| \leq 4b\frac{1+2b}{1-2b}, \quad \left|\frac{\partial}{\partial y_2}Y_2(y_1,z)\right| \leq 6b\frac{1+2b}{1-2b}.$$

Putting everything together we get for all $y \in D_{\delta(1-2b)}$

$$\left\|Y^t(y)\right\| \leq \sqrt{16\frac{1}{4}}\, b\frac{1+2b}{1-2b}(|y_1|+|y_2|), \tag{B.9}$$

$$\left\|\frac{\partial}{\partial y_i}Y^t(y)\right\| \leq 6b\frac{1+2b}{1-2b}. \tag{B.10}$$

Now put $X^t = Y^t \circ P_t$. By construction X^t satisfies the equation (B.4) on $P_t^{-1}(D_{\delta(1-2b)})$. By (B.5) this region contains $D_{\delta(1-4b)}$. Furthermore we get from (B.9), (B.10) and (B.6), (B.8) the desired estimates

$$\left\|X^t(x)\right\| \leq \sqrt{16\frac{1}{4}}b\frac{(1+2b)^2}{1-2b}(|x_1|+|x_2|) < 5b(|x_1|+|x_2|),$$
$$\left\|\frac{\partial}{\partial x_i}X^t(x)\right\| \leq 6b\frac{(1+2b)^2}{1-2b} < 8b.$$

□

CHAPTER 4

The Kadomcev–Petviashvilli Equation

Introduction

The Schrödinger spectral curve $\mathcal{S}(q)$ associated to $q \in L^2_{\mathbb{R}}(\mathbb{R}/2\pi\mathbb{Z})$ is the set of all points $(\xi, \lambda) \in \mathbb{C}^* \times \mathbb{C}$ for which there is a nontrivial distributional solution $\psi(x)$ in $L^\infty_{\mathrm{loc}}(\mathbb{R})$ of the Schrödinger equation

$$-\frac{d^2}{dx^2}\psi + q(x)\psi = \lambda\psi$$

satisfying

$$\psi(x + 2\pi) = \xi\psi(x)$$

for all $x \in \mathbb{R}$. For generic $q \in L^2_{\mathbb{R}}(\mathbb{R}/2\pi\mathbb{Z})$, the curve $\mathcal{S}(q)$ is a Riemann surface of infinite genus and we showed (Theorem 12.1, Example 1) that it satisfies the Geometric Hypotheses of §5.

Suppose $u(x,t)$, $-\infty < t < \infty$, is a solution to the initial value problem for the Korteweg–deVries equation

$$u_t = 3uu_x - \tfrac{1}{2}u_{xxxx}$$

with initial data $u(x,0) = u_0(x) \in C^\infty_{\mathbb{R}}(\mathbb{R}/2\pi\mathbb{Z})$. It is well-known that

$$\mathcal{S}\big(u(\cdot,t)\big) = \mathcal{S}\big(u_0(\cdot)\big)$$

as subsets of $\mathbb{C}^* \times \mathbb{C}$ for all $-\infty < t < \infty$ (see [**McK1**], or Theorem 13.14). In [**MT1**], this fact was used to prove that every spatially periodic solution of the Korteweg–de Vries equation propagates almost periodically in time. In [**MT2**], the theta function for $\mathcal{S}(q)$ was used to give an "explicit" solution to the initial value problem. The technique of [**MT1**] and [**MT2**] relies on the explicit realization of $\mathcal{S}(q)$, by projection onto the λ plane, as a branched double cover of $\mathbb{C}$. Riemann surfaces of infinite genus that are finite sheeted branched covers of $\mathbb{C}$ have been used to study several other integrable 1 + 1-dimensional partial differential equations. See, for example, [**BKM, EM, McK2**] and [**Sch**].

Let

$$\Gamma = (0, 2\pi)\mathbb{Z} \oplus (\omega_1, \omega_2)\mathbb{Z}$$

where $\omega_1 > 0$, $\omega_2 \in \mathbb{R}$. Recall that the heat curve $\mathcal{H}(q)$ associated to $q \in L^2(\mathbb{R}^2/\Gamma)$ is the set of all points $(\xi_1, \xi_2) \in \mathbb{C}^* \times \mathbb{C}^*$ for which there is a nontrivial distributional solution $\psi(x_1, x_2)$ in $L^\infty_{\mathrm{loc}}(\mathbb{R}^2)$ of the "heat equation"

$$\left(\frac{\partial}{\partial x_1} - \frac{\partial^2}{\partial x_2^2}\right)\psi + q(x_1, x_2)\psi = 0$$

satisfying

$$\psi(x_1 + \omega_1, x_2 + \omega_2) = \xi_1\psi(x_1, x_2), \quad \psi(x_1, x_2 + 2\pi) = \xi_2\psi(x_1, x_2).$$

If $q \in C^\infty(\mathbb{R}^2/\Gamma)$ and $\mathcal{H}(q)$ is smooth, then, by Theorem 15.2, it satisfies the geometric hypotheses of §5. There is no natural realization of a heat curve as a branched finite cover of $\mathbb{C}$. For this reason heat curves are intrinsically more complicated than Schrödinger spectral curves.

For each $u \in L^2(\mathbb{R}^2/\Gamma)$ define the function $I(u)$ by

$$I(u)(x_1, x_2) = \int_0^{x_2} u(x_1, s)\, ds - \frac{1}{2\pi} \int_0^{2\pi} dt \int_0^t u(x_1, s)\, ds.$$

The Kadomcev–Petviashvilli equation is

$$\text{(KP)} \qquad u_t = 3uu_{x_2} - \tfrac{1}{2} u_{x_2x_2x_2} - \tfrac{3}{2} I(u_{x_1x_1}).$$

If one differentiates both sides of (KP) with respect to x_2 one recovers the standard KPII equation (see, for example, [**K**])

$$(u_t - 3uu_{x_2} + \tfrac{1}{2} u_{x_2x_2x_2})_{x_2} + \tfrac{3}{2} u_{x_1x_1} = 0.$$

Suppose $u = u(x_1, x_2, t)$ is a solution of the initial value problem for the (KP) equation with intial data $u_0 \in C^\infty_{\mathbb{R}}(\mathbb{R}^2/\Gamma)$. As above, there is an associated family $\mathcal{H}\big(u(\cdot, t)\big)$, $-\infty < t < \infty$, of heat curves. By Theorem 13.14,

$$\mathcal{H}\big(u(\cdot, t)\big) = \mathcal{H}(u_0)$$

as subsets of $\mathbb{C}^* \times \mathbb{C}^*$ for all $-\infty < t < \infty$. In this chapter (Theorem 21.1), we use the theta function on $\mathcal{H}(u_0)$, when $u_0 \in C^\omega_{\mathbb{R}}(\mathbb{R}^2/\Gamma)$, to give an explicit formula for the solution $u(x_1, x_2, t)$. This formula is used to show (Corollary 21.3) that spatially periodic solutions of the Kadomcev–Petviashvilli equation propagate almost periodically in time.

19. The Formula for the Solution

We return to the discussion of Sections 13, 14, 15 and want to solve the initial value problem for the periodic KP-equation

$$\text{(KP)} \qquad u_t = 3uu_{x_2} - \tfrac{1}{2} u_{x_2x_2x_2} - \tfrac{3}{2} I(u_{x_1x_1})$$

for given initial data q. It has been shown by I. Krichever [**K**] that for $q \in C^\omega(\mathbb{R}^2/\Gamma)$ the initial value problem is well posed and can be solved for all time. J. Bourgain [**B**] demonstrated the more difficult fact that the initial value problem is well posed on $H^1(\mathbb{R}^2/\Gamma)$. Here we show that for real analytic q the solution is almost periodic in time. This is done by giving a formula for the solution in terms of theta functions associated to heat curves.

Such a formula is well-known in the case that the normalization of the heat curve $\mathcal{H}(q)$ has finite genus. Such potentials q are often called **finite zone potentials**. We recall the procedure to solve the initial value problem for the KP-equation for initial data q that are finite zone potentials (see [**K**, Chapter II], and [**Mu2**, Chapter III(b), §4].

In this case the normalization of $\mathcal{H}(q)$ is the complement of one point P_∞ on a compact Riemann surface $X(q)$. A local coordinate around P_∞ is $\zeta = i/k_2$ (see Theorem 14.1. Let $A_1, B_1, \dots A_g, B_g$ be a canonical homology basis for $X(q)$, and let $\omega_1, \dots \omega_g$ be the holomorphic one forms on $X(q)$ satisfying $\int_{A_i} \omega_j = \delta_{ij}$. Furthermore denote by θ the associated theta function. The expansions of the forms ω_j at P_∞ define vectors $U, V, W \in \mathbb{C}^g$ by

$$\omega_j = U_j\, d\zeta + V_j \zeta\, d\zeta + \tfrac{1}{2} W_j \zeta^2\, d\zeta + O(\zeta^3) \text{ near } P_\infty.$$

Finally

$$\left\{(\xi_1,\xi_2)\in\mathcal{H}(q)\;\middle|\;(\xi_1,\xi_2) \text{ is a smooth point of } \mathcal{H}(q) \text{ and the nontrivial solution } \psi \text{ of } \left(\frac{\partial}{\partial x_1}-\frac{\partial^2}{\partial x_2^2}+q\right)\psi=0 \text{ vanishes at the point } x_1=x_2=0\right\}$$

defines a divisor of D of degree g on $X(q)$. This divisor is nonspecial, so by Riemann's vanishing theorem there is $Z\in\mathbb{C}^g$ such that D is the zero divisor of

$$x\mapsto\theta\left(Z+\int_{P_\infty}^{x}\vec{\omega}\right)$$

on $X(q)$. Then there is a constant c such that

$$u(x_1,x_2,t)=-2\frac{\partial^2}{\partial x_2^2}\log\theta(Ux_2+Vx_1-\tfrac{1}{2}Wt+Z)+c \tag{19.1a}$$

solves KP, and that

$$u(x_1,x_2,0)=q(x_1,x_2). \tag{19.1b}$$

In addition the constant c is zero if $q\in\mathcal{U}(\Gamma)$, that is if $\int_0^{2\pi}q(x_1,x_2)\,dx_2=0$ (see Lemma 13.11).

Observe that for real valued q the heat curve $\mathcal{H}(q)$ has an antiholomorphic involution induced by $(k_1,k_2)\mapsto(-\overline{k_1},-\overline{k_2})$. The local coordinate i/k_2 is real with respect to this involution. If the canonical homology basis is compatible with the antiholomorphic involution then the vectors U, V, W are real. Also Z is then invariant under the induced antiholomorphic involution on the Jacobian of $X(q)$

The purpose of this chapter is to generalize formula (19.1a)–(19.1b) to initial data for which the heat curve has infinite genus. More precisely we assume from now on that $q\in C^\infty(\mathbb{R}^2/\Gamma)$ and that $\int_0^{2\pi}q(x_1,x_2)\,dx_2=0$. For simplicity we again assume that $\mathcal{H}(q)$ is smooth. The results of the Sections 5, 6, 7 and 15 show that $\mathcal{H}(q)$ has a canonical homology basis A_b, B_b indexed by the set $\Gamma^\#_+$ of all $b\in\Gamma^\#$ with $b_2>0$, a basis $(\omega_b)_{b\in\Gamma^\#_+}$ of the Hilbert space of square integrable holomorphic one forms with

$$\int_{A_b}\omega_c=\delta_{bc}$$

such that the thetafunction associated to the Riemann period matrix

$$\mathcal{R}_{bc}=\int_{B_b}\omega_c$$

converges on the Banachspace

$$B=\left\{(z_b)_{b\in\Gamma^\#_+}\;\middle|\;\sum\frac{|z_b|}{|\log t_b|}<\infty\right\}.$$

Here, t_b are positive constants satisfying (15.1).

To generalize formula (19.1) we have to define analogues of the vectors U, V, W and the divisor D (resp. the associated vector Z). First we discuss the vectors U, V, W.

Consider a marked Riemann surface $(X;A_1,B_1,\dots)$ that satisfies the geometric hypotheses of §5 with $m=1$. By Proposition 6.12 the restriction $w_j(z)\,dz$ of the differential ω_j to the regular piece can be written in the form

$$w_j(z) = w_{j,\mathrm{com}}(z) + \sum_{s\in S} w_{j,s}(z)$$

where

$$w_{j,s}(z) = -\frac{1}{2\pi i}\int_{|\zeta-s|=r(s)} \frac{w_j(\zeta)}{\zeta - z}\,d\zeta,$$
$$w_{j,\mathrm{com}}(z) = -\frac{1}{2\pi i}\int_{\partial K} \frac{w_j(\zeta)}{\zeta - z}\,d\zeta.$$

Furthermore each $w_{j,s}(z)$ and $w_{j,\mathrm{com}}(z)$ is holomorphic outside a bounded subset of $\mathbb{C}$ and decays as $z \to \infty$. Therefore we can consider the expansions of the forms $w_{j,s}(z)\,dz$ and $w_{j,\mathrm{com}}(z)\,dz$ at infinity with respect to the variable i/z

$$\begin{aligned}
&w_{j,s}(z)\,dz \\
&\quad = w_{j,s}^{(1)}\Big(-\frac{dz}{z}\Big) + w_{j,s}^{(2)}\Big(-i\frac{dz}{z^2}\Big) + w_{j,s}^{(3)}\Big(\frac{dz}{z^3}\Big) + w_{j,s}^{(4)}\Big(i\frac{dz}{z^4}\Big) + O(1/|z|^5),\\
&w_{j,\mathrm{com}}(z)\,dz \\
&\quad = w_{j,\mathrm{com}}^{(1)}\Big(-\frac{dz}{z}\Big) + w_{j,\mathrm{com}}^{(2)}\Big(-i\frac{dz}{z^2}\Big) + w_{j,\mathrm{com}}^{(3)}\frac{dz}{z^3} + iw_{j,\mathrm{com}}^{(4)}\frac{dz}{z^4} + O(1/|z|^5)
\end{aligned}$$

with constants $w_{j,s}^{(i)}$, $w_{j,\mathrm{com}}^{(i)}$. By Proposition 6.12 $w_{j,s}$ decays quadratically if $s \neq s_1(j)$, $s_2(j)$ so that in this case $w_{j,s}^{(1)} = 0$. Also by Proposition 6.12

$$w_{j,s_1(j)}^{(1)} + w_{j,s_2(j)}^{(1)} = 0 \quad\text{and}\quad w_{j,\mathrm{com}}^{(1)} = 0.$$

For the next terms we have

PROPOSITION 19.1. *Let $(X; A_1, B_1, \dots)$ be a marked Riemann surface that satisfies the geometric hypotheses* (GH1)–(GH6) *of Section* 5 *with $m = 1$. Then the sums*

$$U_j = w_{j,\mathrm{com}}^{(2)} + \sum_{s\in S} w_{j,s}^{(2)},\quad V_j = w_{j,\mathrm{com}}^{(3)} + \sum_{s\in S} w_{j,s}^{(3)},\quad W_j = w_{j,\mathrm{com}}^{(4)} + \sum_{s\in S} w_{j,s}^{(4)}$$

converges absolutely. Furthermore there is a numerical constant const *independent of j such that for all $j \geq 1$*

$$\Big|U_j + \frac{1}{2\pi}(s_1(j) - s_2(j))\Big| \leq \mathrm{const},\quad \Big|V_j + \frac{1}{2\pi i}(s_1(j)^2 - s_2(j)^2)\Big| \leq \mathrm{const}$$
$$\Big|W_j - \frac{1}{2\pi}(s_1(j)^3 - s_2(j)^3)\Big| \leq \mathrm{const}$$

PROOF.

$$\begin{aligned}
w_{j,s}(z) &= \frac{1}{2\pi i}\int_{|\zeta-s|=r(s)} \frac{w_j(\zeta)}{z-\zeta}\,d\zeta = \frac{1}{2\pi i}\int_{|\zeta-s|=r(s)} \frac{1}{z}\cdot\frac{w_j(\zeta)}{1-\zeta/z}\,d\zeta,\\
&= \sum_{n=1}^{4} \frac{1}{2\pi i}\frac{1}{z^n}\int_{|\zeta-s|=r(s)} \zeta^{n-1} w_j(\zeta)\,d\zeta + O\Big(\frac{1}{|z|^5}\Big).
\end{aligned}$$

Therefore

$$w_{j,s}^{(n)} = \frac{-(i)^{1-n}}{2\pi i}\int_{|\zeta-s|=r(s)} \zeta^{n-1} w_j(\zeta)\,d\zeta \tag{19.2}$$

Similarly

$$w^{(n)}_{j,\mathrm{com}} = \frac{-(i)^{1-n}}{2\pi i}\int_{\partial K} \zeta^{n-1} w_j(\zeta)\,d\zeta \tag{19.3}$$

For $s \neq s_1(j),\ s_2(j)$ we have

$$\begin{aligned} w^{(n)}_{j,s} &= \frac{-(i)^{1-n}}{2\pi i}\int_{|\zeta-s|=4r(s)} \left[s^{n-1} + \binom{n-1}{1} s^{n-2}(\zeta-s) + \cdots (\zeta-s)^{n-2}\right] w_j(\zeta)\,d\zeta, \\ &= \frac{-(i)^{1-n}}{2\pi i}\sum_{r=1}^{n-1} \binom{n-1}{r} s^{n-r}\int_{|\zeta-s|=4r(s)} (\zeta-s)^r w_j(\zeta)\,d\zeta. \end{aligned}$$

Therefore by the estimate of Proposition 6.12 we get for $n \leq 4$

$$w^{(n)}_{j,s}| \leq \mathrm{const}\cdot \sum_{r=1}^{n-1} |s|^{n-r} r(s)^r \big\| w_j\,dz\big|_{\mathcal{A}(s)}\big\|_2.$$

So for $n \leq 5$ and $s \neq s_1(j),\ s_2(j)$

$$|w^{(n)}_{j,s}| \leq \mathrm{const}\, r(s)|s|^{n-2} \big\| w_j\,dz\big|_{\mathcal{A}(s)}\big\|_2.$$

Similarly, for $s = s_\mu(j)$

$$\left| w^{(n)}_{j,s} - \frac{-(i)^{1-n}}{2\pi i}\int_{|\zeta-s|=r(s)} \frac{(-1)^{\mu+1}\zeta^{n-1}}{2\pi i(\zeta-s)}\,d\zeta\right| \\ \leq \mathrm{const}\, r(s)|s|^{n-2}\left\|\left(w_j - \frac{(-1)^{\mu+1}}{2\pi i}\frac{1}{z-s}\right)dz\Big|_{\mathcal{A}(s)}\right\|_2$$

so that for $n \geq 5$

$$\left| w^{(n)}_{j,s_\mu(j)} + (i)^{1-n}\frac{(-1)^{\mu+1}}{2\pi i} s_\mu(j)^{n-1}\right| \\ \leq \mathrm{const}\, r(s)|s|^{n-2}\left\|\left(w_j - \frac{(-1)^{\mu+1}}{2\pi i}\frac{1}{z-s}\right)dz\Big|_{\mathcal{A}(s)}\right\|_2.$$

Therefore by (6.15b)

$$\sum_{\substack{s\in S\\ s\neq s_1(j),s_2(j)}} |w^{(n)}_{j,s}| + \left| w^{(n)}_{j,s_1(j)} + w^{(n)}_{j,s_2(j)} - (i)^{1-n}\left(\frac{1}{2\pi i}s_1(j)^{n-2} - \frac{1}{2\pi i}s_2(j)^{n-2}\right)\right| \\ \leq \mathrm{const}\sum_i \left(r_1(i)|s_1(i)|^{n-1} + r_2(i)|s_2(i)|^{n-1}\right)(\Omega^j_i + \delta_{ij}\aleph_j) \tag{19.4}$$

where, as in (6.18)

$$\Omega^j_i = \begin{cases} \big\|\omega_j\big|_{Y'_i}\big\| & \text{for } i \neq j \\ \Big\|\big(\omega_j - (\phi_j)_*\big(\frac{dz_1}{2\pi i z_1}\big)\big)\big|_{Y'_i}\Big\| & \text{for } i = j \end{cases}$$

and $\aleph_j$ is as in Lemma 6.3. In paricular the sequence of the $\aleph_j$ is bounded in j. By Theorem 6.4 the norm of $(\Omega^j_i)_{i\geq g+1}$ is bounded in j. Thus the right-hand side of (19.4) is finite and bounded uniformly in j if

$$\sum_{s\in S} r(s)^2|s|^{2n-4}$$

is finite. By (GH5)(ii)

$$\sum_{s\in S} r(s)^2|s|^{2n-4} \le \sum_{s\in S} \frac{1}{|s|^{2d+4-2n}} < \infty$$

for $n \le 4$. In a similar way one sees that $|w_{j,\mathrm{com}}^{(n)}|$ is bounded uniformly in j for all $n \le 4$. □

REMARK 19.2. Let $q \in C^\infty(\mathbb{R}^2/\Gamma)$ and $\mathcal{H}(q)$ be the normalization of the associated heat curve. If $\mathcal{H}(q)$ has finite genus then the definition of U, V, W in Proposition 19.1 agrees with the one used in formula (19.1).

If $\mathcal{H}(q)$ is smooth one has for every $\gamma > 0$

$$t_b \le \mathrm{const}\big(\| |b|^{\gamma+2}|\widehat{q}(b)| \|_1\big) \cdot \frac{1}{|b|^{2\gamma}}$$

(see Theorem 14.2 and 15.2). In particular $\sum t_b^\beta < \infty$ for $\beta > 1/\gamma$. In the formula for the solution of the periodic KP-equation we use the theta function on the subspace generated by U, V, W. We will apply Proposition 4.13, so it is useful to note the estimate

$$\begin{gathered} |U_b| t_b^{1-2\beta'/2k}, \quad |V_b| t_b^{1-2\beta'/2k}, \\ |W_b| t_b^{1-2\beta'/2k} \le \mathrm{const}\big(\| \, |b|^{\gamma+2}|\widehat{q}(b)| \, \|_1\big) \cdot \frac{1}{|b|^{\gamma 1-2\beta'/k-3}} \end{gathered} \tag{19.5}$$

which follows directly from the proposition above.

20. Approximations

In this section we formalize what it means for two marked Riemann surfaces that fulfill the geometric hypotheses of §5 to be close to each other. For notational simplicity, we consider only the single sheet case $m = 1$. The definition is such that the corresponding period matrices and theta functions are also close. Then we show that every surface fulfilling the hypotheses of §5 can be approximated by surfaces of finite genus. This approximation result is in turn used to show that (19.1a) is always a solution of the differentiated version (20.16) of the KP equation.

DEFINITION 20.1. Recall that $H(t) = \{(z_1, z_2) \in \mathbb{C}^2 \mid z_1 z_2 = t \text{ and } |z_1|, |z_2| \le 1\}$ is a model handle. Let $K \ge 2$, $1 - 28/15K \ge \sqrt{t}$ and let $\widehat{Y} \subset H(t)$ be diffeomorphic to an annulus and contain $\{(z_1, z_2) \in H(t) \mid |z_1|, |z_2| \le 1 - 1/K\}$. Similarly, let $\widehat{Y}' \subset H(t')$ be diffeomorphic to an annulus and contain $\{(z_1, z_2) \in H(t') \mid |z_1|, |z_2| \le 1 - 1/K\}$. A diffeomorphism

$$f \colon \widehat{Y} \to \widehat{Y}'$$

is said to be K-quasiconformal with distortion at most ε if, firstly, f is holomorphic on $|z_1| \ge 1 - 28/15K$ and on $|z_2| \ge 1 - 28/15K$, secondly, $\big||f(z_1, z_2)_\mu / z_\mu| - 1\big| < \min\{\varepsilon, 1/15K\}$ for at least one point of $\widehat{Y}$ for each of $\mu = 1, 2$ and, thirdly, the pull-back

$$f^*\left(\frac{dz_1}{z_1}\right) = a(z_1, z_2)\frac{dz_1}{z_1} + b(z_1, z_2)\frac{d\overline{z}_1}{\overline{z}_1}$$

obeys

$$\begin{aligned}|a(z_1, z_2) - 1| &\le \min\left\{\varepsilon, \frac{1}{150K}\right\}(|z_1| + |z_2|),\\ |b(z_1, z_2)| &\le \min\left\{\varepsilon, \frac{1}{150K}\right\}(|z_1| + |z_2|),\\ \left|\frac{\partial}{\partial \bar{z}_1} a(z_1, z_2)\right| &\le \min\left\{\varepsilon, \frac{1}{150K}\right\}(|z_1| + |z_2|).\end{aligned}$$

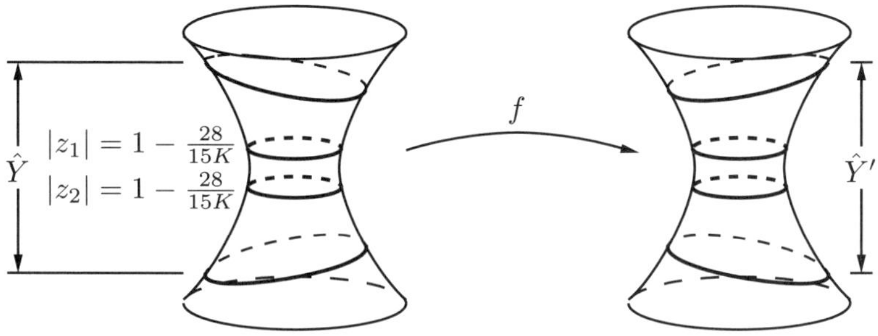

A diffeomorphism F between two marked Riemann surfaces X and X' of genus g is called K-quasiconformal with distortion at most ε if there are $t_1, \dots, t_r, t_1', \dots, t_r' > 0$, sets

$$\left\{(z_1, z_2) \in H(t) \,\middle|\, \left(1 - \frac{1}{K}\right)^{-1} t \le |z_1| \le 1 - \frac{1}{K}\right\} \subset \widehat{Y}_j \subset H(t_j),$$
$$\left\{(z_1, z_2) \in H(t) \,\middle|\, \left(1 - \frac{1}{K}\right)^{-1} t \le |z_1| \le 1 - \frac{1}{K}\right\} \subset \widehat{Y}_j' \subset H(t_j')$$

and maps $\phi_j \colon H(t_j) \to X$, $\phi_j' \colon H(t_j') \to X'$, $1 \le j \le r$ that are biholomorphic onto their images such that

- ${\phi_j'}^{-1} \circ F \circ \phi_j \restriction \widehat{Y}_j$ is K-quasiconformal with distortion at most ε for $1 \le j \le r$
- The sets $\phi_j\big(H(t_j)\big)$ are pairwise disjoint and F induces a biholomorphic map between
$$X \setminus \bigcup_{j=1}^{r} \phi_j(\widehat{Y}_j) \quad \text{and} \quad X' \setminus \bigcup_{j=1}^{r} \phi_j'(\widehat{Y}_j')$$
- $\phi_j(\{(z_1, z_2) \in H(t_j) \mid |z_1| = \sqrt{t_j}\})$ is homologous to a linear combination of A_i, $1 \le i \le g$.

REMARK. Recall that, by definition, a smooth map $U \colon S \to S'$ between Riemann surfaces is quasiconformal with Beltrami coefficient at most ε if, for every $x \in S$, there exists a holomorphic coordinate z around x and a holomorphic coordinate u around $U(x)$ such that $u(z) = u^{-1} \circ U \circ z$ obeys

$$|u_{\bar{z}}(z)| \le \varepsilon |u_z(z)|.$$

Observe that an K-quasiconformal diffeomorphism with distortion at most $\varepsilon < \frac{1}{4}$ is quasiconformal with Beltrami coefficient at most 4ε. We choose this stronger definition, which restricts the second as well as first derivatives of F, so as to be able to convert L^2 bounds on differential forms into pointwise bounds.

DEFINITION 20.2. Let $X = X^{\mathrm{com}} \cup X^{\mathrm{reg}} \cup X^{\mathrm{han}}$ and $X' = X'^{\mathrm{com}} \cup X'^{\mathrm{reg}} \cup X'^{\mathrm{han}}$ be marked Riemann surfaces fulfilling (GH1)–(GH5) with the number of sheets m of X^{reg} and m' of X'^{reg} both being one. In what follows we mark all the objects associated to X' with a prime.

Furthermore let X_0 resp. X_0' be such that

$$\begin{aligned} &X^{\mathrm{com}} \subset X_0, && X'^{\mathrm{com}} \subset X_0', \\ &\partial X_0 \subset X^{\mathrm{reg}}, && \partial X_0' \subset X'^{\mathrm{reg}}, \\ &Y_j \subset X_0 \quad \text{if} \quad Y_j \cap X_0 \neq \varnothing, && Y_j' \subset X_0' \quad \text{if} \quad Y_j' \cap X_0' \neq \varnothing. \end{aligned}$$

Let $0 < \varepsilon < \frac{1}{8}, K \geq 2$. We say that the pair (X, X_0) is (ε, K)—close to (X', X_0') if the following holds.

(i) The set $\mathcal{J} = \{j \mid Y_j \subset X_0\}$ agrees with $\{j \mid Y_j' \subset X_0'\}$. There are compact, simply connected sets $\widehat{D}(s_\mu(j)), j \in \mathcal{J}, \mu = 1, 2$ obeying

$$\begin{gathered} D(s_\mu(j)) \cup D'(s_\mu'(j)) \subset \widehat{D}(s_\mu(j)) \subset \{z \in \mathbb{C} \mid |z - s_\mu(j)| < r_\mu(j)\}, \\ \Phi(\partial \widehat{D}(s_\mu(j))) \subset \phi_j(\{(z_1, z_2) \in H(t_j) \mid \tau_\mu(j) \leq |z_\mu| \leq 2\tau_\mu(j)\}), \\ \Phi'(\partial \widehat{D}(s_\mu(j))) \subset \phi_j'(\{(z_1, z_2) \in H(t_j')\}\{\tau_\mu'(j) \leq |z_\mu| \leq 2\tau_\mu'(j)\}) \end{gathered}$$

and a diffeomorphism

$$F \colon X_0 \to X_0'$$

such that

$$\Phi'^{-1} \circ F = \Phi^{-1} \text{ on } \left\{ x \in X_0 \;\middle|\; \Phi^{-1}(x) \notin \bigcup_{\substack{j \in \mathcal{J} \\ \mu = 1,2}} \widehat{D}(s_\mu(j)) \right\}.$$

For each $j \in \mathcal{J}$

$$F \circ \phi_j(Y_j^{(\circ)}) = \phi_j'({Y^{(\circ)}}'_j)$$

and ${\phi'}_j^{-1} \circ F \circ \phi_j \restriction Y_j^{(\circ)}$ is 2-quasiconformal with distortion at most ε. The restriction of F to X^{com} is a K-quasiconformal diffeomorphism between X^{com} and X'^{com} with distortion at most ε. Furthermore

$$F_*(A_j) = A_j'$$

for all $1 \leq j \leq g = g'$.

(ii) For all $j \in \mathcal{J}$ and $\mu = 1, 2$

$$s_\mu(j) = s_\mu'(j), \quad r_\mu(j) = r_\mu'(j), \quad R_\mu(j) = R_\mu'(j)$$

(iii)

$$\begin{gathered} \|\mathfrak{A}\| \leq K, \quad \|\mathfrak{A}'\| \leq K, \\ \sum_{s \in S} \frac{1}{|s|^{d - 4\delta - 2}} \leq K, \quad \sum_{s' \in S'} \frac{1}{|s'|^{d' - 4\delta' - 2}} \leq K \end{gathered}$$

and, for all j,

$$\begin{gathered} \mathcal{O}^j \leq K^2, \quad \mathcal{O}'^j \leq K^2, \\ \aleph_j \leq K, \quad \aleph_j' \leq K, \\ (\tau_1(j)^2 + \tau_2(j)^2) \ln \frac{\tau_1(j)\tau_2(j)}{t_j} \leq \frac{K}{|\ln t_j|^2}. \end{gathered}$$

Here, $\mathfrak{A}$, $\mathcal{O}^j$ and $\aleph_j$ were defined just before Lemma 6.3. All the other data were defined in (GH1)–(GH5).

(iv) For $j \notin \mathcal{J}$

$$\mathcal{O}^j \le \varepsilon^2, \quad \mathcal{O}'^j \le \varepsilon^2,$$
$$t_j \le \varepsilon, \quad t'_j \le \varepsilon.$$

Furthermore,

$$\left\| (\mathfrak{A}_{i,j})_{\substack{i\notin\mathcal{J}\\ j\in\mathcal{J}}} \right\| \le \varepsilon, \quad \left\| (\mathfrak{A}'_{i,j})_{\substack{i\notin\mathcal{J}\\ j\in\mathcal{J}}} \right\| \le \varepsilon,$$
$$\left\| (\mathfrak{A}_{i,j})_{\substack{i>g\\ j\notin\mathcal{J}}} \right\| \le \varepsilon, \quad \left\| (\mathfrak{A}'_{i,j})_{\substack{i>g\\ j\notin\mathcal{J}}} \right\| \le \varepsilon$$

and, for $j \in \mathcal{J}$

$$\left\| (\widetilde{\Omega}_i^j)_{i\notin\mathcal{J}} \right\|_2 \le \varepsilon, \quad \left\| (\widetilde{\Omega}_i'^j)_{i\notin\mathcal{J}} \right\|_2 \le \varepsilon$$

where $\widetilde{\Omega}$ was defined in (6.13).

(v) There exists a $\gamma > 0$, a collar T of $\Phi^{-1}(\partial X^{\rm com})$ that is contained in $\Phi^{-1}(X_0 \cap X^{\rm reg})$ and a curve Γ in $\Phi^{-1}(X^{\rm reg} \cap X_0)$ of length at most K such that

- for every $j > g$, the points $s_1(j)$ and $s_2(j)$, can be connected by a curve in $\{z \in G \mid \operatorname{dist}(z,T)^2/1 + |z|^2 \ge \gamma/K$. So can the points $s'_1(j)$ and $s'_2(j)$.
- for any holomorphic function w on T obeying $\int_{\Phi^{-1}(\partial X^{\rm com})} w(\zeta)\, d\zeta = 0$

$$\left| \frac{1}{2\pi i} \int_{\Phi^{-1}(\partial X^{\rm com})} \frac{w(\zeta)}{\zeta - z}\, d\zeta \right| \le \frac{\gamma}{\operatorname{dist}(z,T)^2} \| w|_T \|_2.$$

- $\Phi(\Gamma)$ decomposes X into a compact connected component $X(\Gamma)$ containing $X^{\rm com} \cup \Phi(T)$ and a noncompact component such that for all $j \ge g+1$ either $Y_j^{(\circ)} \subset X(\Gamma)$ or $Y_j^{(\circ)} \cap X(\Gamma) = \varnothing$.
- Let $\mathcal{J}(\Gamma) = \{ i \in \mathcal{J} \mid Y_i^{(\circ)} \subset X(\Gamma) \}$.

$$\left\| (\mathfrak{A}_{i,k})_{\substack{i\notin\mathcal{J}(\Gamma)\\ k\ge g+1}} \right\| < \frac{1}{8}, \quad \left\| (\mathfrak{A}'_{i,k})_{\substack{i\notin\mathcal{J}(\Gamma)\\ k\ge g+1}} \right\| < \frac{1}{8},$$

$$\sum_{\substack{\mu=1,2\\ i\ge g+1\\ i\notin\mathcal{J}(\Gamma)}} \frac{R_\mu(i)^2}{\operatorname{dist}(s_\mu(i),T)^4} < \frac{1}{2^{10}\pi^2\gamma^2}, \quad \sum_{\substack{\mu=1,2\\ i\ge g+1\\ i\notin\mathcal{J}(\Gamma)}} \frac{R'_\mu(i)^2}{\operatorname{dist}(s'_\mu(i),T)^4} < \frac{1}{2^{10}\pi^2\gamma^2}$$

- $$\operatorname{length}(\Gamma) \sup_{\substack{j\ge g+1\\ z\in\Gamma}} \left(\sum_{\mu=1,2} \frac{3r_\mu(j)}{|z - s_\mu(j)|^2} \aleph_j + \frac{1}{2\pi} \frac{1}{|z - s_1(j)|} + \frac{1}{2\pi} \frac{1}{|z - s_2(j)|} \right) \le K,$$

$$\operatorname{length}(\Gamma) \sup_{\substack{j\ge g+1\\ z\in\Gamma}} \left(\sum_{\mu=1,2} \frac{3r'_\mu(j)}{|z - s'_\mu(j)|^2} \aleph'_j + \frac{1}{2\pi} \frac{1}{|z - s'_1(j)|} + \frac{1}{2\pi} \frac{1}{|z - s'_2(j)|} \right) \le K,$$

$$\operatorname{length}(\Gamma) \sup_{\substack{j\ge g+1\\ j\notin\mathcal{J}_c\\ z\in\Gamma}} \left(\sum_{\mu=1,2} \frac{3r_\mu(j)}{|z - s_\mu(j)|^2} \aleph_j + \frac{1}{2\pi} \frac{1}{|z - s_1(j)|} + \frac{1}{2\pi} \frac{1}{|z - s_2(j)|} \right) \le \varepsilon,$$

$$\operatorname{length}(\Gamma) \sup_{\substack{j\geq g+1 \\ j\notin \mathcal{J}_c \\ z\in\Gamma}} \left(\sum_{\mu=1,2} \frac{3r'_\mu(j)}{|z-s'_\mu(j)|^2}\aleph'_j + \frac{1}{2\pi}\frac{1}{|z-s'_1(j)|} + \frac{1}{2\pi}\frac{1}{|z-s'_2(j)|} \right) \leq \varepsilon$$

- $$\operatorname{dist}(\Gamma, T)^2 \geq 4\gamma \operatorname{length}(\Gamma),$$
- $$\operatorname{length}(\Gamma)^2 \sum_k \sup_{\mu=1,2} \frac{36 r_\mu(k)^2}{\operatorname{dist}(s_\mu(k),\Gamma)^4} \leq \frac{1}{16}$$

(vi) For each $1 \leq i \leq g$ there exist L_i, δ_i with $L_i/\delta_i \leq K^2$ and a quasiconformal diffeomorphism u_i, with Beltrami coefficient bounded by $\frac{1}{2}$, from $\mathcal{U} = \mathbb{R}/L_i\mathbb{Z}\times[0,\delta_i]$ into X^{com} with $u_i(\mathbb{R}/L_i\mathbb{Z}\times\{0\}) = B_i$.

First we observe that every marked Riemann surface $X = X^{\mathrm{com}} \cup X^{\mathrm{reg}} \cup X^{\mathrm{han}}$ is close to one of finite genus.

PROPOSITION 20.3. *Let $X = X^{\mathrm{com}} \cup X^{\mathrm{reg}} \cup X^{\mathrm{han}}$ be a marked Riemann surface with $m = 1$ fulfilling* (GH1)–(GH5) *such that*

$$\sup_j \left(\tau_1(j)^2 + \tau_2(j)^2\right) |\ln t_j|^2 \ln \frac{\tau_1(j)\tau_2(j)}{t_j} < \infty.$$

Then there is $K > 0$ such that for every compact subset Z of X and any $\varepsilon > 0$ there is

- *a submanifold X_0 of X with boundary,*
- *a marked Riemann surface X' of genus genus(X_0) and*
- *a compact submanifold $X'_0 \subset X'$.*

such that X_0 contains Z and (X', X'_0) is (ε, K)—close to (X, X_0) and the K-quasiconformal diffeomorphism F is biholomorphic on X^{com}.

PROOF. By Lemma 6.9(b) there is a collar T of $\Phi^{-1}(\partial X^{\mathrm{com}})$ and a constant γ such that the second bullet of condition (v) in Definition 20.2 is satisfied. By Lemma 6.1, there exists a curve Γ in $\phi^{-1}(X^{\mathrm{reg}})$ satisfying the third, fourth, sixth and seventh bullets of (v). By Lemma 6.3 and (GH5)(ii) there is $K \geq 0$ such that the conditions (iii) in Definition 20.2 are fulfilled for X. It is clearly also possible to choose K large enough that the first bullet and the first half of the fifth bullet of (v) are satisfied. Again possibly enlarging K, it is possible to choose L_i, δ_i, $\mu_i > 0$, $1 \leq i \leq g$, and quasiconformal diffeomorphisms u_i, $1 \leq i \leq g$ satisfying condition (vi).

By Lemma 6.3 and (GH2)(iv) it is possible to choose n so that condition (iv) and the second part of the fifth bullet of (v) are satisfied for $j > n$. Choose X_0 such that $Z \subset X_0$ and $Y_j \subset X_0$ for all $g < j \leq n$. This can be done by putting $X_0 = X(\Gamma')$ for a suitable Γ' as in Lemma 6.1. Put $\mathcal{J} = \{j \mid Y_j \subset X_0\}$. We define X' as the Riemann surface obtained by glueing X_0 to $\mathbb{C}\setminus\Big(K \cup \bigcup_{\substack{j\in\mathcal{J} \\ \mu=1,2}} \operatorname{int} D\big(s_\mu(j)\big)\Big)$ along $X_0 \cap X^{\mathrm{reg}}$ resp. $\Phi^{-1}(X_0 \cap X^{\mathrm{reg}})$ using Φ as a glueing map. Furthermore, define X'_0 to be the part of X' corresponding to X_0 in this construction and $F\colon X_0 \to X'_0$ to be the identity map. Conditions (i) and (ii) are trivially satisfied and the required bounds on primed quantities are inherited from the corresponding bounds on unprimed quantities. □

THEOREM 20.4. *Let $X = X^{\mathrm{com}} \cup X^{\mathrm{reg}} \cup X^{\mathrm{han}}$ and $X' = X'^{\mathrm{com}} \cup X'^{\mathrm{reg}} \cup X'^{\mathrm{han}}$ be marked Riemann surfaces with $m = m' = 1$ fulfilling* (GH1)–(GH5) *and let*

$X_0 \subset X$, $X_0' \subset X'$ be compact submanifolds with boundary in X and X' respectively. Assume that (X, X_0) and (X', X_0') are (ε, K)–close, with ε smaller than some strictly positive universal constant. Let $F\colon X_0 \to X_0'$ be a diffeomorphism as in part (i) of Definition 20.2. *Then there is a numerical constant* const *independent of ε, K, X and X' such that*

(a) *For all $j \in \mathcal{J}$*

$$|t_j - t_j'| \leq \text{const}\, \varepsilon t_j.$$

(b) *Define $\mathcal{J}_c = \{i \mid Y_i \subset X_0\} \cup \{1, \ldots, g\}$. For $i, j \in \mathcal{J}_c$*

$$\left| \int_{B_i} \omega_j - \int_{B_i'} \omega_j' \right| \leq \text{const}\, \varepsilon K^4.$$

(c) *For every compact subset $\mathcal{K}$ in the universal covering $\pi\colon \widetilde{X}_0 \to X_0$ there is a constant $C_{\mathcal{K}}$ which depends only on $\mathcal{K}$, not on X', such that for all $x_1, x_2 \in \mathcal{K}$ and all $j \in \mathcal{J}_c = \mathcal{J}$*

$$\left| \int_{x_1}^{x_2} (\omega_j - F^* \omega_j') \right| \leq \varepsilon K^6 C_{\mathcal{K}}.$$

(d) *Consider the quantities U, V, W of Proposition* 19.1. *Then for $j \in \mathcal{J}_c$*

$$|U_j - U_j'| \leq \text{const}\, \varepsilon K^5, \quad |V_j - V_j'| \leq \text{const}\, \varepsilon K^5, \quad |W_j - W_j'| \leq \text{const}\, \varepsilon K^6$$

and for $j \notin \mathcal{J}_c$

$$\left| U_j + \frac{1}{2\pi}\big(s_1(j) - s_2(j)\big) \right| \leq \text{const}\, \varepsilon K^3, \quad \left| V_j + \frac{1}{2\pi i}\big(s_1(j)^2 - s_2(j)^2\big) \right| \leq \text{const}\, \varepsilon K^3,$$

$$\left| W_j - \frac{1}{2\pi}\big(s_1(j)^3 - s_2(j)^3\big) \right| \leq \text{const}\, \varepsilon K^4.$$

To prepare for the proof we first note the following

Lemma 20.5 (Modification of Lemma 6.8). *Let $\sqrt{t} < a < A < 1$. Let f be a differentiable function on a neighbourhood of the annulus $\{z \in \mathbb{C} \mid t \leq |z| \leq 1\}$. Let C_1 and C_2 be curves (without self-intersection) of winding number one in the outer annulus $\{z \in \mathbb{C} \mid A \leq |z| \leq 1\}$ and inner annulus $\{z \in \mathbb{C} \mid t \leq |z| \leq t/A\}$ respectively. Suppose that*

$$\int_{C_1} f(\zeta) \frac{d\zeta}{\zeta} = 0.$$

Then, for all $t/a \leq |z| \leq a$

$$|f(z)| \leq \frac{|z|}{2\pi(A-a)} \int_{C_1} \left| f(\zeta) \frac{d\zeta}{\zeta} \right| + \frac{t}{2\pi(A-a)|z|} \int_{C_2} \left| f(\zeta) \frac{d\zeta}{\zeta} \right|$$
$$+ \frac{1}{2\pi} \left| \int_{R_{12}} \frac{f_{\bar z}(\zeta)}{\zeta - z}\, d\zeta \wedge d\bar\zeta \right|$$

where R_{12} is the region between C_1 and C_2.

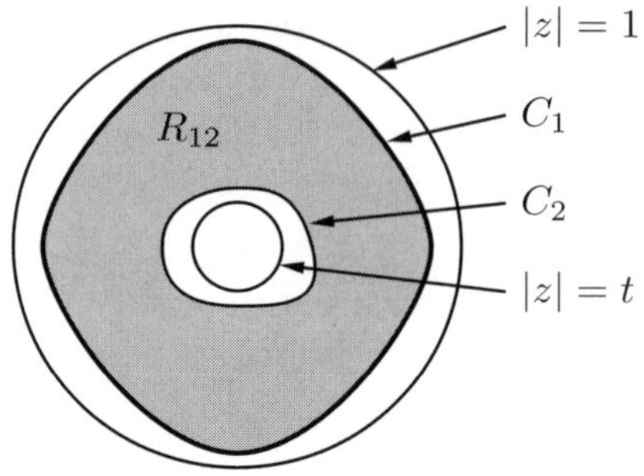

PROOF. By Cauchy's integral formula

$$\begin{aligned} f(z) &= \frac{1}{2\pi i}\left(\int_{C_1} f(\zeta)\frac{d\zeta}{\zeta - z} - \int_{C_2} f(\zeta)\frac{d\zeta}{\zeta - z} + \int_{R_{12}} \frac{f_{\bar z}(\zeta)}{\zeta - z}\, d\zeta \wedge d\bar\zeta\right) \\ &= \frac{1}{2\pi i}\left(\int_{C_1} f(\zeta)\frac{d\zeta}{\zeta} + z\int_{C_1} \frac{1}{\zeta - z} f(\zeta)\frac{d\zeta}{\zeta} - \int_{C_2} \frac{\zeta}{\zeta - z} f(\zeta)\frac{d\zeta}{\zeta} \right. \\ &\qquad\qquad \left. + \int_{R_{12}} \frac{f_{\bar z}(\zeta)}{\zeta - z}\, d\zeta \wedge d\bar\zeta\right). \end{aligned}$$

For $t/a \le |z| \le a$ we have

$$\frac{1}{|\zeta - z|} \le \frac{1}{A - a} \quad \text{if } \zeta \in C_1,$$
$$\left|\frac{\zeta}{\zeta - z}\right| \le \frac{t}{|z|(A - a)} \quad \text{if } \zeta \in C_2.$$

As $\int_{C_1} f(\zeta)\frac{d\zeta}{\zeta} = 0$ we get the desired estimate. □

For the proof of Theorem 20.4, we define

$$\Omega_i^j = \left\| \left(\omega_j - \delta_{ij}\frac{1}{2\pi i}(\phi_j)_*\left(\frac{dz_1}{z_1}\right)\right)\Big|_{Y_i^{(\circ)}} \right\|_2$$

and we define $\Omega_i'^j$ analogously. Recall that $Y_i^{(\circ)}$ is the cylinder in Y_i bounded by the curves $\Phi\big(\{z \in \mathbb{C} \mid |z - s_\mu(i)| = R_\mu(i)\}\big)$, $\mu = 1, 2$. Furthermore write

$$\Phi^*(\omega_j) = w_j(z)\, dz, \quad \Phi'^*(\omega_j') = w_j'(z)\, dz.$$

We plan to mimic the proof of Theorem 6.4. We put, for $i \in \mathcal{J}$, $j \in \mathcal{J}_c$

$$D_i^j = 2\pi \max_{\mu = 1,2} R_\mu(i) \sup_{|z - s_\mu(i)| = R_\mu(i)} |w_j(z) - w_j'(z)|$$

and define $Y_i^{(\circ\circ)}$ to be the cylinder in Y_i bounded by the curves

$$\Phi\big(\{z \in \mathbb{C} \mid |z - s_\mu(i)| = 4r_\mu(i)\}\big), \quad \mu = 1, 2$$

LEMMA 20.6. *There is a numerical constant* const *independent of X and X' such that for $i \in \mathcal{J}$, $j \in \mathcal{J}_c$*

$$\big\|(\omega_j - F^*\omega_j')\big|_{Y_i^{(\circ\circ)}}\big\|_2 \le D_i^j + \varepsilon \operatorname{const} K(\Omega_i'^j + \delta_{ij}). \tag{20.1}$$

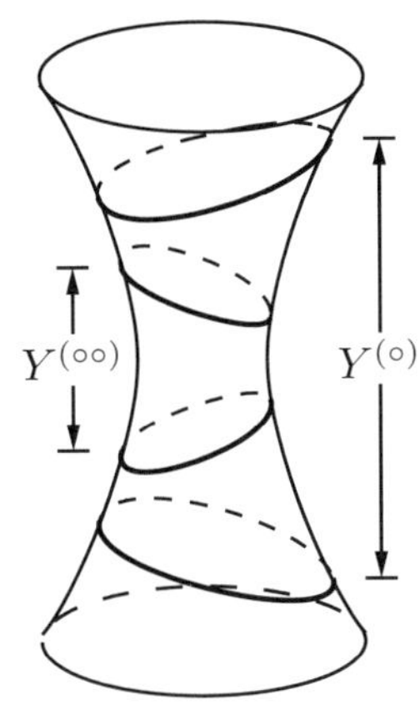

More precisely, if one writes on the model handle $H(t_i)$

$$\phi_i^*(\omega_j - F^*\omega_j') = \alpha_j^{(i)}(z_1, z_2)\frac{dz_1}{z_1} + \beta_j^{(i)}(z_1, z_2)\frac{d\bar{z}_1}{\bar{z}_1}.$$

Then on $\phi_i^{-1}(Y_i^{(\circ\circ)})$

$$(20.2)\qquad \begin{aligned} |\alpha_j^{(i)}(z_1, z_2)| &\le \frac{3}{\pi}(|z_1| + |z_2|)D_i^j + \text{const}\frac{\varepsilon K}{|\ln t_i|}(\Omega_i'^j + \delta_{ij}), \\ |\beta_j^{(i)}(z_1, z_2)| &\le \varepsilon\text{const}(|z_1| + |z_2|)(\Omega_i'^j + \delta_{ij}) \end{aligned}$$

$\alpha_j^{(i)}$ *is holomorphic outside* $\{(z_1, z_2) \in H(t_i) \mid |z_1| \le 2\tau_i$ *and* $|z_2| \le 2\tau_i\}$ *and* $\beta_j^{(i)}$ *is zero outside this set.*

PROOF. Clearly (20.1) follows from (20.2), so it suffices to prove (20.2). To simplify the notation we delete the sub and superscripts i and j whenever convenient, write $\widehat{F}$ for $\phi_i'^{-1} \circ F \circ \phi_i$, z for the variable z_1 on $H(t_i)$ and $u = u(z)$ for the first component of $\widehat{F}(z, t_i/z)$. With this notation

$$(20.3)\qquad \frac{du}{u} = a(z)\frac{dz}{z} + b(z)\frac{d\bar{z}}{\bar{z}}$$

On $\phi_i^{-1}\big(\{x \in X^{\text{reg}} \mid \Phi^{-1}(x) \notin \widehat{D}\big(s_1(i)\big) \cup \widehat{D}\big(s_2(i)\big)\}\big)$ the function $a(z)$ is holomorphic and $b(z)$ is zero. By part (i) of Definition 20.2 and Definition20.1

$$d\ln\frac{u}{z} = \frac{du}{u} - \frac{dz}{z} = [a(z) - 1]\frac{dz}{z} + b(z)\frac{d\bar{z}}{\bar{z}}$$

and, on $Y^{(\circ)}$,

$$\left|[a(z) - 1]\frac{dz}{z}\right| \le \varepsilon[|dz_1| + |dz_2|], \qquad \left|b(z)\frac{d\bar{z}}{\bar{z}}\right| \le \varepsilon[|dz_1| + |dz_2|].$$

Consequently $\ln u/z$ varies by at most $4\pi\varepsilon$ over the handle. Since, by the second requirement of Definition 20.1, we have that $1 - \varepsilon \le |u/z| \le 1 + \varepsilon$ for at least one $z \in Y^{(\circ)}$ there is a $\varphi \in \mathbb{R}$ such that

$$(20.4)\qquad \left|\frac{u}{z} - e^{i\varphi}\right| \le \text{const}\,\varepsilon.$$

Part (a) of Theorem 20.4 follows from (20.4) and its analogue, $|u_2/z_2 - e^{-i\varphi}| \le \text{const}\,\varepsilon$ since

$$|t_i - t_i'| = |z_1z_2 - u_1u_2| \le |z_1e^{i\varphi} - u_1|\,|z_2| + |u_1||z_2e^{-i\varphi} - u_2| \le \text{const}\,\varepsilon|z_1z_2| + \text{const}|u_1|\varepsilon|z_2| \le \text{const}\,\varepsilon|z_1||z_2| = \text{const}\,\varepsilon t_i. \tag{20.5}$$

Write

$$\phi_i'^*\omega_j' = g'(u)\frac{du}{u}, \quad \phi_i^*\omega_j = g(z)\frac{dz}{z}.$$

By Lemma 6.8, scaled to the handle $H(4t)$ as in Proposition 6.16,

$$\left|g'(u) - \frac{\delta_{ij}}{2\pi}\right| \le \text{const}\left(|u| + \frac{t'}{|u|}\right)\Omega_i'^j, \tag{20.6a}$$

$$\left|g(z) - \frac{\delta_{ij}}{2\pi}\right| \le \text{const}\left(|z| + \frac{t}{|z|}\right)\Omega_i^j \tag{20.6b}$$

for $3t' \le |u| \le \frac{1}{3}$ and $3t \le |z| \le \frac{1}{3}$. Now

$$\phi_i^*\omega_j - \phi_i^*F^*\omega_j' = g(z)\frac{dz}{z} - g'(u)\left(a(z)\frac{dz}{z} + b(z)\frac{d\bar z}{\bar z}\right) = \alpha(z)\frac{dz}{z} + \beta(z)\frac{d\bar z}{\bar z}$$

with

$$\alpha(z) = g(z) - g'(u)a(z), \quad \beta(z) = -g'(u)b(z).$$

Clearly $\beta(z)$ is zero on $\{(z_1, z_2) \in H(t_i) \mid |z_1| > 2\tau_1 \text{ or } |z_2| > 2\tau_2\}$. By (20.6a), part (i) of Definition 20.2 and Definition 20.1

$$|\beta(z)| \le \text{const}\,\varepsilon\left(|z| + \frac{t}{|z|}\right)(\Omega_i'^j + \delta_{ij}).$$

To bound $|\alpha(z)|$ we apply Lemma 20.5 to $f(z) = \alpha(z)$,

$$C_1 = \{z \in H(t_i) \mid |g_{i1}(z_1) - s_1(i)| = R_1(i)\},$$
$$C_2 = \{z \in H(t_i) \mid |g_{i2}(z_2) - s_2(i)| = R_2(i)\}.$$

On C_1, $\beta(z) = 0$ so that the hypothesis $\int_{C_1} c\alpha(z)/z\,dz = \int_{C_1} \phi_i^*\omega_j - \phi_i^*F^*\omega_j' = 0$ is satisfied. The bounds on the first two terms in the conclusion of Lemma 20.5 are

$$\frac{3|z|}{\pi}\int_{C_1}\left|\alpha(\zeta)\frac{d\zeta}{\zeta}\right| \le \frac{3|z|}{\pi}\int_{|\xi - s_1(i)| = R_1(i)} |w_j(\xi) - w_j'(\xi)|\,|d\xi| \le \frac{3|z|}{\pi}D_i^j,$$
$$\frac{3t_j}{\pi|z|}\int_{C_2}\left|\alpha(\zeta)\frac{d\zeta}{\zeta}\right| \le \frac{3t_j}{\pi|z|}\int_{|\xi - s_2(i)| = R_2(i)} |w_j(\xi) - w_j'(\xi)|\,|d\xi| \le \frac{3t_j}{\pi|z|}D_i^j.$$

To bound the third term, we observe that

$$\alpha_{\bar z} = -(g_u'u_{\bar z}a + g'a_{\bar z}) = -\left(\frac{u}{\bar z}g_u'ab + g'a_{\bar z}\right)$$

by (20.3), since $g(z)$ and $g'(u)$ are holomorphic. By Cauchy's estimate and (20.6a)

$$|g_u'| \le \text{const}\left(1 + \frac{t'}{|u|^2}\right)\Omega_i'^j$$

for $6t' \le |u| \le \frac{1}{6}$. By (GH3) and Definition 20.2(i), the support of b and $a_{\bar z}$ is contained in $\{z \in \mathbb{C} \mid t/2\tau_2 \le |z| \le 2\tau_1\} \subset \{z \in \mathbb{C} \mid 6t' \le |u| \le \frac{1}{6}\}$. Therefore,

$$\int_{R_{12}} \left| \frac{\alpha_{\bar z}}{\zeta - z} \, d\zeta \wedge d\bar\zeta \right| \le \varepsilon \mathrm{const}(\Omega_i'^j + \delta_{ij}) \int_{t/2\tau_2 \le |\zeta| \le 2\tau_1} \left[\left(|\zeta| + \frac{t}{|\zeta|} \right)^2 \frac{1}{|\zeta|} + \left(|\zeta| + \frac{t}{|\zeta|} \right) \right] \left| \frac{d\zeta \wedge d\bar\zeta}{\zeta - z} \right|$$
$$\le \varepsilon \mathrm{const}(\Omega_i'^j + \delta_{ij}) \int_{t/2\tau_2}^{2\tau_1} dr \left[r + \frac{t}{r} + \frac{t^2}{r^3} \right] \int_{-\pi}^{\pi} d\phi \frac{r}{\big||z| - re^{i\phi}\big|}.$$

Since

$$\int_{-\pi}^{\pi} d\phi \frac{1}{\big||z/r| - e^{i\phi}\big|} \le \mathrm{const}\big(1 + \big|\ln \big||z| - r\big|\big| + |\ln r|\big)$$

we get, using Definition 20.2(iii)

$$\int_{R_{12}} \left| \frac{\alpha_{\bar z}}{\zeta - z} \, d\zeta \wedge d\bar\zeta \right| \le \varepsilon \mathrm{const}(\Omega_i'^j + \delta_{ij})(\tau_1^2 + \tau_2^2) \ln \frac{\tau_1 \tau_2}{t^2}$$
$$\le \varepsilon \mathrm{const}(\Omega_i'^j + \delta_{ij}) \frac{K}{|\ln t|}. \qquad \square$$

We shall first prove Theorem 20.4 under the additional hypotheses that $X^{\mathrm{com}} = X'^{\mathrm{com}} = \varnothing$ and $\|\mathfrak{A}\|, \|\mathfrak{A}'\| < \frac{1}{4}$.

LEMMA 20.7. *Assume that* $X^{\mathrm{com}} = X'^{\mathrm{com}} = \varnothing$ *and* $\|\mathfrak{A}\|, \|\mathfrak{A}'\| < \frac{1}{4}$. *Then*

$$\begin{aligned} &\|\Omega^j\|_2, \|\Omega'^j\|_2 \le 3K && \text{for all } j, \\ &\|\Omega^j\|_2, \|\Omega'^j\|_2 \le 3\varepsilon K && \text{for } j \notin \mathcal{J}, \\ &\|(\Omega_i^j)_{i\notin\mathcal{J}}\|_2, \|(\Omega_i'^j)_{i\notin\mathcal{J}}\|_2 \le 9\varepsilon K && \text{for all } j, \\ &\|D^j\|_2 \le \mathrm{const}\, \varepsilon K^2 && \text{for } j \in \mathcal{J}. \end{aligned}$$

PROOF. By the inequality following (6.17)

$$\|\Omega^j\|_2 \le 2\big(\sqrt{\mathcal{O}^j} + \big\|\big(\mathfrak{A}_{ij}\aleph_j\big)_{i>g}\big\|\big)$$

and one has the same inequality for the primed objects. So the assertions of the first two lines of the lemma follow from parts (iii) and (iv) of Definition 20.2.

Inequality (6.16a) yields for $j \in \mathcal{J}$

$$\big\|(\Omega_i^j)_{i\notin\mathcal{J}}\big\|_2 \le \|(\widetilde\Omega_i^j)_{i\notin\mathcal{J}}\|_2 + \aleph_j \big\|(\mathfrak{A}_{i,j})_{i\notin\mathcal{J}}\big\|_2 + \left\|(\mathfrak{A}_{i,k})_{\substack{i\notin\mathcal{J}\\ k\in\mathcal{J}}}\right\| \|\Omega^j\|_2 + \left\|(\mathfrak{A}_{i,k})_{\substack{i\notin\mathcal{J}\\ k\notin\mathcal{J}}}\right\| \big\|(\Omega_k^j)_{k\notin\mathcal{J}}\big\|_2.$$

So the third assertion of the Lemma follows from the preceeding ones and part (iv) of Definition 20.2.

As in Proposition 6.5 we write

$$w_j(z) = \sum_{s\in S} w_{j,s}(z) \tag{20.7a}$$

with

$$w_{j,s}(z) = -\frac{1}{2\pi i} \int_{|\zeta - s| = r(s)} \frac{w_j(\zeta)}{\zeta - z} \, d\zeta.$$

Define $w'_{j,s'}$ in the same way so that

$$w'_j(z) = \sum_{s' \in S'} w'_{j,s'}(z) \tag{20.7b}$$

By estimate (6.9) in Proposition 6.5, we have for $i, j \in \mathcal{J}$, $k \notin \mathcal{J}$

$$\begin{aligned} 2\pi \max_{\mu=1,2} R_\mu(i) \sup_{|z-s_\mu(i)|=R_\mu(i)} & \left| w_{j,s_1(k)}(z) + w_{j,s_2(k)}(z) \right| \\ & \leq 24\pi \max_{\mu,\tau=1,2} R_\mu(i) r_\tau(k) \frac{1}{|s_\mu(i) - s_\tau(k)|^2} \Omega_k^j \\ & \leq \mathfrak{A}_{i,k} \Omega_k^j \end{aligned} \tag{20.8a}$$

and similarly

$$2\pi \max_{\mu=1,2} R_\mu(i) \sup_{|z-s_\mu(i)|=R_\mu(i)} |w'_{j,s'_1(k)}(z) + w'_{j,s'_2(k)}(z)| \leq \mathfrak{A}'_{i,k} \Omega'^j_k. \tag{20.8b}$$

Furthermore, by Lemma 6.9(c), for $k \in \mathcal{J}$

$$|w_{j,s_\tau(k)}(z) - w'_{j,s_\tau(k)}(z)| \leq \frac{3r_\tau(k)}{|z - s_\tau(k)|^2} \|(\omega_j - F^*\omega'_j)|_{Y_k^{\circ\circ}}\|_2$$

if $|z - s_\mu(i)| = R_\mu(i)$ for some $i \in \mathcal{J}$ and $\mu = 1, 2$. Therefore

$$\begin{aligned} 2\pi \max_{\mu=1,2} R_\mu(i) \sup_{|z-s_\mu(i)|=R_\mu(i)} & \left| \sum_{\tau=1}^{2} w_{j,s_\tau(k)}(z) - w'_{j,s_\tau(k)}(z) \right| \\ & \leq \mathfrak{A}_{i,k} \left\| (\omega_j - F^*\omega'_j) \right|_{Y_k^{\circ\circ}} \Big\|_2. \end{aligned} \tag{20.9}$$

Inserting (20.7), (20.8), (20.9) in the definition of D_i^j we get

$$D_i^j \leq \sum_{k \in \mathcal{J}} \mathfrak{A}_{i,k} \left\| (\omega_j - F^*\omega'_j) \right|_{Y_k^{\circ\circ}} \Big\|_2 + \sum_{k \notin \mathcal{J}} (\mathfrak{A}_{i,k} \Omega_k^j + \mathfrak{A}'_{i,k} \Omega'^j_k). \tag{20.10}$$

In this inequality we insert the first statement of Lemma 20.6 and get, for $i, j \in \mathcal{J}$

$$D_i^j \leq \sum_{k \in \mathcal{J}} \mathfrak{A}_{i,k} D_k^j + \sum_{k \notin \mathcal{J}} (\mathfrak{A}_{i,k} \Omega_k^j + \mathfrak{A}'_{i,k} \Omega'^j_k) + \varepsilon \text{const}\, K \sum_{k \in \mathcal{J}} \mathfrak{A}_{i,k} (\Omega'^j_k + \delta_{jk}).$$

As $\|\mathfrak{A}\| < \frac{1}{4}$ we get

$$\begin{aligned} \|D^j\|_2 & \leq 2\|\mathfrak{A}\| \left\| (\Omega_k^j\,)_{k \notin \mathcal{J}} \right\|_2 + 2\|\mathfrak{A}'\| \left\| (\Omega'^j_k)_{k \notin \mathcal{J}} \right\|_2 + \varepsilon \text{const}\, K \|\mathfrak{A}\| (\|\Omega'^j\| + 1) \\ & \leq \text{const}\, \varepsilon K^2 \end{aligned}$$

by the first and third lines of this lemma. □

Lemmas 20.6 and 20.7 combined give pointwise bounds on $(\omega_j - F^*\omega'_j)\big|_{Y_i^{\circ\circ}}$ for $i, j \in \mathcal{J}$. Pointwise bounds on the regular pieces are given by

LEMMA 20.8. *Assume that* $X^{\text{com}} = X'^{\text{com}} = \varnothing$ *and* $\|\mathfrak{A}\|, \|\mathfrak{A}'\| < \frac{1}{4}$.

(a) *Define*

$$U_R = \{z \in \mathbb{C} \mid |z - s| \geq R(s), |z - s'| \geq R(s') \text{ for all } s \in S, s' \in S'\},$$
$$U_r = \{z \in \mathbb{C} \mid |z - s| \geq 4r(s), |z - s'| \geq 4r(s') \text{ for all } s \in S, s' \in S'\}.$$

There is a universal constant independent of X, X' etc. such that for all $j \in \mathcal{J}$, $z \in U_R$

$$|w_j(z) - w'_j(z)| \le \text{const}\,\frac{\varepsilon K^3}{1+|z|^2}.$$

If $z \in U_r \setminus U_R$ then there exists an $s \in S$ and/or an $s' \in S'$ such that $|z-s| \le R(s)$ and/or $|z-s'| \le R(s')$. Then, for all $j \in \mathcal{J}$

$$|w_j(z) - w'_j(z)| \le \text{const}\,\varepsilon K^2\left(\frac{r(s)}{|z-s|^2} + \frac{r'(s')}{|z-s'|^2}\right) + \text{const}\,\frac{\varepsilon K^3}{1+|z|^2}.$$

(b) *For $j \notin \mathcal{J}$*

$$\left| w_j(z) - \frac{1}{2\pi i}\frac{1}{z-s_1(j)} + \frac{1}{2\pi i}\frac{1}{z-s_2(j)}\right| \le \text{const}\,\frac{\varepsilon K^2}{1+|z|^2}$$

if $|z-s| \ge R(s)$ for all $s \in S$. If for some $s \in S$, $4r(s) \le |z-s| \le R(s)$ then

$$\left| w_j(z) - \frac{1}{2\pi i}\frac{1}{z-s_1(j)} + \frac{1}{2\pi i}\frac{1}{z-s_2(j)}\right| \le \text{const}\,\varepsilon K\left(\frac{r(s)}{|z-s|^2}\right) + \text{const}\,\frac{\varepsilon K^2}{1+|z|^2}.$$

Similar bounds apply to $w'_j(z)$.

(c) *For $i \in \mathcal{J}$, $j \notin \mathcal{J}$*

$$\left|\frac{\phi_i^*(\omega_j)}{dz_1/z_1}\right| \le \text{const}\,\varepsilon K(|z_1|+|z_2|) \quad \textit{if } |z_1|,|z_2| \le \tfrac14,$$
$$\left|\frac{\phi_i^* F^*\omega'_j}{dz_1/z_1}\right| \le \text{const}\,\varepsilon K(|z_1|+|z_2|) \quad \textit{if } |z_1|,|z_2| \le \tfrac14.$$

PROOF. (a) By Lemma 6.9(c) and Proposition 6.5

$$\begin{aligned}
&|w_j(z) - w'_j(z)| \\
&\le \sum_{\substack{k\in\mathcal{J}\\ \mu=1,2}} |w_{j,s_\mu(j)}(z) - w'_{j,s_\mu(j)}(z)| + \sum_{\substack{k\notin\mathcal{J}\\ \mu=1,2}} |w_{j,s_\mu(j)}(z)| + \sum_{\substack{k\notin\mathcal{J}\\ \mu=1,2}} |w'_{j,s'_\mu(j)}(z)| \\
&\le \sum_{\substack{k\in\mathcal{J}\\ \mu=1,2}} \frac{3r_\mu(k)}{|z-s_\mu(k)|^2}\big\|(\omega_j - F^*\omega'_j)|_{Y_k^{\circ\circ}}\big\|_2 + \sum_{\substack{k\notin\mathcal{J}\\ \mu=1,2}} \frac{3r_\mu(k)}{|z-s_\mu(k)|^2}\Omega_k^j + \sum_{\substack{k\notin\mathcal{J}\\ \mu=1,2}} \frac{3r'_\mu(k)}{|z-s'_\mu(k)|^2}\Omega_k'^j \\
&\le \text{const}\,\varepsilon K^2\left(\sum_{s\in S}\frac{r(s)}{|z-s|^2} + \sum_{s'\in S'}\frac{r'(s')}{|z-s'|^2}\right)
\end{aligned}$$

by Lemmas 20.6 and 20.7. As in (6.3), but using part (iii) of Definition 20.2,

$$\sum_{s\in S}\frac{r(s)}{|z-s|^2} + \sum_{s'\in S'}\frac{r'(s')}{|z-s'|^2} \le \text{const}\,\frac{K}{1+|z|^2}.$$

The claim follows.

(b) follows from Proposition 6.5 and Lemma 20.7.

(c) The first line follows from Lemma 6.8, applied to the scaled handle $H(4t_i)$, and Lemma 20.7. To prove the second line observe that

$$\phi_i^* F^*\omega'_j = g'(u_1)\left(a(z_1)\frac{dz_1}{/}z_1 + b(z_1)\frac{d\bar z_1}{\bar z_1}\right)$$

with $g'(u_1)$ estimated in (20.6a). Definition 20.2 (i) and Definition 20.1 provide bounds on $a(z_1)$ and $b(z_1)$. □

PROOF OF THEOREM 20.4 (Simple single sheet case). We now prove Theorem 20.4 under the additional hypotheses that $X^{\rm com} = X'^{\rm com} = \varnothing$ and $\|\mathfrak{A}\|, \|\mathfrak{A}'\| < \frac{1}{4}$. Part (a) has been proven in (20.5). For part (b), observe that, for each $i \in \mathcal{J}$, the cycle B_i can be represented as the union of

$$h_i = \phi_i\big(\{(z_1, z_2) \in H(t_i) \mid z_1 > 0\}\big) \cap Y_i^{(\circ)}$$

and $\Phi(b_i)$, where b_i is a path in $\{z \in \mathbb{C} \mid |z - s| \geq R(s)$ for $s \in S, |z' - s'| \geq R'(s')$ for $s' \in S'\}$ with the property that

$$\text{length}\{z \in b_i \mid |z| \leq \rho\} \leq \text{const}\, \rho \qquad \text{for all } \rho > 0.$$

By the first statement of Lemma 20.8(a)

$$\left| \int_{\Phi(b_i)} \omega_j - \int_{\Phi'(b_i)} \omega_j' \right| \leq \int_{b_i} \big|(w_j(z) - w_j'(z))\, dz\big| \leq \text{const}\, \varepsilon K^3.$$

By the second statement of Lemma 20.8(a)

$$\left| \int_{h_i \setminus Y_i^{(\circ\circ)}} (\omega_j - F^*\omega_j') \right| \leq \text{const}\, \varepsilon K^2 \sum_{\mu=1}^{2} \int_{r_\mu(j)}^{R_\mu(j)} r_\mu(j) \frac{dt}{t^2} + \text{const}\, \varepsilon K^3 \leq \text{const}\, \varepsilon K^3.$$

Furthermore, by the pointwise estimate of Lemma 20.6 and the estimate on D_i^j of Lemma 20.7

$$\left| \int_{h_i \cap Y_i^{(\circ\circ)}} (\omega_j - F^*\omega_j') \right| \leq \text{const}\, \varepsilon K^2.$$

This proves part (b) of the Theorem. Part (c) is proven in the same way.

We prove the bound of part (d) on $|W_j - W_j'|$, $j \in \mathcal{J}$. The remaining bounds are proven similarly. Recall from Proposition 19.1 and (19.2) that

$$W_j = \sum_{s \in S} w_{j,s}^{(4)}$$

where

$$w_{j,s}^{(n)} = -\frac{i^{1-n}}{2\pi i} \int_{|\zeta - s| = r(s)} \zeta^{n-1} w_j(\zeta)\, d\zeta.$$

For $j \in \mathcal{J}$ write

$$W_j - W_j' = \sum_{\substack{i \in \mathcal{J} \\ \mu = 1,2}} [w_{j,s_\mu(i)}^{(4)} - w_{j,s_\mu(i)}'^{(4)}] + \sum_{\substack{i \notin \mathcal{J} \\ \mu = 1,2}} w_{j,s_\mu(i)}^{(4)} - \sum_{\substack{i \notin \mathcal{J} \\ \mu = 1,2}} w_{j,s'_\mu(i)}'^{(4)}.$$

If $s = s_\mu(i)$ for some $i \in \mathcal{J}$ then

$$\begin{aligned}
|w_{j,s}^{(n)} - w_{j,s}'^{(n)}| &\leq \frac{1}{2\pi} \left| \int_{|\zeta - s| = 4r(s)} \zeta^{n-1} [w_j(\zeta) - w_j'(\zeta)]\, d\zeta \right| \\
&= \frac{1}{2\pi} \left| \int_{|\zeta - s| = 4r(s)} [\zeta^{n-1} - s^{n-1}][w_j(\zeta) - w_j'(\zeta)]\, d\zeta \right| \\
&\leq r(s)(n-1)[|s| + 4r(s)]^{n-2} \int_{|\zeta - s| = 4r(s)} |w_j(\zeta) - w_j'(\zeta)||d\zeta| \\
&\leq \text{const}_n \varepsilon K^3 r(s) |s|^{n-2} \int_{|\zeta - s| = 4r(s)} \left(\frac{r(s)}{|\zeta - s|^2} + \frac{1}{1 + |\zeta|^2} \right) |d\zeta| \\
&\leq \text{const}_n \varepsilon K^3 r(s) |s|^{n-2} [1 + r(s)|s|^{-2}] \\
&\leq \text{const}_n \varepsilon K^3 r(s) |s|^{n-2}
\end{aligned}$$

by Lemma 20.8(a). Similarly, by Lemma 6.9(a) followed by Lemma 20.7 if $s = s_\mu(i)$ and $s' = s'_\mu(i)$ for $i \notin \mathcal{J}$ then

$$\begin{aligned}|w^{(n)}_{j,s}| &\le \frac{1}{2\pi}\left|\int_{|\zeta-s|=r(s)} \zeta^{n-1} w_j(\zeta)\,d\zeta\right| \\ &= \frac{1}{2\pi}\left|\int_{|\zeta-s|=r(s)} [\zeta^{n-1}-s^{n-1}] w_j(\zeta)\,d\zeta\right| \\ &\le \left\|[\zeta^{n-1}-s^{n-1}]w_j(\zeta)\big|_{r<|\zeta-s|<2r}\right\|_2 \\ &\le \mathrm{const}_n r(s)|s|^{n-2}\Omega^j_i \\ &\le \mathrm{const}_n \varepsilon K r(s)|s|^{n-2}\end{aligned}$$

and

$$|w'^{(n)}_{j,s'}| \le \mathrm{const}_n \varepsilon K r(s)|s|^{n-2}.$$

Consequently

$$|W_j - W'_j| \le \mathrm{const}\,\varepsilon K^3[\Sigma r(s)|s|^2 + \sum r'(s')|s'|^2\Sigma] \le \mathrm{const}\,\varepsilon K^4$$

by (GH5) part (ii) and Definition 20.2 part (iii). □

We now wish to prove Theorem 20.4, allowing X^{com} to be nonempty and deleting the simplifying assumption $\|\mathfrak{A}\| < \frac{1}{4}$. We start with three general Lemmata.

LEMMA 20.9. *Let $U\colon S \to S'$ be a quasiconformal diffeomorphism with Beltrami coefficient at most $\varepsilon < 1$. Let ω' be any form on S'. Then*

$$\|U^*\omega'\|_{L^2(S)} \le \sqrt{\frac{1+\varepsilon}{1-\varepsilon}}\|\omega'\|_{L^2(S')}.$$

Write $U^*\omega' = \alpha + \beta$ with α of type $(1,0)$ and β of type $(0,1)$. Then, if ω' is of type $(1,0)$,

$$\|\beta\|_{L^2(S)} \le \frac{\varepsilon}{\sqrt{1-\varepsilon^2}}\|\omega'\|_{L^2(S')}.$$

PROOF. Locally, let $\omega' = a(u)du + b(u)\,d\bar{u}$. Then

$$\begin{aligned}U^*\omega' \wedge \overline{*U^*\omega'} &= i\{[|a|^2+|b|^2][|u_z|^2+|u_{\bar z}|^2] + 4\,\mathrm{Re}[a\bar{b}u_z u_{\bar z}]\}\,dz\wedge d\bar{z}, \\ U^*(\omega' \wedge \overline{*\omega'}) &= i[|a|^2+|b|^2][|u_z|^2-|u_{\bar z}|^2]\,dz\wedge d\bar{z}.\end{aligned}$$

The first claim follows from

$$\begin{gathered}[|a|^2+|b|^2][|u_z|^2-|u_{\bar z}|^2] \ge [|a|^2+|b|^2]|u_z|^2(1-\varepsilon^2), \\ [|a|^2+|b|^2][|u_z|^2+|u_{\bar z}|^2] + 4|a\bar{b}u_z u_{\bar z}| \le [|a|^2+|b|^2]|u_z|^2(1+\varepsilon^2+2\varepsilon).\end{gathered}$$

For the second observe that

$$\beta \wedge \overline{*\beta} = i|au_{\bar z}|^2\,dz\wedge d\bar{z} = \left|\frac{u_{\bar z}}{u_z}\right|^2\left[1-\left|\frac{u_{\bar z}}{u_z}\right|^2\right]^{-1} U^*(\omega' \wedge \overline{*\omega'}). \quad \square$$

LEMMA 20.10. *Let S be a Riemann surface with boundary ∂S and canonical homology basis $A_1,\dots,A_g,B_1,\dots,B_g$. Suppose that α and β are differential forms* *of type* $(1,0)$ *and* $(0,1)$ *respectively on* S *such that*

$$d(\alpha+\beta)=0,$$

$$\int_{A_i}(\alpha+\beta)=0 \quad \text{for } i=1,\dots,g, \quad \int_{\Delta}(\alpha+\beta)=0 \quad \text{for all components } \Delta \text{ o f}\partial S.$$

Then

$$\|\alpha+\beta\|_{L^2(S)} \le \sqrt{2}\|\beta\|_{L^2(S)} + \int_{\partial S}|\alpha+\beta|.$$

PROOF. Put

$$\omega=\alpha+\beta.$$

We have

$$*\omega = *\alpha + *\beta = -i\alpha + i\beta, \quad *\overline{\omega} = i\overline{\alpha} - i\bar{\beta} = i\overline{\omega} - 2i\bar{\beta}.$$

Hence

$$\|\omega\|^2_{L^2(S)} = \int_S \omega\wedge *\overline{\omega} = i\int_S \omega\wedge\overline{\omega} - 2i\int_S \omega\wedge\bar{\beta}.$$

We apply Lemma 2.8, the Riemann period relations, with $\omega=\omega$ and $\eta=\overline{\omega}$ to bound the first term. Both ω and η are closed. By hypothesis the integral of ω around each component of ∂S is zero. Therefore there is a single-valued C^∞ function f, defined on a neighbourhood of ∂S, such that $\omega=df$ on that neighbourhood and f has a zero on each component of ∂S. Hence by Lemma 2.8 and the vanishing of the A periods,

$$\int_S \omega\wedge\overline{\omega} = \sum_{k=1}^{g}\left(\int_{A_k}\omega\int_{B_k}\overline{\omega} - \int_{B_k}\omega\int_{A_k}\overline{\omega}\right) + \int_{\partial S} f\overline{\omega} = \int_{\partial S} f\overline{\omega}.$$

Consequently

$$\left|\int_S \omega\wedge\overline{\omega}\right| \le \left[\int_{\partial S}|\omega|\right]^2.$$

Since $\alpha\wedge\bar{\beta}=0$ we have

$$-i\int_S \omega\wedge\bar{\beta} = -i\int_S \beta\wedge\bar{\beta} = \int_S \beta\wedge *\bar{\beta} = \|\beta\|^2_{L^2(S)}.$$

Therefore

$$\|\omega\|^2_{L^2(S)} \le \left[\int_{\partial S}|\omega|\right]^2 + 2\|\beta\|^2_{L^2(S)}$$

which implies that

$$\|\omega\|_{L^2(S)} \le \sqrt{2}\|\beta\|_{L^2(S)} + \int_{\partial S}|\omega|. \quad \square$$

LEMMA 20.11. *Let* $\mathcal{U}$ *be either* $\mathbb{R}/L\mathbb{Z}\times[0,\delta]$ *or* $[0,L]\times[0,\delta]$ *with the natural complex structures. Denote by* $\mathcal{U}_t$ *the subset of* $\mathcal{U}$ *consisting of those points whose second component is* t. *Let* $u\colon\mathcal{U}\to\mathcal{X}$ *be a quasiconformal diffeomorphism into a Riemann surface* $\mathcal{X}$ *whose Beltrami coefficient is bounded by* $\mu<1$. *Let* ω *be a closed one form on* $u(\mathcal{U})$. *In the event that* $\mathcal{U}=[0,L]\times[0,\delta]$, *assume that* $\omega(u(0,t))/dt$ *and* $\omega(u(L,t))/dt$ *are bounded in absolute value by* C_B. *Then*

$$\left|\int_{u(\mathcal{U}_0)}\omega\right| \le \sqrt{\frac{L}{\delta}}\sqrt{\frac{1+\mu}{1-\mu}}\|\omega|_{u(\mathcal{U})}\|_2 + \begin{cases}0 & \text{if } \mathcal{U}=\mathbb{R}/L\mathbb{Z}\times[0,\delta]\\ \delta C_B & \text{if } \mathcal{U}=[0,L]\times[0,\delta]\end{cases}$$

PROOF. Let $\omega' = u^*\omega = S(s,t)\,ds + T(s,t)\,dt$ be the pull-back of ω by u. By Stokes' theorem, for every $\tau \in [0,\delta]$

$$\begin{aligned}\int_{u(\mathcal{U}_0)} \omega &= \int_{\mathcal{U}_0} \omega' \\ &= \int_{\{0\}\times[0,\tau]} \omega' + \int_{\mathcal{U}_\tau} \omega' - \int_{\{L\}\times[0,\tau]} \omega' \\ &= \int_0^L S(s,\tau)\,ds + \int_0^\tau T(0,t)\,dt - \int_0^\tau T(L,t)\,dt.\end{aligned}$$

Averaging over τ

$$\int_{u(\mathcal{U}_0)} \omega = \frac{1}{\delta}\int_0^\delta \int_0^L S(s,\tau)\,ds\,d\tau + \frac{1}{\delta}\int_0^\delta \left[\int_0^\tau T(0,t)\,dt - \int_0^\tau T(L,t)\,dt\right] d\tau.$$

If $\mathcal{U} = \mathbb{R}/L\mathbb{Z} \times [0,\delta]$, the second term is exactly zero, while if $\mathcal{U} = [0,L]\times[0,\delta]$, it is bounded by

$$\frac{1}{\delta}\int_0^\delta \int_0^\tau 2C_B\,dt\,d\tau = \frac{1}{\delta}\int_0^\delta 2C_B\tau\,d\tau = C_B\delta.$$

By the Cauchy–Schwarz inequality the first term is bounded by

$$\begin{aligned}\frac{1}{\delta}\int_0^\delta \int_0^L S(s,\tau)\,ds\,d\tau &\le \frac{1}{\delta}\|S\|_{L^2(\mathcal{U})}\sqrt{L\delta} \\ &\le \sqrt{\frac{L}{\delta}}\|\omega'\|_{L^2(\mathcal{U})} \\ &\le \sqrt{\frac{L}{\delta}}\sqrt{\frac{1+\mu}{1-\mu}}\|\omega\|_{L^2(u(\mathcal{U}))}\end{aligned}$$

by Lemma 20.9. □

For $j \in \mathcal{J}_c = \mathcal{J} \cup \{1,\dots,g\}$ and $i \in \mathcal{J}$, define

$$M_i^j = \left\|(\omega_j - F^*\omega_j')\big|_{Y_i^{\circ\circ}}\right\|_2, \quad M_{\text{com}}^j = \left\|(w_j - w_j')\,dz\big|_T\right\|_2,$$
$$D_\Gamma^j = \sup_{z\in\Gamma} |w_j(z) - w_j'(z)|.$$

LEMMA 20.12.

$$\begin{aligned}&\|\Omega^j\|_2, \|\Omega'^j\|_2 \le \operatorname{const} K^2 \quad \textit{for all } j \ge 1, \\ &\|\omega_j\|_{L^2(X^{\text{com}})}, \|\omega_j'\|_{L^2(X'^{\text{com}})} \le \operatorname{const} K^2 \quad \textit{for all } j \ge 1, \\ &\|\Omega^j\|_2, \|\Omega'^j\|_2 \le \operatorname{const} \varepsilon K \quad \textit{for } j \notin \mathcal{J}_c, \\ &\|(\Omega_i^j)_{i\notin\mathcal{J}}\|_2, \quad \|(\Omega_i'^j)_{i\notin\mathcal{J}}\|_2 \le \operatorname{const} \varepsilon K^2 \quad \textit{for all } j \ge 1, \\ &\|M^j\|_2 \le \operatorname{const} \varepsilon K^3 \quad \textit{for } j \in \mathcal{J}_c, \\ &\left\|(\omega_j - F^*\omega_j')\big|_{X(\Gamma)}\right\|_2 \le \operatorname{const} \varepsilon K^3 \quad \textit{for } j \in \mathcal{J}_c.\end{aligned}$$

PROOF. We first prove the first two lines for $1 \le j \le g$. Let u_j be the quasi-conformal map of Definition 20.2 part (vi). By the Riemann period relations and Lemma 20.11

$$\|\omega_j\|^2 = \operatorname{Im} \int_{B_j} \omega_j \le \sqrt{\frac{L_j}{\delta_j}}\sqrt{\frac{3/2}{1/2}}\|\omega_j\| \le \sqrt{3}K\|\omega_j\|.$$

Therefore

$$\|\omega_j\| \le \sqrt{3}K$$

which give the first two lines for ω_j. To get the first two lines for ω_j', it suffices to replace ω_j by ω_j', B_j by B_j' and u_j by $u_j \circ F$ in the above argument. Note that, since u_j and F are quasiconformal with Beltrami coefficents at most $\frac{1}{2}$ and $\frac{1}{4}$ respectively, the composition $u_j \circ F$ is quasiconformal with Beltrami coefficent at most $\frac{6}{7}$.

Next we prove the first three lines for $j \ge g+1$. The bounds on the first line follow from (6.24) and Definition 20.2 parts (iii, v) via

$$\|\overline{\Omega}^j\| \le |\widetilde{\Omega}_\Gamma^j| + 2\big(\|\widetilde{\Omega}^j\| + \|(\mathfrak{A}_{ij}\aleph_j)_{i>g}\|\big) \le K + 2(K+K^2) \le 4K^2$$

$$\|\underline{\Omega}^j\| \le 3|\widetilde{\Omega}_\Gamma^j| + 2\big(\|\widetilde{\Omega}^j\| + \|(\mathfrak{A}_{ij}\aleph_j)_{i>g}\|\big) \le 3K + 2(K+K^2) \le 6K^2$$

and the analogous bounds on the primed quantities. We have used $\|\widetilde{\Omega}^j\| = \sqrt{\mathcal{O}^j} \le K$. The third line follows similarly using part (iv) of Definition 20.2.

To get the second line we apply (6.22) to give

$$\|\omega_j\|_{L^2(X^{\text{com}})} \le \|\omega_j\|_{L^2(X(\Gamma))} \le |\widetilde{\Omega}_\Gamma^j| + \tfrac{1}{2}\|\Omega^j\| \le K + 4K^2 \le 5K^2.$$

We now move on to the fourth line for $j \in \mathcal{J}_c$. By (6.18), for $j \in \mathcal{J}$ and the analogous inequality for $1 \le j \le g$

$$\begin{aligned}\big\|(\Omega_i^j)_{i\notin\mathcal{J}}\big\|_2 &\le \Big\|(\mathfrak{A}_{i,k})_{\substack{i\notin\mathcal{J}\\ k\in\mathcal{J}}}\Big\|\,\big\|(\Omega_k^j)_{k\in\mathcal{J}}\big\|_2 + \Big\|(\mathfrak{A}_{i,k})_{\substack{i\notin\mathcal{J}\\ k\notin\mathcal{J}}}\Big\|\,\big\|(\Omega_k^j)_{k\notin\mathcal{J}}\big\|_2 \\ &\qquad + \begin{cases} 0 & \text{if } 1 \le j \le g \\ \big\|(\widetilde{\Omega}_i^j)_{i\notin\mathcal{J}}\big\|_2 + |\aleph_j|\big\|(\mathfrak{A}_{i,j})_{i\notin\mathcal{J}}\big\|_2 & \text{if } j \in \mathcal{J}\end{cases} \\ &\le 8\varepsilon K^2 + \tfrac{1}{4}\big\|(\Omega_k^j)_{k\notin\mathcal{J}}\big\|_2 + \varepsilon + K\varepsilon.\end{aligned}$$

Here $i,k \notin \mathcal{J}$ includes $i,k = \text{com}$. For the second term we used the fact that $i \notin \mathcal{J}$ implies $Y_i^{(\circ)} \cap X(\Gamma) = \varnothing$. The desired result follows.

That just leaves the last two lines. Recall that, by (20.1) of Lemma 20.6

$$M_i^j \le D_i^j + \varepsilon \,\text{const}\, K(\Omega_i'^j + \delta_{ij}) \tag{20.11}$$

for all $i \in \mathcal{J}$, $j \in \mathcal{J}_c$. As in (6.16), (6.18) and (20.10)

$$D_i^j \le \mathfrak{A}_{i,\text{com}} M_{\text{com}}^j + \sum_{k\in\mathcal{J}} \mathfrak{A}_{i,k} M_k^j + \sum_{k\notin\mathcal{J}} \mathfrak{A}_{i,k}\Omega_k^j + \sum_{k\notin\mathcal{J}} \mathfrak{A}'_{i,k}\Omega_k'^j, \tag{20.12}$$

$$D_\Gamma^j \le \mathfrak{A}_{\Gamma,\text{com}} M_{\text{com}}^j + \sum_{k\in\mathcal{J}} \mathfrak{A}_{\Gamma,k} M_k^j + \sum_{k\notin\mathcal{J}} \mathfrak{A}_{\Gamma,k}\Omega_k^j + \sum_{k\notin\mathcal{J}} \mathfrak{A}'_{\Gamma,k}\Omega_k'^j \tag{20.13}$$

where we recall that

$$\mathfrak{A}_{i,\text{com}} = \sup_{\mu=1,2} \frac{4\pi\gamma R_\mu(i)}{\text{dist}(s_\mu(i),T)^2}$$

and define

$$\mathfrak{A}_{\Gamma,\text{com}} = \frac{\gamma}{\text{dist}(\Gamma,T)^2}, \qquad \mathfrak{A}_{\Gamma,k} = \sup_{\mu=1,2} \frac{6r_\mu(k)}{\text{dist}(s_\mu(k),\Gamma)^2}.$$

As in (6.20)

$$(M_{\text{com}}^j)^2 + \sum_{\substack{i\ge g+1\\ Y_i^{(\circ)}\subset X(\Gamma)}} (M_i^j)^2 \le \big\|(\omega_j - F^*\omega_j')\big|_{X(\Gamma)}\big\|_2^2.$$

Let β be the $(0,1)$ part of $\omega_j - F^*\omega'_j$. It is the same as the $(0,1)$ part of $-F^*\omega'_j$ and vanishes on $X^{\mathrm{reg}} \cap X_0$. By Lemma 20.10, Lemma 20.9 and Lemma 20.6

$$\begin{aligned}(20.14)\quad \left\|(\omega_j - F^*\omega'_j)|_{X(\Gamma)}\right\|_2 &\le 2\|\beta\|_{L^2(X(\Gamma))} + \int_\Gamma |w_j - w'_j|\,|dz| \\ &\le 2\left[\|\beta\|^2_{L^2(X^{\mathrm{com}})} + \sum_{\substack{i\ge g+1\\ Y_i^{(\circ)}\subset X(\Gamma)}} \|\beta\|^2_{L^2(Y_i^{(\circ)})}\right]^{1/2} + \int_\Gamma |w_j - w'_j|\,|dz| \\ &\le \mathrm{const}\,\varepsilon 2\left[\|\omega'_j\|^2_{L^2(X'^{\mathrm{com}})} + \sum_{\substack{i\ge g+1\\ Y_i'^{(\circ)}\subset X'(\Gamma)}} (\Omega_i'^j + \delta_{ij})^2\right]^{1/2} + \mathrm{length}(\Gamma) D^j_\Gamma \\ &\le \mathrm{length}(\Gamma) D^j_\Gamma + \mathrm{const}\,\varepsilon K^2\end{aligned}$$

by the first two lines of the current lemma.

As in §6, we use $\underline{V}$ to denote the vector having components V_{com} and V_i with $i \ge g+1, Y_i^{(\circ)} \subset X(\Gamma)$ and $\overline{V}$ to denote the vector having components V_i with $i \in \mathcal{J}$, $Y_i'^{(\circ)} \cap X(\Gamma) = \varnothing$. In this notation, the conclusion of the last paragraph is that

$$\|\underline{M}^j\| \le \mathrm{length}(\Gamma) D^j_\Gamma + \mathrm{const}\,\varepsilon K^2.$$

By (20.11) and the first line of the current lemma

$$\|\overline{M}^j\| \le \|\overline{D}^j\| + \mathrm{const}\,\varepsilon K^3.$$

Definition 20.2(v) implies that $\|\overline{\mathfrak{A}V}\| \le \frac14 \|V\|$. So, by (20.12) and (20.13)

$$\|\overline{D}^j\| \le \tfrac14 \|M^j\| + \mathrm{const}\,\varepsilon K^3, \quad \mathrm{length}(\Gamma) D^j_\Gamma \le \tfrac12 \|M^j\| + \mathrm{const}\,\varepsilon K^2.$$

Hence

$$\|M^j\| \le \mathrm{const}\,\varepsilon K^3, \quad \left\|(\omega_j - F^*\omega'_j)|_{X(\Gamma)}\right\|_2 \le \mathrm{const}\,\varepsilon K^3. \quad \square$$

LEMMA 20.13. (a) *Define*

$$U_R = \left\{ z \in \Phi^{-1}(X^{\mathrm{reg}}) \;\middle|\; \frac{\mathrm{dist}(z,T)^2}{1+|z|^2} \ge \frac{\gamma}{K} \text{ and } |z-s| \ge R(s), \; |z-s'| \ge R(s') \forall s \in S, s' \in S' \right\},$$

$$U_r = \left\{ z \in \Phi^{-1}(X^{\mathrm{reg}}) \;\middle|\; \frac{\mathrm{dist}(z,T)^2}{1+|z|^2} \ge \frac{\gamma}{K} \text{ and } |z-s| \ge 4r(s), \; |z-s'| \ge 4r(s') \forall s \in S, s' \in S' \right\}.$$

There is a universal constant independent of X , X' etc. such that for all $j \in \mathcal{J}_c, z \in U_R$

$$|w_j(z) - w'_j(z)| \le \mathrm{const}\frac{\varepsilon K^4}{1+|z|^2}.$$

If $z \in U_r \setminus U_R$ then there exists an $s \in S$ and/or an $s' \in S'$ such that $|z-s| \le R(s)$ and/or $|z-s'| \le R(s')$. Then, for all $j \in \mathcal{J}_c$

$$|w_j(z) - w'_j(z)| \le \mathrm{const}\,\varepsilon K^3\left(\frac{r(s)}{|z-s|^2} + \frac{r'(s')}{|z-s'|^2}\right) + \mathrm{const}\frac{\varepsilon K^4}{1+|z|^2}.$$

(b) *For* $j \notin \mathcal{J}_c$

$$\left| w_j(z) - \frac{1}{2\pi i}\frac{1}{z - s_1(j)} + \frac{1}{2\pi i}\frac{1}{z - s_2(j)} \right| \le \text{const}\frac{\varepsilon K^2}{1+|z|^2}$$

if $\operatorname{dist}(z,T)^2/1+|z|^2 \ge \gamma/K$ *and* $|z-s| \ge R(s)$ *for all* $s \in S$. *If for some* $s \in S$, $4r(s) \le |z-s| \le R(s)$ *then*

$$\left| w_j(z) - \frac{1}{2\pi i}\frac{1}{z - s_1(j)} + \frac{1}{2\pi i}\frac{1}{z - s_2(j)} \right| \le \text{const}\,\varepsilon K\left(\frac{r(s)}{|z-s|^2}\right) + \text{const}\frac{\varepsilon K^2}{1+|z|^2}.$$

Similar bounds apply to $w'_j(z)$.

(c) *For* $i \in \mathcal{J}$, $j \notin \mathcal{J}_c$

$$\left|\frac{\phi_i^*(\omega_j)}{dz_1/z_1}\right| \le \text{const}\,\varepsilon K(|z_1|+|z_2|) \qquad \text{if } |z_1|,|z_2| \le \tfrac14,$$

$$\left|\frac{\phi_i^* F^* \omega'_j}{dz_1/z_1}\right| \le \text{const}\,\varepsilon K(|z_1|+|z_2|) \qquad \text{if } |z_1|,|z_2| \le \tfrac14.$$

PROOF. (a) By Lemma 6.9(c) and Proposition 6.12

$$\begin{aligned}
|w_j(z) - w'_j(z)| &\le |w_{j,\text{com}}(z) - w'_{j,\text{com}}(z)| + \sum_{\substack{k\in\mathcal{J}\\ \mu=1,2}} |w_{j,s_\mu(j)}(z) - w'_{j,s_\mu(j)}(z)| \\
&\quad + \sum_{\substack{k\notin\mathcal{J}\\ \mu=1,2}} |w_{j,s_\mu(j)}(z)| + \sum_{\substack{k\notin\mathcal{J}\\ \mu=1,2}} |w'_{j,s'_\mu(j)}(z)| \\
&\le \frac{\gamma}{\operatorname{dist}(z,T)^2} M^j_{\text{com}} + \sum_{\substack{k\in\mathcal{J}\\ \mu=1,2}} \frac{3r_\mu(k)}{|z-s_\mu(k)|^2} M^j_k \\
&\quad + \sum_{\substack{k\notin\mathcal{J}\\ \mu=1,2}} \frac{3r_\mu(k)}{|z-s_\mu(k)|^2}\Omega^j_k + \sum_{\substack{k\notin\mathcal{J}\\ \mu=1,2}} \frac{3r'_\mu(k)}{|z-s'_\mu(k)|^2}\Omega'^j_k \\
&\le \text{const}\,\varepsilon K^3 \left(\frac{\gamma}{\operatorname{dist}(z,T)^2} + \sum_{s\in S}\frac{r(s)}{|z-s|^2} + \sum_{s'\in S'}\frac{r'(s')}{|z-s'|^2}\right)
\end{aligned}$$

by Lemma 20.12. As in (6.3)

$$\frac{\gamma}{\operatorname{dist}(z,T)^2} + \sum_{s\in S}\frac{r(s)}{|z-s|^2} + \sum_{s'\in S'}\frac{r'(s')}{|z-s'|^2} \le \text{const}\frac{K}{1+|z|^2}.$$

The claim now follows from part (iii) of Definition 20.2.

(b) follows from Proposition 6.12 and Lemma 20.12.

(c) The first line follows from Lemma 6.8, applied to the scaled handle $H(4t_i)$, and Lemma 20.12. To prove the second line observe that, in the notation of Lemma 20.6, $\phi_i^* F^* \omega'_j = g'(u_1)\big(a(z_1)dz_1/z_1 + b(z_1)d\bar z_1/\bar z_1\big)$ with $g'(u_1)$ estimated in (20.6a) and (20.4). Definition 20.2(i) and Definition 20.1 provide bounds on $a(z_1)$ and $b(z_1)$. □

PROOF OF THEOREM 20.4. Part (a) has been proven in (20.5). For part (b), first consider $i \in \mathcal{J}$. The cycle B_i can be represented as the union of $h_i \cup \Phi(b_i)$ with $\sum_{i'<i} c_{i,i'} A_{i'}$ where

$$h_i = \phi_i\big(\{(z_1,z_2) \in H(t_i) \mid z_1 > 0\}\big) \cap Y_i^{(\circ)}$$

and b_i is a path in U_R with the property that

$$\text{length}\left\{z \in b_i \;\middle|\; |z| \le \rho\right\} \le \text{const}\,\rho \qquad \text{for all } \rho > 0.$$

The integral of $\omega_j - F^*\omega_j'$ over each A–cycle $A_{i'}$ is zero. By the first statement of Lemma 20.13(a)

$$\left| \int_{\Phi(b_i)} \omega_j - \int_{\Phi'(b_i)} \omega_j' \right| \le \int_{b_i} \left|\left(w_j(z) - w_j'(z)\right) dz\right| \le \text{const}\,\varepsilon K^4.$$

By the second statement of Lemma 20.13(a)

$$\left| \int_{h_i \setminus Y_i^{(\circ\circ)}} (\omega_j - F^*\omega_j') \right| \le \text{const}\,\varepsilon K^3 \sum_{\mu=1}^{2} \int_{r_\mu(j)}^{R_\mu(j)} r_\mu(j) \frac{dt}{t^2} + \text{const}\,\varepsilon K^4 \le \text{const}\,\varepsilon K^4.$$

Furthermore, by the pointwise estimate of Lemma 20.6, the bound on D_i^j arising from Lemma 20.13(a) and the estimate on $\Omega_i'^j$ of Lemma 20.12,

$$\left| \int_{h_i \cap Y_i^{(\circ\circ)}} (\omega_j - F^*\omega_j') \right| \le \text{const}\,\varepsilon K^4.$$

This proves part (b) of the Theorem when $i \in \mathcal{J}$.

Now fix any $1 \le i \le g$. By Definition 20.2(vi), there exist L_i, δ_i obeying $\sqrt{L_i/\delta_i} \le K$ and a quasicomformal diffeomorphism u_i, with Beltrami coefficient bounded by $\frac{1}{2}$, from $\mathcal{U} = \mathbb{R}/L_i\mathbb{Z} \times [0, \delta_i]$ into X^{com} with $u_i(\mathcal{U}_0) = B_i$. Then, by Lemmas 20.11 and 20.12

$$\left| \int_{B_i} \omega_j - F^*\omega_j' \right| \le \sqrt{\frac{L_i}{\delta_i}} \sqrt{3/2/1/2}\varepsilon K^3 \le \text{const}\,\varepsilon K^4.$$

We now prove part (c). Recall that, by part (i) and the first bullet of part (v) of Definition 20.2 and by Definition 20.1, there is a cover

$$X^{\text{com}} \cup \Phi\left\{ z \in G \;\middle|\; \frac{\text{dist}(z, T)^2}{1 + |z|^2} \le \frac{\gamma}{K} \right\} \subset \bigcup_{i=1}^{N} D_i \cup \bigcup_{\ell=1}^{r} H_\ell \subset X(\Gamma)$$

sets

$$\left\{ (z_1, z_2) \in H(t_\ell) \;\middle|\; \left(1 - \frac{1}{K}\right)^{-1} t_\ell \le |z_1| \le 1 - \frac{1}{K} \right\} \subset \widehat{Y}_\ell \subset H(t_\ell) \quad 1 \le \ell \le r,$$

$$\left\{ (z_1, z_2) \in H(t_\ell') \;\middle|\; \left(1 - \frac{1}{K}\right)^{-1} t_\ell' \le |z_1| \le 1 - \frac{1}{K} \right\} \subset \widehat{Y}_\ell' \subset H(t_\ell') \quad 1 \le \ell \le r$$

and biholomorphic maps

$$\Phi_i\colon \{z \in \mathbb{C} \mid |z| < 1\} \to D_i \quad 1 \le i \le N, \qquad \psi_\ell\colon H(t_\ell) \to H_\ell \quad 1 \le \ell \le r$$

and a corresponding family of coordinate patches containing X'^{com} such that

$$\Phi_i'^{-1} \circ F \circ \Phi_i \text{ is biholomorphic} \quad 1 \le i \le N,$$

$$\psi_\ell'^{-1} \circ F \circ \psi_\ell \restriction \widehat{Y}_\ell \text{ is } K\text{-quasiconformal of distortion at most } \varepsilon \quad 1 \le \ell \le r.$$

Furthermore we may choose this cover such that x_1 is joined to by x_2 by a curve that is a union of a finite number (depending only on $\mathcal{K}$) of pieces with the image under the universal covering π of each piece being of one of the four following types:

- the image under some ϕ_k of a line segment in Y_k.

- the image under Φ of a line segment in U_R.
- the image under some Φ_i of a line segment in $\{z \in \mathbb{C} \mid |z| \le \frac{1}{2}\}$.
- the image under some ψ_ℓ of a line segment in $\{(z_1, z_2) \in H(t_\ell) \mid |z_1|, |z_2| \le 1 - 20/15K\}$

Pieces of the first two types were treated in part (b).

Pieces of the third type are bounded using Lemma 6.9(a). Because $F \circ \Phi_i$ is holomorphic on the unit disk, the pullback

$$\Phi_i^*(\omega_j - F^*\omega_j') = w_{i,j}(z)\, dz$$

is a holomorphic form with L^2 norm bounded by $\left\|(\omega_j - F^*\omega_j')\big|_{X(\Gamma)}\right\|_2 \le \text{const}\, \varepsilon K^3$. Hence, by Lemma 6.9(a), $|w_{i,j}(z)| \le \text{const}\, \varepsilon K^3$ on $\{z \in \mathbb{C} \mid |z| \le 1/2\}$ and the integral of $\Phi_i^*(\omega_j - F^*\omega_j')$ along any line segment in $\{z \in \mathbb{C} \mid |z| \le 1/2\}$ obeys a similar bound.

Pieces of the fourth type are bounded using a variant of Lemma 20.6 similarly to pieces of the first type. By way of preparation we make some preliminary bounds. Because $F \circ \psi_\ell$ is holomorphic on $|z_1| > 1 - 28/15K$ and on $|z_2| > 1 - 28/15K$, the pullback

$$\psi_\ell^*(\omega_j - F^*\omega_j') = \alpha_{\ell,j}(z_1)\frac{dz_1}{z_1} + \beta_{\ell,j}(z_1)\frac{d\bar{z}_1}{z_1}$$

restricted to these two neighbourhoods is a holomorphic form (that is, $\beta_{\ell,j} = 0$ and $\alpha_{\ell,j}$ is holomorphic) with L^2 norm bounded by $\left\|(\omega_j - F^*\omega_j')\big|_{X(\Gamma)}\right\|_2 \le \text{const}\, \varepsilon K^3$. Define

$$D_{\ell,j} \equiv \sup_{|z_1| = 1 - 17/15K} |\alpha_{\ell,j}(z)| + \sup_{|z_2| = 1 - 17/15K} |\alpha_{\ell,j}(z)|.$$

By the Cauchy integral formula, for $|z_1| = 1 - 17/15K$,

$$\alpha_{\ell,j}(z_1) = \frac{1}{2\pi i}\int_{|\zeta| = 1 - 16/15K} \frac{\alpha_{\ell,j}(\zeta)}{\zeta - z_1}\, d\zeta - \frac{1}{2\pi i}\int_{|\zeta| = 1 - 18/15K} \frac{\alpha_{\ell,j}(\zeta)}{\zeta - z_1}\, d\zeta.$$

By Lemma 6.9(a),

$$\frac{1}{2\pi i}\int_{|\zeta| = 1 - 16/15K} \frac{\alpha_{\ell,j}(\zeta)}{\zeta - z_1}\, d\zeta \le \text{const}\, \varepsilon K^{3/2} \left\|(\omega_j - F^*\omega_j')\big|_{X(\Gamma)}\right\|_2 \le \text{const}\, \varepsilon K^{9/2}.$$

Applying a similar argument for the integral over $|\zeta| = 1 - 18/15K$ and then twice more for $|z_2| = 1 - 17/15K$

$$D_{\ell,j} \le \text{const}\, \varepsilon K^{9/2}.$$

As the image under ψ_ℓ of the circle $|z_1| = \sqrt{t_\ell}$ is homologous to a finite linear combination of the cycles A_j, $1 \le j \le g$, we have that

$$\int_{|z_1| = \sqrt{t_\ell}} \psi_\ell^*(\omega_j - F^*\omega_j') = 0 \quad \text{for all } j \ge 1,$$

$$\int_{|z_1| = \sqrt{t_\ell}} \psi_\ell^*\omega_j = 0 \quad \text{for all } j > g,$$

$$\left|\int_{|z_1| = \sqrt{t_\ell}} \psi_\ell^*\omega_j\right| \le C_K' \quad \text{for all } 1 \le j \le g.$$

In Lemma 20.12 it was proven that

$$\left\|\psi_\ell^*\omega_j\big|_{\hat{Y}_\ell}\right\|_2 \le \left\|\omega_j\big|_{X(\Gamma)}\right\|_2 \le \text{const}\, K^2.$$

Similarly,

$$\big\|(\psi'_\ell)^*\omega'_j|_{\widehat{Y}'_\ell}\big\|_2 \le \text{const}\, K^2.$$

Using the above preliminary estimates, we have, as in Lemma 20.6,

$$|\alpha_{\ell,j}(z_1)| \le K\left(|z_1| + \left|\frac{t_\ell}{z_1}\right|\right) D_{\ell,j} + \text{const}\,\varepsilon K^{5/2}\frac{1}{t_\ell}\left(\big\|(\psi'_\ell)^*\omega'_j|_{\widehat{Y}'_\ell}\big\|_2 + C'_{\mathcal{K}}\right)$$

$$|\beta_{\ell,j}(z_1)| \le \text{const}\,\varepsilon K^{3/2}\left(|z_1| + \left|\frac{t_\ell}{z_1}\right|\right)\left(\big\|(\psi'_\ell)^*\omega'_j|_{\widehat{Y}'_\ell}\big\|_2 + C'_{\mathcal{K}}\right)$$

for $|z_1|$, $|t_\ell/z_1| \le 1 - 20/15K$. Thus, on this domain,

$$|\alpha_{\ell,j}(z_1)| \le \text{const}_{\mathcal{K}}\varepsilon K^{11/2}, \quad |\beta_{\ell,j}(z_1)| \le \text{const}_{\mathcal{K}}\varepsilon K^{5/2}$$

and part (c) follows.

We prove the bound of part (d) on $|W_j - W'_j|$, $j \in \mathcal{J}$. The remaining bounds are proven similarly. From Proposition 19.1 and (19.2)–(19.3) and the Cauchy integral formula we have that

$$W_j = w^{(4)}_{j,\Gamma} + \sum_{\substack{s\in S\\ s \text{ outside } \Gamma}} w^{(4)}_{j,s}$$

where

$$w^{(n)}_{j,s} = -\frac{i^{1-n}}{2\pi i}\int_{|\zeta-s|=r(s)} \zeta^{n-1} w_j(\zeta)\,d\zeta, \quad w^{(n)}_{j,\Gamma} = -\frac{i^{1-n}}{2\pi i}\int_\Gamma \zeta^{n-1} w_j(\zeta)\,d\zeta.$$

The proof now continues as in the simple single sheet case. □

COROLLARY 20.14 (Solutions of the KP equation). *Let* $X = X^{\text{com}} \cup X^{\text{reg}} \cup X^{\text{han}}$ *be a marked Riemann surface that fulfils* (GH1)–(GH5) *with one regular sheet* $(m = 1)$. *Assume that*

$$\sup_j \left(\tau_1(j)^2 + \tau_2(j)^2\right)|\ln t_j|^2 \ln\frac{\tau_1(j)\tau_2(j)}{t_j} < \infty$$

and that

$$\lim_{j\to\infty}\frac{1}{|\log t_j|}\left(s_1(j)^n - s_2(j)^n\right) = 0 \quad \textit{for } n = 1, 2, 3.$$

Then there is a constant c *such that for every* $e \in B$

$$u(x_1, x_2, t) = -2\frac{\partial^2}{\partial x_2^2}\ln\theta\big(Ux_2 + Vx_1 - \tfrac{1}{2}Wt + e\big) + c \tag{20.15}$$

solves the KP equation

$$\left(u_t - 3uu_{x_2} + \tfrac{1}{2}u_{x_2x_2x_2}\right)_{x_2} + \tfrac{3}{2}u_{x_1x_1} = 0 \tag{20.16}$$

whenever $\theta\big(Ux_2 + Vx_1 - \frac{1}{2}Wt + e\big) \ne 0$.

PROOF. Fix $R > 0$. By the holomorphicity of θ, it suffices to prove that there exists a constant c such that the expression (20.15) satisfies the KP equation for all e, t, x_1 and x_2 obeying $\|Ux_2\|$, $\|Vx_1\|$, $\|Wt\|$, $\|e\| < R$.

We denote by $\mathcal{R}$ the period matrix $\mathcal{R}_{ij} = \int_{B_i}\omega_j$ of X, so $\theta(z) = \theta(z, \mathcal{R})$. For $\mathcal{J} \subset \{1, 2, \dots\}$ let $\theta_{\mathcal{J}}(z, \mathcal{R})$ be the truncated theta function

$$\theta_{\mathcal{J}}(z, \mathcal{R}) = \sum_{\substack{n\in\mathbb{Z}^\infty\\ n_j = 0 \text{ if } j\notin\mathcal{J}}} e^{2\pi i\langle z,n\rangle} e^{\pi\langle n,\mathcal{R}n\rangle}.$$

Now fix $\varepsilon > 0$. Since the series for $\theta(z, \mathcal{R})$ converges uniformly on B_{5R} there is $N > 0$ such that for all $\mathcal{J} \subset \mathbb{N}$ with $\{1, 2, \dots, N\} \subset \mathcal{J}$

$$|\theta(z, \mathcal{R}) - \theta_{\mathcal{J}}(z, \mathcal{R})| < \varepsilon \qquad \text{for } z \in B_{5R}.$$

Furthermore there is δ such that for $\{1, 2, \dots, N\} \subset \mathcal{J}$

(i) $|\theta(z, \mathcal{R}) - \theta_{\mathcal{J}}(z, \mathcal{R}')| < \varepsilon$ for $z \in B_{5R}$ if $|\mathcal{R}_{ij} - \mathcal{R}'_{ij}| < \delta$ for $i, j \in \mathcal{J}$,

(ii) $|\theta(z, \mathcal{R}) - \theta_{\mathcal{J}}(z', \mathcal{R})| < \varepsilon$ if $z, z' \in B_{5R}$ with $|z_j - z'_j| < \delta$ for $j \in \mathcal{J}$.

Now, by Proposition 20.3 and Theorem 20.4, there exist

a compact submanifold(with boundary) X_0 in X containing the image of $\mathcal{K}$
a marked Riemann surface $X' = X'^{\text{com}} \cup X'^{\text{reg}} \cup X'^{\text{han}}$ of genus genus(X_0)
a compact submanifold $X'_0 \subset X'$ and a diffeomorphism $F\colon X_0 \to X'_0$

such that

$\mathcal{J} = \{j \mid Y_j \subset X_0\}$ contains $\{1, 2, \dots, N\}$
the period matrix $\mathcal{R}'$ of X' fulfills $|\mathcal{R}'_{ij} - \mathcal{R}_{ij}| < \delta$ for all $i, j \in \mathcal{J}$
$\|Ux_2 + Vx_1 - \frac{1}{2}Wt - U'x_2 - V'x_1 + \frac{1}{2}W't\| < \delta$ for all all t, x_1 and x_2 obeying $\|Ux_2\|$, $\|Vx_1\|$, $\|Wt\| < R$.

Consequently, for any e, t, x_1 and x_2 obeying $\|Ux_2\|$, $\|Vx_1\|$, $\|Wt\| < R$ and $\|e\| < 2R$

$$\left|\theta\big(Ux_2 + Vx_1 - \tfrac{1}{2}Wt + e; \mathcal{R}\big) - \theta\big(U'x_2 + V'x_1 - \tfrac{1}{2}W't + e; \mathcal{R}'\big)\right| < \varepsilon.$$

By [**Mu2**, p. 3.239], for each such X' there is a constant c' such that

$$u'(x_1, x_2, t) = -2\frac{\partial^2}{\partial x_2^2} \ln \theta\big(U'x_2 + V'x_1 - \tfrac{1}{2}W't + e; \mathcal{R}'\big)$$

solves

$$\left(u'_t - 3u'u'_{x_2} + \tfrac{1}{2}u'_{x_2x_2x_2}\right)_{x_2} + \tfrac{3}{2}u'_{x_1x_1} = 3c'u'_{x_2x_2}.$$

We can choose a sequence of ε's and approximating Riemann surfaces X' such that the corresponding c''s converge to some value c, which is possibly infinite. Then $u' + c'$ converges to u.

If c is finite, then, by the Cauchy integral formula, we get the desired equation. Assume now that c is infinite. Then $u_{x_2x_2} = 0$. That is

$$\frac{\partial^4}{\partial x_2^4} \ln \theta(e + Ux_2\ ) = 0 \qquad \text{for all } e \in B.$$

Therefore, for each $e \in B$, the line

$$\{e + x_2U \mid x_2 \in \mathbb{C}\}$$

is either completely contained in the theta-divisor Θ, or does not meet it at all. So, for all smooth points e of Θ the vector U lies in the tangent space $T_e\Theta$ of Θ at e. Therefore U also lies in the kernel of the second derivative of θ at any such point. That is,

$$U \in \ker H(e) \qquad \text{for all } e \in \Theta_{\text{reg}}$$

As $\ker H(z) = 0$ for z in a dense subset of Θ_{reg} (Corollary 9.9 and Lemma 9.4(a), this implies that $U = 0$. Then u is a constant and the KP equation is trivially satisfied. □

REMARK. In a similar way one can show that for any marked Riemann surface $X = X^{\mathrm{com}} \cup X^{\mathrm{reg}} \cup X^{\mathrm{han}}$ that fulfils (GH1)–(GH5) with one regular sheet ($m = 1$) and with

$$\sup_j \big(\tau_1(j)^2 + \tau_2(j)^2\big) |\ln t_j|^2 \ln \frac{\tau_1(j)\tau_2(j)}{t_j} < \infty$$

any point x and any local coordinate ζ about this point, formula (20.15) gives a solution of the KP equation for U, V , W defined by

$$\omega_j = U_j \, d\zeta + V_j \zeta \, d\zeta + \tfrac{1}{2} W_j \zeta^2 \, d\zeta + O(\zeta^3) \text{ near } x$$

as in the beginning of §19.

21. Real Periodic Potentials

The main result in this section is

THEOREM 21.1. *Let $q \in L^2(\mathbb{R}^2/\Gamma)$ be a real analytic potential satisfying*

$$\int_0^{2\pi} q(x_1, x_2)\, dx_2 = 0 \quad \textit{for all } x_1 \in \mathbb{R}.$$

Assume that its associated heat curve $\mathcal{H}(q)$ is smooth. Let θ be the theta function of $\mathcal{H}(q)$ on the torus

$$\mathcal{T} = (\mathbb{R}/\mathbb{Z})^\infty$$

with metric

$$d(\mathrm{x}, \mathrm{y}) = \inf_{\mathbf{n} \in \mathbb{Z}^\infty} \|\mathrm{x} - \mathrm{y} - \mathbf{n}\|.$$

Then there exist $e \in \mathcal{T}$ and $U, V, W \in \mathbb{R}^\infty$ such that

$$u(x_1, x_2, t) = -2 \frac{\partial^2}{\partial x_2^2} \ln \theta\big(e + U x_2 + V x_1 - \tfrac{1}{2} W t\big)$$

is a C^∞ function on $\mathbb{R}^3$ which solves the KP *equation*

$$\text{(KP)} \qquad u_t = 3 u u_{x_2} - \tfrac{1}{2} u_{x_2 x_2 x_2} - \tfrac{3}{2} I(u_{x_1 x_1})$$

and satisfies the initial condition

$$u(x_1, x_2, 0) = q(x_1, x_2).$$

Here, as before,

$$I(u)(x_1, x_2) = \int_0^{x_2} u(x_1, s)\, ds - \frac{1}{2\pi} \int_0^{2\pi} dt \int_0^t u(x_1, s)\, ds.$$

REMARK 21.2. In the event that $\mathcal{H}(q)$ is singular, the results of the Theorem apply to the normalization of $\mathcal{H}(q)$.

Theorem 21.1 and Remark 21.2 show that the initial value problem for periodic (KP) with real analytic initial data can be solved for all time.¡ This had been shown by Krichever [**K**] and refined by Bourgain [**B**]. In addition the Theorem gives qualitative information on the solution. Namely, by Proposition 4.14.

COROLLARY 21.3. *Let $q \in L^2(\mathbb{R}^2/\Gamma)$ be a real analytic potential for which $\int_0^{2\pi} q(x_1, x_2)\, dx_2 = 0$ for all $x_1 \in \mathbb{R}$. Then the solution of* (KP) *with initial data q is almost periodic in time.*

In preparation for the proof of Theorem 21.1, we investigate the structure of $\mathcal{H}(q)$ for **real** potentials q. Clearly $\mathcal{H}(q)$ is invariant under the antiholomorphic involution

$$\sigma\colon \mathbb{C}^* \times \mathbb{C}^* \to \mathbb{C}^* \times \mathbb{C}^*, \quad (\xi_1, \xi_2) \mapsto (\bar{\xi}_1^{-1}, \bar{\xi}_2^{-1}).$$

In [**K**, Chapter I, §2], Krichever describes the structure of $\mathcal{H}(q)$ in more detail.

THEOREM 21.4 ([**K**, Chapter I, §2]). *Let $q \in L^2(\mathbb{R}^2/\Gamma)$ be a real analytic potential. Let*

$$\mathcal{H}_A(q) = \{(\xi_1, \xi_2) \in \mathcal{H}(q) \mid \xi_1, \xi_2 \in \mathbb{R}\}.$$

Then $\mathcal{H}_A(q)$ is the disjoint union of connected components a_0 and a_b, $b \in \Gamma^\#$, $b_2 > 0$. For each $b \in \Gamma^\#$, $b_2 > 0$, either a_b is diffeomorphic to a circle or a_b is a point. In the latter case a_b is an ordinary double point of $\mathcal{H}(q)$. There is holomorphic map

$$K_2\colon \mathcal{H}(q) \setminus \Bigg(\bigcup_{\substack{b \in \Gamma^\# \\ b_2 > 0}} a_b \Bigg) \to \mathbb{C}$$

such that

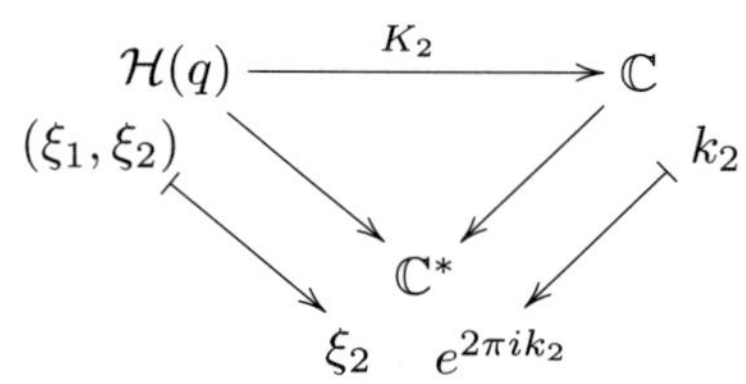

commutes. K_2 is biholomorphic to its image and $K_2(a_0)$ is the imaginary axis. The image of K_2 is the complement of a set of disjoint "cuts" $c_b(q)$, $b \in \Gamma^\#$, $b_2 \neq 0$. Each cut $c_b(q)$ is either a compact interval or a point on the line $\{k_2 \in \mathbb{C} \mid \operatorname{Re} k_2 = -\frac{1}{2}b_2\}$. The cut $c_b(q)$ is the reflection of $c_{-b}(q)$ across the imaginary axis. For each fixed b_2, the cuts $c_b(q)$ are ordered along the line $\{k_2 \in \mathbb{C} \mid \operatorname{Re} k_2 = -\frac{1}{2}b_2\}$ according to b_1.

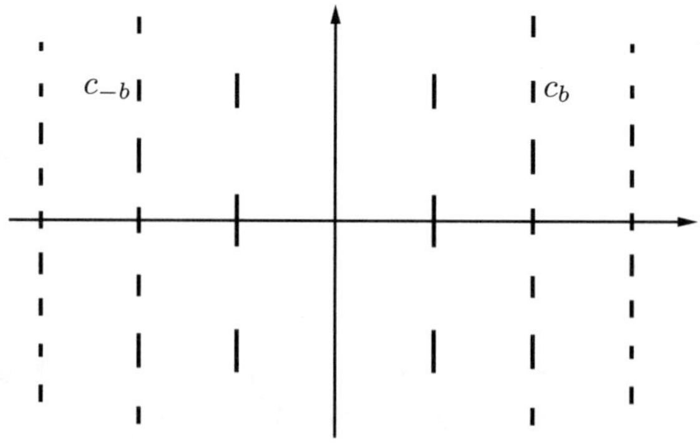

Let $\mathcal{M}(q)$ be the Riemann surface obtained from $\mathbb{C}$ by gluing, for each $b \in \Gamma^\#$, $b_2 > 0$, the cut c_b to the cut c_{-b} using translation by b_2. Then K_2 induces a biholomorphic map from $\mathcal{H}(q)$ to $\mathcal{M}(q)$ that maps a_b to the circle defined by c_b.

Let

$$D = \left\{ (\xi_1, \xi_2) \in \mathcal{H}(q) \;\middle|\; \begin{array}{l} \exists \psi(x_1, x_2) \neq 0 \text{ obeying } \left(\frac{\partial}{\partial x_1} - \frac{\partial^2}{\partial x_2^2}\right)\psi + q\psi = 0, \\ \psi(x_1 + \omega_1, x_2 + \omega_2) = \xi_1 \psi(x_1, x_2), \\ \psi(x_1, x_2 + 2\pi) = \xi_2 \psi(x_1, x_2) \text{ and } \psi(0,0) = 0 \end{array} \right\}.$$

Then $D \subset \bigcup_{\substack{b \in \Gamma^\# \\ b_2 > 0}} a_b$ *and each* a_b *contains exactly one point of* D*, counted with multiplicity.*

LEMMA 21.5. *Let* $\beta \geq 4$. *Assume that* $q \in L^2(\mathbb{R}^d/\Gamma)$ *obeys* $\widehat{q}(0) = 0$ *and* $\||b|^\beta \widehat{q}(b)\|_1 < \infty$. *Then there is a constant* const, *depending only on* $\||b|^\beta \widehat{q}(b)\|_1$ *such that for every* $d \in \Gamma^\#$ *with* $d_2 \neq 0$ *and* $|d| >$ const

$$|v_d - s_d| \leq \frac{\text{const}}{|d|^\beta}, \qquad |w_d - s_d| \leq \frac{\text{const}}{|d|^\beta}$$

where $s_d = \widehat{\phi}_d(0,0)$ *was defined in Theorem* 14.2.

PROOF. Define, as in the proof of Theorem 14.2,

$$x_1(k_1, k_2) = P_0(k) - \mathcal{D}_{1,1}(k_1, k_2) = ik_1 + k_2^2 - \mathcal{D}_{1,1}(k_1, k_2),$$
$$x_2(k_1, k_2) = P_d(k) - \mathcal{D}_{2,2}(k_1, k_2) = i(k_1 + d_1) + (k_2 + d_2)^2 - \mathcal{D}_{2,2}(k_1, k_2). \quad \square$$

The functions $\mathcal{D}(k)_{i,j}$ were given in Proposition 14.5, where it was also shown that $k \in T_0 \cap T_d$ is on $\widehat{\mathcal{H}}(q)$ if and only if

$$x_1(k) x_2(k) = h(k) \quad \text{where } h(k) = \big(\widehat{q}(-d) - \mathcal{D}_{2,1}(k)\big)\big(\widehat{q}(d) - \mathcal{D}_{1,2}(k)\big).$$

We shall shortly show that, for $|d| \geq$ const, the equation

$$\frac{\partial}{\partial k_1}\big(x_1(k)x_2(k) - h(k)\big) = 0$$

has precisely two solutions in $T_0 \cap T_d$. These are the points v_d and w_d.

In Proposition 14.5 and Lemmas 14.6 and 14.8, it was proven that, for $k \in T_0 \cap T_d$ and $m, n \geq 0$,

$$\left| \frac{\partial^{n+m}}{\partial^n k_1 \partial^m k_2} \mathcal{D}_{i,j} \right| \leq \begin{cases} \frac{\text{const}}{[1+|z_d|]^{2-m}}, & \text{if } i = j, \\ \text{const} \frac{|d|^m}{[1+|d|]^\beta}, & \text{if } i \neq j, \end{cases}$$

with the constant depending only on n, m and $\||b|^\beta \widehat{q}(b)\|_1$. Consequently,

$$\frac{\partial x_1}{\partial k_1} = i - \frac{\partial \mathcal{D}_{1,1}}{\partial k_1} = i + O(|z_d|^{-2}), \qquad \frac{\partial x_2}{\partial k_1} = i - \frac{\partial \mathcal{D}_{2,2}}{\partial k_1} = i + O\big(|z_d|^{-2}\big)$$

and

$$\begin{aligned} \frac{\partial}{\partial k_1}(x_1 x_2 - h) &= \left(i - \frac{\partial \mathcal{D}_{2,2}}{\partial k_1}\right) x_1 + \left(i - \frac{\partial \mathcal{D}_{1,1}}{\partial k_1}\right) x_2 - \frac{\partial h}{\partial k_1} \\ &= \big(i + O(|z_d|^{-2})\big) x_1 + \big(i + O(|z_d|^{-2})\big) x_2 + O(|d|^{-2\beta}) \end{aligned}$$

First, substitute $k_1 = k_1(x_1, x_2)$ and $k_2 = k_2(x_1, x_2)$ in $h(k)$, $\partial \mathcal{D}_{2,2}/\partial k_1$, $\partial \mathcal{D}_{2,2}/\partial k_1$ and $\partial h/\partial k_1$ and think of

$$x_1 x_2 = h, \quad \left(i - \frac{\partial \mathcal{D}_{2,2}}{\partial k_1}\right) x_1 + \left(i - \frac{\partial \mathcal{D}_{1,1}}{\partial k_1}\right) x_2 - \frac{\partial h}{\partial k_1} = 0,$$

as two equations in the two unknowns x_1 and x_2. As in the proof of Theorem 14.2

$$\left|\frac{\partial}{\partial x_1}\frac{\partial \mathcal{D}_{1,1}}{\partial k_1}\right| \le \frac{\text{const}}{1+|d_2 z_d|}, \qquad \left|\frac{\partial}{\partial x_1}\frac{\partial \mathcal{D}_{2,2}}{\partial k_1}\right| \le \frac{\text{const}}{1+|d_2 z_d|}$$

$$\left|\frac{\partial h}{\partial x_1}\right| \le \frac{\text{const}}{1+|d|^{2\beta-1}}, \qquad \left|\frac{\partial}{\partial x_1}\frac{\partial h}{\partial k_1}\right| \le \frac{\text{const}}{1+|d|^{2\beta-1}}$$

so that, by the implicit function theorem, the second equation has a unique solution $x_2 = x_2(x_1)$ and this solution obeys

$$x_2 = -\frac{i - \partial \mathcal{D}_{2,2}/\partial k_1}{i - \partial \mathcal{D}_{1,1}/\partial k_1} x_1 + \frac{1}{i - \partial \mathcal{D}_{1,1}/\partial k_1}\frac{\partial h}{\partial k_1}.$$

Substituting this in the first equation gives

$$x_1(ax_1 + b) - h = 0$$

with

$$\begin{aligned} a &= -\frac{i - \partial \mathcal{D}_{2,2}/\partial k_1}{i - \partial \mathcal{D}_{1,1}/\partial k_1} = -1 + O(|z_d|^{-2}), \\ b &= \frac{1}{i - \partial \mathcal{D}_{1,1}/\partial k_1}\partial h/\partial k_1 = O(|d|^{-2\beta}), \\ h &= O(|d|^{-2\beta}). \end{aligned}$$

By Rouchés theorem, this has the same number of solutions as $-x^2 = 0$, namely two. The solutions obey

$$x_1 = \frac{1}{2a}\{-b \pm \sqrt{b^2 + 4ah}\}$$

and hence

$$|x_1|, |x_2| \le \frac{\text{const}}{[1+|d|]^\beta}.$$

In the proof of Theorem 14.2 we showed that the point s_d obeyed

$$|x_1|, |x_2| \le \frac{1}{[1+|d|]^{2\beta-1}}$$

Hence, in terms of the coordinates (x_1, x_2), the points v_d, w_d are at most a distance $\text{const}/[1+|d|]^\beta$ away from s_d. Since we also showed in Theorem 14.2 that

$$\frac{\partial k_2}{\partial x_1} = -\frac{1}{2d_2}(\mathbb{1} + O(|d_2 z_d|^{-1})), \qquad \frac{\partial k_2}{\partial x_2} = \frac{1}{2d_2}(\mathbb{1} + O(|d_2 z_d|^{-1}))$$

the corresponding distance in terms of k_2 is at most $\text{const}/[1+|d|]^\beta$.

We give a decomposition of $\mathcal{H}(q)$ based on Theorem 21.4 that is slightly different from the one used in §15. Define, as in §15, for $d \in \Gamma^\#$, $d_2 > 0$,

$$\tau_d = \frac{1}{|z_d|^{13}}, \qquad r_d = \frac{2}{|d_2 z_d^{14}|}, \qquad R_d = \frac{1}{6|d_2 z_d|}$$

and also set

$$\hat{t}_d = \frac{|v_d - w_d|}{9R_d^2}.$$

Fix β sufficiently large. Then there is a constant ρ depending only on $\||b|^\beta \widehat{q}\|_1$ such that, for $d \in \Gamma^\#$, $d_2 > 0$, $|d| > \rho$, the circle of radius R_d and centre $v_d + w_d/2$ lies completely between the ellipse with focii v_d, w_d and semiaxes $|v_d - w_d|/4(1/\sqrt{\hat{t}_d} \pm$

$\sqrt{\hat{t}_d}$) and the ellipse with focii v_d, w_d and semiaxes $|v_d - w_d|/4(1/2\sqrt{\hat{t}_d} \pm 2\sqrt{\hat{t}_d})$. Also, if ρ is chosen large enough (depending only on $\||b|^\beta \widehat{q}\|_1$), then for $|d| > \rho$

- no two of the above ellipses intersect and
- the conditions of (GH2) are fulfilled.

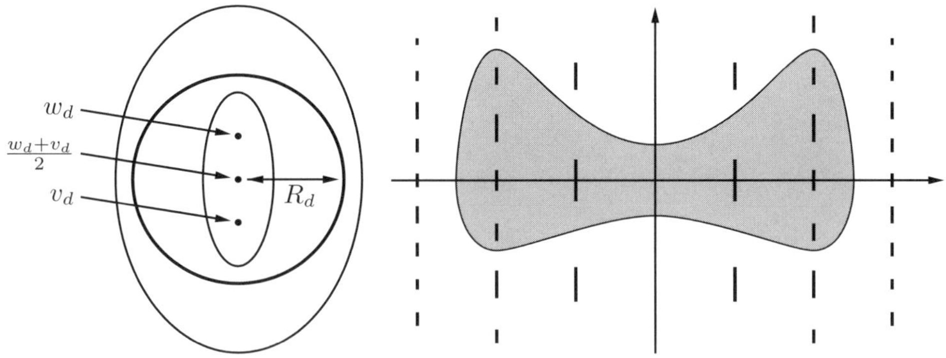

Choose a closed σ–invariant curve around the origin with inner and outer radii ρ and 2ρ that avoids all of the ellipses above. Define K_2^{com} to be the interior of this curve and $\mathcal{H}(q)^{\text{com}}$ to be the closure of the inverse image of K_2^{com} under the map K_2. Furthermore, let G be the complement of the interior of K_2^{com} and the union over all d with $v_d + w_d/2 \notin K_2^{\text{com}}$ of the interior of the ellipse with focii v_d, w_d and semiaxes $|v_d - w_d|/4(\tau_d/\sqrt{\hat{t}_d} \pm \tau_d\sqrt{\hat{t}_d})$. Define $\Phi\colon G \to \mathcal{H}(q)$ as the inverse of the map K_2.

We set

$$t_d = \hat{t}_d \quad \text{if } \frac{v_d + w_d}{2} \notin K_2^{\text{com}}.$$

If $v_d + w_d/2 \in K_2^{\text{com}}$ choose t_d close to 1 such that the ellipses with with focii v_d, w_d and semiaxes $|v_d - w_d|/4(1/\sqrt{t_d} \pm \sqrt{t_d})$ do not overlap. Define

$$P_{d,1}\colon H(t_d) \to \mathbb{C}, \qquad (z_1, z_2) \mapsto \frac{v_d + w_d}{2} + \frac{v_d - w_d}{4\sqrt{t_d}}(z_1 + z_2)$$

$$P_{d,2}\colon H(t_d) \to \mathbb{C}, \qquad (z_1, z_2) \mapsto \frac{v_{-d} + w_{-d}}{2} + \frac{v_{-d} - w_{-d}}{4\sqrt{t_d}}(z_1 + z_2).$$

Then $P_{d,1}$ maps the "centre of the handle"

$$\{(z_1, z_2) \in H(t_d) \mid |z_1| = |z_2| = \sqrt{t_d}\}$$

as a two-fold cover on the line segment $[v_d, w_d]$ and it maps the "edge of the handle"

$$\{(z_1, z_2) \in H(t_d) \mid |z_1| = 1\}$$

to the ellipse with focii v_d and w_d and semiaxes of lengths $|v_d - w_d|/4(1/\sqrt{t_d} \pm \sqrt{t_d})$. Similarly, $P_{d,2}$ maps the "centre of the handle" as a two-fold cover on the line segment $[v_{-d}, w_{-d}]$ and maps the "edge of the handle" to the ellipse with focii v_{-d} and w_{-d} and semiaxes of lengths $|v_d - w_d|/4(1/\sqrt{t_d} \pm \sqrt{t_d})$.

Let $\widehat{\sigma}\colon \mathbb{C} \to \mathbb{C}$, $k_2 \mapsto -\bar{k}_2$ be reflection in the imaginary axis. Then for each $(z_1, z_2) \in H(t)$

$$P_{d,2}(\bar{z}_2, \bar{z}_1) = \widehat{\sigma} \circ P_{d,1}(z_1, z_2).$$

We define

$$\begin{aligned} P_d \colon H(t_d) \setminus \{(z_1, z_2) \in H(t_d) \mid |z_1| = |z_2| = \sqrt{t_d}\} &\to \mathbb{C}, \\ (z_1, z_2) &\mapsto \begin{cases} P_{d,1}(z_1, z_2), & \text{if } |z_1| > \sqrt{t_d}, \\ P_{d,2}(z_1, z_2), & \text{if } |z_2| > \sqrt{t_d}. \end{cases} \end{aligned}$$

There is a unique holomorphic

$$\phi_d \colon H(t_d) \to \mathcal{H}(q)$$

such that

$$\begin{array}{ccc} H(t_d) \setminus \{(z_1, z_2) \in H(t_d) \mid |z_1| = |z_2| = \sqrt{t_d}\} & \longrightarrow & \mathcal{H}(q) \setminus \bigcup_c a_c \\ & {\scriptstyle P_d} \searrow \quad \swarrow {\scriptstyle K_2} & \\ & \mathbb{C} & \end{array}$$

commutes. As in §15 one verifies that this decomposition satisfies (GH1)–(GH6).

Theorem 21.4 also shows that the heat curves for real potentials all have the same topological structure. To make this more precise, we fix a ball $\mathcal{P}$ in the space of real analytic potentials $q \in L^2(\mathbb{R}^2/\Gamma)$ with x_2–average zero and construct the "universal family" $\mathcal{H}$ of heat curves over $\mathcal{P}$. More precisely, put

$$\mathcal{H} = \big\{ \big((\xi_1, \xi_2); q\big) \in \mathbb{C}^* \times \mathbb{C}^* \times \mathcal{P} \;\big|\; (\xi_1, \xi_2) \in \mathcal{H}(q) \big\}.$$

By Lemma 13.6 and Theorem 13.8, $\mathcal{H}$ is an analytic subvariety of $\mathbb{C}^* \times \mathbb{C}^* \times \mathcal{P}$. Denote by $h \colon \mathcal{H} \to \mathcal{P}$ the projection $\big((\xi_1, \xi_2); q\big) \mapsto q$. The fiber $h^{-1}(q)$ is the heat curve $\mathcal{H}(q)$.

Fix $\rho > 0$ such that the decomposition constructed above works for all $q \in \mathcal{P}$. We can choose the decomposition of the heat curves $\mathcal{H}(q)$, $q \in \mathcal{P}$ in such a way that

(i) the restriction of h to

$$\mathcal{H}^{\text{com}} = \big\{ \big((\xi_1, \xi_2); q\big) \in \mathcal{H} \;\big|\; (\xi_1, \xi_2) \in \mathcal{H}(q)^{\text{com}} \big\}$$

is a locally trivial differentiable fibre bundle over the complement of

$$\Delta = \{ q \in \mathcal{P} \mid \mathcal{H}(q)^{\text{com}} \text{ is singular} \}.$$

The the arithmetic genus of $\mathcal{H}(q)^{\text{com}}$ is a constant over $\mathcal{P}$. We denote this number by g.

(ii) the restriction of h to $\partial \mathcal{H}^{\text{com}}$ is a trivial fibre bundle with fibres S^1.

(iii) for all real valued $q \in \mathcal{P}$

- σ maps each of the pieces $\mathcal{H}(q)^{\text{com}}$, $\mathcal{H}(q)^{\text{reg}}$, $\mathcal{H}(q)^{\text{han}}$ onto itself,
- the fixed point set of σ on $\mathcal{H}(q)^{\text{com}}$ consists of one interval $a_0 \cap \mathcal{H}(q)^{\text{com}}$ and g ovals $a_1, \ldots, a_g$ which represent the cycles $A_1, \ldots, A_g$,
- $\Phi^{-1} \circ \sigma \circ \Phi$ is the map $z \mapsto -\bar{z}$,
- for $b \in \Gamma^\sharp$ obeying $b_2 > 0, v_b + w_b/2 \notin K_2^{\text{com}}$ the involution on the model handle $H(t_b)$ is given by

$$\phi_b^{-1} \circ \sigma \circ \phi_b \colon H(t_b) \to H(t_b), \quad (z_1, z_2) \mapsto \left(\frac{t_b}{\bar{z}_1}, \frac{t_b}{\bar{z}_2} \right) = (\bar{z}_2, \bar{z}_1).$$

The fixed point set of σ on $\mathcal{H}(q)$ consists of the curve a_0, the ovals $a_1, \dots, a_g$ and the ovals

$$a_b = \phi_b\{(z_1, z_2) \in H(t_b) \mid |z_1| = |z_2| = t_b^{1/2}\}$$

and is illustrated in the figures below

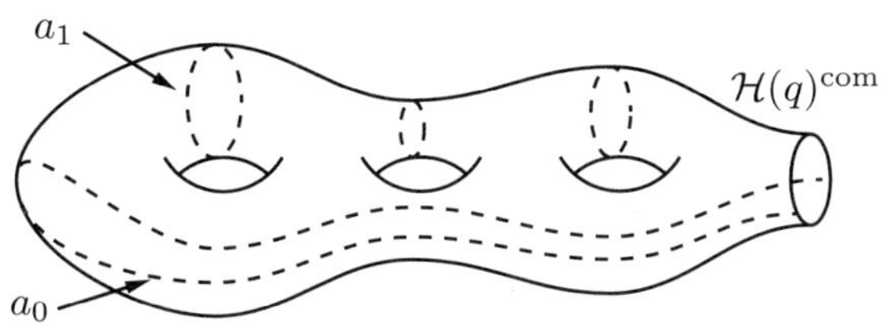

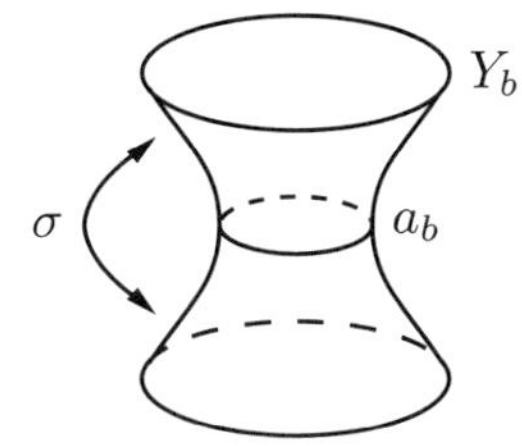

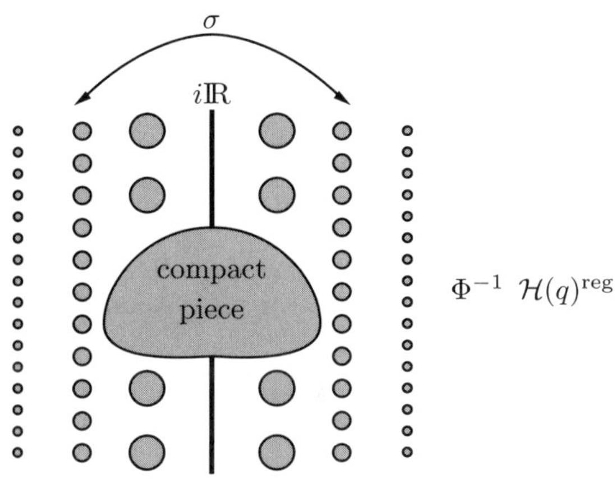

COROLLARY 21.6. *The Riemann period matrix $\mathcal{R}$, for the heat curve $\mathcal{H}(q)$, is pure imaginary. The vectors U, V, W are real.*

PROOF. The A–cycles are invariant under the antiholomorphic involution σ. This implies that $\sigma^*\omega_b = \overline{\omega}_b$. As σ is orientation reversing, $\sigma B_c = -B_c$. Hence

$$\mathcal{R}_{c,b} = \int_{B_c} \omega_b = \int_{-B_c} \overline{\omega}_b = -\overline{\mathcal{R}}_{c,b}.$$

By (19.2)–(19.3) and the fact that $w_b(-\bar{\xi}) = -\overline{w_b(\xi)}$ we have that $w_{b,s}^{(n)} + w_{b,-\bar{s}}^{(n)}$ and $w_{b,\rm com}^{(n)}$ are real. The reality of U, V, W then follows from Proposition 19.1. $\square$

We denote by $\mathcal{P}_{\mathbb{R}}$ the set of real valued potentials in $\mathcal{P}$ and set

$$\mathcal{H}_r = \big\{ \big((\xi_1, \xi_2); q \big) \in \mathcal{H} \bigm| (\xi_1, \xi_2) \in \mathbb{R} \times \mathbb{R}, q \in \mathcal{P}_{\mathbb{R}} \big\}.$$

The maps K_2 of Theorem 21.4 define a map

$$K_2 \colon h^{-1}(\mathcal{P}_{\mathbb{R}}) \setminus \mathcal{H}_r \to \mathbb{C} \times \mathcal{P}_{\mathbb{R}}.$$

Furthermore, there are real analytic maps

$$v_b, w_b \colon \mathcal{P}_{\mathbb{R}} \to \mathbb{C}$$

such that $\operatorname{Re} v_b = \operatorname{Re} w_b = -\frac{1}{2} b_2$, $\operatorname{Im} v_b \le \operatorname{Im} w_b$ and $c_b(q)$ is the line segment joining $v_b(q)$ and $w_b(q)$. Observe that, for $q \in \mathcal{P}_{\mathbb{R}}$, the heat curve $\mathcal{H}(q)$ is singular if and only if $v_b(q) = w_b(q)$ for some b. In this case $v_b(q)$ corresponds to an ordinary double point of $\mathcal{H}(q)$.

We define, for each $b \in \Gamma^\#$ with $b_2 > 0$,

$$\mathcal{A}_b = \bigcup_{q \in \mathcal{P}_\mathbb{R}} a_b(q) \times \{q\}.$$

This is a connected component of $\mathcal{H}_r$ and $\mathcal{A}_b \cap \mathcal{H}(q) = a_b$.

Now fix $\hat{e} = (\hat{e}_b)_{b \in \Gamma^\#, b_2 > 0}$ such that $\hat{e}_b \in \mathbb{R}$ for all b and

$$\lim_{b \to \infty} \frac{\hat{e}_b}{\log |b|} = 0.$$

For each $q \in \mathcal{P}$ for which $\mathcal{H}(q)$ is smooth $\hat{e}$ lies in the Banach space associated to $\mathcal{H}(q)$. If $q \in \mathcal{P}_\mathbb{R}$ and $\mathcal{H}(q)$ has an ordinary double point at a_c then we consider $\hat{e}$ as a vector in the Banach space with the variable e_c deleted. Put

$$\mathcal{N} = \left\{ ((\xi_1, \xi_2); q) \in \mathcal{H} \,\middle|\, \theta\left(\hat{e} + \int_\infty^{(\xi_1, \xi_2)} \vec{\omega} \right) = 0 \right\}.$$

By Theorem 7.11 the restriction of h to $\mathcal{N} \cap \mathcal{H}^{\mathrm{com}}$ is a (possibly ramified) covering of degree g over $\mathcal{P} \setminus \Delta$. We use this picture to prove

PROPOSITION 21.7. *Let $q \in C^\omega(\mathbb{R}/\Gamma)$ be a real valued potential with $\int_0^{2\pi} q(x_1, x_2)\, dx_2 = 0$ and let $\hat{e} \in B$ be a real vector in the Banach space*

$$B = \left\{ (e_b)_{b \in \Gamma_+^\sharp} \,\middle|\, \lim_{b \to \infty} \frac{e_b}{|\ln t_b|} = 0 \right\}.$$

Then

$$\theta\left(\hat{e} + \int_\infty^x \vec{\omega} \right)$$

has exactly one zero $\widehat{x}_b$ on a_b, $b \in \Gamma^\#$, $b_2 > 0$ and no other zeroes.

PROOF. Observe that $\theta(\hat{e}) \neq 0$ by Corollary 21.6 and Proposition 4.12. We may assume that q is in the ball $\mathcal{P}$ discussed above. Put

$$\mathcal{S} = \{ q \in \mathcal{P}_\mathbb{R} \setminus \Delta \mid \mathcal{N} \cap \mathcal{A}_b \cap \mathcal{H}(q) \text{ consists of exactly one point for each } b \in \Gamma^\#, \\ b_2 > 0 \text{ with } a_b \subset \mathcal{H}^{\mathrm{com}} \}.$$

If $q \in \mathcal{P}$ is small, then by Theorem 7.11 (choosing $\mathcal{H}(q)^{\mathrm{com}} = \varnothing$)

$$\mathcal{N} \cap \mathcal{H}(q) = \bigcup_{\substack{b \in \Gamma^\# \\ b_2 > 0}} \mathcal{N} \cap Y_b$$

and, for each $b \in \Gamma^\#$, $b_2 > 0$, $\mathcal{N} \cap Y_b$ consists of one point. If q is real-valued, the anti-holomorphic involution σ maps $\mathcal{N}$ and Y_b to themselves. Thus, if q is small and real-valued, $\mathcal{N} \cap Y_b$ is a fixed point of σ, i.e., $\mathcal{N} \cap Y_b \subset a_b$. This shows that all sufficiently small $q \in \mathcal{P}_\mathbb{R} \setminus \Delta$ are contained in $\mathcal{S}$.

Now fix $q \in \mathcal{P}_\mathbb{R} \setminus \Delta$. Let $0 = t_0 < t_1 < t_2 < \cdots < t_m$ be the points in $[0, 1]$ for which $tq \in \Delta$. Put $t_{m+1} = 1$.

Next we claim that if $tq \in \mathcal{S}$ for some $t \in (t_{n-1}, t_n)$, $1 \leq n \leq m+1$ then $tq \in \mathcal{S}$ for all $t \in (t_{n-1}, t_n)$. The points of $\mathcal{N} \cap \mathcal{H}(tq)^{\mathrm{com}}$ move continuously with t. Suppose that $t \in (t_{n-1}, t_n)$ and $b \in \Gamma^\#$, $b_2 > 0$ are such that the cardinality of $\mathcal{A}_b \cap \mathcal{H}(tq) \cap \mathcal{N}$ jumps. In other words, suppose that as t' passes through t, a point

$z(t') \in \mathcal{H}(t'q) \cap \mathcal{N}$ leaves $\mathcal{A}_b$. If t' is such that $z(t') \notin \mathcal{A}_b$ then, by σ–invariance, $\sigma z(t')$ is a second point of $\mathcal{H}(t'q) \cap \mathcal{N}$ near $\mathcal{A}_b$. So there must be at least one point of multiplicity two in $\mathcal{A}_b \cap \mathcal{H}(tq) \cap \mathcal{N}$. This is impossible at any point t with $tq \in \mathcal{S}$ because then all the g points of $\mathcal{H}(tq)^{\mathrm{com}} \cap \mathcal{N}$ lie in different connected components of $\mathcal{H}_r(tq)$. This shows that $\{t \in (t_{n-1}, t_n) \mid tq \in \mathcal{S}\}$ is both open and closed in (t_{n-1}, t_n).

Next we show that if $tq \in \mathcal{S}$ for all $t \in (t_{n-1}, t_n)$ for some $1 \le n \le m$ then there is $t' \in (t_n, t_{n+1})$ with $t'q \in \mathcal{S}$. Since $h\colon \mathcal{N} \cap \mathcal{H}^{\mathrm{com}} \to \mathcal{P}$ is a g–fold covering over $\mathcal{P} \setminus \Delta$ there exist holomorphic maps from the punctured disk $\dot{D} = \{z \in \mathbb{C} \mid 0 < |z| < \varepsilon\}$

$$f_i\colon \dot{D} \to \mathcal{N} \cap \mathcal{H}^{\mathrm{com}}, \quad i = 1, \ldots, k$$

and $\alpha_i \in \mathbb{Z}$, $\alpha_i \ge 1$ such that the diagrams

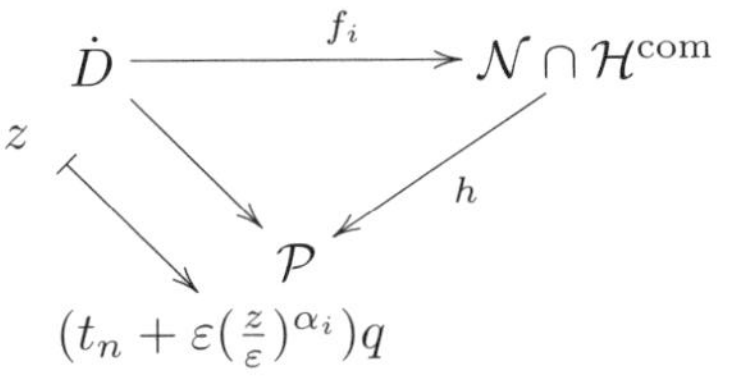

commute, $\alpha_1 + \cdots + \alpha_k = g$ and $\mathcal{N} \cap \mathcal{H}^{\mathrm{com}} \cap h^{-1}\big((t_n + \dot{D})q\big)$ is the union of the images of the f_i. The f_i have removeable singularities at $z = 0$ and hence can be analytically continued to maps

$$\bar{f}_i\colon \{z \in \mathbb{C} \mid |z| < \varepsilon\} \to \mathcal{H}^{\mathrm{com}}.$$

If $\alpha_i \ge 2$ for some i then there would be $b \ne c \in \Gamma^{\#}$ such that the points of $\mathcal{A}_b \cap \mathcal{H}(tq) \cap \mathcal{N}$ and $\mathcal{A}_c \cap \mathcal{H}(tq) \cap \mathcal{N}$ would have the same limit as $t \to t_n-$. This is impossible since $\mathcal{A}_b \cap \mathcal{H}(t_n q)$ and $\mathcal{A}_c \cap \mathcal{H}(t_n q)$ are separated. So $k = g$ and $\alpha_1 = \cdots = \alpha_k = 1$. Since the construction is continuous and σ–invariant $f_i\big((-\varepsilon, \varepsilon)\big)$ is completely contained in some $\mathcal{A}_{b_i}$. By hypothesis, for $i \ne j$, $f_i(-\varepsilon)$ and $f_j(-\varepsilon)$ lie in different components of $\mathcal{H}_r\big((t_n - \varepsilon)q\big)$. Thus $b_i \ne b_j$ for $i \ne j$ and $(t_n + \varepsilon)q \in \mathcal{S}$.

Combining the last three paragraphs, it follows that tq lies in $\mathcal{S}$ for $0 < t \le 1$ and $t \ne t_1, \ldots, t_m$. In particular $q \in \mathcal{S}$.

So $\theta(\hat{e} + \int_\infty^{\widehat{x}} \vec{\omega})$ has exactly one zero $\widehat{x}_b$ on each a_b such that $a_b \subset \mathcal{H}(q)^{\mathrm{com}}$ and no other zeroes on $\mathcal{H}(q)^{\mathrm{com}}$. For $\mathcal{H}(q) \setminus \mathcal{H}(q)^{\mathrm{com}}$ the statement follows immediately from the σ–invariance of $\mathcal{N}$ and Theorem 7.11. □

Now, let $q \in C^\omega(\mathbb{R}^2/\Gamma)$ be a real potential with $\int_0^{2\pi} q(x_1, x_2)\, dx_2 = 0$ for which the associated heat curve $\mathcal{H}(q)$ is smooth. Choose a real vector $\hat{e}$ in the Banach space

$$B = \left\{ z \in \mathbb{C}^\infty \;\middle|\; \lim_{b \to \infty} \frac{z_b}{|\log t_b|} = 0 \right\}$$

of $\mathcal{H}(q)$ and let $\widehat{x}_b$ be the zero of $\theta(\hat{e} + \int_\infty^x \vec{\omega})$ on the oval a_b as in Proposition 21.7. Furthermore, let $y_b \in a_b$ be the unique point of $D \cap a_b$ specified in Theorem 21.4. Then, for $j \ge g + 1$,

$$\left| \int_{\widehat{x}_b}^{y_b} (\phi_b)_* \left(\frac{1}{2\pi i} \frac{dz_1}{z_1} \right) \right| \le 1$$

so by Lemma 6.8 and Theorem 6.4, the sequence $(\int_{\widehat{x}_b}^{y_b} \omega_b)$ is bounded. Therefore, $(y_1, y_2, \dots)$ is a divisor of index zero in the sense of §8. We put

$$e = \hat{e} - \sum_{\substack{b \in \Gamma^{\#} \\ b_2 > 0}} \int_{\widehat{x}_b}^{y_b} \vec{\omega}.$$

Observe that, again by Corollary 21.6 and Proposition 4.12, $\theta(e) \neq 0$. Then, by Proposition 8.5(a) the points $y_1, y_2, \dots$ are the zeroes of $\theta(e + \int_\infty^x \vec{\omega})$ on $\mathcal{H}(q)$.

Now define U, V, W as in §19. It follows from Theorem 4.6, Corollary 21.6, Proposition 4.12 and (19.5) that

$$(x_1, x_2, t) \mapsto \theta\big(e + x_2 U + x_1 V - \tfrac{1}{2} t W\big)$$

is a nowhere vanishing C^∞ function of (x_1, x_2, t). Hence

$$u(x_1, x_2, t) = -2 \frac{\partial^2}{\partial x_2^2} \ln \theta\big(e + U x_2 + V x_1 - \tfrac{1}{2} W t\big)$$

is well-defined everywhere. We can make our main result more precise.

THEOREM 21.1′. *In the situation above, $u(x_1, x_2, t)$ solves* (KP) *and*

$$u(x_1, x_2, 0) = q(x_1, x_2).$$

In §20, we showed that there is a constant c such that $u(x_1, x_2, t) + c$ is a solution of the (KP) equation. It remains to verify the initial condition and $c = 0$.

We first verify that if $u + c$ obeys the initial condition then $c = 0$. So assume that $u + c$ obeys the initial condition. Then, as $\int_0^{2\pi} q(x_1, x_2)\, dx_2 = 0$ we have

$$\begin{aligned} 2\pi c &= 2 \int_\xi^{\xi + 2\pi} \frac{d^2}{dx_2^2} \ln \theta(e + U x_2)\, dx_2 \\ &= 2 \frac{d}{dx_2} \ln \theta(e + U x_2)\Big|_{x_2 = \xi + 2\pi} - 2 \frac{d}{dx_2} \ln \theta(e + U x_2)\Big|_{x_2 = \xi} \\ &= 2 \frac{d}{dx_2} \ln \frac{\theta(e + 2\pi U + U x_2)}{\theta(e + U x_2)}\bigg|_{x_2 = \xi}. \end{aligned}$$

Hence

$$\theta(e + 2\pi U + U\xi) = e^{a + \pi c \xi} \theta(e + U\xi).$$

Since, for real ξ, θ is real and bounded we have $c = 0$.

We have chosen the vector e by an algorithm that, in the finite genus case, ensures that the initial condition is satisfied. To prove it in the general case, we approximate $\mathcal{H}(q)$ by heat curves of finite genus. By convention, if θ' is the θ function for an approximating finite genus heat curve $\mathcal{H}(q')$ and $z \in B$, then $\theta'(z)$ is evaluated by ignoring all components z_b of z for which a_b' is a point. The following two theorems prove that approximation by finite genus heat curves is possible.

THEOREM 21.8 ([**K**, Chapter I, §3]). *Let $q \in C^\omega(\mathbb{R}^2/\Gamma)$ be a real potential with $\int_0^{2\pi} q(x_1, x_2)\, dx_2 = 0$. Then, for any $n > 0$, $\varepsilon > 0$, there exists a real-valued finite zone potential q' with*

$$\sum_{b \in \Gamma^{\#}} (1 + |b|^n) |\widehat{q}(b) - \widehat{q'}(b)| < \varepsilon.$$

THEOREM 21.9. *Fix a real* $q \in C^\omega(\mathbb{R}^2/\Gamma)$. *There is a constant* K *such that the following holds. Let* $Z \subset \mathcal{H}(q)$ *be a compact subset containing* $\mathcal{H}(q)^{\mathrm{com}}$ *and let* $\varepsilon > 0$. *Then, there is* $\delta > 0$ *such that for all* q' *with* $\sum_b |b|^4 |\widehat{q}(b) - \widehat{q'}(b)| < \delta$ *there are compact submanifolds with boundary* $X_0 \subset \mathcal{H}(q), X_0' \subset \mathcal{H}(q')$ *with* $Z \subset X_0$ *and a diffeomorphism* $F\colon X_0 \to X_0'$ *such that*

(i) $\big(\mathcal{H}(q), X_0\big)$ *is* (ε, K)*–close to* $\big(\mathcal{H}(q'), X_0'\big)$ *via* F

(ii) *the antiholomorphic involutions preserve* X_0 *and* X_0' *and* F *is compatible with the antiholomorphic involutions. In formulae:* $\sigma(X_0) = X_0$, $\sigma'(X_0') = X_0'$ *and* $F \circ \sigma = \sigma' \circ F$

(iii) *Let* $z_b' \in a_b'$ *be the zeroes of* $\theta'\big(e + \int_\infty^{z'} \vec{\omega}'\big)$ *on* $\mathcal{H}(q')$. *Then*
$$\big|\phi_b^{-1}(z_b) - \phi_b^{-1}\big(F^{-1}(z_b')\big)\big| \le \varepsilon\sqrt{t_b} \quad \text{for all } b \text{ such that } Y_b \subset X_0.$$

(iv) *Let* $y_b' \in a_b'$ *be the unique point of* $D' \cap a_b'$ *of Theorem 21.4. Then*
$$\big|\phi_b^{-1}(y_b) - \phi_b^{-1}\big(F^{-1}(y_b')\big)\big| \le \varepsilon\sqrt{t_b} \quad \text{for all } b \text{ such that } Y_b \subset X_0.$$

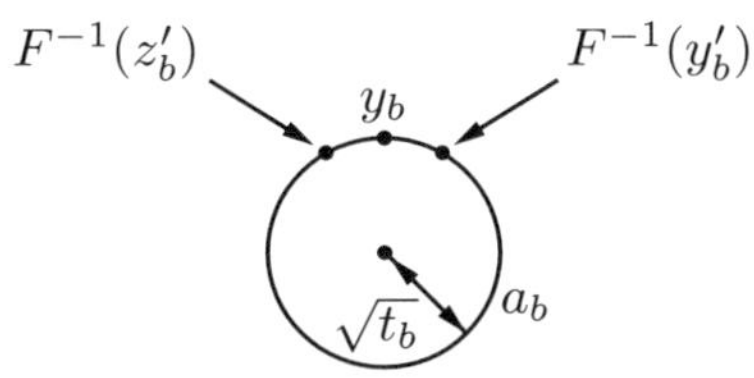

Before we prove Theorem 21.9, we use it to give the

PROOF OF THEOREM 21.1$'$. Approximate q by finite zone potentials with respect to the norm $\sum_b |b|^4 |\widehat{q}(b)|$. If q' is an approximating potential as in Theorem 21.8, then
$$q'(x_1, x_2) = -\frac{\partial^2}{\partial x_2^2} \log \theta'(e' + U'x_2 + V'x_1)$$
where
$$e' = e - \sum_b \int_{z_b'}^{y_b'} \vec{\omega}'$$
since
$$\theta'\left(e' + \int_\infty^{y'} \vec{\omega}'\right)$$
vanishes precisely on $\{y_b'\}$. By Proposition 8.1, Proposition 6.16 and Theorem 21.9, e' converges to e in B. Furthermore by Theorem 21.9(i) and Theorem 20.4 the theta function
$$\theta'(e' + U'x_2 + V'x_1)$$
converges to
$$\theta(e + Ux_2 + Vx_1)$$
uniformly for all x_1, x_2 in any bounded set by the method of Corollary 20.14 (solutions of the KP equation). By analyticity this proves that $u(x_1, x_2, 0) = q(x_1, x_2)$. That u obeys the KP equation was proven before. □

For the proof of Theorem 21.9 we need

LEMMA 21.10. *Let $v, w, v', w' \in \mathbb{C}$, $0 < t,\ t' < 1$ and let*

$$P_1 \colon H(t) \to \mathbb{C}, \qquad (z_1, z_2) \mapsto \frac{v+w}{2} + \frac{1}{4\sqrt{t}}(v-w)(z_1+z_2),$$
$$P_1' \colon H(t') \to \mathbb{C}, \qquad (z_1, z_2) \mapsto \frac{v'+w'}{2} + \frac{1}{4\sqrt{t'}}(v'-w')(z_1+z_2).$$

If $t \le (4/15 \times 9)^2$ set $K = 2$. Otherwise, let $K > 2$ obey $t < (1 - 28/15K)^2(1 + 1/K)^{-4}$. Assume that $|v - v'| < \varepsilon|v - w|$, $|w - w'| < \varepsilon|v - w|$ and $|t - t'| < \varepsilon t$. There is a universal constant const *such that, if $\varepsilon < 1/\mathrm{const}\, K^6$, then all of the following hold. There are*

$$\left\{(z_1,z_2) \in H(t) \;\middle|\; \frac{t}{1-\mathrm{const}\,K\varepsilon} \le |z_1| \le 1 - \mathrm{const}\,K\varepsilon\right\} \subset \widehat{Y} \subset H(t),$$
$$\left\{(z_1,z_2) \in H(t') \;\middle|\; \frac{t'}{1-\mathrm{const}\,K\varepsilon} \le |z_1| \le 1 - \mathrm{const}\,K\varepsilon\right\} \subset \widehat{Y}' \subset H(t')$$

and a K–quasiconformal diffeomorphism $f \colon \widehat{Y} \to \widehat{Y}'$ of distortion at most $\mathrm{const}\, K^5 \varepsilon$ *such that*

$$P_1' \circ f(z_1, z_2) = P_1(z_1, z_2)$$

for $(z_1, z_2) \in \widehat{Y}$ with $|z_1| \ge 1 - 28/15K$ or $|z_2| \ge 1 - 28/15K$ ($= 1/15$ for $K = 2$) and

$$f(z_1, z_2) = \sqrt{\frac{t'}{t}}(z_1, z_2)$$

for $(z_1, z_2) \in \widehat{Y}$ with $|z_1|, |z_2| \le (1 - 28/15K)(1 + 1/K)^{-1}$ ($= 2/45$ for $K = 2$). Furthermore

$$|f(z_1, z_2) - (z_1, z_2)| \le \mathrm{const}\, K\varepsilon.$$

If the line through v and w coincides with that through v' and w' then f commutes with the antiholomorphic involutions $(z_1, z_2) \mapsto (\bar{z}_2, \bar{z}_1)$ on $H(t)$ and $H(t')$.

REMARK. Be careful to distinguish between ε and ε. We shall use Lemma 21.10 to construct K–quasiconformal diffeomorphisms of distortion at most ε in the proof of Theorem 21.9. There, we first pick K. Then, for any desired $\varepsilon > 0$, we pick ε such that $\mathrm{const}\, K^5\varepsilon < \varepsilon$ and $\mathrm{const}\, K\varepsilon < 1/K$.

PROOF. Let $\widehat{F}_{t,K} \colon \mathbb{C} \to [0,1]$ be a C^∞ function which takes the value one inside the ellipse with focii $\pm\sqrt{t}$ and semiaxes $\frac{1}{2}(1 - 28/15K)(1 + 1/K)^{-1} \pm t/2(1 - 28/15K)^{-1}(1 + 1/K)$, which vanishes outside the ellipse with focii $\pm\sqrt{t}$ and semiaxes $\frac{1}{2}(1 - s28/15K) \pm t/2(1 - 28/15K)^{-1}$, and which is invariant under reflection about the real axis. Hence $\widehat{F}_{t,K}(\bar{z}) = \widehat{F}_{t,K}(z)$. Because the distance between the two ellipses is

$$\frac{1}{2}\left(1-\frac{28}{15K}\right)+\frac{t}{2}\left(1-\frac{28}{15K}\right)^{-1}-\frac{1}{2}\left(1-\frac{28}{15K}\right)\left(1+\frac{1}{K}\right)^{-1}-\frac{t}{2}\left(1-\frac{28}{15K}\right)^{-1}\left(1+\frac{1}{K}\right)$$
$$=\frac{1}{2}\left(1-\frac{28}{15K}\right)\left(1+\frac{1}{K}\right)^{-1}\frac{1}{K}-\frac{t}{2}\left(1-\frac{28}{15K}\right)^{-1}\frac{1}{K}$$
$$\geq\frac{1}{2}\left(1-\frac{28}{15K}\right)\left(1+\frac{1}{K}\right)^{-1}\frac{1}{K}-\tfrac{1}{2}\left(1-\frac{28}{15K}\right)\left(1+\frac{1}{K}\right)^{-4}\frac{1}{K}$$
$$\geq\frac{1}{2}\frac{1}{15}\frac{2}{3}\frac{1}{K}\left[1-\left(1+\frac{1}{K}\right)^{-3}\right]$$
$$\geq\frac{\text{const}}{K^2}$$

it is possible to construct $\widehat{F}_{t,K}$ so that

$$\sup_{x,y}|\partial^\alpha\widehat{F}_{t,K}(x,y)|\leq\text{const}_{|\alpha|}(\text{const}K^2)^{|\alpha|}.$$

Define $F\colon\mathbb{C}\to\mathbb{C}$ by

$$F(z)=z+\left[\frac{v'+w'-v-w}{2}+\frac{2z-v-w}{v-w}\frac{v'-w'-v+w}{2}\right]\widehat{F}_{t,K}\left(\sqrt{t}\frac{2z-v-w}{v-w}\right).$$

We will define f in terms of F. Before doing so, we derive some properties of F. Observe that, for each $a>\sqrt{t}$, the image under the map

$$(z_1,z_2)\in H(t)\mapsto\sqrt{t}\frac{2P_1(z_1,z_2)-v-w}{v-w}=\frac{z_1+z_2}{2}\in\mathbb{C}$$

of the circle $|z_1|=a$ and of the circle $|z_2|=a$ is the ellipse with focii $\pm\sqrt{t}$ and semiaxes $\frac{1}{2}(a\pm t/a)$. Whenever $\widehat{F}(\sqrt{t}2z-v-w/v-w\ )=1$, and this includes all $z=P_1(z_1,z_2)$ with $|z_1|,|z_2|=t/|z_1|\leq(1-28/15K)(1+1/K)^{-1}$,

$$\frac{2F(z)-v'-w'}{v'-w'}=\frac{2z-v'-w'}{v'-w'}+\frac{v'+w'-v-w}{v'-w'}+\frac{2z-v-w}{v-w}\frac{v'-w'-v+w}{v'-w'}$$
$$=\frac{2z-v-w}{v'-w'}+\frac{2z-v-w}{v-w}\frac{v'-w'-v+w}{v'-w'}$$
$$=\frac{2z-v-w}{v-w}.$$

Consequently, denoting $2z-v-w/v-w=s$,

$$F\left(\frac{v+w}{2}+s\frac{v-w}{2}\right)=\frac{v'+w'}{2}+s\frac{v'-w'}{2}$$

for all $\sqrt{t}|s|$ inside the ellipse with focii $\pm\sqrt{t}$ and semiaxes $\frac{1}{2}(1-28/15K)(1+1/K)^{-1}$ $\pm\,t/2(1+1/K)(1-28/15K)^{-1}$. This includes $s=\pm1$, so

$$F(v)=v',\quad F(w)=w'$$

and F intertwines reflection about the line through v and w with reflection about the line through v' and w', at least in a neighbourhood of the line segment joining v and w.

For general s

$$F\left(\frac{v+w}{2}+s\frac{v-w}{2}\right)=\frac{v'+w'}{2}+s\frac{v'-w'}{2}$$
$$+\left[\frac{v'+w'-v-w}{2}+s\frac{v'-w'-v+w}{2}\right][\widehat{F}(\sqrt{t}s)-1]$$
$$=\frac{v'+w'}{2}+\frac{v'-w'}{2}\left\{s+\left[\frac{v'+w'-v-w}{v'-w'}+s\left(1-\frac{v-w}{v'-w'}\right)\right]\right.$$
$$\left.\times[\widehat{F}(\sqrt{t}\,s)-1]\right\}.$$

Note that, by hypothesis, $|(v'+w'-v-w)/(v-w)|, |1-(v'-w')/(v-w)|\le 2\varepsilon$ and that, by construction, $|d\widehat{F}_{t,K}(\sqrt{t}s)/ds|\le \text{const}\,K^2$. Hence, as long as we have chosen the const in $\varepsilon\le 1/\text{const}\,K^2$ sufficiently large, the map from s to $s+\big[(v'+w'-v-w)/(v'-w')+s(1-(v-w)/(v'-w'))\big][\widehat{F}_{t,K}(\sqrt{t}\,s)-1]$, and consequently the map from z to $F(z)$, is globally bijective. Furthermore, if the line through v and w coincides with the line through v' and w' then $(v-w)/(v'-w')$ and $(v'+w'-v-w)/(v'-w')$ are real so that

$$F\left(\frac{v+w}{2}+\bar{s}\frac{v-w}{2}\right)$$
$$=\overline{\frac{v'+w'}{2}+\frac{v'-w'}{2}\left\{s+\left[\frac{v'+w'-v-w}{v'-w'}+s\left(1-\frac{v-w}{v'-w'}\right)\right][\widehat{F}_{t,K}(\sqrt{t}\,s)-1]\right\}}$$

and F commutes with relection in the line joining v and w.

Now define $f\colon \widehat{Y}=\{(z_1,z_2)\in H(t)\mid F\circ P_1(z_1,z_2)\in \text{range}\,P_1'\}\to H(t')$ so that

$$P_1'\circ f=F\circ P_1$$

P_1' is a 2 to 1 map, so for each (z_1,z_2) in the domain of f we have two possible values of $(u_1,u_2)=f(z_1,z_2)$. Choose $|u_1|>|u_2|$ if and only if $|z_1|>|z_2|$. Note the argument of $\widehat{F}_{t,K}$ in $F\circ P_1$ is $\sqrt{t}2P_1(z_1,z_2)-v-w/v-w=\frac{1}{2}(z_1+z_2)$ and that for $|z_1|,|z_2|\le(1-28/15K)(1+1/K)^{-1}$ we have $\widehat{F}_{t,K}(z_1+z_2/2)=1$. For all such (z_1,z_2)

$$F\circ P_1(z_1,z_2)=\frac{v'+w'}{2}+\frac{z_1+z_2}{2\sqrt{t}}\frac{v'-w'}{2}=P_1'\left(\sqrt{\frac{t'}{t}}z_1,\sqrt{\frac{t'}{t}}z_2\right).$$

Thus

$$f(z_1,z_2)=\sqrt{\frac{t'}{t}}(z_1,z_2)$$

on $\{(z_1,z_2)\in H(t)\mid |z_1|,|z_2|\le(1-28/15K)(1+1/K)^{-1}\}$. On the other hand, in the event that $\max\{|z_1|,|z_2|\}\ge(1-28/15K)$, we have $\widehat{F}_{t,K}(z_1+z_2/2)=0$. For all such (z_1,z_2)

$$P_1'\circ f=F\circ P_1=P_1.$$

Now write $f(z_1, z_2) = (u_1, u_2)$. By definition

$$
\begin{aligned}
&\frac{v' + w'}{2} + \frac{v' - w'}{2} \frac{u_1 + u_2}{2\sqrt{t'}} \\
&= \frac{v' + w'}{2} + \frac{v' - w'}{2} \left\{ \frac{z_1 + z_2}{2\sqrt{t}} + \left[\frac{v' + w' - v - w}{v' - w'} + \frac{z_1 + z_2}{2\sqrt{t}} \frac{v' - w' - v + w}{v' - w'} \right] \right. \\
&\qquad\qquad \left. \times \left[\widehat{F}\left(\frac{z_1 + z_2}{2} \right) - 1 \right] \right\}
\end{aligned}
$$

so that

$$
\begin{aligned}
u_1 + u_2 &= \sqrt{\frac{t'}{t}} \left\{ z_1 + z_2 + 2\sqrt{t} \left[\frac{v' + w' - v - w}{v' - w'} + \frac{z_1 + z_2}{2\sqrt{t}} \frac{v' - w' - v + w}{v' - w'} \right] \right. \\
&\qquad\qquad \left. \times \left[\widehat{F}_{t,K}\left(\frac{z_1 + z_2}{2} \right) - 1 \right] \right\} \\
&= \sqrt{\frac{t'}{t}} (z_1 + z_2 + U)
\end{aligned}
$$

with

$$
U = 2\sqrt{t} \left[\frac{v' + w' - v - w}{v' - w'} + \frac{z_1 + z_2}{2\sqrt{t}} \frac{v' - w' - v + w}{v' - w'} \right] \left[\widehat{F}_{t,K}\left(\frac{z_1 + z_2}{2} \right) - 1 \right]
$$

bounded by

$$
|U| \le 6\sqrt{t}\varepsilon \left(1 + \frac{z_1 + z_2}{2\sqrt{t}} \right) \le 12\varepsilon
$$

and, more generally, obeying

$$
|\partial^\alpha U| \le \mathrm{const}_{|\alpha|} \varepsilon (\mathrm{const}\, K^2)^{|\alpha|}
$$

We now solve for u_1 in terms of z_1

$$
\begin{aligned}
&u_1 + \frac{t'}{u_1} = \sqrt{\frac{t'}{t}} \left(z_1 + \frac{t}{z_1} + U \right) \\
&\implies u_1^2 - \sqrt{\frac{t'}{t}} \left(z_1 + \frac{t}{z_1} + U \right) u_1 + t' = 0 \\
&\implies u_1 = \frac{1}{2} \left\{ \sqrt{\frac{t'}{t}} \left(z_1 + \frac{t}{z_1} + U \right) \pm \sqrt{\frac{t'}{t} \left(z_1 + \frac{t}{z_1} + U \right)^2 - 4t'} \right\} \\
&\implies u_1 = \frac{1}{2} \left\{ \sqrt{\frac{t'}{t}} \left(z_1 + \frac{t}{z_1} + U \right) \pm \sqrt{\frac{t'}{t}} \sqrt{\left(z_1 - \frac{t}{z_1} \right)^2 + 2U \left(z_1 + \frac{t}{z_1} \right) + U^2} \right\}.
\end{aligned}
$$

To satisfy $|u_1| > |u_2|$ for $|z_1| > |z_2|$, we take the $+$ sign, so

$$
\begin{aligned}
u_1 = \sqrt{\frac{t'}{t}} &\left(z_1 + \frac{U}{2} \right) \\
&+ \sqrt{\frac{t'}{t}} \left(z_1 - \frac{t}{z_1} \right) \left[\sqrt{1 + \left[2U \left(z_1 + \frac{t}{z_1} \right) + U^2 \right] \left(z_1 - \frac{t}{z_1} \right)^{-2}} - 1 \right].
\end{aligned}
$$

We already know that $U = 0$ if $|z_1|, |z_2| \leq (1 - 28/15K)(1 + 1/K)^{-1}$. Otherwise,

$$\left|z_1 - \frac{t}{z_1}\right| = |z_1 - z_2| \geq \left(1 - \frac{28}{15K}\right)\left(1 + \frac{1}{K}\right)^{-1} - \sqrt{t}$$
$$\geq \left(1 - \frac{28}{15K}\right)\left(1 + \frac{1}{K}\right)^{-2}\frac{1}{K} \geq \frac{1}{15}\frac{4}{9}\frac{1}{K}$$

and

$$|u_1 - z_1| \leq \text{const}\, \varepsilon + \text{const}\, KU \leq \text{const}\, K\varepsilon.$$

Similarly $|u_2 - z_2| \leq \text{const}\, K\varepsilon$, so we have verified

$$|f(z_1, z_2) - (z_1, z_2)| \leq \text{const}\, K\varepsilon.$$

Finally, we must show that $f^* du_1/u_1$ satisfies the K–quasiconformal bounds corresponding to distortion at most $\text{const}\, K^5\varepsilon$. If $|z_1|, |z_2| \leq (1 - 28/15K) \times (1 + 1/K)^{-1}$

$$f^* \frac{du_1}{u_1} = \frac{dz_1}{z_1}.$$

On the other hand, if, for example, $|z_1| \geq (1 - 28/15K)(1 + 1/K)^{-1}$ then

$$\begin{aligned}
f^* \frac{du_1}{u_1} &= f^* \frac{d(u_1 + u_2)}{u_1 - u_2} \\
&= \sqrt{\frac{t'}{t}} \frac{d(z_1 + z_2) + dU}{u_1 - u_2} \\
&= \sqrt{\frac{t'}{t}} \frac{z_1 - z_2}{u_1 - u_2} \frac{d(z_1 + z_2)}{z_1 - z_2} + \sqrt{\frac{t'}{t}} \frac{dU}{u_1 - u_2} \\
&= \sqrt{\frac{t'}{t}} \frac{z_1 - z_2}{u_1 - u_2} \frac{dz_1}{z_1} + \sqrt{\frac{t'}{t}} \frac{dU}{u_1 - u_2} \\
&= a\frac{dz_1}{z_1} + b\frac{d\bar{z}_1}{\bar{z}_1}
\end{aligned}$$

where

$$a = \sqrt{\frac{t'}{t}} \frac{z_1 - z_2}{u_1 - u_2} + \sqrt{\frac{t'}{t}} \frac{z_1 U_{z_1}}{u_1 - u_2}, \qquad b = \sqrt{\frac{t'}{t}} \frac{\bar{z}_1 U_{\bar{z}_1}}{u_1 - u_2}.$$

Hence it remains only to show that, when $|z_1| \geq (1 - 28/15K)(1 + 1/K)^{-1}$, each of

$$\begin{aligned}
a - 1 &= \sqrt{\frac{t'}{t}} - 1 + \sqrt{\frac{t'}{t}} \frac{z_1 - u_1 - z_2 + u_2}{u_1 - u_2} + \sqrt{\frac{t'}{t}} \frac{z_1 U_{z_1}}{u_1 - u_2}, \\
b &= \sqrt{\frac{t'}{t}} \frac{\bar{z}_1 U_{\bar{z}_1}}{u_1 - u_2}, \\
\frac{\partial a}{\partial \bar{z}_1} &= \sqrt{\frac{t'}{t}} (z_1 - z_2) \frac{\partial}{\partial \bar{z}_1} \frac{1}{u_1 - u_2} + \sqrt{\frac{t'}{t}} z_1 \frac{\partial}{\partial \bar{z}_1} \frac{U_{z_1}}{u_1 - u_2} \\
&= -\sqrt{\frac{t'}{t}} \frac{z_1 - z_2}{(u_1 - u_2)^2} \frac{\partial}{\partial \bar{z}_1}(u_1 - u_2) - \sqrt{\frac{t'}{t}} \frac{z_1 U_{z_1}}{(u_1 - u_2)^2} \frac{\partial}{\partial \bar{z}_1}(u_1 - u_2) \\
&\qquad + \sqrt{\frac{t'}{t}} \frac{z_1}{u_1 - u_2} U_{z_1 \bar{z}_1}
\end{aligned}$$

is bounded by const $K^5\varepsilon(|z_1|+|z_2|)$. In fact, as

$$|z_1|, |z_2| \le 1,$$

$$|z_1| \ge \left(1 - \frac{28}{15K}\right)\left(1+\frac{1}{K}\right)^{-1} \ge \frac{1}{15}\frac{2}{3}$$

$$|z_1 - z_2| \ge |z_1 - \sqrt{t}| \ge \begin{cases} \frac{1}{15}\left(\frac{2}{3}-\frac{4}{9}\right), & \text{if } t \le \left(\frac{4}{15\times 9}\right)^2, \\ \frac{1-28/15K}{1+1/K} - \frac{1-28/15K}{(1+1/K)^2}, & \text{if } t \ge \left(\frac{4}{15\times 9}\right)^2, \end{cases}$$

$$\ge \begin{cases} \frac{1}{15}\frac{2}{9}, & \text{if } t \le \left(\frac{4}{15\times 9}\right)^2, \\ \frac{1}{15}\frac{4}{9}\frac{1}{K}, & \text{if } t \ge \left(\frac{4}{15\times 9}\right)^2, \end{cases}$$

$$|u_1 - z_1| \le \text{const}\, K\varepsilon,$$

$$|u_2 - z_2| \le \text{const}\, K\varepsilon,$$

$$|u_1 - u_2| \ge |z_1 - z_2| - \text{const}\, K\varepsilon \ge \frac{\text{const}}{K}$$

$$\left|\sqrt{\frac{t'}{t}} - 1\right| \le \text{const}\, \varepsilon$$

we have

$$|a-1| \le \text{const}\, K^2\varepsilon + \text{const}\, K|U_{z_1}|,$$

$$|b| \le \text{const}\, K|U_{\bar z_1}|,$$

$$\left|\frac{\partial a}{\partial \bar z_1}\right| \le \text{const}\, K^2(1+|U_{z_1}|)\left|\frac{\partial}{\partial \bar z_1}(u_1-u_2)\right| + \text{const}\, K|U_{z_1\bar z_1}|.$$

Applying $|\partial^\alpha U| \le \text{const}_{|\alpha|}\, \varepsilon(\text{const}\, K^2)^{|\alpha|}$ yields

$$|a-1| \le \text{const}\, K^2\varepsilon + \text{const}\, K^3\varepsilon,$$

$$|b| \le \text{const}\, K^3\varepsilon,$$

$$\left|\frac{\partial a}{\partial \bar z_1}\right| \le \text{const}\, K^2\left|\frac{\partial}{\partial \bar z_1}(u_1-u_2)\right| + \text{const}\, K^5\varepsilon.$$

We used $|U_{z_1}| \le \text{const}\, K^2\varepsilon \le \text{const}$ in the last line. Finally

$$\left|\frac{\partial}{\partial \bar z_1}(u_1-u_2)\right| = \left|\left(1+\frac{t'}{u_1^2}\right)\frac{\partial u_1}{\partial \bar z_1}\right| = \left|\left(1+\frac{t'}{u_1^2}\right)\frac{u_1}{\bar z_1}b\right|$$

$$= \left|(u_1+u_2)\frac{b}{\bar z_1}\right| \le \text{const}\, K^3\varepsilon. \qquad \square$$

Proof of Theorem 21.9. Fix $q \in C^\infty(\mathbb{R}^2/\Gamma)$ and $\rho > 0$ such that the decomposition after Theorem 21.4 works for all $q' \in C^\infty(\mathbb{R}^2/\Gamma)$ with $\||b|^\beta \hat{q}'\|_1 < 2\||b|^\beta \hat{q}\|_1$. Then there exist $K > 0$ and $\gamma > 0$, a tubular neighbourhood T and a contour Γ, such that all of Definition 20.2 parts (iii) and (vi) as well as all parts of (v) except the second half of the fifth bullet, are satisfied for all $q' \in C^\infty(\mathbb{R}^2/\Gamma)$ with $\||b|^\beta \hat{q}'\|_1 < 2\||b|^\beta \hat{q}\|_1$. Further increase K, if necessary, so that $t_b < (1-28/15K)^2(1+1/K)^{-4}$ for all $b \in \Gamma^\#$, $b_2 > 0$ and all $q' \in C^\infty(\mathbb{R}^2/\Gamma)$ with $\||b|^\beta \hat{q}'\|_1 < 2\||b|^\beta \hat{q}\|_1$. Now let $0 < \varepsilon < 1/150$ and pick ε such that const $K^6\varepsilon < \varepsilon$ with the const being that of Lemma 21.10. Choose a contour ∂G_0 in $\mathbb{C}$ such that (iv) and the second half of the fifth bullet of (v) are satisfied. If q' is sufficiently close to q then v_b' can be made arbitrarily close to v_b and w'_b can be made arbitrarily close to w_b for all b with $v_b + w_b/2$ inside ∂G_0. In particular, $|v - v'| < \varepsilon|v_b - w_b|$, $|w_b - w'_b| < \varepsilon|v_b - w_b|$ and $|t_b - t'_b| < \varepsilon t_b$ can be satisfied for all b with $v_b + w_b/2$ inside ∂G_0. Define $\mathcal{H}(q')_0$ to be the compact part of $\mathcal{H}(q')$ bounded by $\Phi'(\partial G_0)$. If q' is close to q we define

$$F \colon \mathcal{H}(q)_0 \to \mathcal{H}(q')_0$$

by

$$F(x) = K_2'^{-1} \circ K_2(x) \quad \text{for } x \notin \phi_b\{(z_1, z_2) \in H(t_b) \mid |z_1|, |z_2| \leq \tfrac{1}{2}\} \qquad \forall b \in \Gamma^{\#},\ b_2 > 0,$$

$$F(x) = \phi'_b \circ f_b \circ \phi_b(x) \quad \text{for } x \in \phi_b\{(z_1, z_2) \in H(t_b) \mid |z_1|, |z_2| \leq \tfrac{1}{2}\}$$

where K_2 was defined in the discussion following Lemmma 21.5 and f_b is defined in Lemma 21.10. Then, if q' is sufficiently close to q (i), (ii) hold.

Parts (iii) and (iv) of Theorem 21.9 are obvious as z_b, y_b depend continuously on q. $\square$

Bibliography

[Ac1] R. Accola, *Some classical theorems on open Riemann surfaces*, Bull. Amer. Math. Soc. (N.S.) **73** (1967), 13–26.

[Ac2] ———, *The bilinear relation on open Riemann surfaces*, Trans. Amer. Math. Soc. **96** (1960), 143–161.

[Ah] L. Ahlfors, *Normalintegrale auf offenen Riemannschen Flächen*, Ann. Acad. Sci. Fennicae. Ser. A. I. Math.-Phys. **1947**, (1947). no. 35 (German).

[AS] L. Ahlfors and L. Sario, *Riemann surfaces*, Princeton Univ. Press, Princeton, NJ, 1960.

[An] A. Andreotti, *On a theorem of Torelli*, Amer. J. Math. **80** (1958), 801–828.

[AdC] E. Arbarello and C. de Concini, *On a set of equations characterizing Riemann matrices*, Ann. of Math. (2) **112** (1984), 119–140.

[AC] ———, *Another proof of a conjecture of S. P. Novikov on periods of abelian integrals on Riemann surfaces*, Duke Math. J. **54** (1987), 163–178.

[BKM] D. Bättig, T. Kappeler, and B. Mityagin, *On the Korteweg–de Vries equation: frequencies and initial value problem*, Pacific J. Math. **181** (1997), 1–55.

[BEKT] A. Bobenko, N. Ercolani, H. Knörrer, and E. Trubowitz, *Density of heat curves in the moduli space*, Panoramas of Mathematics (Warsaw, 1992/1994), Banach Center Publ., vol. 34, Polish Acad. Sci., Warsaw, 1995, pp. 19–27,

[B] J. Bourgain, *On the Cauchy problem for the Kadomtsev–Petviashvili equation*, Geom. Funct. Anal. **3** (1993), 315–341.

[EM] N. Ercolani and H. McKean, *Geometry of* KdV(4): *Abel sums, Jacobi variety and the theta function in the scattering case*, Invent. Math. **99** (1990), 483–544.

[FK] H. M. Farkas and I. Kra, *Riemann surfaces*, Grad. Texts in Math., vol. 71, Springer-Verlag, New York-Berlin, 1980.

[FKT] J. Feldman, H. Knörrer, and E. Trubowitz, *The perturbatively stable spectrum of a periodic Schrödinger operator*, Invent. Math. **100** (1990), 259–300.

[GG] M. Golubitsky and V. Guillemin, *Stable mappings and their singularities*, Grad. Texts in Mathe., vol. 14, Springer-Verlag, New York-Berlin, 1973.

[GH] P. Griffiths and J. Harris, *Principles of algebraic geometry*, Wiley, 1978.

[H] M. Heins, *Riemann surfaces of infinite genus*, Ann. Math. **55** (1952), 296–317.

[KT] H. Knörrer and E. Trubowitz, *A directional compactification of the complex Bloch variety*, Comment. Math. Helv. **65** (1990), 114–149.

[HW] G. H. Hardy and E. M. Wright, *An introduction to the theory of numbers*, Oxford University Press, Oxford, 1979.

[K] I. Krichever, *Spectral theory of two-dimensional periodic operators and its applications*, Russian Math. Surveys **44** (1989), 145–225.

[Kru] S. L. Krushkal, *Quasiconformal mappings and Riemann surfaces*, John Wiley & Sons, New York-Toronto, Ont.-London, 1979.

[LG] P. Lelong and L. Gruman, *Entire functions of several complex variables*, Springer-Verlag, New York–Berlin, 1986.

[McK1] H. McKean, *Integrable systems and algebraic curves*, Global Analysis, (Proc. Biennial Sem. Canad. Math. Congr., Calgary, Alta., 1978) (M. Grmela and J. E. Marsden, eds.) (1978), Lecture Notes in Math., vol. 755, Springer Verlag, 1979, pp. 83–200.

[McK2] ———, *Boussinesq's equation on the circle*, Comm. Pure Appl. Math. **34** (1981), 599–691.

[MT1] H. P. McKean and E. Trubowitz, *Hill's operator and hyperelliptic function theory in the presence of infinitely many branch points*, Comm. Pure Appl. Math. **29** (1976), 143–226.

[MT2] ———, *Hill's surfaces and their theta functions*, Bull. Amer. Math. Soc. (N.S.) **84**, no. 6 (1978), 1042–1085.

[Me1] F. Merkl, *A Riemann–Roch theorem for infinite genus Riemann surfaces*, Invent. Math. **139** (2000), 393–438.

[Me2] ———, *An asymptotic expansion for Bloch functions on Riemann surfaces of infinite genus and almost periodicity of the Kadomcev–Petviashvilli flow*, Math. Phys. Anal. Geom. **2** (1999), no. 3, 245–278.

[Mu1] D. Mumford, *Tata lectures on theta.* I, Progr. Math., vol. 28, Birkhäuser Boston, Inc., Boston, MA, 1983.

[Mu2] ———, *Tata lectures on theta.* II. *Jacobian theta functions and differential equations*, Progr. Math., vol. 43. Birkhäuser Boston, Inc., Boston, MA, 1984.

[Na] M. Nakai, *On Evans potential*, Proc. Japan Acad. **38** (1962), 624–629.

[Ne] R. Nevanlinna, *Quadratisch integrierbare differentiale auf einer Riemannschen Mannigfaltigkeit*, Ann. Acad. Sci. Fennicae. Ser. A. I. Math.-Phys. **1941**, (1941). no. 1.

[PT] J. Pöschel and E. Trubowitz, *Inverse spectral theory*, Academic Press, New York-London, 1987.

[RS] M. Reed and B. Simon, *Methods of modern mathematical physics.* I. *Functional analysis*, Academic Press, New York–London, 1972.

[R] H. Royden, *The Riemann–Roch theorem*, Comment. Math. Helv. **34** (1960), 37–51.

[SN] L. Sario and M. Nakai, *Classification theory of Riemann surfaces*, Grundlehren Math. Wiss., vol. 164, Springer-Verlag, New York–Berlin, 1970.

[Sch] M. Schmidt, *Integrable systems and Riemann surfaces of infinite genus*, Mem. Amer. Math. Soc. **122** (1996), no. 581.

[S] T. Shiota, *Characterization of Jacobian varieties in terms of soliton equations*, Invent. Math. **83** (1986), 333–382.

[T] K. Tahara, *On the hyperelliptic Riemann surfaces of infinite genus with absolutely convergent Riemann's theta functions*, Nagoya Math. J. **33** (1968), 57–73.

[W] H. Wu, *The equidistribution theory of holomorphic curves*, Ann. of Math. Stud. vol. 64, Princeton University Press, Princeton, NJ, 1970.